Geographical Analysis and Modeling of Urban Heat Island Formation

Geographical Analysis and Modeling of Urban Heat Island Formation

Editors

Yuji Murayama
Ruci Wang

MDPI

Basel • Beijing • Wuhan • Barcelona • Belgrade • Novi Sad • Cluj • Manchester

Editors

Yuji Murayama
University of Tsukuba
Tsukuba, Japan

Ruci Wang
Chiba University
Chiba, Japan

Editorial Office
MDPI
St. Alban-Anlage 66
4052 Basel, Switzerland

This is a reprint of articles from the Special Issue published online in the open access journal *Remote Sensing* (ISSN 2072-4292) (available at: https://www.mdpi.com/journal/remotesensing/special_issues/geographical_UHI).

For citation purposes, cite each article independently as indicated on the article page online and as indicated below:

Lastname, A.A.; Lastname, B.B. Article Title. *Journal Name* **Year**, *Volume Number*, Page Range.

ISBN 978-3-0365-9274-9 (Hbk)
ISBN 978-3-0365-9275-6 (PDF)
doi.org/10.3390/books978-3-0365-9275-6

Contents

About the Editors

Yuji Murayama

Yuji Murayama is a professor emeritus and adjunct researcher at the University of Tsukuba, Japan. His research interests are urban and transport geography, geographical information science, and spatial analysis. He has edited and published many books including "*Spatio-Temporal Analysis of Urbanization Using GIS and Remote Sensing in Developing Countries*" (2021, MDPI), "*Urban Development in Asia and Africa: Geospatial Analysis of Metropolises*" (2017, Springer Nature), "*Progress in Geospatial Analysis*" (2012, Springer), "*Spatial Analysis and Modeling in Geographical Transformation Process: GIS-Based Applications*" (2011, Springer), "*Recent Advances in Remote Sensing and GIS in Sub-Saharan Africa*" (2010, Nova Publishers), and "*Japanese Urban System*" (2000, Springer).

Ruci Wang

Ruci Wang is an assistant professor at the Center for Environmental Remote Sensing, Chiba University. She obtained her Ph.D. from the University of Tsukuba in 2020. Her research interests are urbanization, urban heat islands, and future scenario simulation.

remote sensing

MDPI

Editorial

Editorial: Special Issue on Geographical Analysis and Modeling of Urban Heat Island Formation

Yuji Murayama [1,*] and Ruci Wang [1,2]

1 Faculty of Life and Environmental Sciences, University of Tsukuba, 1-1-1 Tennodai, Tsukuba 305-8572, Japan; wang.ruci.fw@u.tsukuba.ac.jp
2 Center for Environmental Remote Sensing (CEReS), Chiba University, 1-33, Yayoicho, Inage-ku, Chiba 263-8522, Japan
* Correspondence: mura@geoenv.tsukuba.ac.jp

Abstract: This Special Issue focuses on the data, methods, techniques, and empirical outcomes of urban heat island studies from a time and space perspective. We showcase research papers, empirical studies, conceptual or analytic reviews, and policy-related tasks to help achieve urban sustainability. We are interested in target methodologies and datasets capturing urban heat island phenomena, including novel techniques for urban heat island monitoring and forecasting with the integration of remote sensing and GIS, the spatial relationship between urban heat island intensity and land use/cover distribution in metropolitan areas, the geographical patterns and processes of urban heat island phenomena in large cities, spatial differences in urban heat island intensity between developing and developed countries, urban heat island disaster mitigation and adaptation for future urban sustainability, and prediction and scenario analysis of urban heat island formation for policy and planning purposes.

Keywords: urban remote sensing; land surface temperature; urbanization; sustainable cities; impervious surface; spatial analysis

Citation: Murayama, Y.; Wang, R. Editorial: Special Issue on Geographical Analysis and Modeling of Urban Heat Island Formation. *Remote Sens.* **2023**, *15*, 4474. https://doi.org/10.3390/rs15184474

Received: 25 June 2023
Revised: 24 August 2023
Accepted: 5 September 2023
Published: 12 September 2023

1. Introduction

The urban heat island (UHI) phenomenon, which is related to rapid urbanization, has attracted considerable attention from academic scholars and governmental policymakers because of its profound influence on citizens' daily lives [1]. The UHI effect has negative human impacts, including indirect economic loss, poor air quality, reduced comfort, imbalanced public health, and increased mortality rates [2,3]. The temperature difference between the center and the periphery is expanding, especially in large cities, which may result from land use/cover composition changes and increasing anthropogenic heat sources [4]. According to a United Nations estimate, nearly 54% of the world's population currently resides in urban regions, and by 2050, that number is expected to rise to 66% [5]. Urbanization is expected to add another 2.5 billion people to the global population by 2050, with Asia and Africa accounting for more than 90% of the growth. If traditional city planning continues without considering environmental factors, living conditions may be seriously degraded.

Therefore, the monitoring and modeling of urban heat island formation is important for management and sustainable development, especially in developing countries. This Special Issue focuses on the data, methods, techniques, and empirical outcomes of urban heat island studies from a geographical perspective, i.e., a time and space viewpoint. A total of 14 articles and 1 review paper are included in this Special Issue, all contributing to the field of sustainable urban development. The included studies highlight four points of importance:

(1) The spatial relationship between urban heat island intensity and land use/cover distribution in metropolitan areas;

(2) Geographical patterns and processes of urban heat island formation in large cities based on empirical studies;

(3) Spatial differences in urban heat island intensity between developing and developed counties;

(4) Useful methodologies and datasets for capturing urban heat island phenomena.

This Special Issue discusses the latest developments in these subjects, providing a review of recent geographical research on UHI effects. In the editorial, we first examine current UHI trends and discuss the impact factors of the land surface temperature (LST) in various case studies. The selected papers highlight the regional climatic parameters, topography, size, and population of each city, as well as urban materials and the distribution of green spaces, all of which affect changes in UHI intensity. Finally, we emphasize the significance and contribution of urban environmental studies and discuss sustainable development prospects for future UHI studies.

2. Current Trends in UHI Formation

With the rapid pace of industrialization and urbanization in recent decades, the UHI effect has gradually harmed our daily lives. Therefore, scientists and planning authorities have increasingly focused on mitigating the impact of UHIs by allocating land use/cover distribution and considering air ventilation in urban center planning. As a result, a growing number of UHI studies have been conducted.

Most UHI mitigation research concentrates on urban landscapes and building design in an effort to avoid intensive development that causes the loss of green spaces and an increase in impervious surfaces, which cause overheating in urban centers [6,7]. Considering that spatial structure and urban growth are not consistent across cities, long-term spatiotemporal monitoring should be carried out for various types of cities [8].

Twenty-one cities from four nations were targeted as case studies in this Special Issue. All included studies have one thing in common: population growth and urban expansion exacerbated UHI phenomena. However, water and green spaces were found to lessen the effects, especially in the central area.

Each study examined the spatial influence on UHIs by employing a different methodology. For example, Wang et al. created 12 functional construction land zones based on various social and economic indicators to examine how they contributed to the changes in the urban thermal environment [9]. We can deepen our understanding of urban thermal warming mechanisms by exploring diverse functional land zones. Another study proposed new macro-perspectives for reducing UHI phenomena by redistributing land use/cover. Zheng et al. attempted to detect cooling effects and scales using Landsat 8 Operational Land Imager/Thermal Infrared Sensor (OLI/TIRS) and Sentinel-2 data [10]. They demonstrated the application of a distance–LST scatter diagram and a multiple linear regression method, taking two inner city lakes as study objects. According to their findings, a high density of green spaces, combined with dispersed, modest structures, may aid in extending the cooling effect of inner lakes.

Shi et al. compared daytime and nighttime LSTs to explore the general spatial distribution of urban thermal environments [11]. They provided a valuable method to characterize the UHI effect more effectively and illustrate its evolution during the day. Moreover, Zhou et al. investigated the driving factors based on the temporal and spatial variation in LSTs in Zhengzhou [12]. Their findings showed a positive correlation between human activity and LSTs, accelerating the UHI effect. Although the cooling effect of vegetation and water was superior to that of topography, the role of albedo on LSTs confirmed the geographical variation.

Investigations into how the design and layout of the landscape may impact the LST on the city scale are vital. Sarif et al. examined the influence of land indices on the dynamics of LSTs from the city center to the periphery to debate the directional profiling of LSTs [13]. In Prayagraj City, the LST distribution was lower in the forested regions than in the built-up areas, bare soils, and sands. Meng et al. highlighted that various land cover patterns are considerably influenced by the spatial distribution of LST [14]. Karunaratne et al. conducted a study on the temperate mountain valley metropolis of Kathmandu, Nepal [15]. Similar to other studies, where the mean LST tended to shift in an east–south–north–west pattern, consistent with urban growth, the mean LST tended to be higher in the city's core. In Kathmandu, the LST pattern is influenced by both valley wind and urban heat island circulation. The valley wind is affected by the heat island circulation in a specific way, demonstrating that during the daytime, the valley wind speed slows down, and the LST difference reduces between the metropolitan region and the mountain slopes. On the other hand, mountain wind speeds rise at night, and the LST differential between them tends to increase.

Derdouri et al. examined the relationship between land use/cover changes and surface UHI in large cities [4]. In a systematic review, the authors attempted to synthesize, contrast, and critically evaluate numerous empirical studies conducted between 2001 and 2020, including regional characteristics, data sources, techniques for classifying land use/cover and quantifying surface UHI, and mechanisms of interaction between surface UHI intensity and land use/cover. Finally, by discussing spatial and temporal changes in land use/cover, they suggested concrete alleviation actions. Such a study can support decision-making and pave the way for future academic research, particularly in vulnerable cities that have not received considerable attention to date.

3. Prospects of UHI Formation

We can identify two directions for future UHI research: mitigating the impact of UHI formation and adapting to UHI effects on sustainability.

Over the past two decades, UHI-related studies have shown remarkable progress [4]. However, case studies of UHIs are more than just a distinction between developed and developing cities because architecture and urban design vary among cities. Many researchers have discovered that urban patches with varying densities of vegetation significantly impact LST formation, although this phenomenon has not been investigated scientifically in detail.

Zhang et al. discussed the relationship between urban vegetation components and LST distribution in Xuzhou City, China [16]. Their findings demonstrate that essential aspects in controlling the thermal environment include spatial distribution features such as patch proportion, natural connection degree, predominance degree, shape complexity, and aggregation degree of areas with a high vegetation density. The distribution, scale, and heat-reducing properties of different landscapes should be analyzed to capture the future trends in UHI patterns. In addition to water and wetlands, surface and roof materials should be re-investigated for their cooling effects.

One of the primary concerns with UHIs in geographical studies is that climate change adaptation may be more costly in urban compared to non-urban locations, owing to the increase in UHI intensity. Therefore, future UHI research is expected to evaluate the urban thermal security pattern and suggest future planning strategies that provide a favorable layout based on sustainable development goals to mitigate the consequences of UHIs.

Sismanidis et al. explored the differences in the seasonal hysteresis of surface urban heat island intensity (SUHII) between climates [17]. They offer a thorough typology of the daytime and nighttime SUHII hysteresis loops. The analysis results reveal that the seasonal hysteresis of the SUHII exhibits twisted, flat, and triangle-like patterns, in addition to concave up and down forms. Furthermore, Hu et al. proposed a regional heat island network based on circuit theory simulation [18]. They discussed the locational characteristics of UHI patches and the spatial patterns of collaborative optimization in Wuhan City, China.

With the acceleration in urbanization, urban areas continue to spread out, with a decreasing distance between urban core areas. As a result, urban agglomeration or conurbation has developed with accompanying UHI formation. An integrated research framework to assess the spatial effects of multiple environmental circumstances on habitat quality was developed by Liu et al. [19]. By highlighting the connections and interactions between various environmental challenges in urban agglomerations and ecosystems, the authors discussed the importance of the designed multidimensional sustainability and co-benefits Liu et al. also investigated urban agglomeration, taking the Pearl River Delta, China, as the study area [20]. Compared with cities with low urbanization rates, the authors showed that the effect of land cover and socioeconomic determinants on the daytime LST was more significant in highly urbanized cities.

Integrating machine learning algorithms with remote sensing data is an important topic that has received considerable attention. Applying regression analysis and machine learning algorithms, Garzón et al. evaluated modeling techniques to assess the impact of various elements on surface UHIs [21]. In this paper, an attempt was made to illustrate the applicability of machine learning algorithms in the surface mapping of UHI intensities by quantifying surface UHIs using different contributing parameters.

4. Contributions to Future UHI Studies

To summarize this editorial, we chart the progress in related UHI studies. The UHI phenomenon is prevalent in various cities. An effective urban design reduces UHI formation while simultaneously achieving the objectives of sustainable development. As is customary, remote sensing serves as the primary data source for the analysis of the correlation between UHI intensity and urban dispersion. However, a considerable debate continues about whether the data sources are reliable enough to accurately reflect the features of cities (e.g., 2D or 3D building data). Do we need to focus on gathering actual big datasets for each building (such as building type and building height), or does the suitable size of the urban area suffice? These and other concerns are addressed, in part, in this editorial (Section 3), although they remain challenges to be solved in the future.

For researchers and city planners, we hope that this Special Issue will inspire novel concepts and methods that can lead to theoretical comprehension and practical application with respect to UHI formation and effects.

Author Contributions: This editorial was prepared by Y.M. and R.W. and reviewed by Y.M. All authors have read and agreed to the published version of the manuscript.

Funding: This research was funded by the Japan Society for the Promotion of Science (JSPS) (grants 21K01027 and 21F21003).

Acknowledgments: The authors thank the editors and anonymous reviewers for their valuable comments and suggestions.

Conflicts of Interest: The authors declare no conflict of interest.

References

1. Kaloustian, N.; Diab, Y. Effects of Urbanization on the Urban Heat Island in Beirut. *Urban Clim.* **2015**, *14*, 154–165. [CrossRef]
2. Shishegar, N. The Impact of Green Areas on Mitigating Urban Heat Island Effect: A Review. *Int. J. Environ. Sustain.* **2014**, *9*, 119–130. [CrossRef]
3. Tan, J.; Zheng, Y.; Tang, X.; Guo, C.; Li, L.; Song, G.; Zhen, X.; Yuan, D.; Kalkstein, A.J.; Li, F.; et al. The Urban Heat Island and Its Impact on Heat Waves and Human Health in Shanghai. *Int. J. Biometeorol.* **2010**, *54*, 75–84. [CrossRef]
4. Derdouri, A.; Wang, R.; Murayama, Y.; Osaragi, T. Understanding the Links between LULC Changes and SUHI in Cities: Insights from Two-Decadal Studies (2001–2020). *Remote Sens.* **2021**, *13*, 3654. [CrossRef]
5. United Nations Raises Projected World Population. Available online: https://www.prb.org/resources/united-nations-raises-projected-world-population/ (accessed on 13 September 2021).
6. Aleksandrowicz, O.; Vuckovic, M.; Kiesel, K.; Mahdavi, A. Current Trends in Urban Heat Island Mitigation Research: Observations Based on a Comprehensive Research Repository. *Urban Clim.* **2017**, *21*, 1–26. [CrossRef]
7. Wang, R.; Hou, H.; Murayama, Y.; Morimoto, T. A Three-Dimensional Investigation of Spatial Relationship between Building Composition and Surface Urban Heat Island. *Buildings* **2022**, *12*, 1240. [CrossRef]
8. Yang, L.; Yu, K.; Ai, J.; Liu, Y.; Lin, L.; Lin, L.; Liu, J. The Influence of Green Space Patterns on Land Surface Temperature in Different Seasons: A Case Study of Fuzhou City, China. *Remote Sens.* **2021**, *13*, 5114. [CrossRef]
9. Wang, H.; Li, B.; Yi, T.; Wu, J. Heterogeneous Urban Thermal Contribution of Functional Construction Land Zones: A Case Study in Shenzhen, China. *Remote Sens.* **2022**, *14*, 1851. [CrossRef]

10. Zheng, Y.; Li, Y.; Hou, H.; Murayama, Y.; Wang, R.; Hu, T. Quantifying the Cooling Effect and Scale of Large Inner-City Lakes Based on Landscape Patterns: A Case Study of Hangzhou and Nanjing. *Remote Sens.* **2021**, *13*, 1526. [CrossRef]
11. Shi, W.; Hou, J.; Shen, X.; Xiang, R. Exploring the Spatio-Temporal Characteristics of Urban Thermal Environment during Hot Summer Days: A Case Study of Wuhan, China. *Remote Sens.* **2022**, *14*, 6084. [CrossRef]
12. Zhou, S.; Liu, D.; Zhu, M.; Tang, W.; Chi, Q.; Ye, S.; Xu, S.; Cui, Y. Temporal and Spatial Variation of Land Surface Temperature and Its Driving Factors in Zhengzhou City in China from 2005 to 2020. *Remote Sens.* **2022**, *14*, 4281. [CrossRef]
13. Sarif, M.O.; Gupta, R.D.; Murayama, Y. Assessing Local Climate Change by Spatiotemporal Seasonal LST and Six Land Indices, and Their Interrelationships with SUHI and Hot–Spot Dynamics: A Case Study of Prayagraj City, India (1987–2018). *Remote Sens.* **2023**, *15*, 179. [CrossRef]
14. Meng, Q.; Liu, W.; Zhang, L.; Allam, M.; Bi, Y.; Hu, X.; Gao, J.; Hu, D.; Jancsó, T. Relationships between Land Surface Temperatures and Neighboring Environment in Highly Urbanized Areas: Seasonal and Scale Effects Analyses of Beijing, China. *Remote Sens.* **2022**, *14*, 4340. [CrossRef]
15. Karunaratne, S.; Athukorala, D.; Murayama, Y.; Morimoto, T. Assessing Surface Urban Heat Island Related to Land Use/Land Cover Composition and Pattern in the Temperate Mountain Valley City of Kathmandu, Nepal. *Remote Sens.* **2022**, *14*, 4047. [CrossRef]
16. Zhang, Y.; Wang, Y.; Ding, N. Spatial Effects of Landscape Patterns of Urban Patches with Different Vegetation Fractions on Urban Thermal Environment. *Remote Sens.* **2022**, *14*, 5684. [CrossRef]
17. Sismanidis, P.; Bechtel, B.; Perry, M.; Ghent, D. The Seasonality of Surface Urban Heat Islands across Climates. *Remote Sens.* **2022**, *14*, 2318. [CrossRef]
18. Hu, C.; Li, H. Reverse Thinking: The Logical System Research Method of Urban Thermal Safety Pattern Construction, Evaluation, and Optimization. *Remote Sens.* **2022**, *14*, 6036. [CrossRef]
19. Liu, F.; Murayama, Y.; Masago, Y. Spatial Influence of Multifaceted Environmental States on Habitat Quality: A Case Study of the Three Largest Chinese Urban Agglomerations. *Remote Sens.* **2023**, *15*, 921. [CrossRef]
20. Liu, W.; Meng, Q.; Allam, M.; Zhang, L.; Hu, D.; Menenti, M. Driving Factors of Land Surface Temperature in Urban Agglomerations: A Case Study in the Pearl River Delta, China. *Remote Sens.* **2021**, *13*, 2858. [CrossRef]
21. Garzón, J.; Molina, I.; Velasco, J.; Calabia, A. A Remote Sensing Approach for Surface Urban Heat Island Modeling in a Tropical Colombian City Using Regression Analysis and Machine Learning Algorithms. *Remote Sens.* **2021**, *13*, 4256. [CrossRef]

remote sensing

MDPI

Review

Understanding the Links between LULC Changes and SUHI in Cities: Insights from Two-Decadal Studies (2001–2020)

Ahmed Derdouri [1], Ruci Wang [2,*], Yuji Murayama [2] and Toshihiro Osaragi [1]

1 School of Environment and Society, Tokyo Institute of Technology, 2 Chome-12-1 Ookayama, Meguro City, Tokyo 152-8550, Japan; derdouri.a.aa@m.titech.ac.jp (A.D.); osaragi.t.aa@m.titech.ac.jp (T.O.)
2 Faculty of Life and Environmental Sciences, University of Tsukuba, 1-1-1 Tennodai, Tsukuba City 305-8572, Ibaraki, Japan; mura@geoenv.tsukuba.ac.jp
* Correspondence: wang.ruci.fw@u.tsukuba.ac.jp

Abstract: An urban heat island (UHI) is a serious phenomenon associated with built environments and presents threats to human health. It is projected that UHI intensity will rise to record levels in the following decades due to rapid urban expansion, as two-thirds of the world population is expected to live in urban areas by 2050. Nevertheless, the last two decades have seen a considerable increase in the number of studies on surface UHI (SUHI)—a form of UHI quantified based on land surface temperature (LST) derived from satellite imagery—and its relationship with the land use/cover (LULC) changes. This surge has been facilitated by the availability of freely accessible five-decade archived remotely sensed data, the use of state-of-art analysis methods, and advancements in computing capabilities. The authors of this systematic review aimed to summarize, compare, and critically analyze multiple case studies—carried out from 2001 to 2020—in terms of various aspects: study area characteristics, data sources, methods for LULC classification and SUHI quantification, mechanisms of interaction coupled with linking techniques between SUHI intensity with LULC spatial and temporal changes, and proposed alleviation actions. The review could support decision-makers and pave the way for scholars to conduct future research, especially in vulnerable cities that have not been well studied.

Keywords: urban heat island (UHI); land use land cover (LULC); land surface temperature (LST); spatiotemporal changes; SUHI-contributing factors; satellite imagery; literature review

Citation: Derdouri, A.; Wang, R.; Murayama, Y.; Osaragi, T. Understanding the Links between LULC Changes and SUHI in Cities: Insights from Two-Decadal Studies (2001–2020). *Remote Sens.* **2021**, *13*, 3654. https://doi.org/10.3390/rs13183654

Academic Editor: Zina Mitraka

Received: 11 August 2021
Accepted: 10 September 2021
Published: 13 September 2021

1. Introduction

Since the start of the industrial revolution in Great Britain circa the 1780s [1], substantial urban expansion and population growth have been observed in industrialized countries. To put things in perspective, there were fewer than 50 cities with over 100,000 residents in 1800, there were approximately 900 cities in 1950 [2], and there were thousands of cities in 2019 [3]. In fact, by the end of the last century, almost 370 cities had over one million inhabitants worldwide, which increased to 584 in 2018, and it is estimated to reach 706 cities by the end of 2030 [4]. Though these accelerated trends have contributed to economic growth and social development in many parts of the world, they have also led to several environmental issues at different scales. Urban heat islands (UHIs) are perhaps the most evident and most documented manifestation of these radical anthropogenic activities.

UHI, also known as an "urban heat sink" or an "oasis effect" [5], refers to the phenomenon that occurs in urban areas (UAs) that involves an excessive increase in interrelated air, subsurface, and surface temperatures compared to those observed in underdeveloped surroundings [6]. The common term "heat island"—reportedly coined by British climatologist Gordon Manley in 1958 [7]—was given its name because the resulting spatial shape of the isotherms creates one or more island-like features [8]. The UHI phenomenon was first documented by Luke Howard over 200 years ago in his study of London's climate, where he found fluctuations in temperatures measured in the city and its rural surroundings during daytimes and nighttimes—the city had 3.7 °F warmer nights and 0.34 °F colder days than the countryside [9]. Statistically, UHIs have been documented in over 1400 cities across the globe in all continents without exception [10].

Urbanization has been highlighted in numerous studies as the leading culprit behind UHIs. Economic growth usually drives new UAs to accommodate new industrial and commercial hubs and ultimately build connecting transportation networks and new residences, especially for a floating population hoping for stable life away from numerous hardships and the lack of opportunities associated with rural areas. Perhaps, urbanization leads to social stability and economic prosperity; on the other hand, land cover conversion to urban uses has detrimental effects on the balance of the natural environment. Natural surfaces, including vegetation and water bodies, contribute to the balance of energy heat fluxes because they are excellent solar radiation absorbers. Vegetation uses a significant amount of the absorbed radiation through evapotranspiration to release water vapor that subsequently helps to cool the air in their proximity [11].

Additionally, vegetation reduces surface temperatures by offering a shading layer that shields land surfaces from direct sun radiation [12]. Water bodies provide cooling effects, much as vegetation does, as they provide a source for moisture capable of lowering nearby ambient temperatures. By contrast, impervious surfaces (ISs) such as concrete (buildings) and asphalt (streets) halt the interchange of heat between different environmental components because of their low reflectivity and capacity to absorb solar radiation [13], leading to heat imbalance and, consequently, local climate change. Ultimately, the alteration of previous surfaces to ISs introduces perturbations into the balance of the local climate, resulting in UHIs. Moreover, the heat emitted by traffic, industries, factories, and air conditioners contributes to increasing the local temperature [14]. Additionally, the amount of airflow is reduced as narrow streets and tall buildings trap heat, thereby intensifying the heat island effect [14].

In the short and long term, UHIs have severe implications for many areas of life on earth, including socioeconomic and environmental issues. Air pollution may increase because of UHIs, daytime temperatures become warmer, and nighttime cooling becomes less effective [15]. These alterations lead to discomfort and an increase in human premature mortality rates due to excessive heat. In fact, extreme heat is a primary contributor to the rise in weather-related human mortality [16–18]. Between 1991 and 2018, 37% of the world's heat-related fatalities may have been linked to human-caused global warming [19]. Additionally, because of UHIs, urban rain islands can form, resulting in greater precipitation during the flood season in flood-prone places and leading to waterlogging at a regional scale [20]. In Jinan City, for instance, it has been proven that the URI effect is spatially correlated with that of the UHI, resulting in an increased frequency and severity of short-duration precipitation episodes [21]. During the rainy season, areas with intensive construction receive more rain. With so much rain concentrated in urbanized areas, which are characterized by low surface infiltration capacities, the city has become more and more vulnerable to floods.

The UHI phenomenon is also an obstacle to achieving sustainable development. In 2015, the United Nations provided the 17 Sustainable Development Goals (SDGs), four of which are directly or indirectly related to UHIs. UHIs comprise one of the significant reasons for apparent temperature increases, which have the potential to be particularly serious for heat-associated deaths (SDG 3 (Good Health and Well-Being)), and people who are living in high-temperature areas also have increased electricity bills (SDG 7 (Affordable and Clean Energy)). Additionally, high-rise and -density buildings lead to UHIs and can affect the quality of eco-environments in living areas (SDG 11 (Sustainable Cities and Communities)). UHIs can also influence urban microclimates, and in the long run, their effects could indicate global warming (SDG 13 (Climate Action)). According to the United Nations, 60% of the world's people (around 5 billion) will live in UAs by 2050. The increase in urban populations will be the cause of urban agglomeration, and then UHIs will become more obvious. Therefore, the proper awareness and analysis of the relationship between

UHIs and land use land cover (LULC) changes, as well as how to relieve the UHI effect by urban planning, are crucial for achieving the SDGs.

Broadly, UHIs can be categorized into two types: surface UHIs (SUHIs) and atmospheric UHIs (AUHIs). Several aspects differentiate the two UHI forms, as summarized by Sharma and Joshi [22]: (i) day/night prominence: during daytimes and nighttimes, SUHIs are prominent, while AUHIs are mainly noticeable at night [23]; (ii) seasonal variations: during wintertime, AUHIs are more intense than SUHIs and vice versa in the summertime [24,25]; (iii) UHI intensity (UHIi): SUHI intensity generally tends to be much higher than AUHI intensity [26]; (iv) data sources: while AUHI measurements are mostly retrieved from ground-based meteorological stations network and field surveys, SUHI measurements are quantified using airborne or satellites thermal data [27,28]; and (v) ideal observation time: nighttimes under a clear sky and calm conditions are the best times for investigating AUHIs, but thermal data taken during the daytimes are more suitable for investigating SUHIs [29].

Furthermore, Oke distinguished two types of AUHIs [30]: the urban canopy layer (UCL) and the urban boundary layer (UBL). A UCL is formed by microscale activities occurring in urban streets—often referred to as "canyons"—that connect buildings situated under the roofline [31]. By contrast, a UBL is a localized mesoscale phenomenon that is influenced by the nature of the urban topography above the roofline [31]. Fixed and mobile instruments can be used to observe UBLs and UCLs. While the most common devices used to monitor UCLs are fixed screens and automobiles, UBLs can be observed using fixed tower sodars or aircrafts and tetroons [32]. However, these instruments and techniques are time- and budget-consuming, especially when focusing on city-, metropolitan-, or mega-city-scale and regional studies. Moreover, the availability of historical or up-to-date measurements is another obstacle that hinders investigating historical trends of AUHIs. Conversely, using satellite-derived data, spatiotemporal studies of SUHI distribution may be conducted in a cost-effective and time-saving manner [33] at local and regional scales.

The use of remotely sensed data to study urban climate started in the 1970s as a result of successful observational satellite programs such as Television InfraRed Observational Satellite (TIROS), Nimbus, Environmental Science Services Administration Satellite (ESSA), and Landsat, resulting in the launch of several satellites—notably TIROS-1 (1960), Nimbus series (1964–1978), ESSA-1 (1966), and Landsat series (since 1972). Reportedly, Rao may have been the first to show that satellite imagery can be used to investigate urban climate [34]. Through his study, he investigated surface temperature trends in cities across the Mid-Atlantic coast of the US using the infrared radiometer (IR) data of TIROS-I. Subsequently, other studies using similar approaches followed. For instance, Matson et al. used NOAA-5 satellite data to identify nocturnal UHIs in the Midwest and Northwestern US [35], and Price investigated UHIs in New York City and the New England region [36]. Taking advantage of the significant advancements in computing software and hardware in the last two decades, researchers have been able to concisely investigate SUHIs and their driving factors vis-à-vis LULC in different regions of the world, relying on sophisticated developed techniques to retrieve land surface temperatures (LSTs) and extract LULC information from freely accessible historical satellite data covering almost half a century of data. Particularly, Landsat series data have been extensively used to evaluate the connection between LULC and LST.

Several review studies have been published in recent years that summarize our knowledge on SUHIs from multiple perspectives, including LST retrieval methods from thermal sensors [37], exploring factors amplifying its intensity [38], and mitigation strategies [39]. Deilami et al., for instance, focused on how spatial factors (e.g., LULC and urban form), dynamics, and temporal variations (e.g., yearly and seasonal) impact SUHIs. The authors reviewed studies published between 1965 and 2017 [38]. In a broad yet comprehensive review, Zhou et al. explored several aspects based on the literature published from 1972 to 2018 [10]. The authors investigated the popular thermal sensors and methods used to retrieve SUHIs from them in addition to the main drivers of SUHI variations. While most review studies have focused on a worldwide geographic scope, others have focused on a concise geographic scale limited to a region or even a country. Kotharkar et al., for instance, concentrated their review on South Asian cities because the region hosts approximately 15% of the world's population [40]. The authors reviewed published papers between 1973 and 2017, with a primary focus on investigating empirical measurements of UHIs, observation, modeling, their impact, and proposed mitigation strategies.

In this review paper, however, we focus on how spatial and temporal changes of different factors, in particular those related to LULC, have impacted SUHIs in cities across the globe from the turn of the 21st century till the present. In addition to the characteristics of the reviewed literature in terms of annual trends, sources, and geographical coverage, a detailed key points analysis that concerns five critical topics was conducted: (1) the primary data sources and methods employed for extracting LULC and retrieving LST information, (2) the methods used for evaluating LULC and SUHIs, (3) the most common factors deemed critical in affecting SUHI magnitude across time and space, (4) an overview of the proposed mitigation strategies, and finally (5), the main limitations and future directions.

With that in mind, the overall structure of this review takes the form of five sections, excluding this introductory section. Section 2 is concerned with the followed review methodology, specifically the selection criteria of the reviewed literature and the main considered databases. Section 3 presents a general synopsis of the findings of the review, focusing on the synthesis of the literature and geographical coverage of the reviewed studies. In Section 4, a detailed analysis of the obtained results focusing on seven key elements is presented: (i) the main satellite data sources, (ii) the methods of LULC extraction, (iii) the methods of SUHI quantification, (iv) the assessment methods of the relationship between LULC and SUHIs, (v) the main factors affecting UHIs, (vi) the proposed mitigation strategies, and lastly (vii), the future challenges and areas for further research. Section 5 summarizes the findings.

2. Review Methodology

In this review, we concentrated on the literature focusing on connections between LULC spatiotemporal changes and UHIs, specifically SUHIs extracted from satellite images. For that reason and because satellite images have become more accessible, we concentrated on studies published in the last two decades—precisely between January 2001 and November 2020. The management of remotely sensed data has become easy, thanks to the advancement of GIS in terms of software and hardware (computer performance). Three databases were explored, namely: Scopus, Web of Science, and Google Scholar. Furthermore, only peer-reviewed journal articles written in English were retained. Conference papers, book chapters, reports, and other types were excluded.

Regarding the followed inclusion strategy, we employed a four-step approach (Figure 1) to screen the existing literature. First, we identified published papers based on a broad search query. Considering the varying terminology employed by researchers, we built search queries by combining possible variants of key terms including: "UHI", "urban heat island", "land use", "LULC", "spatiotemporal", and "satellite image". Second, we conducted an initial assessment of the resulting papers based on their titles and abstracts. We excluded unrelated papers, including health- and energy-focused and climatology-related papers—notably those concerning simulation and numeric modeling. Third, a deeper screening was conducted to further refine the resulting papers from the previous screening based on the content of each paper, following the filtering out of duplicated papers collected from the three considered indexing databases. Through this deep screening, we ensured that the included manuscripts satisfied the following requirements:

- At least two-date satellite images were used to analyze changes of LULC and SUHIs.
- The authors explicitly assessed the relationship between the different LULC classes and SUHIs.
- The study was city-focused; district-level and regional-level studies were excluded.
- For multicity studies, we considered each city as a case study.

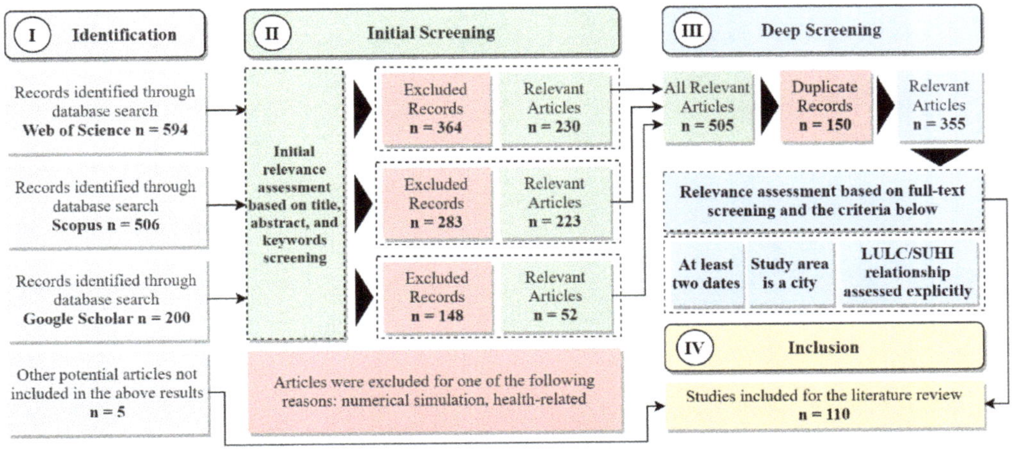

Figure 1. Methodological flowchart.

This process yielded a total of 110 eligible papers out of 1300 (duplicates included) initially found, based on selection queries. Three other studies that fit our inclusion criteria were added because they were not included in the query-based search results. Relevant attributes for each retained manuscript—including author(s), title, source, year of publication, and keywords—were collected. Additionally, we gathered the characteristics of the target area(s), namely: country, climate region, type (i.e., landlocked or coastal), and size (via population and/or spatial extent). Information related to spatiotemporal LULC changes was also recorded. We focused on study period, data sources, considered seasons, day/night, and LULC extraction method(s). A detailed list of all included studies with a brief description of used methods, the characteristics of the study area(s), and results is presented in the Supplementary Materials.

3. General Findings

This section presents a descriptive statistical analysis of the gathered literature. Section 3.1 describes the statistical results regarding the yearly growth and publication sources of the selected papers. Section 3.2 provides insights regarding the investigated cities in terms of their geographical distribution concerning their physical characteristics (e.g., climate and topography). Section 3.3 describes the general characteristics of the reviewed publications in terms of the selected study areas and study periods.

3.1. Literature Synopsis: Trends and Source

Figure 2 shows the yearly counts of published papers investigating the relationship between LULC changes and SUHIs. Two noteworthy findings emerge from these counts. First, the usage of satellite images to investigate UHIs vis-à-vis spatiotemporal LULC changes is still in its initial phase, which is reflected by the short period and a small number of annually published studies (25 at most). It should be noted that the lower number of studies was possibly due to the selection criteria we adopted, as we required at least two dates to analyze LULC changes and, subsequently, SUHIs. The reader is referred to a more inclusive review by Zhou et al. regarding a longer coverage time range and looser inclusion criteria [10]. Second, a generally increasing tendency in the annual numbers can be seen, especially since 2010, suggesting a rising interest in this research topic. This could be attributed to the fact that medium-spatial-resolution Landsat data have been made freely accessible since then. Furthermore, the significant optimization of computing capabilities (e.g., big data storage and short execution/processing timing) in the last two decades, together with the significant and constant improvement in GIS and remote sensing methods, has attracted a growing number of researchers to contribute to this important topic.

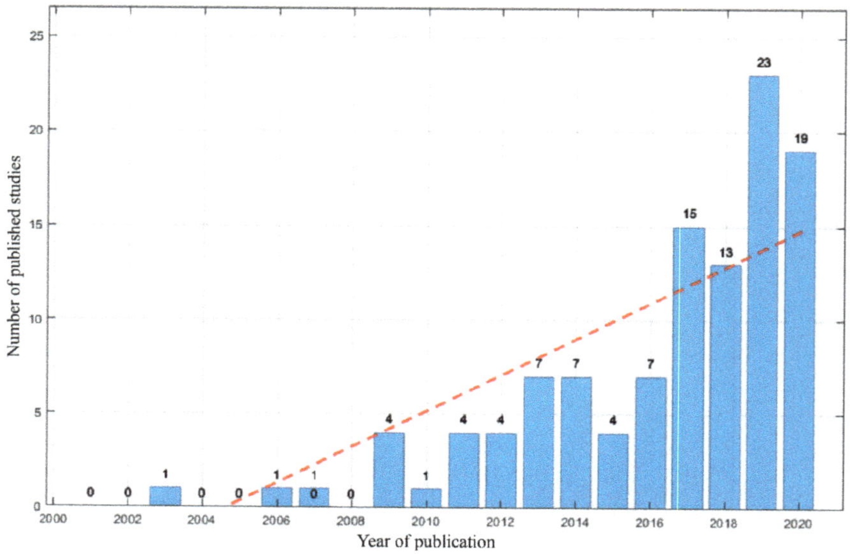

Figure 2. Annual trends of the published literature regarding studies' investigations of SUHI/LULC changes.

In terms of publication sources, the selected literature appeared in 61 journals. A vast majority of studies were published in *Sustainability* (10.8%) and Urban Climate (9.6%). Other top leading journals included the International Journal of Remote Sensing, Remote Sensing, and Sustainable Cities and Society, with 7.2% each. About 6% of articles were published in the *International Journal of Applied Earth Observation* and *Geoinformation* and *ISPRS Journal of Photogrammetry* and *Remote Sensing*.

3.2. Geographical Coverage and Cities Characteristics

Figure 3 displays the geographical distribution of the investigated studies. It should be noted that the number of studies does not correspond to the number of cities, as several considered studies analyzed multiple cities at once, notably [41], where the authors investigated the interlacement between LULC changes and the magnitude of UHIs in 10 Indian cities.

From a geographical standpoint, SUHI–LULC links investigated in 133 cities of 27 countries worldwide were reported. Figure 3 illustrates the spatial distribution of the total numbers of the considered studies in every country. Investigations focusing on Asian cities were predominant (83.2%). Few studies focused on cities in Africa (6.6%), South America (3.6%), North America (2.9%), Europe (2.2%), and Oceania (1.5%).

In terms of climatological characteristics, the spatial distribution and percentage of investigated cities based on their zones on the Köppen–Geiger climate system—according to a recent update published in [42]—are illustrated in Figure 3 and Table 1. Most investigated cities (42%) are characterized by a temperate climate. Specifically, those located in humid subtropical (Cfa) and dry-winter humid subtropical (Cwa) were found to represent 23.4% and 12.4% of all investigated cities, respectively. Most of these cities are located in Asia, such as Shanghai, Fuzhou, and Wuhan for Cfa and Guangzhou, Shenzhen, and Hanoi for Cwa. These climates are generally characterized by mild-to-warm summers and cool-to-cold winters [43]. Desert-climate cities were also found to be common target areas, as about 25% of all cities are located in such climate regions, most likely because UHIs thrive in such climates due to limited vegetation cover and water resources. Researchers have mainly focused on cities with hot arid (BWh) and hot semiarid (BSh) climates, with each of them representing 10% of the total considered cities. Another 23% of the examined cities belong to the tropical climates, with a focus on cities enjoying Aw (tropical savanna) climate (14%). Constant high temperatures typically characterize a tropical climate. Continental-climate cities were found to be the least investigated, with a share of only 10%, and approximately 7% of these studies focused on the Dwa (hot summer continental) climate. This climate has warm months that average between 10 and 20 °C, and its cold months are at or below 0 °C.

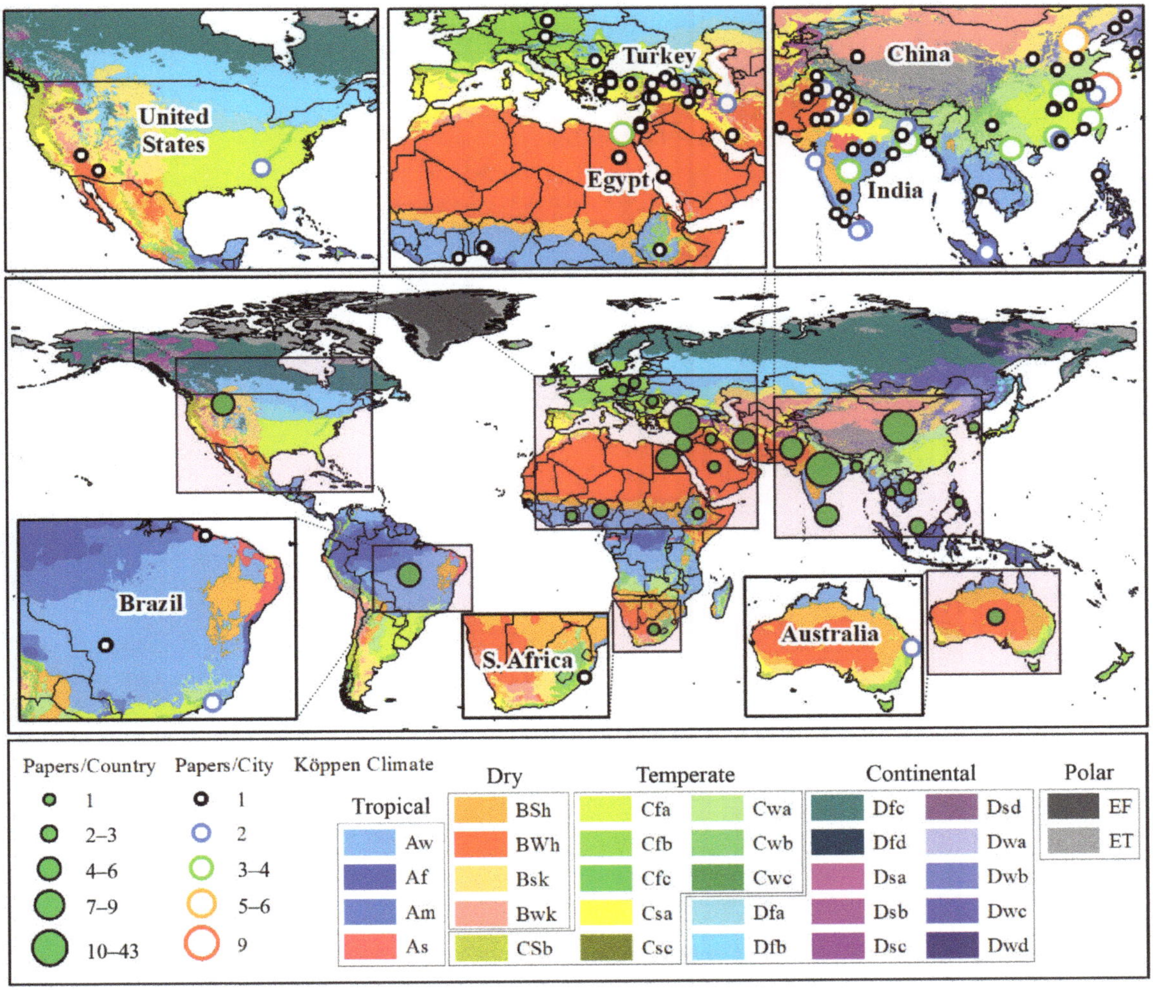

Figure 3. Spatial distribution of case studies aggregated based on countries where cities are located with regard to the world's Köppen climate classification according to [42] (licensed under CC BY 4.0). The circle size on the left indicates the range of papers count per country.

Overall, cities with hot summers (BSh, BWh, and Dwa), and humid/dry winters (Cfa, Cwa, and Aw) were found to be the most investigated cities. The focus on cities under these specific climates is attributed to many reasons. First, increasing SUHI magnitudes directly impacts climate change and urban expansion, especially since most cities under these climate regimes are situated in developing countries. Second, in contrast to tropical and continental cities, satellite data covering desert cities are not ruined by high percentages of clouds during most seasons of the year. This specific issue, which is discussed in detail in Section 4.7.1, is the most challenging obstacle facing studies investigating SUHIs in general and particularly those focusing on connections between SUHIs and LULC over wide areas and long time.

Table 1. Share of studies regarding the climate regions classified based on Köppen climate classification.

Temperate					Arid				Tropical			Continental	
41.6%					24.8%				23.4%			10.2%	
Cfa	Cwa	Csa	Csb	Cwb	BSh	BWh	BSk	BWk	Aw	Am	Af	Dwa	Dfa
23.4%	12.4%	4.4%	0.72%	0.72%	10.2%	10.2%	3.6%	0.7%	13.9%	5.1%	4.4%	7.3%	0.7%

3.3. General Characteristics: Study Periods and Target Areas

The use of historical remotely sensed data going back to the 1970s has made it possible to study decadal LULC spatiotemporal changes, as well as their impact on LST. Over half of the studies used research durations of more than 10 years, as shown in Figure 4. Selecting such lengthy periods helped provide distinct contrasts between past and current trends.

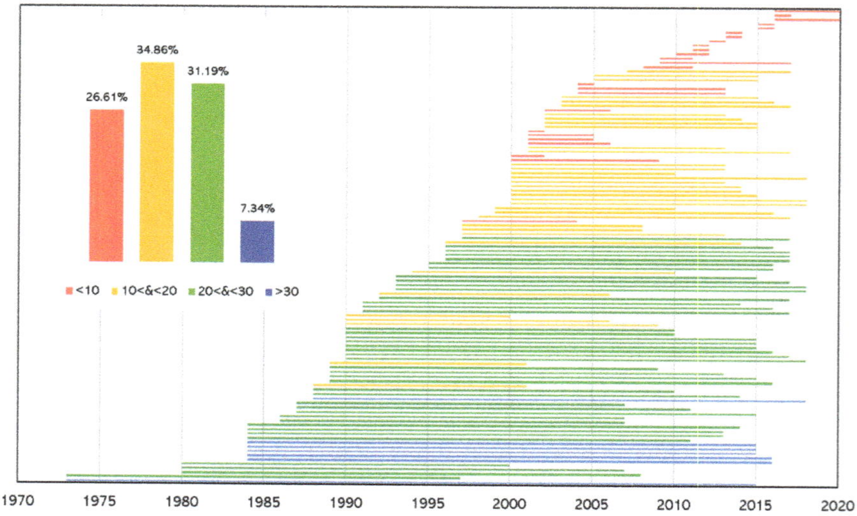

Figure 4. Timeline of study periods considered in the reviewed literature along with the percentage of studies based on different decadal time intervals.

To accommodate explosive population increases in developing countries while ensuring faster economic growth, urbanization rates in such countries are generally faster than those observed in developed countries. Previous studies have affirmed that urban expansion is one of the leading drivers of SUHI development in UAs. These facts are reflected in the number of target areas in developing countries presented in Figure 5, illustrating the number of reviewed studies per country categorized according to their development indexes [44]. In detail, compared to 7.3% in developed countries, 92.7% of investigations were reported in developing nations. Chinese and Indian cities have been the most investigated. Shanghai was ranked first, with nine publications [20,45–52], followed by Delhi [22,53–56] and Beijing [48,57–60], with five publications each.

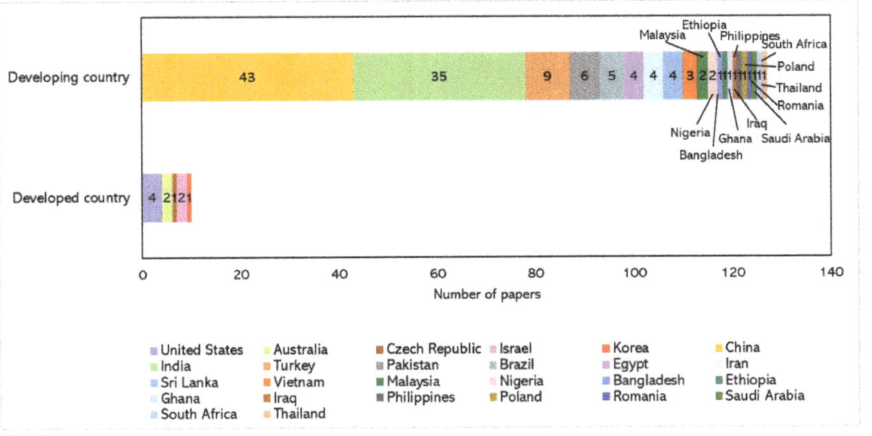

Figure 5. The number of studies that focused on cities in developed and developing nations.

4. Content Key Focuses

This section focuses on the key elements used in the literature to study the interlacements between LULC changes and SUHIs. Firstly, we identify and compare the main data sources used in terms of offered advantages and limitations (Section 4.1). Secondly, we detail the most common techniques of LULC extraction (Section 4.2) and SUHI calculation (Section 4.3). Thirdly, qualitative and quantitative methods for analyzing the association of LULC and UHI changes in cities are detailed (Section 4.4). Next, we provide an overview of the factors affecting SUHIs (Section 4.5). Finally, the mitigation strategies recommended in the considered literature (Section 4.6), followed by the limitations and future directions (Section 4.7), are summarized.

4.1. Data Sources

Table 2 shows the most common satellite data sources used to extract LULC changes and/or LST retrieval. Landsat series satellite data were found to have been dominant among researchers (95.5%), followed by MODIS (10.9%) and ASTER (4.5%). Only one study relied on HJ-1B data (0.9%). Other sources of imagery were found to include high-resolution satellites mainly used for LULC classification validation purposes such as IKONOS [61], SPOT [62,63], IRS LISS-III [64], and Sentinel 2A [65]. Additionally, Table 2 shows the percentage of studies based on the considered study periods. Landsat data are often used for all study period ranges—specifically for periods between 10 and 20 years (37%). MODIS data, on the other hand, are generally used in medium-study-period studies (50%). Regarding high-resolution imagery, they were used extensively in studies focusing on medium-period studies (43%). It should be noted that this imagery type is generally used for validation of LULC classification rather than LULC changes analysis.

Landsat series satellite data were used in approximately 95.9% of the considered literature for LULC extraction, LST retrieval, or (mostly) for both. More than half of reviewed studies (77.6%) were found to have used Landsat series data to investigate medium-to-very-long (more than 10 years) LULC changes between 1980 and 2020 (Table 2). As a cooperation program between the National Aeronautics and Space Administration (NASA) and the United States Geological Survey (USGS), the series has been operating since 1972 and consists of eight satellites that can image the earth in moderate resolution. They are considered to comprise the longest-running optical remote-sensing-based satellite constellation for monitoring the earth. The Landsat Multispectral Scanner (MSS) was used on Landsat 1–5. For Landsats 4 and 5, the Thematic Mapper (TM) sensor was also mounted. Landsat 6 was an upgraded version of its predecessors, carrying MSS and the Enhanced Thematic Mapper (ETM). However, it did not reach orbit. Landsat 7 was launched in

1999 with the ETM plus (ETM+) on board. While the satellite is active, its resulting scenes have missed approximately 22% of the data due to failure in the Scan Line Corrector since May 2003. The primary sensors on Landsat 8 are the Operational Land Imager (OLI) and the Thermal Infrared Sensor (TIRS). Though LULC information could be extracted from early images, temperature information retrieval was made possible in July 1982 when the TM sensor was introduced to Landsat 4 and others onward.

Table 2. Share of studies regarding the satellite data source used/study period considered categorized based on imagery resolution.

Spatial Resolution	Satellite/Sensor and Total Studies (%)			Use Per Study Period Length	
				Classification *	Total (%)
High resolution	All sources (13.6%) including IKONOS, SPOT, GeoEye, and QuickBird			Short Medium Long Very long	28.6 42.9 21.4 7.1
Medium resolution	Landsat Series (95.5%)	ASTER (4.5%)	HJ-1B (0.9%)	Short Medium Long Very long	22.4 31.8 37.4 8.4
Low resolution	MODIS (10.9%)			Short Medium Long Very long	25.0 50.0 25.0 0.0

* Study period length is classified based on the following: Short = [>10]; Medium [≥10 and <20]; Long [≥20 and <30]; and Very long [≥30].

Landsat data are popular among researchers for many reasons. First, the data have been provided at no cost since January 2009, when the USGS made all Landsat data free to the public (Landsat 7 data were made free in October 2008). Before that, a single scene's costs varied between $20 and $4000 [66]. This has been beneficial, especially for studies focusing on developing nations characterized by fast urbanization trends and population growth. Second, Landsat data are considered to comprise the genuine global archive that has constantly been updated for almost half a century following a strategy of regular imagery acquisition rather than the limited acquisition on images of interest or ready for purchase [66]. Third, the characteristics of the latest satellites and their imagery, mainly those related to spatial resolution, revisiting cycle, and swath dimensions (Table A1) (which are described as moderate), are convenient for most LULC and SUHI studies focusing on cities. Generally, one scene per city is needed. However, depending on the geographic location of some cities with regard to the path and row of the satellite, mosaicking multiple scenes may be required (see [67]). Fourth, the extensive use of Landsat imagery has been an incentive to develop well-documented techniques for optimal use.

MODIS: The second commonly used satellite imagery for LULC/SUHI studies is that of MODIS (10.9%). In general, MODIS data have been used in SUHI studies focusing on medium study periods (between 10 and 20 years) (Table 2) and/or seasonal variations. MODIS is an acronym that stands for Moderate Resolution Imaging Spectroradiometer. It is a NASA Earth Observing System (EOS) sensor that flies on NASA's Terra (1999) and Aqua (2002) satellites. Terra's orbit crosses the equator from north to south in the morning, and Aqua crosses the equator from south to north in the afternoon, providing worldwide coverage every 1–2 days. The EOS satellites have a scanning pattern of 55 degrees and orbit at the height of 705 km, with a swath width of 2330 km.

ASTER: ASTER imagery has been used in 4.5% of studies, making it the third most popular source. ASTER is another sensor mounted on the *Aqua* satellite launched into orbit in 1999. Given its capabilities, ASTER is usually used for nighttime analysis [54]. Satellite datasets of Terra ASTER level-1B contain radiometrically calibrated and geometrically registered data for all ASTER channels [54].

HJ-1B: Another imagery source used in the considered literature is HJ-1B, an optical minisatellite (2008) that forms the HJ-1 (Huan Jing = Environment) constellation along with another two small satellites—optical HJ-1A (2008) and radar HJ-1C (2012). The instruments onboard HJ-1B include two charge-coupled device (CCD) cameras and an infrared (IRS) camera whose spatial resolution range is between 30 and 300 m [68]. Although the minisatellite constellation is primarily designed for environmental monitoring and disaster risk assessment such as floods and forest fires [68,69], HJ-1B imagery was used to assess the relationship of LULC and SUHIs in [70].

4.2. LULC Classification: Types, Methods, and Indices

In this subsection, we shed light on the procedure of extracting LULC information from satellite data, focusing on (i) regularly adopted LULC classification schemes, (ii) LULC extraction methods including supervised and unsupervised methods, and (iii) indices used in tandem with LULC classification to identify the composition or configuration of target areas.

4.2.1. LULC Types

In the considered literature, authors often opted for different LULC classifications depending on several factors, including the geographical settings of the study area, the main focus of the study, and the quality of remotely sensed data. While the terminology used to depict each varies from one study to another, it was observed that the four classes most often present in most cases were UA, vegetation, water, and bare land.

In most cases, LULC types are defined based on the authors' or experts' knowledge of the geographic settings of the targeted area. In [71], for instance, the authors selected LULC classes as a result of consultation with an expert familiar with the city of Erbil to better assess the associations between LULC and LST. Another approach for determining LULC classes is to rely on the classification scheme provided by governmental or scientific agencies. For example, several studies focusing on Chinese cities used the land use classification system by the China National Committee of Agricultural Divisions dated back to 1984 [72] or by the Chinese Academy of Sciences (as in [73]). In comparative studies focusing on cities with similar characteristics (e.g., climate and landscape configuration) that are not necessarily located within the same country or region, authors have preferred to use a unified classification system such as the USGS 24-category land use categories. Fan et al., for example, used USGS 24 to compare the impacts of the spatiotemporal variations of LULC on UHIs in five desert cities located on three different continents (North America, Africa, and Asia) [5]. Additionally, there have been instances where authors used ready LULC classes based on inventories published by scientific agencies. An example of such an occurrence could be found in the study on the Polish city of Poznań [74], where the authors used the Europe-focused CORINE Land Cover inventory, with a total of 44 classes, updated and maintained regularly within the Copernicus Programme [75]. The same classification scheme was also followed in a comparative study of the effects of SUHIs in seven big Turkish cities [76].

A few studies were found to combine LULC types into few representative classes, as in [77], where the authors combined multiple LULC types into what they called "functional zones". For instance, a functional zone labeled as "dense green space" was created as a result of the combination of several types of forests (indigenous, plantations, and thicket), water bodies, and wetlands. Besides the high multiplicity of LULC types and the small spatial extent of certain types, the main rationale behind this approach, according to the authors, was the similarity of the properties of the combined types. Huang and Wang opted for a similar approach by defining "urban functional zones" (UFZs), which are abstracted from typical LULC types to depict human activities [78]. On the other hand, it was observed that authors generally tended to expand typical LULC types to more representative classes to evaluate their impacts on SUHIs. For instance, vegetation classes were classified into several types based on their density [79,80] or types [79]. In the same vein, urban development types were expanded to further understand the impact of different types of urban expansion on the worsening of UHIs.

In addition to typical LULC classes such as IS, vegetation, water, and cropland, Tran et al. classified urbanized areas into different categories to reflect their actual development types [80]: (i) infill, which refers to newly developed constructions surrounded by older built-up areas; (ii) extension, depicting new constructions intersecting with older built-up areas; and (iii) leapfrog development, describing new built-up areas separated from old UAs. The authors found that infill development exhibits the highest LST, followed by extension and (lastly) by leapfrog development, which is attributed to the high-LST LULC types surrounding infill and the good planning policies followed when constructing leapfrog UAs, which require proximity to public spaces such as parks and water bodies. It is evident that selecting convenient LULC classes for SUHI studies is a crucial step that might provide new insight and improve the understanding of the influence of different LULC types on SUHIs.

4.2.2. Extraction Methods

Pixel-, subpixel-, and object-based classifications have been used in revised studies using parametric and nonparametric methods. Though parametric methods, including the maximum likelihood classifier (MLC) and iterative self-organizing data analysis (ISODATA), are the most used, nonparametric machine-learning algorithms were found to have gained ground in recent SUHI-related publications. Such methods include support vector machine (SVM; [61,76,81–83]), random forest (RF; [81–85]), k-nearest neighbor (kNN; [81–83]), and neural networks (NNs; [81–83]). Nonparametric techniques present the advantage of being able to be used without a priori assumption on data distribution. This advantage has enabled researchers to evaluate several algorithms to determine which one has the highest accuracy scores. For instance, in [83], the authors compared four different classification techniques: SVM, kNN, RF, and NN. SVM outperformed all in terms of overall accuracy and kappa statistic. A similar approach and identical set of algorithms were employed in [81,82].

Regarding accuracy, studies relying on supervised methods have typically used a confusion matrix, also known as an error matrix, which encompasses statistics comparing the count of real samples and the corresponding predicted ones. Overall accuracy is the generally used metric to assess classification results, which correspond to the percentage of correctly predicted samples of all classes. Two other metrics reflecting the accuracies of different classes have also been calculated: the user's accuracy and the producer's accuracy [31,41,80,86–99]. Another measure used in tandem with the aforementioned accuracy metrics or alone is the kappa statistic [100], the values of which range between 0 and 1. The lowest values indicate slight to nonagreement between two datasets. In detail, according to Monserud and Leemans, values below 0.4 indicate poor or very poor agreement, values between 0.4 and 0.55 indicate fair agreement, values between 0.55 and 0.7 indicate good agreement, values between 0.7 and 0.85 indicate very good agreement, and values above 0.85 indicate excellent agreement between images [101].

For most studies, the minimum required for the overall classification accuracy was 85%, following the recommendations of [102,103]. More recent studies have applied an even stricter accuracy threshold of 90% for both overall accuracy and kappa statistic, as suggested in [104]. These demanding requirements may be attributed to the decreasing spatial resolution of recent remotely sensed data, along with the advancements of classification techniques. Nevertheless, it was difficult to reach such high accuracies in some cases. Fan et al., for example, employed an object-based classifier to perform eight-class LULC classification in five desert cities [5]. The authors succeeded in achieving accuracies superior to 85% for all three target years in the considered cities except for those of the Indian city of Jodhpur—which ranged between 80% and 82.57%. They attributed this relatively low accuracy to the complex spatial distribution of different LULC classes alongside the similarity of spectral responses in several of these classes. For this reason, unsupervised classification is usually first conducted by using satellite data to determine the spectral separability of the LULC classes [54]. Unsupervised methods include ISODATA alone or as a hybrid method with a supervised algorithm [105]. The initial use of a hybrid approach initially allows authors to gain insight into the number of major spectral differences [77]. Other studies have solely relied on an unsupervised approach [92,99,106].

Samples for validation purposes have been collected through field surveys using GPS [54,93,107,108] and aerial photos [51,72], as well as being extracted from the commercial imagery of high-resolution satellites such as IKONOS [58,109], SPOT [51,52,63,72,107,110,111], QuickBird [20,58], GeoEye [111], and GaoFen-1 [112] provided by private companies. However, researchers are moving away from these methods for several reasons. First, fieldwork is a time- and budget-consuming task that requires hefty effort. Second, high-resolution images are costly, which has led to issues related to the availability of historical data in many regions in the world. In recent years, however, the free-of-charge Google Earth has offered a reliable alternative for these traditional sources to extract balanced validation points using sampling techniques such as stratified random sampling [76,99,113–115].

4.2.3. Indices as LULC Proxy

The use of indices calculated from different sensors bands or other indices was found to be common in the reviewed literature. Authors have used these indices to further assess the relationship between LST and different characteristics of target areas including biophysical properties, landscape composition and/or configuration, and (less frequently), an alternative approach to the supervised/unsupervised methods for extracting LULC types. Table A2 lists the most employed indices in the reviewed studies, as well as the proportion of each category of indices. Broadly, the employed indices were found to fall into five categories: vegetation, built-up, water, bare land, and landscape metrics.

There are several vegetation-related indices. However, the normalized difference vegetation index (NDVI) and fractional vegetation cover (FVC) remain the most employed indices, accounting for 52.5% and 8.2%, respectively, of SUHI research. The NDVI has been solely used to extract emissivity values for LST retrieval (Section 4.3.1) in several studies; as a result, these studies were not included in this analysis. A few other studies relied on the soil-adjusted vegetation index (SAVI), transformed difference vegetation index (TDVI), and enhanced vegetation index (EVI), which were used together only in 4.5% of reviewed studies.

The built-up indices were found to be second in terms of frequency of use in the reviewed studies. While several new indices, including the enhanced built-up and bareness index (EBBI), index-based built-up index (IBI), and normalized difference impervious surface index (NDISI)—accounting for 2.7%, 0.9%, and 0.9%, respectively—have been introduced in recent studies, the normalized difference built-up index (NDBI) was employed in almost 24.5% of studies to represent built-up areas. Water was mostly represented by the modified difference water index (NDWI) in about 7.3% of studies, followed by the modified normalized difference water index (MNDWI) with a share of 6.4%. Of the studies, only 0.8% relied on the land surface water index (LSWI) and normalized difference moisture index (NDMI) each. Two indices were used as proxies of bare lands: the normalized difference bareness index (NDBaI) and dry bare-soil index (DBSI), which were found to account for, respectively, 4.9% and 2.1% of studies. Landscape metrics, both compositional and configurational, were employed in approximately 5.7% of studies.

4.3. SUHI Calculation Methods

In this subsection, we shed light on the approaches employed in the considered literature to quantify SUHIs. The approaches roughly fall into two categories: the first consists of methods using retrieved LST as a proxy, and the second is based on the calculation of SUHI intensity defined as LST differences between UAs and less developed areas (e.g., rural). Numerous studies combined both approaches. Though LST retrieval is a critical step in the process, a detailed description of the different methods used is beyond the scope of this review. Instead, the reader is referred to a published review by Li et al. for a comprehensive summary of the methodologies and algorithms developed for LST retrieval [116].

4.3.1. LST as a Proxy of SUHI

LST is regarded as the land surface's radiative skin temperature [117], which is considered a key element in the physics of the land surface through energy and water exchanges with the atmosphere [109]. Though LST is believed to be more adequate for quantifying UHIs at urban canopy layers [6], SUHIs tend to follow the same patterns as AUHIs according to several previous studies [118–120]. In most of the considered literature, LST was used as the sole proxy of SUHIs to analyze their spatial variability concerning different LULC classes, which can be attributed to the fact that LST is highly correlated with surface properties [121]. Chen et al., for instance, classified the retrieved LST information into five categories (very low, low, medium, high, and very high), reflecting its magnitude, based on the deviation from the mean value [112]. The relationships between LST and different LULC data are further detailed in the following section. In this section, an overview of the methods of LST retrieval from satellite thermal data is presented as a critical step in the reviewed literature.

The retrieval of LST from satellite thermal data is a complex but critical procedure in SUHI studies. Its complexity resides not only in the multiple processes that are involved, including radiometric sensor calibrations and the required adjustments for air and surface emissivity [109] but also in the number of parameters that need to be accurately known, such as emissivity transmittance and atmospheric temperature [114].

In general, efforts to retrieve surface temperature (ST) from the thermal data of satellite sensors have been made since the late 1960s, when Anding and Kauth introduced a method known as the split window (SW) that was capable of estimating sea surface temperature [122]. Since then, numerous attempts were made to extend the SW technique to obtain LST before the algorithm was extended to be able to retrieve it [123,124]. A wide range of other algorithms and techniques have been additionally proposed to retrieve LST.

According to Li et al., LST-retrieval methods fall into three broad categories depending on whether land surface emissivity (LSE) and atmospheric quantities are known [116]: (i) retrieval methods with known LSE including single-channel (SC) algorithms, multichannel (MC) algorithms, and multiangle algorithms; (ii) retrieval methods with unknown LSE including the classification-based emissivity method (CBM) and NDVI-based emissivity methods (NBEMs); and (iii) retrieval methods with unknown emissivity and unknown atmospheric quantities. Among these methods, SC algorithms, characterized by their simplicity, were found to have been widely used in the reviewed literature. As their name suggests, they are used to extract LST from the thermal data of sensors with a single TIR channel (e.g., Landsat 4–5 (TM) and 7 (ETM+)).

Various variants of the single-channel method have been proposed, notably the algorithm (often referred to as the mono-window algorithm (MWA)) specifically developed for the thermal band of Landsat 5 TM channel 6 by Qin et al. [125], which was applied in multiple reviewed studies [14,94,112,114,126,127]. The MWA generally requires three parameters: emissivity, transmittance, and effective mean atmospheric temperature [112]. Jiménez-Muñoz and Sobrino developed a universal single-channel (USC) algorithm capable of retrieving LST from any TIR channel requiring two parameters, such as atmospheric moisture content and emissivity [128]. Similar to the MWA, the USC algorithm has been applied in the considered literature across different cities [45,111]. SC algorithms, however, are limited in their applicability due to a variety of criteria that have been deemed difficult, if not impossible, to meet—notably, the a priori knowledge of the emissivity of each pixel [116] and detailed information about atmospheric profiles during the satellite overpass of a given target area [51,116]. In contrast to SC algorithms, MC algorithms, known as split-window algorithms, have been employed to retrieve LST from sensors with multiple adjacent thermal bands, including Landsat 8 TIRS, ASTER, and MODIS without the need for accurate atmospheric profile data at acquisition time.

4.3.2. SUHI Intensity

SUHIi is the second most common approach that has been used to quantify SUHI magnitudes in cities. Generally, SUHIi is the difference between retrieved LST means calculated in UAs and their surroundings, often underdeveloped, including suburban or rural areas. The boundary that has to be established to distinguish between the two regions is problematic to define. Broadly, two methods have been employed in the literature.

The first heuristic approach relies on city features such as traffic rings, administrative boundaries, and buffer zones. In a Shanghai-focused study [51], the authors divided the target area into three sublevels based on the inner and outer traffic rings surrounding the city: (i) the city proper located within the inner traffic ring, (ii) a periurban area within the inner and outer traffic rings, and (iii) a region beyond the outer traffic ring, which is a rural area with low-to-moderately dense residential areas, farmland, and natural surroundings. After considering this layout, the authors calculated seasonal and interannual variations of two SUHI intensity indices: the first was between the city center and surrounding rural areas, and the other was between urban fringe and surrounding rural areas. A similar approach has been employed in other cities such as Jinan city [107] and Fuzhou city [72]. Another way to delineate urban and rural areas found in the literature consists of using administrative boundaries [129]. Buffer zoning while centering the city core is an alternative approach found in the literature to extract urban/rural areas. Rasul et al. defined rural areas with a 10 km buffer zone beyond the city core [71]. However, the authors did not report how the core was delineated. It should be noted that this approach was applied to an arid city with outskirts characterized by quasi similar land cover classes. Nonetheless, these heuristic approaches might not be suitable for studies focusing on the decadal monitoring of SUHIi due to major changes in landscapes.

Other studies have opted to rely on LULC classification as a way to differentiate urban- and nonurban-labeled pixels. Ultimately, the LST means of areas classed as nonurban are subtracted from areas identified as nonurban, including vegetation (grass, forest, etc.) and bare land. Numerous studies were found to have followed this approach [130,131]. Following the same concept, other studies employed ISs as proxies of UAs, and rural areas have been represented by green space (GS) in most cases. This approach was applied, for instance, by Estoque and Murayama to calculate the SUHIi in the tropical mountain of Baguio, Philippines [12]. The authors calculated the SUHIi as the mean LST difference between ISs and GSs. Furthermore, in an attempt to replicate the concept of the inter-zone UHI intensity comparison proposed in [132], which involves an initial classification of the landscape into urban or local climate zones followed by inter-zone temperature differences, the authors distinguished two types of GS in addition to IS: a GS1 that includes forest and shrubland and a GS2 that includes grassland and cultivated land. Subsequently, the SUHI magnitude between the IS and other zones (GS, GS1, and GS2) was calculated. The same approach was applied in Kandy City, Sri Lanka [83].

4.4. Relationship Assessment of LULC and SUHIs

Various methods have been employed to assess the relationship between LULC changes and SUHI changes. Some studies used more than one method for the analysis, either for comparison or validation purposes.

Regression analysis tops the list with approximately 20.9% of manuscripts. Ordinary least square (OLS) regression has been used more frequently to characterize the relationship between LST and LULC based on several indices presented in Section 4.2.3. Though popular among researchers, OLS estimators present a major limitation; according to Deilami et al. [133], they do not account for spatial variability, which leads to issues associated with spatial autocorrelation and nonstationarity. Due to this limitation, the geographically weighted regression (GWR) was used in several studies as a potential alternative to counter OLS limitations. In GWR models, the spatial variability between the response variable (LST) and explanatory variables (e.g., indices and other factors detailed in Section 4.5) is taken into consideration. A GWR model generates estimates for every point in space based on local linear regression estimators that depend on a subset of information from neighboring points [134]. Although it was found that GWR models outperform OLS models in multiple studies [108,133–135], both procedures have their own benefits. In contrast to GWR, which is more effective at the local level, OLS has been found to aid in quantifying the impacts of various factors of SUHIs at the regional level [133,134].

Chart analysis, which relies on the zonal statistics method and is presented in the form of charts to illustrate the relationship between LULC and SUHIs over time, was found to come second (19.8%). Among the selected papers, 11 analyzed the mean LST for each LULC category and compared the LST change in different years. Overall, the trend of mean LST was found to be increasing in all LULC categories, and the built-up always had the highest LST among all LULC categories [54]. Moreover, vegetation coverage is an essential factor that can help cool cities and reduce the impact of LST. However, the benefit of vegetation has not been given sufficient attention. For example, the percentage of vegetation was found to have decreased from 6.3% to 1.9% over 25 years in Shiraz city, Iran [136]. At the same time, the mean LST of vegetation changed from 35.1 °C in 1993 to 39.5 °C in 2018, which proved that the cooling power of vegetation was affected by the total quantity. More accurate analyses would help to better identify the impact of different LULC types on LST. For instance, some authors plotted the space lattice of the LST, NDVI, and LULC, and the results showed that the LST values of water and green spaces were lower than those of built-up and cropland spaces [137].

Correlation analysis was found to be the third most common method (8%), and it has often been used together with regression analysis. We distinguish three types of correlations used in reviewed studies: (i) Pearson correlation, (ii) Kendall rank correlation, and (iii) canonical correlation analysis. Pearson correlation was used in most of the studies [11,45,52,97,99]. Only two studies opted for the other two types. Lo and Quattrochi used both canonical and Pearson correlations analyses to investigate the relationship between LST spatial patterns and those of volatile organic compounds (VOC) and nitrogen oxide (NOx) emissions in Atlanta, US [31]. Kendall rank correlation was used in [5] to detect the monotonic dependence between SUHIs and relative urban-rural vegetation differences—a metric that measures the difference of NDVI means in urban and rural areas—concerning the populations and UAs of five desert cities. The authors argued that Kendall's coefficient offers resistance against outliers and missing values in addition to its capability to identify any type of monotonic relationship—not just linear dependence, as measured by the Pearson coefficient.

Calculating the contribution index (CI) was found to be another common approach (8%) used to evaluate the impacts of different LULC types on LST variations. The CI measures the thermal contribution of various LULC categories on the LST by multiplying the proportion of a given LULC type to the entire study area by the difference between the mean LST of the LULC type in the question and that of the whole study area. Positive CI values of an LULC type indicate a direct influence on enhancing SUHIs, whereas negative values indicate a negative effect on SUHIs [138]. The index was first employed by Chen et al. to assess the impact of 10-year LUCL spatiotemporal changes on SUHIs at a regional-scale level in the Pearl River Delta located in Guangdong Province and a city-scale level in Shenzhen city [87]. It was then used in multiple studies in different research areas, including Shenyang, China [94]; Ethekwini, South Africa [77]; Wuhan, China [139]; Malda, India [138]; and Delhi, India [55]. Based on the obtained CI values of different LULC types, Pramanik and Punia further calculated another index called the landscape index (LI), defined as the quotient of the CI of SUHI sink (i.e., croplands, vegetation, and water bodies) and source (i.e., built-up, fallow, and bare lands) landscapes [55].

Though fewer, other noteworthy methods such as the ANOVA test [61], the crossover comparison method [12,83], grid-level analysis [11], centroid movement analysis [45,140], temperature vegetation index space [55,141], and hot-spot analysis [80] have also been used in studies.

4.5. Factors Affecting SUHIs

Various factors contribute to variations in SUHI magnitude over space and time. While the examined literature mainly focused on the impacts of LULC types and their interannual dynamics, the effects of additional variables were also investigated. Figure 6 shows the main factors that were found to have been used in the reviewed literature, as well as the most common combinations between them. These variables fall into four broad categories: (i) LULC types and their spatiotemporal dynamics, (ii) landscape composition and configuration, (iii) terrain characteristics, and (iv) socioeconomic factors. In the following section, we provide a detailed explanation of the main factors included in each category.

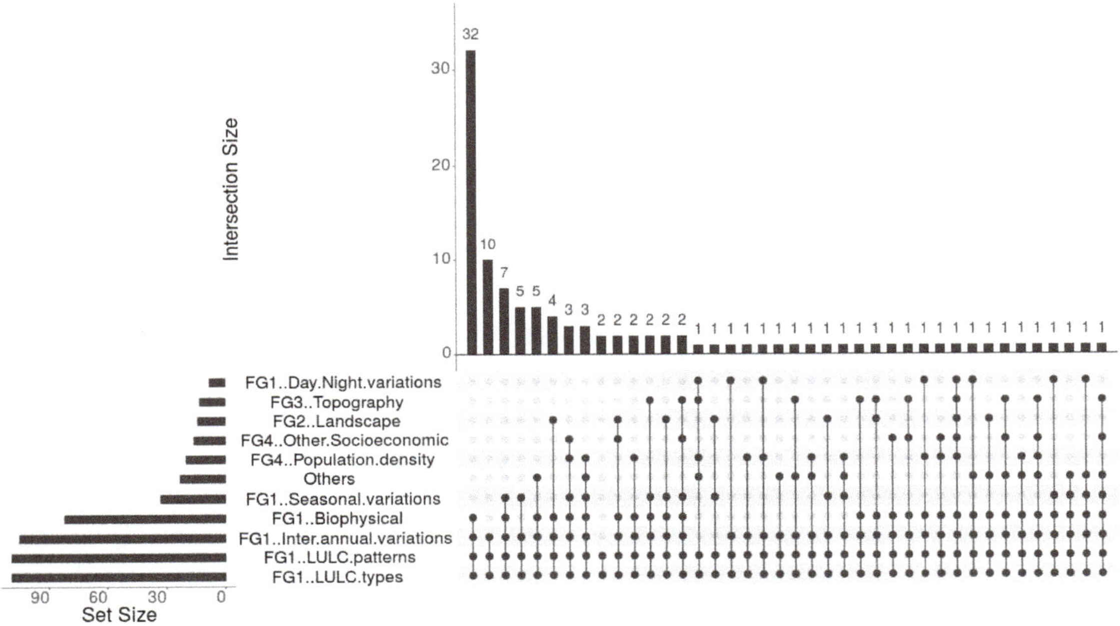

Figure 6. The main SUHI-contributing factors along with the frequency of their combinations, often considered when analyzing SUHI evolutions in the reviewed literature. FG refers to the main group each factor belongs to as follows: FG1: LULC types and their spatiotemporal dynamics; FG2: landscape composition and configuration; FG3: terrain characteristics; and FG4: socioeconomic factors.

4.5.1. LULC Types and Their Spatiotemporal Dynamics

The conversion of land cover to urban land uses across space and over time has been singled out in most reviewed studies as the primary cause for the increase in recorded LST values in UAs and ultimately high SUHI intensity in cities across the globe, regardless of their climates or geographic settings. In the tropical Malaysian capital city of Kuala Lumpur, for instance, Amanollahi et al. found that the LSTs of areas previously covered by forest and light vegetation before being transformed into UAs as a result of a 16-year urban sprawl (1990–2006) had increased by 4 and 3 °C, respectively [142]. Similar observations were reported in [97] for the temperate Indian municipality of English Bazar, where the LULC dynamics of 23 years (1991–2014) resulted in the radical transformation of multiple land cover classes into the built-up areas of water, agricultural lands, and Mango orchards, which led to increases in LST values by 1.9, 1.4, and 1.5 °C, respectively. In Beijing, a continental-climate city, Guo et al. noticed an 8% increase in ISs from 2005 to 2015, resulting in the mean LST of ISs being about 2 °C higher than that recorded in green areas [59]. In an arid climate, El-Hattab et al. found an increase of approximately 1.6 °C in Cairo from

1984 to 2015 as a result of a heavy urban expansion and vegetation cover loss [143]. In mountain cities, identical tendencies were also observed. Increases in SUHIi were reported for instance, in Baguio (+0.7 °C between 1987 and 2015) [12] and Kandy (+2.3 °C between 1996 and 2017) [83].

The resulting LULC spatial structure has been found to predominantly affect the spatial patterns of the recorded LST and to intensify SUHI effects [80]. By examining various studies carried out in different cities across the globe, we compiled a list of the LULC types with considerable influence on SUHIs including ISs/UAs, vegetation, water bodies, and seasonal variations.

- IS/UA: UAs, along with pavements and road networks, form ISs. A substantial body of research points to the fact that ISs have strong warming effects in cities [20], regardless of the reigning climate (tropical, desert, etc.) or the geographical settings (topography, elevation, etc.). While some studies have focused on ISs, others have separately investigated the effects of UAs, roads, and pavements on UHI. However, both sets of studies concluded that the highest LST values are exhibited by either UAs or ISs [11,14,41,45,72,74,80,97,111,112,115,126,133]. Regarding association, it has been found that both ISs and UAs have strong and significant linear positive correlations with LST [80,87,99,112] during all seasons [97]. This is attributed to the fact ISs absorb more solar radiation and have greater thermal capacities and conductivities, allowing heat to be retained during the day and released at night [143]. Ultimately, urban expansion exacerbates this process. Tran et al. estimated that a 1% increase in UAs in the Hanoi inner city could raise the surface temperature anywhere between 0.075 and 0.108 °C [80]. Urban expansion, however, does not only concern the size of UAs (i.e., footprint areas), due not only to their low albedo roofing materials (e.g., concrete and asbestoses)—as observed in [11,33,106,113]—but also the buildings' heights [56,76], UA density [112], porosity (defined as the ratio of total open space to total built-up areas [133]), and UA development level [74]. Another IS component, which consists of pavements such as parking lots and harbor jetty covered with dark materials such as asphalt, was found to be a significant contributor to UHIs [33]. Roads, on the other hand, have been found to increase SUHI impacts in two ways: first through their paved surfaces that absorb shortwave radiation and store heat throughout the day and release it slowly at night [99] (similarly to other pavements) and second via emissions produced by traffic passing through them.
- Vegetation: In contrast to ISs, vegetation absorbs solar radiation and removes a great amount of stored heat via evapotranspiration—a process that releases water vapor into the ambient air and subsequently contributes to cooling surrounding areas [11,12]. The relationship between LST and vegetation cover is complex. It depends on many considerations related to the study area (e.g., seasonal variations and landscape topography) and the characteristics of the vegetation cover itself (e.g., the nature of species, heights, and density). Numerous studies in the reviewed literature reported a negative linear relationship between LST and vegetation, as quantified through multiple indices (refer to Section 4.2.3 for a detailed list), most notably the NDVI [11,45,53,87,88,98,111,126,136,144–149] and FVC [53]. Rotem-Mindali et al. even found an exponential relationship between the NDVI and LST [150]. A few other studies reported negligible correlation due to various possible causes. Rasul et al., for instance, found a weak yet significant inverse relationship between LST and the NDVI during the summer season in the city of Erbil, Iraq, which is characterized by a temperate climate [71]. After considering seasonal variations in the Chinese city of Jinan, Meng and Liu concluded that FVC is negatively correlated with LST during all seasons except for winter [107]. Likewise, Wang et al. argued that variations of seasons were a possible cause of their obtained weak correlation between LST and the NDVI in Shanghai [20]. After exploring the reviewed case studies, however, it became evident that a fast rate of urban expansion to the detriment of vegetation cover leads to the weakening of the impact of vegetation on LST in comparison to the influence

exerted by ISs. This can be concluded from the weak correlation between LST and vegetation when compared with LST and ISs, as reported in [20,96].

- Water: As with vegetation, water bodies have a cooling effect due to their ability to absorb heat and release it in the form of water vapor, leading to lower ambient temperatures in their vicinities. In an investigation of the Chinese city of Wuhan, in which water accounts for over 25% of the total area of the central district—including two rivers (the Yangtze River and Han River), East Lake (the largest urban lake in China), and dozens of other lakes—Wu et al. found that both the area and the spatial distribution of water bodies contributed to significant distributions in the effects caused by SUHIs [70]. Moreover, the findings reported in multiple studies have affirmed that water bodies exhibit the lowest LSTs [11,20,41,45,47,60,73,78,79,88,90,93,97,98,111,112,115,138,139,151–153], along with vegetation cover. Furthermore, water has generally been found to have a negative correlation with LST [20,111,154], except for a few circumstances due to various reasons primarily related to the climate characteristics of cities. For instance, in [99], the authors attributed the lack of a significant relationship between LST variations and water bodies' changes that occurred in desert city Phoenix between 2000 and 2014 to the scarcity of large and evenly distributed open-air water bodies in the target area, along with possible LULC classification errors. In comparison with vegetation, water usually has a less significant association with LST [20].
- Seasonal variations: Seasonal variations have significant impacts on the spatial and temporal distribution of SUHIs. While the bulk of research has focused on interannual variations, seasonal fluctuations were also taken into account in several manuscripts, particularly in those with short study durations (less than 10 years). An illustration of such investigations was reported in [107], where the authors analyzed SUHI variations in all seasons for the Chinese city of Jinan. Based on LST data derived from Landsat images acquired between 1992 and 2011, they calculated two SUHIi indices (while considering rural areas) based on the two traffic rings surrounding Jinan urban center. Their findings showed that both SUHIi indices were stronger during summer (0.98–1.75 °C) and spring (0.40–0.85 °C) and weaker during autumn (0.16–0.37 °C) and winter (from −0.05 to −0.03 °C). These results are aligned with those reported in [52] for Shanghai. In the tropical Indian city of Delhi, Sharma and Joshi also found that summer had the maximum SUHIi (16.2 °C), followed by monsoon and spring seasons with SUHIi values of 13.8 and 12 °C, respectively [22]. On the other hand, the post-monsoon and winter seasons exhibited the lowest SUHIi values of 10.5 and 7.4 °C, respectively. The dominant factors impacting SUHI levels depend on seasonal changes. For instance, Zhang et al. determined that water, vegetation, and developed lands are the major drivers during all seasons except for summer in Shanghai [52]. These results partly agree with those reported in [77], where the authors found that the functional zone "dense green spaces" (including different types of forests, water bodies, and wetlands) in Ethekwini, South Africa, had a major heat contribution in the autumn, winter, and spring seasons. On the other hand, Song et al. reported that ISs had higher mean LSTs during all seasons except for winter in the Chinese temperate city of Hangzhou [155].

4.5.2. Landscape Composition and Configuration

In addition to LULC's different types and their dynamics, several studies have analyzed the relationships between urban climate and landscape patterns. A wide range of metrics has been developed to characterize these patterns and relate them with ecological processes [156]. As described in Section 4.2.3, these metrics can be categorized into two general groups: compositional and configurational. Metrics for landscape composition assess the existence and quantity of various patch types in the landscape without specification on its spatial characteristics, whereas those for landscape configuration measure the spatial distribution of patches within the landscape [50]. An illustration of such investigation was reported in [50], the authors analyzed the impacts of landscape structure on SUHIs in the

Shanghai metropolitan area in early spring (March 2001) and summer (July 2001). Eight landscape metrics—compositional (PLAND, SHEI, and SHDI) and configurational (ED, PD, LSI, CI, and CONTAGION)—were selected to investigate the relationship between LST and landscape patterns with regard to the LULC types of residential, public facilities, industrial, traffic, green land, water, agricultural, and other land uses including under-construction sites, cemeteries, and historic relics. High correlations were found between LST and landscape metrics for residential and urban green, with significant fluctuations observed due to alterations in SUHI spatial configurations induced by seasonal changes affecting vegetation patterns. Furthermore, residential areas' contributions to LST were found to be significantly influenced by their morphological characteristics. High-rise residential areas were found to have lower LSTs than low-rise residential areas for many reasons. First, low-rise residential zones have more horizontal active surfaces than high-rises. Second, smaller structures with lower thermal inertia produce shorter shadows than high ones. Third, when wind speed rises with height, the lower boundary layer's aerodynamics vary considerably. As a result, high-rise structures have a greater aerodynamic conductance than low-rise ones. Finally, the temperature of the lower boundary layer drops with height. The top of the high rise is cooler than the bottom. The proposed methodology was further expanded in [78], where the authors selected 2D and 3D building metrics related to shape, arrangement, composition, and distribution (in addition to landscape metrics) to evaluate the association between SUHIs and 2D/3D urban morphology in Wuhan, China. Two functional zones were considered: built-up (including residential, industrial, commercial, open space, and public services) and non-built-up (including urban green areas, agricultural lands, forests, rivers, and lakes). The findings of the study can be summarized as follows: (1) during the summer daylight, four LULC classes, namely buildings, other ISs (e.g., pavements, open areas, and squares), grass/shrubs, and trees, have significant effects on intraurban LST variation at a fine-scale; (2) the proportion of trees is the primary element affecting LST cooling via evapotranspiration and shade casting; and (3) during summer daylight, 3D urban morphology has an effect on LST, although the correlations are not as tight as those for 2D urban morphology. It was found that the most significant 3D metrics are the mean height (MH) and the sky view factor (SVF)—a metric measuring the extent of 3D open space that ranges from 0 (no sky in sight) to 1 (no obstacles in sight). A negative correlation was found between MH and LST in summer, indicating that high-rise buildings may help counteract SUHIs throughout the day. The SVF impact on LST is complex because it depends on the surrounding environment and relates to enhancing air circulation in the case of a higher SVF and reducing incoming solar radiation in the case of a lower SVF.

4.5.3. Terrain Characteristics

Generally speaking, most of the investigated cities in the reviewed literature are located in flat areas with slight or barely noticeable topographic variations. For that reason, few researchers have considered topography-related factors. On the other hand, researchers who focused on mountain cities or those dotted with varied landscapes concluded the importance of topography in understanding the spatial variability of LST. In a study that focused on exploring the spatial variations of LST between 2000 and 2010 in the tropical Malaysian capital Kuala Lumpur, Amanollahi et al. concluded that its elevated landscape plays a significant role in stabilizing the heat island in the city in two ways [142]. First, the expansion of ISs has resulted in fragmented vegetation cover—mainly consisting of forests in the study area—which has led to increases in surface temperatures; second, forests located at high elevations were recorded to have the lowest temperatures. The rise of LST due to fragmented forests had been somewhat regulated by the low LST recorded in high-elevated forests, resulting in the highest standard deviation of LST values of any LULC type. Second, the Titiwangsa mountain range with peaks over 2500 m above sea level (asl), located to the east, northeast, and north of the city, acts as a dam that stops winds coming from the north and east from blowing in the directions toward the city which, according to the authors, stabilizes the heat island in Kuala Lumpur. These findings relatively correspond to those reported in [157], in which the author partly attributed the lowest LST values observed in forests to their highest altitudes in the Czech city of Brno. Even in cities with a dry climate and relatively low landscape, such as Cairo, topography has been deemed a key factor influencing LST [106]. The author found that the mean LST at the highest region, El-Mokatam Plateau at 500 m asl, was 37 °C in comparison to the 39 °C value recorded at 100 m—which indicates an LST increase of 0.5 °C for each 100 m elevation. In the tropical mountain city of Baguio, the Philippines, Estoque and Murayama investigated the effects of topography on UHIs through additional parameters including elevation, slope, aspect, and hill shade [12]. Empirical analysis showed that these parameters, together with those of landscape composition, were capable of explaining a great deal of LST spatial variability in three different years of 1987, 2001, and 2015.

4.5.4. Socioeconomic Factors

Though relatively limited in occurrence, socioeconomic factors have also been considered as potential variables that contribute to elevated LST in cities, especially from the perspective that anthropogenic activities of residential, industrial, and/or commercial natures are responsible for enhancing surface temperatures.

- Population density (PD): PD has been introduced in several studies as a factor impacting SUHIs. For example, Zhang et al. investigated the links between PD and SUHIs in Nanchang, China [115]. The authors found significant positive correlations between PD and mean LST in 2000 and 2013, concluding that as PD increases, LST also increases. This result is consistent with that found in other studies carried out in different cities such as Wuhan [129], Fuzhou [72], Brisbane [133], Shanghai [45], Zhengzhou [73], and Hefei [158]. However, a PD-induced LST increase does not concern the total number of people located in a given area as much as it is related to the socioeconomic activities carried out by people daily in houses or places of work (e.g., industrial centers) [115]. This was confirmed in [129], where the authors concluded that, in contrast to PD variations that have been found to be somewhat correlated with LST, Wuhan population changes from 2000 to 2009 had no direct relationship with LST.
- Other socioeconomic factors: Our investigation shows that socioeconomic factors are often overlooked, generally because of a lack of data. Nevertheless, several researchers introduced such variables and assessed their impact on SUHI development. These include (i) emissions such as those of VOC and NOx [31], waste gas emissions [46], and carbon dioxide (CO_2) [115]; (ii) electricity [76,96]; (iii) employment density [61,133]; (iv) night light [45]; (v) gross domestic product (GDP, [46]); and (vi) house rent [159].

4.6. *An Overview of Proposed Mitigation Strategies*

In light of the findings of the reviewed studies, the magnitude of SUHIs can be said to be associated with rapid urban sprawl to the detriment of green areas. Several suggestions have been proposed in the literature to alleviate SUHI impacts. We present a compiled list of the main strategies suggested:

- Promoting greenery: Implementing policies encouraging more green areas [11,73,99,115], preferably within the urban premise and beyond, is one of the most suggested SUHI mitigation strategies given that increases in SUHI magnitude are highly associated with depletion of vegetation cover. Within UAs, greening concepts need to be implemented in both the horizontal and vertical directions [82]. Kleerekoper et al. described four forms of vegetation that can be fostered: parks, trees along streets, green in private gardens, and green roofs or facades [160]. Regarding the cooling effect, Wong et al. reported that ground greenery often lowers the surface temperature by 2–9 °C, whereas roofs or buildings walls covered with green layers reduce surface temperature by approximately 17 °C [161]. Additionally, the type of planted vegetation makes a great difference in cooling effects. For instance, Zhang compared the cooling effects of five regionally common shrubs in Guangzhou, China [162]. According to the author, only one vegetation type, *Murraya exotica L.*, showed excellent cooling effects. Furthermore, various studies have emphasized other parameters deemed critical for maximizing SUHI mitigation gains, including park shape [110,161], park size [110,161], and plant placement [161]. Thus, such considerations need to be considered by urban planners prior to implementing a strategy for optimal outcomes in the long run. Beyond UAs, the use of greenbelts surrounding cities is an effective way to combat SUHIs. In [94], the authors demonstrated the cooling effects of the greenbelt surrounding the Chinese city of Shenyang based on an investigation of LULC changes on SUHIs from 1986 to 2007, although this effect had started to fade because of urban sprawl. In desert cities, greenbelts are highly recommended, as seen in [106], where the authors recommended expanding greenbelts to protect new urban communities in the Cairo metropolitan area against SUHIs amplified by air pollution caused by dust and suspended aerosols.

- Safeguarding water bodies: Similar to vegetation, the reviewed studies reported that changes in water bodies had had substantial effects on SUHI mitigation [50,87,163], concluding that alleviating SUHIs necessitates safeguarding water bodies. In Wuhan city, Wu et al. found considerable spatial variations in SUHI effects, which they attributed to water bodies' distribution [70]. In a study carried out on the tropical city of Kuala Lumpur, Amanollahi et al. recommended increasing the number of retention ponds and adding new vegetation areas [142]. Among various advanced materials and techniques, Cai et al. suggested using waterscapes in the city of Fuzhou to counter the impacts of SUHIs [72]. While it has been reported that the impacts of water bodies on SUHIs are generally less effective than those of vegetation cover [164], combining both approaches would be a good strategy to reduce SUHI impacts.
- Using cool roofing/paving materials: Though the increase in green and blue areas may be feasible in cities under moderate climates (i.e., temperate, tropical, and continental), such measures are difficult, if not impossible, to implement in desert cities due to scarcity of water resources, as reported in the case of Phoenix [99]. Reducing ISs' properties to absorb solar radiation by using reflective materials [11,33,165] is an alternative option in cities with harsh and moderate climates alike. That can be conducted using highly emissive materials to prevent heat retention and by painting roofs and pavements with white paint [11].
- Other notable measures: Regarding LULC changes, the aforementioned measures have been the most suggested ones in the reviewed literature to alleviate SUHIs. A few studies suggested other ways such as (i) relying more and more on renewable energy (e.g., solar and wind) at the expense of fossil-fuel-based energy to reduce carbon emissions [11] and (ii) promoting incentive programs such as "carbon credits" targeting polluting companies to reduce emitted anthropogenic gases [11].

In summary, mitigation measures depend on the characteristics of the target area, mainly in terms of the reigning climate, local topography, size, and geographical settings. It is worth noting that this review only covers commonly suggested mitigation strategies in the considered literature. Policy and technology responses developed for the alleviation of UHI impacts are not discussed. For more information, one may refer to a study by Kleerekoper et al. [160] or a recent and more focused review conducted by Degirmenci et al. [39].

4.7. Limitations and Future Directions

In this subsection, the limitations or obstacles facing studies focusing on the associations between LULC spatiotemporal changes and SUHIs are discussed. This is followed by the proposed future paths that such a topic may be directed toward.

4.7.1. Limitations

Though studies focusing on the historical trends of SUHIs in relation to LULC changes have helped our understanding, several critical limitations have been pinpointed in the reviewed literature and need to be addressed in future research. These limitations concern the used data, executed methods, or a principal combination of both.

Perhaps the biggest challenge that limits such studies is the incapacity to validate satellite-derived LSTs using in situ measurements collected from field surveys or historical data collected via climatological station networks. The lack of such networks in targeted areas and the time- and budget-consuming fieldwork are the main obstacles that have hindered the validation process of satellite-derived LSTs in numerous case studies. The deployment of reliable networks usually comes with huge expenses, as it requires continuous maintenance. Even when available, in situ records may be incomplete due to various reasons, such as the forced relocation of stations (as reported in [144]), which makes them inadequate for studies focusing on the long-period monitoring of SUHI and LULC changes. In addition, high costs or privacy barriers may hinder researchers from accessing such valuable data. Nevertheless, numerous researchers have been able to confirm small differences between SUHIs and AUHIs, as illustrated in [54], where the authors found an error of 2–3 °C between satellite-derived mean LSTs and average temperatures based on in situ measurements. Error fluctuations were found to vary depending on the LULC type, while vegetation averaged 28.7 °C based on field measurements and 30.8 °C based on estimation via satellite data; UAs averaged 30.1 °C based on field measurements, and satellite estimations indicated 32.7 °C. In the city of Zhengzhou, quasi-identical results were described by Min et al., who found differences ranging between 2.7 and 4.7 °C, based on historical meteorological data [73]. Similarly, using records from four climatological stations, Nguyen et al. found that these differences varied between 0.3 and 3.34 °C, with higher differences observed in UAs and lower differences observed in rural areas [96]. A different approach was used in [22], where the authors conducted a field survey intending to collect over 200 ground observations across Delhi during the spring season. A high correlation (0.89), along with a relatively high coefficient of multiple determination ($r^2 = 0.79$) and quasi-equal standard deviation (0.72 °C), was reported between the two sets, thus indicating that the LST estimated from satellite thermal sensor data was as precise as those measured in situ. These findings are partly aligned with those reported in [97], in which the authors found a significant correlation between air temperatures (measured at 137 sites) and satellite-estimated LST during the winter (January 2014) and pre-monsoon (April 2014) seasons in the city of Malda. Aside from validation using meteorological station data or field survey measurements, LST data retrieved from satellite sensors have been used for comparison against each other. This was described in [126], where the authors used the MODIS LST product to validate LST estimated from Landsat imagery in different years (2006, 2009, and 2016). Empirical analysis showed the existence of moderate-to-strong correlations ranging from 0.44 and 0.55 between the LST estimated from Landsat (5 TM and 8 TIRS) and MODIS. According to Swain et al. [166], the refined MODIS LST products were reliable because they had already been validated using ground observations; they cited [167] as a reference paper where the authors demonstrated that their accuracy was better than 1 K in 39 out of 47 cases, with all 47 cases having a root mean squares of differences less than 0.7 K. A similar approach was used by Chaka and Oda to validate Landsat 8 TIRS data in the city of Hawassa, Southern Ethiopia [79].

A second limitation observed in the reviewed literature is related to the limited number of dates that were considered for analyzing the links between SUHIs and LULC changes, especially those monitoring long periods. This might hinder the provision of a true picture of the trends of the impacts of different LULC changes on SUHIs in cities during such a long timeframe, specifically those related to seasonality [22] or the occurrence of rare abnormal events (e.g., drought and cold waves) that might lead to misleading interpretations. A study by Feng et al. in Xiamen city, China, revealed that a cold wave that occurred in 1992 may have been the cause behind water bodies exhibiting the highest LST that year, despite being one of the coolest LULC types in other years (1987, 1997, and 2007) [121]. Moreover, in [111], the authors partly attributed the high temperature recorded in 2015 observed in the mountain city of Yan'an city, China, to the significant climatic impact of the global El Niño phenomenon. Another rare event impacting SUHIs was described in [166], where the authors reported an unusual increase in MODIS-derived mean LST recorded in 2009 in agricultural lands, which resulted in a small difference between rural areas and UAs. These anomalies were attributed to the severe drought that affected the region in 2009, which impacted the vegetation cover and soil moisture content. From this perspective, while it requires massive efforts and resources, relying on imagery data with higher temporal frequency to monitor interannual [83] and seasonal [22] SUHI variations is highly recommended because it would improve our understanding of the impacts of LULC spatial and temporal changes and prevent misleading interpretations. An example of such investigations was illustrated in [168]; using 507 Landsat images, the authors analyzed historic LULC changes—which occurred in Atlanta between 1984 and 2011—and explored their effects on thermal landscape patterns.

The third challenge that faces SUHI/LULC studies is related to the quality of archived remotely sensed data. It is evident that multiple data sources are available, as presented in Section 4.1, which have allowed researchers to carry out studies in different cities across the globe. However, a serious limitation that impedes the use of satellite data is the high percentage of cloud cover, notably in tropical regions. This issue was illustrated in [114], where the authors described cloud coverage as a "curse" for optical remote sensing following their struggle to find Landsat data of the Kuala Lumpur metropolitan city with minimal cloud cover. Subsequently, the authors could finally collect images during the summer seasons of five years between 1997 and 2013, covering percentages ranging from 4% to 24%. In the same study area, Amanollahi et al. found the same issue regarding the collection of cloud-free Landsat imagery, leading the authors to somewhat limit their investigation of SUHIs to only two dates (1990 and 2006) [142]. Though the two collected images were over 95% cloud-free, the authors noticed that 90% of clouds were clustered above the vegetated area, leading the authors to classify them as vegetation. Clouds not only contribute to less accurate LULC classification [142] but also affect the estimation of LST. This was further detailed in [169], where the authors used over 80 Landsat images to monitor SUHI evolution in the tropical metropolitan area of Rio de Janeiro between 1984 and 2015. It was found that in addition to the limited number of collected cloud-free imagery, LST changes were affected by cloud-induced noise from undetectable clouds, cloud shadows, and aerosols.

4.7.2. Future Directions

The findings of this study indicate that the majority of SUHI research has been concentrated in South Asian nations (Section 3.2), namely China and India, with less attention paid to cities in developing countries in other continents, particularly Africa and South America. Nine studies were interested in cities of six African countries—Cairo [5,106,113,143], Ethekwini [77], Hawassa [79], Akure [153], Osogbo [130], and Accra [170]. Only Brazilian cities, including Rio de Janeiro [8,169], Cuiaba-Varzea Grande [171], and Paço do Lumiar [172], were investigated in South America. More SUHI studies targeting cities in the African and South American continents are highly recommended.

There is no doubt that SUHIs are heavily impacted by the spatial and temporal dynamics of LULC. The increase witnessed in SUHI magnitude in several cities is attributed to IS horizontal expansion to the detriment of previous green and blue areas as a result of population growth and rising anthropogenic activities. As demonstrated in Section 4.5, however, other critical factors, such as seasonal variations, landscape composition and configuration, terrain characteristics in hilly and mountainous cities, and socioeconomic variables, highly contribute to the development of SUHIs. Nonetheless, it has been observed that only a few studies incorporated such important variables into their analyses. While acquiring data associated with some variables would be challenging in certain target areas, it is important to consider the changes of as many variables as possible while investigating SUHIs.

In the considered literature, most studies focused on exploring the past trends of SUHIs in regard to LULC changes. However, recent studies, although few (four publications), have started to pay more attention to simulating future trends based on past ones. From this perspective, Tran et al. estimated future urban climate patterns based on predicted LULC changes in the Hanoi inner city [80]; the authors employed a nonparametric regression to predict future LST values in 2023 based on past LST spatial variations with respect to spatiotemporal changes witnessed within five LULC types (i.e., urban, vegetation, cropland, water, and bare land). The examined two scenarios regeared whether future development growth would be low or high. From 2015 to 2023, the hotter LST zones (≥40 °C) were found to be likely to grow, while the cooler LST zones (≤38.5 °C) were found to tend to decline. From low-to-high growth scenarios, the same pattern was found to persist, albeit less pronounced. Guo et al. simulated two possible 2025 future scenarios in which LULC and LST in Beijing fluctuate depending on whether or not the city's population is controlled [59]. The authors used the conversion of land use and its effect at the small regional extent (CLUE-S) model to simulate LST based on land use demand that was estimated using a linear extrapolation of population and land use type. The population-controlled scenario had a projected reduction in future IS demand of 7.69%, resulting in a reduction of 1.1 °C in the average LST. Another investigation of the sort was reported by Wang et al. for the Chinese city of Nanjing [137], in which the authors simulated LULC patterns using the cellular automata-Markov chain model and estimated LST of the years 2030 and 2050 based on past trends (2000–2018); they subsequently defined areas of high and moderate risk depending on estimated LST values. The model predicted that high LST risk regions would increase, assuming urbanization is maintained from 2018 to 2050. A similar approach using the multilayer perceptron-Markov chain model was applied to Lahore city, Pakistan, to detect how future LULC changes would impact LST [173]. The findings suggested that if vegetation cover decreases by 3% in the next 15 years, there would be a rise of roughly 2 °C by 2035. The results of these studies are more relevant in the context that they provide concrete information to planners and decision-makers regarding future trends. Thus, future studies of this kind are strongly recommended.

5. Summary and Concluding Remarks

Since the turn of the century, research focused on historical and current SUHI trends with respect to LULC spatiotemporal changes in cities has significantly increased. This review shows that although case studies are growing, they have been geographically skewed. These studies have focused on South Asian cities, but other cities in emerging African or Latin American nations with fast urbanization trends have been less studied. Thus, more studies are highly required, especially in these cities, because gaining an accurate picture of SUHIs' long-term impacts is a crucial first step toward minimizing their consequences.

The increasing trend in SUHI studies is attributed to the free access to a multidecadal archive of satellite imagery data and the development of robust techniques for LULC extraction, LST retrieval, and the relationship assessment between the two. In terms of the used data, the Landsat series was the main source in most reviewed studies due to its long-term archived data dated back to the 1970s. On the other hand, MODIS and ASTER data were primarily used to assess SUHIs regarding day and night variations. Several methods to extract LULC from these datasets, notably the MLC, have been employed. In contrast to LULC extraction, LST retrieval is a complicated task due to the lack of a universal method that applies to all thermal sensors and the demanding nature of existing methods. Regression analysis was found to have been the most common way to evaluate LULC changes and LST.

SUHIs are strongly influenced by the spatial and temporal dynamics of LULC. The rise in SUHI magnitude seen in many cities is primarily linked to the horizontal expansion of ISs at the expense of formerly green and blue regions as a consequence of population growth and increased human activity. However, other important factors such as seasonal changes, landscape composition and configuration, topographical characteristics (especially in hilly and mountainous cities), and socioeconomic determinants all play significant roles in the formation of SUHIs.

Specific mitigation measures to alleviate SUHIs are target area-specific, as they depend on geographical settings and climatic conditions. Nonetheless, four common strategies have been proposed in the revised literature: promoting greenery via planting and vertical greening, safeguarding waterbodies, utilizing solar-blocking materials for roofing and pavements, and adopting optimal urban designs.

The most significant limitation of SUHI studies concerns the inability to verify satellite-derived LSTs with in situ measurements from field surveys or historical data from climatological station networks. Another obstacle is related to the high cloud cover in remote sensing data, particularly in tropical cities. In addition, it was found that few papers included other critical factors (e.g., socioeconomic variables, topography, and landscape metrics) in their analysis to assess SUHI evolution.

In terms of prospects for SUHI research, the addition of other characteristics such as geography and socioeconomic variables will provide more insight into how SUHIs are evolving. This will enable the more realistic modeling of future SUHI trends based on historical patterns. The results of such research will be useful to planners and decision-makers because they will provide specific information regarding future trends.

Supplementary Materials: The following are available online at https://www.mdpi.com/article/10.3390/rs13183654/s1. Table S1: Descriptive list of the reviewed literature sorted chronologically based on year of publication (YoP).

Author Contributions: Conceptualization, A.D. and R.W.; methodology, A.D. and R.W.; writing—original draft preparation, A.D. and R.W.; writing—review and editing, A.D., R.W., Y.M. and T.O.; visualization, A.D. and R.W.; supervision, Y.M. and T.O.; funding acquisition, Y.M. All authors have read and agreed to the published version of the manuscript.

Funding: This research was partly supported by JSPS grant 21K01027.

Institutional Review Board Statement: Not applicable.

Informed Consent Statement: Not applicable.

Data Availability Statement: Not applicable.

Acknowledgments: The authors would like to thank the editors and the anonymous reviewers for their valuable comments that improved the final version of this article. The first author would also like to extend his gratitude to the Yoshida Scholarship Foundation for the funding received to undertake his PhD.

Conflicts of Interest: The authors declare no conflict of interest.

Appendix A

Table A1. Characteristics of the satellite data sources used in SUHI studies.

Thermal Sensors		MSS/TM/ETM+/TIRS	MODIS	ASTER
Carrier Satellite		Landsat Series	Aqua and Terra	Terra
Resolution	Spatial	30 m [1]	250 m (bands 1–2) 500 m (bands 3–7) 1000 m (bands 8–36)	VNIR: 15 m SWIR: 30 m TIR: 90 m
	Temporal	16 days	1–2 days	1–16 days
Coverage	Swath	L4: 170 × 183 km L5: Width (185 km) L7: Width (185 km) L8: 185 × 180 km	2330 × 10 km	60 × 60 km

Table A1. *Cont.*

Thermal Sensors		MSS/TM/ETM+/TIRS	MODIS	ASTER
	Temporal	L4: 1982–2001 L5: 1984–2013 L7: 1999–ongoing L8: 2013–ongoing	Terra: since 1999 Aqua: since 2002	1999–ongoing (Terra)
Sensors		L4: MSS and TM L5: MSS and TM L7: ETM+ L8: OLI and TIRS	Terra MODIS (1999) Aqua MODIS (2002)	Terra ASTER (1999)

[1] The spatial resolution of thermal infrared band 6 of L4 is 120 m.

Table A2. List of the most common indices employed in the reviewed literature.

Index Category		Index	Name	Reference	Studies Share
Biophysical	Vegetation	NDVI	Normalized Difference Vegetation Index	[174]	58.2%
		FVC	Fractional Vegetation Cover	[175]	9.1%
		SAVI	Soil-Adjusted Vegetation Index	[176]	2.7%
		TDVI	Transformed Difference Vegetation Index	[177]	0.9%
		EVI	Enhanced Vegetation Index	[178]	0.9%
	Built-Up	NDBI	Normalized Difference Built-Up Index	[179]	32.9%
		IBI	Index-Based Built-Up Index	[180]	1.2%
		EBBI	Enhanced Built-Up and Bareness Index	[181]	3.7%
		NDISI	Normalized Difference Impervious Surface Index	[182]	1.2%
		DBI	Dry Built-up Index	[183]	1.2%
	Water	NDWI	Modified Difference Water Index	[184,185]	9.8%
		MNDWI	Modified Normalized Difference Water Index	[186]	8.5%
		LSWI	Land Surface Water Index	[187]	1.2%
		NDMI	Normalized Difference Moisture Index	Used in [85,154]	1.2%
	Bare Land	NDBaI	Normalized Difference Bareness Index	[188]	7.3%
		DBSI	Dry Bare-Soil Index	[183]	2.4%
Landscape Composition		PLAND	Percentage of Landscape area	[189]	8.5%
		SHEI	Shannon's Evenness Index		
		SHDI	Shannon's Diversity Index		
Landscape Configuration		ED	Edge Density		
		PD	Patch Density		
		LSI	Landscape Shape Index		
		CI	Clumpiness Index		
		CONTAG	Contagion		
		COHESION	Patch Cohesion Index		

References

1. Hobsbawm, E. *The Age of Revolution: 1789–1848*, 1st ed.; Vintage: New York, NY, USA, 1996; p. 27. ISBN 978-0-679-77253-8.
2. Davis, K. The Origin and Growth of Urbanization in the World. *Am. J. Sociol.* **1955**, *60*, 429–437. [CrossRef]
3. UNSD. *Demographic Yearbook 2019*; UNSD: New York, NY, USA, 2021; ISBN 978-92-1-148351-2.
4. United Nations. *The World's Cities in 2018—Data Booklet*; United Nations: New York, NY, USA, 2018; ISBN 978-92-1-047610-2.
5. Fan, C.; Myint, S.W.; Kaplan, S.; Middel, A.; Zheng, B.; Rahman, A.; Huang, H.-P.; Brazel, A.; Blumberg, D.G. Understanding the Impact of Urbanization on Surface Urban Heat Islands-A Longitudinal Analysis of the Oasis Effect in Subtropical Desert Cities. *Remote Sens.* **2017**, *9*, 672. [CrossRef]
6. Oke, T.R. The Heat Island of the Urban Boundary Layer: Characteristics, Causes and Effects. In *Wind Climate in Cities*; Cermak, J.E., Davenport, A.G., Plate, E.J., Viegas, D.X., Eds.; NATO ASI Series; Springer: Dordrecht, The Netherlands, 1995; pp. 81–107, ISBN 978-94-017-3686-2.
7. Manley, G. On the Frequency of Snowfall in Metropolitan England. *Q. J. R. Meteorol. Soc.* **1958**, *84*, 70–72. [CrossRef]
8. Lucena, A.J.; Rotunno Filho, O.C.; França, J.R.A.; Peres, L.F.; Xavier, L.N.R. Urban Climate and Clues of Heat Island Events in the Metropolitan Area of Rio de Janeiro. *Theor. Appl. Climatol.* **2013**, *111*, 497–511. [CrossRef]
9. Howard, L. *The Climate of London: Deduced from Meteorological Observations*, 1st ed.; Cambridge University Press: Cambridge, UK, 2012; ISBN 978-1-108-04951-1.
10. Zhou, D.; Xiao, J.; Bonafoni, S.; Berger, C.; Deilami, K.; Zhou, Y.; Frolking, S.; Yao, R.; Qiao, Z.; Sobrino, J.A. Satellite Remote Sensing of Surface Urban Heat Islands: Progress, Challenges, and Perspectives. *Remote Sens.* **2019**, *11*, 48. [CrossRef]

11. Kikon, N.; Singh, P.; Singh, S.K.; Vyas, A. Assessment of Urban Heat Islands (UHI) of Noida City, India Using Multi-Temporal Satellite Data. *Sustain. Cities Soc.* **2016**, *22*, 19–28. [CrossRef]
12. Estoque, R.C.; Murayama, Y. Monitoring Surface Urban Heat Island Formation in a Tropical Mountain City Using Landsat Data (1987–2015). *ISPRS J. Photogramm. Remote Sens.* **2017**, *133*, 18–29. [CrossRef]
13. Oke, T.R. The Energetic Basis of the Urban Heat Island. *Q. J. R. Meteorol. Soc.* **1982**, *108*, 1–24. [CrossRef]
14. Singh, P.; Kikon, N.; Verma, P. Impact of Land Use Change and Urbanization on Urban Heat Island in Lucknow City, Central India. A Remote Sensing Based Estimate. *Sustain. Cities Soc.* **2017**, *32*, 100–114. [CrossRef]
15. US EPA. Heat Island Impacts. Available online: https://www.epa.gov/heatislands/heat-island-impacts (accessed on 17 July 2021).
16. Gasparrini, A.; Guo, Y.; Hashizume, M.; Lavigne, E.; Zanobetti, A.; Schwartz, J.; Tobias, A.; Tong, S.; Rocklöv, J.; Forsberg, B.; et al. Mortality Risk Attributable to High and Low Ambient Temperature: A Multicountry Observational Study. *Lancet* **2015**, *386*, 369–375. [CrossRef]
17. Hsiang, S.; Kopp, R.; Jina, A.; Rising, J.; Delgado, M.; Mohan, S.; Rasmussen, D.J.; Muir-Wood, R.; Wilson, P.; Oppenheimer, M.; et al. Estimating Economic Damage from Climate Change in the United States. *Science* **2017**, *356*, 1362–1369. [CrossRef]
18. Ye, X.; Wolff, R.; Yu, W.; Vaneckova, P.; Pan, X.; Tong, S. Ambient Temperature and Morbidity: A Review of Epidemiological Evidence. *Environ. Health Perspect.* **2012**, *120*, 19–28. [CrossRef] [PubMed]
19. Vicedo-Cabrera, A.M.; Scovronick, N.; Sera, F.; Royé, D.; Schneider, R.; Tobias, A.; Astrom, C.; Guo, Y.; Honda, Y.; Hondula, D.M.; et al. The Burden of Heat-Related Mortality Attributable to Recent Human-Induced Climate Change. *Nat. Clim. Chang.* **2021**, *11*, 492–500. [CrossRef] [PubMed]
20. Wang, H.; Zhang, Y.; Tsou, J.Y.; Li, Y. Surface Urban Heat Island Analysis of Shanghai (China) Based on the Change of Land Use and Land Cover. *Sustainability* **2017**, *9*, 1538. [CrossRef]
21. Zhao, Y.; Xia, J.; Xu, Z.; Zou, L.; Qiao, Y.; Li, P. Impact of Urban Expansion on Rain Island Effect in Jinan City, North China. *Remote Sens.* **2021**, *13*, 2989. [CrossRef]
22. Sharma, R.; Joshi, P.K. Identifying Seasonal Heat Islands in Urban Settings of Delhi (India) Using Remotely Sensed Data—An Anomaly Based Approach. *Urban Clim.* **2014**, *9*, 19–34. [CrossRef]
23. Yuan, F.; Bauer, M.E. Comparison of Impervious Surface Area and Normalized Difference Vegetation Index as Indicators of Surface Urban Heat Island Effects in Landsat Imagery. *Remote Sens. Environ.* **2007**, *106*, 375–386. [CrossRef]
24. Imhoff, M.L.; Zhang, P.; Wolfe, R.E.; Bounoua, L. Remote Sensing of the Urban Heat Island Effect across Biomes in the Continental USA. *Remote Sens. Environ.* **2010**, *114*, 504–513. [CrossRef]
25. Papanastasiou, D.K.; Kittas, C. Maximum Urban Heat Island Intensity in a Medium-Sized Coastal Mediterranean City. *Theor. Appl. Climatol.* **2012**, *107*, 407–416. [CrossRef]
26. Grimmond, S. Urbanization and Global Environmental Change: Local Effects of Urban Warming. *Geogr. J.* **2007**, *173*, 83–88. [CrossRef]
27. Tran, H.; Uchihama, D.; Ochi, S.; Yasuoka, Y. Assessment with Satellite Data of the Urban Heat Island Effects in Asian Mega Cities. *Int. J. Appl. Earth Obs. Geoinf.* **2006**, *8*, 34–48. [CrossRef]
28. Voogt, J.A.; Oke, T.R. Thermal Remote Sensing of Urban Climates. *Remote Sens. Environ.* **2003**, *86*, 370–384. [CrossRef]
29. Roth, M.; Oke, T.R.; Emery, W.J. Satellite-Derived Urban Heat Islands from Three Coastal Cities and the Utilization of Such Data in Urban Climatology. *Int. J. Remote Sens.* **1989**, *10*, 1699–1720. [CrossRef]
30. Oke, T.R. The Distinction between Canopy and Boundary-layer Urban Heat Islands. *Atmosphere* **1976**, *14*, 268–277. [CrossRef]
31. Lo, C.P.; Quattrochi, D.A. Land-Use and Land-Cover Change, Urban Heat Island Phenomenon, and Health Implications: A Remote Sensing Approach. *Photogramm. Eng. Remote Sens.* **2003**, *69*, 1053–1063. [CrossRef]
32. Cermak, J.E.; Davenport, A.G.; Plate, E.J.; Viegas, D.X. *Wind Climate in Cities*; Springer Science & Business Media: Berlin/Heidelberg, Germany, 1994; ISBN 978-0-7923-3202-2.
33. Senanayake, I.P.; Welivitiya, W.D.D.P.; Nadeeka, P.M. Remote Sensing Based Analysis of Urban Heat Islands with Vegetation Cover in Colombo City, Sri Lanka Using Landsat-7 ETM+ Data. *Urban Clim.* **2013**, *5*, 19–35. [CrossRef]
34. Rao, P.K. Remote Sensing of Urban Heat Islands from an Environmental Satellite. *Bull. Am. Meteorol. Soc.* **1972**, *53*, 647–648.
35. Matson, M.; McClain, E.P.; McGinnis, D.F., Jr.; Pritchard, J.A. Satellite Detection of Urban Heat Islands. *Mon. Weather Rev.* **1978**, *106*, 1725–1734. [CrossRef]
36. Price, J.C. Assessment of the Urban Heat Island Effect through the Use of Satellite Data. *Mon. Weather Rev.* **1979**, *107*, 1554–1557. [CrossRef]
37. Mohamed, A.A.; Odindi, J.; Mutanga, O. Land Surface Temperature and Emissivity Estimation for Urban Heat Island Assessment Using Medium- and Low-Resolution Space-Borne Sensors: A Review. *Geocarto Int.* **2017**, *32*, 455–470. [CrossRef]
38. Deilami, K.; Kamruzzaman, M.; Liu, Y. Urban Heat Island Effect: A Systematic Review of Spatio-Temporal Factors, Data, Methods, and Mitigation Measures. *Int. J. Appl. Earth Obs. Geoinf.* **2018**, *67*, 30–42. [CrossRef]
39. Degirmenci, K.; Desouza, K.C.; Fieuw, W.; Watson, R.T.; Yigitcanlar, T. Understanding Policy and Technology Responses in Mitigating Urban Heat Islands: A Literature Review and Directions for Future Research. *Sustain. Cities Soc.* **2021**, *70*, 102873. [CrossRef]
40. Kotharkar, R.; Ramesh, A.; Bagade, A. Urban Heat Island Studies in South Asia: A Critical Review. *Urban Clim.* **2018**, *24*, 1011–1026. [CrossRef]

41. Sultana, S.; Satyanarayana, A.N.V. Urban Heat Island Intensity during Winter over Metropolitan Cities of India Using Remote-Sensing Techniques: Impact of Urbanization. *Int. J. Remote Sens.* **2018**, *39*, 6692–6730. [CrossRef]
42. Beck, H.E.; Zimmermann, N.E.; McVicar, T.R.; Vergopolan, N.; Berg, A.; Wood, E.F. Present and Future Köppen-Geiger Climate Classification Maps at 1-Km Resolution. *Sci. Data* **2018**, *5*, 180214. [CrossRef] [PubMed]
43. Simmons, M.T. Climates and Microclimates: Challenges for Extensive Green Roof Design in Hot Climates. In *Green Roof Ecosystems*; Sutton, R.K., Ed.; Ecological Studies; Springer International Publishing: Cham, Switzerland, 2015; pp. 63–80, ISBN 978-3-319-14983-7.
44. IMF. *World Economic Outlook (WEO)*; World Economic and Financial Surveys; International Monetary Fund: Washington, DC, USA, 2020; pp. 159–162.
45. Chen, L.; Jiang, R.; Xiang, W.-N. Surface Heat Island in Shanghai and Its Relationship with Urban Development from 1989 to 2013. *Adv. Meteorol.* **2016**, *2016*, 9782686. [CrossRef]
46. Cui, L.; Shi, J. Urbanization and Its Environmental Effects in Shanghai, China. *Urban Clim.* **2012**, *2*, 1–15. [CrossRef]
47. Du, H.; Zhou, F.; Li, C.; Cai, W.; Jiang, H.; Cai, Y. Analysis of the Impact of Land Use on Spatiotemporal Patterns of Surface Urban Heat Island in Rapid Urbanization, a Case Study of Shanghai, China. *Sustainability* **2020**, *12*, 1171. [CrossRef]
48. Hu, Y.; Hou, M.; Jia, G.; Zhao, C.; Zhen, X.; Xu, Y. Comparison of Surface and Canopy Urban Heat Islands within Megacities of Eastern China. *ISPRS J. Photogramm. Remote Sens.* **2019**, *156*, 160–168. [CrossRef]
49. Li, J.-J.; Wang, X.-R.; Wang, X.-J.; Ma, W.-C.; Zhang, H. Remote Sensing Evaluation of Urban Heat Island and Its Spatial Pattern of the Shanghai Metropolitan Area, China. *Ecol. Complex.* **2009**, *6*, 413–420. [CrossRef]
50. Li, J.; Song, C.; Cao, L.; Zhu, F.; Meng, X.; Wu, J. Impacts of Landscape Structure on Surface Urban Heat Islands: A Case Study of Shanghai, China. *Remote Sens. Environ.* **2011**, *115*, 3249–3263. [CrossRef]
51. Li, Y.-Y.; Zhang, H.; Kainz, W. Monitoring Patterns of Urban Heat Islands of the Fast-Growing Shanghai Metropolis, China: Using Time-Series of Landsat TM/ETM+ Data. *Int. J. Appl. Earth Obs. Geoinf.* **2012**, *19*, 127–138. [CrossRef]
52. Zhang, H.; Qi, Z.-F.; Ye, X.-Y.; Cai, Y.-B.; Ma, W.-C.; Chen, M.-N. Analysis of Land Use/Land Cover Change, Population Shift, and Their Effects on Spatiotemporal Patterns of Urban Heat Islands in Metropolitan Shanghai, China. *Appl. Geogr.* **2013**, *44*, 121–133. [CrossRef]
53. Kant, Y.; Bharath, B.D.; Mallick, J.; Atzberger, C.; Kerle, N. Satellite-Based Analysis of the Role of Land Use/Land Cover and Vegetation Density on Surface Temperature Regime of Delhi, India. *J. Indian Soc. Remote Sens.* **2009**, *37*, 201–214. [CrossRef]
54. Mallick, J.; Rahman, A.; Singh, C.K. Modeling Urban Heat Islands in Heterogeneous Land Surface and Its Correlation with Impervious Surface Area by Using Night-Time ASTER Satellite Data in Highly Urbanizing City, Delhi-India. *Adv. Space Res.* **2013**, *52*, 639–655. [CrossRef]
55. Pramanik, S.; Punia, M. Land Use/Land Cover Change and Surface Urban Heat Island Intensity: Source—Sink Landscape-Based Study in Delhi, India. *Environ. Dev. Sustain.* **2020**, *22*, 7331–7356. [CrossRef]
56. Singh, R.B.; Grover, A.; Zhan, J. Inter-Seasonal Variations of Surface Temperature in the Urbanized Environment of Delhi Using Landsat Thermal Data. *Energies* **2014**, *7*, 1811–1828. [CrossRef]
57. Cai, G.; Du, M.; Xue, Y. Monitoring of Urban Heat Island Effect in Beijing Combining ASTER and TM Data. *Int. J. Remote Sens.* **2011**, *32*, 1213–1232. [CrossRef]
58. Ding, H.; Shi, W. Land-Use/Land-Cover Change and Its Influence on Surface Temperature: A Case Study in Beijing City. *Int. J. Remote Sens.* **2013**, *34*, 5503–5517. [CrossRef]
59. Guo, L.; Liu, R.; Men, C.; Wang, Q.; Miao, Y.; Zhang, Y. Quantifying and Simulating Landscape Composition and Pattern Impacts on Land Surface Temperature: A Decadal Study of the Rapidly Urbanizing City of Beijing, China. *Sci. Total Environ.* **2019**, *654*, 430–440. [CrossRef]
60. Ye, C.; Chen, R.; Li, Y.; Liu, T.; Diao, K.; Li, J. Characterization of Combined Effects of Urban Built-Up and Vegetated Areas on Long-Term Urban Heat Islands in Beijing. *Can. J. Remote Sens.* **2019**, *45*, 634–649. [CrossRef]
61. Kamruzzaman, M.; Deilami, K.; Yigitcanlar, T. Investigating the Urban Heat Island Effect of Transit Oriented Development in Brisbane. *J. Transp. Geogr.* **2018**, *66*, 116–124. [CrossRef]
62. Bokaie, M.; Shamsipour, A.; Khatibi, P.; Hosseini, A. Seasonal Monitoring of Urban Heat Island Using Multi-Temporal Landsat and MODIS Images in Tehran. *Int. J. Urban Sci.* **2019**, *23*, 269–285. [CrossRef]
63. Xiong, Y.; Huang, S.; Chen, F.; Ye, H.; Wang, C.; Zhu, C. The Impacts of Rapid Urbanization on the Thermal Environment: A Remote Sensing Study of Guangzhou, South China. *Remote Sens.* **2012**, *4*, 2033–2056. [CrossRef]
64. Padmanaban, R.; Bhowmik, A.K.; Cabral, P. Satellite Image Fusion to Detect Changing Surface Permeability and Emerging Urban Heat Islands in a Fast-Growing City. *PLoS ONE* **2019**, *14*, e0208949. [CrossRef]
65. Dhar, R.B.; Chakraborty, S.; Chattopadhyay, R.; Sikdar, P.K. Impact of Land-Use/Land-Cover Change on Land Surface Temperature Using Satellite Data: A Case Study of Rajarhat Block, North 24-Parganas District, West Bengal. *J. Indian Soc. Remote Sens.* **2019**, *47*, 331–348. [CrossRef]
66. Wulder, M.A.; Masek, J.G.; Cohen, W.B.; Loveland, T.R.; Woodcock, C.E. Opening the Archive: How Free Data Has Enabled the Science and Monitoring Promise of Landsat. *Remote Sens. Environ.* **2012**, *122*, 2–10. [CrossRef]
67. Melaas, E.K.; Wang, J.A.; Miller, D.L.; Friedl, M.A. Interactions between Urban Vegetation and Surface Urban Heat Islands: A Case Study in the Boston Metropolitan Region. *Environ. Res. Lett.* **2016**, *11*, 054020. [CrossRef]

68. Zhang, W.; Liu, S. Applications of the Small Satellite Constellation for Environment and Disaster Monitoring and Forecasting. *Int. J. Disaster Risk Sci.* **2010**, *1*, 9–16. [CrossRef]
69. Wang, Q.; Wu, C.Q.; Li, Q.; Li, J.S. Chinese HJ-1A/B Satellites and Data Characteristics. *Sci. China Earth Sci.* **2010**, *53*, 51–57. [CrossRef]
70. Wu, H.; Ye, L.-P.; Shi, W.-Z.; Clarke, K.C. Assessing the Effects of Land Use Spatial Structure on Urban Heatislands Using HJ-1B Remote Sensing Imagery in Wuhan, China. *Int. J. Appl. Earth Obs. Geoinf.* **2014**, *32*, 67–78. [CrossRef]
71. Rasul, A.; Balzter, H.; Smith, C. Spatial Variation of the Daytime Surface Urban Cool Island during the Dry Season in Erbil, Iraqi Kurdistan, from Landsat 8. *Urban Clim.* **2015**, *14*, 176–186. [CrossRef]
72. Cai, Y.; Zhang, H.; Zheng, P.; Pan, W. Quantifying the Impact of Land Use/Land Cover Changes on the Urban Heat Island: A Case Study of the Natural Wetlands Distribution Area of Fuzhou City, China. *Wetlands* **2016**, *36*, 285–298. [CrossRef]
73. Min, M.; Zhao, H.; Miao, C. Spatio-Temporal Evolution Analysis of the Urban Heat Island: A Case Study of Zhengzhou City, China. *Sustainability* **2018**, *10*, 1992. [CrossRef]
74. Majkowska, A.; Kolendowicz, L.; Półrolniczak, M.; Hauke, J.; Czernecki, B. The Urban Heat Island in the City of Poznań as Derived from Landsat 5 TM. *Theor. Appl. Climatol.* **2017**, *128*, 769–783. [CrossRef]
75. Land Monitoring Service. CORINE Land Cover—Copernicus Land Monitoring Service. Available online: https://land.copernicus.eu/pan-european/corine-land-cover (accessed on 18 July 2021).
76. Dihkan, M.; Karsli, F.; Guneroglu, N.; Guneroglu, A. Evaluation of Urban Heat Island Effect in Turkey. *Arab. J. Geosci.* **2018**, *11*, 186. [CrossRef]
77. Odindi, J.O.; Bangamwabo, V.; Mutanga, O. Assessing the Value of Urban Green Spaces in Mitigating Multi-Seasonal Urban Heat Using MODIS Land Surface Temperature (LST) and Landsat 8 Data. *Int. J. Environ. Res.* **2015**, *9*, 9–18.
78. Huang, X.; Wang, Y. Investigating the Effects of 3D Urban Morphology on the Surface Urban Heat Island Effect in Urban Functional Zones by Using High-Resolution Remote Sensing Data: A Case Study of Wuhan, Central China. *ISPRS J. Photogramm. Remote Sens.* **2019**, *152*, 119–131. [CrossRef]
79. Chaka, D.S.; Oda, T.K. Understanding Land Surface Temperature on Rift Areas to Examine the Spatial Variation of Urban Heat Island: The Case of Hawassa, Southern Ethiopia. *GeoJournal* **2019**, *86*, 993–1014. [CrossRef]
80. Tran, D.X.; Pla, F.; Latorre-Carmona, P.; Myint, S.W.; Caetano, M.; Kieu, H.V. Characterizing the Relationship between Land Use Land Cover Change and Land Surface Temperature. *ISPRS J. Photogramm. Remote Sens.* **2017**, *124*, 119–132. [CrossRef]
81. Dissanayake, D.M.S.L.B.; Morimoto, T.; Ranagalage, M.; Murayama, Y. Land-Use/Land-Cover Changes and Their Impact on Surface Urban Heat Islands: Case Study of Kandy City, Sri Lanka. *Climate* **2019**, *7*, 99. [CrossRef]
82. Priyankara, P.; Ranagalage, M.; Dissanayake, D.M.S.L.B.; Morimoto, T.; Murayama, Y. Spatial Process of Surface Urban Heat Island in Rapidly Growing Seoul Metropolitan Area for Sustainable Urban Planning Using Landsat Data (1996–2017). *Climate* **2019**, *7*, 110. [CrossRef]
83. Ranagalage, M.; Dmslb, D.; Murayama, Y.; Zhang, X.; Estoque, R.C.; Enc, P.; Morimoto, T. Quantifying Surface Urban Heat Island Formation in the World Heritage Tropical Mountain City of Sri Lanka. *ISPRS Int. J. Geo-Inf.* **2018**, *7*, 341. [CrossRef]
84. Khan, M.S.; Ullah, S.; Sun, T.; Rehman, A.U.; Chen, L. Land-Use/Land-Cover Changes and Its Contribution to Urban Heat Island: A Case Study of Islamabad, Pakistan. *Sustainability* **2020**, *12*, 3861. [CrossRef]
85. Roy, S.; Pandit, S.; Eva, E.A.; Bagmar, M.S.H.; Papia, M.; Banik, L.; Dube, T.; Rahman, F.; Razi, M.A. Examining the Nexus between Land Surface Temperature and Urban Growth in Chattogram Metropolitan Area of Bangladesh Using Long Term Landsat Series Data. *Urban Clim.* **2020**, *32*. [CrossRef]
86. Atasoy, M. Assessing the Impacts of Land-Use/Land-Cover Change on the Development of Urban Heat Island Effects. *Environ. Dev. Sustain.* **2020**, *22*, 7547–7557. [CrossRef]
87. Chen, X.-L.; Zhao, H.-M.; Li, P.-X.; Yin, Z.-Y. Remote Sensing Image-Based Analysis of the Relationship between Urban Heat Island and Land Use/Cover Changes. *Remote Sens. Environ.* **2006**, *104*, 133–146. [CrossRef]
88. Choudhury, D.; Das, K.; Das, A. Assessment of Land Use Land Cover Changes and Its Impact on Variations of Land Surface Temperature in Asansol-Durgapur Development Region. *Egypt. J. Remote Sens. Space Sci.* **2019**, *22*, 203–218. [CrossRef]
89. Ghosh, S.; Chatterjee, N.D.; Dinda, S. Relation between Urban Biophysical Composition and Dynamics of Land Surface Temperature in the Kolkata Metropolitan Area: A GIS and Statistical Based Analysis for Sustainable Planning. *Modeling Earth Syst. Environ.* **2019**, *5*, 307–329. [CrossRef]
90. Grigoraș, G.; Urițescu, B. Land Use/Land Cover Changes Dynamics and Their Effects on Surface Urban Heat Island in Bucharest, Romania. *Int. J. Appl. Earth Obs. Geoinf.* **2019**, *80*, 115–126. [CrossRef]
91. Gupta, N.; Mathew, A.; Khandelwal, S. Spatio-Temporal Impact Assessment of Land Use / Land Cover (LU-LC) Change on Land Surface Temperatures over Jaipur City in India. *Int. J. Urban Sustain. Dev.* **2020**. [CrossRef]
92. Hoan, N.T.; Liou, Y.-A.; Nguyen, K.-A.; Sharma, R.C.; Tran, D.-P.; Liou, C.-L.; Cham, D.D. Assessing the Effects of Land-Use Types in Surface Urban Heat Islands for Developing Comfortable Living in Hanoi City. *Remote Sens.* **2018**, *10*, 1965. [CrossRef]
93. Lakra, K.; Sharma, D. Geospatial Assessment of Urban Growth Dynamics and Land Surface Temperature in Ajmer Region, India. *J. Indian Soc. Remote Sens.* **2019**, *47*, 1073–1089. [CrossRef]
94. Lu, D.; Song, K.; Zang, S.; Jia, M.; Du, J.; Ren, C. The Effect of Urban Expansion on Urban Surface Temperature in Shenyang, China: An Analysis with Landsat Imagery. *Environ. Model. Assess.* **2015**, *20*, 197–210. [CrossRef]

95. Miky, Y.H. Remote Sensing Analysis for Surface Urban Heat Island Detection over Jeddah, Saudi Arabia. *Appl. Geomat.* **2019**, *11*, 243–258. [CrossRef]
96. Nguyen, T.M.; Lin, T.-H.; Chan, H.-P. The Environmental Effects of Urban Development in Hanoi, Vietnam from Satellite and Meteorological Observations from 1999–2016. *Sustainability* **2019**, *11*, 1768. [CrossRef]
97. Pal, S.; Ziaul, S. Detection of Land Use and Land Cover Change and Land Surface Temperature in English Bazar Urban Centre. *Egypt. J. Remote Sens. Space Sci.* **2017**, *20*, 125–145. [CrossRef]
98. Rousta, I.; Sarif, M.O.; Gupta, R.D.; Olafsson, H.; Ranagalage, M.; Murayama, Y.; Zhang, H.; Mushore, T.D. Spatiotemporal Analysis of Land Use/Land Cover and Its Effects on Surface Urban Heat Island Using Landsat Data: A Case Study of Metropolitan City Tehran (1988–2018). *Sustainability* **2018**, *10*, 4433. [CrossRef]
99. Wang, C.; Myint, S.W.; Wang, Z.; Song, J. Spatio-Temporal Modeling of the Urban Heat Island in the Phoenix Metropolitan Area: Land Use Change Implications. *Remote Sens.* **2016**, *8*, 185. [CrossRef]
100. Cohen, J. A Coefficient of Agreement for Nominal Scales. *Educ. Psychol. Meas.* **1960**, *20*, 37–46. [CrossRef]
101. Monserud, R.A.; Leemans, R. Comparing Global Vegetation Maps with the Kappa Statistic. *Ecol. Model.* **1992**, *62*, 275–293. [CrossRef]
102. Anderson, J.R.; Hardy, E.E.; Roach, J.T.; Witmer, R.E. *A Land Use and Land Cover Classification System for Use with Remote Sensor Data*; Professional Paper; U.S. Government Printing Office: Washington, DC, USA, 1976; Volume 964.
103. Townshend, J.R. *Terrain Analysis and Remote Sensing*, 1st ed.; Unwin Hyman: London, UK, 1981; ISBN 978-0-04-551037-5.
104. Lea, C.; Curtis, A.C. *Thematic Accuracy Assessment Procedures: National Park Service Vegetation Inventory*; Natural Resource Report; Version 2.0.; U.S. Department of the Interior, National Park Service, Natural Resource Program Center: Fort Collins, CO, USA, 2010.
105. Lillesand, T.; Kiefer, R.W.; Chipman, J. *Remote Sensing and Image Interpretation*; John Wiley & Sons: Hoboken, NJ, USA, 2015; ISBN 978-1-118-34328-9.
106. Hereher, M.E. Retrieving Spatial Variations of Land Surface Temperatures from Satellite Data—Cairo Region, Egypt. *Geocarto Int.* **2017**, *32*, 556–568. [CrossRef]
107. Meng, F.; Liu, M. Remote-Sensing Image-Based Analysis of the Patterns of Urban Heat Islands in Rapidly Urbanizing Jinan, China. *Int. J. Remote Sens.* **2013**, *34*, 8838–8853. [CrossRef]
108. Zhou, X.; Wang, Y.-C. Dynamics of Land Surface Temperature in Response to Land-Use/Cover Change. *Geogr. Res.* **2011**, *49*, 23–36. [CrossRef]
109. Zhang, Y.; Odeh, I.O.A.; Han, C. Bi-Temporal Characterization of Land Surface Temperature in Relation to Impervious Surface Area, NDVI and NDBI, Using a Sub-Pixel Image Analysis. *Int. J. Appl. Earth Obs. Geoinf.* **2009**, *11*, 256–264. [CrossRef]
110. Yu, Z.; Guo, X.; Zeng, Y.; Koga, M.; Vejre, H. Variations in Land Surface Temperature and Cooling Efficiency of Green Space in Rapid Urbanization: The Case of Fuzhou City, China. *Urban For. Urban Green.* **2018**, *29*, 113–121. [CrossRef]
111. Zhang, X.; Wang, D.; Hao, H.; Zhang, F.; Hu, Y. Effects of Land Use/Cover Changes and Urban Forest Configuration on Urban Heat Islands in a Loess Hilly Region: Case Study Based on Yan'an City, China. *Int. J. Environ. Res. Public Health* **2017**, *14*, 840. [CrossRef]
112. Yang, C.; He, X.; Yan, F.; Yu, L.; Bu, K.; Yang, J.; Chang, L.; Zhang, S. Mapping the Influence of Land Use/Land Cover Changes on the Urban Heat Island Effect-A Case Study of Changchun, China. *Sustainability* **2017**, *9*, 312. [CrossRef]
113. Effat, H.A.; Hassan, O.A.K. Change Detection of Urban Heat Islands and Some Related Parameters Using Multi-Temporal Landsat Images; a Case Study for Cairo City, Egypt. *Urban Clim.* **2014**, *10*, 171–188. [CrossRef]
114. Yusuf, Y.A.; Pradhan, B.; Idrees, M.O. Spatio-Temporal Assessment of Urban Heat Island Effects in Kuala Lumpur Metropolitan City Using Landsat Images. *J. Indian Soc. Remote Sens.* **2014**, *42*, 829–837. [CrossRef]
115. Zhang, X.; Estoque, R.C.; Murayama, Y. An Urban Heat Island Study in Nanchang City, China Based on Land Surface Temperature and Social-Ecological Variables. *Sustain. Cities Soc.* **2017**, *32*, 557–568. [CrossRef]
116. Li, Z.-L.; Tang, B.-H.; Wu, H.; Ren, H.; Yan, G.; Wan, Z.; Trigo, I.F.; Sobrino, J.A. Satellite-Derived Land Surface Temperature: Current Status and Perspectives. *Remote Sens. Environ.* **2013**, *131*, 14–37. [CrossRef]
117. Khan, A.; Chatterjee, S.; Weng, Y. 2 - Characterizing thermal fields and evaluating UHI effects. In *Urban Heat Island Modeling for Tropical Climates*; Khan, A., Chatterjee, S., Weng, Y., Eds.; Elsevier: Amsterdam, The Netherlands, 2021; pp. 37–67, ISBN 978-0-12-819669-4.
118. Göttsche, F.-M.; Olesen, F.-S.; Bork-Unkelbach, A. Validation of Land Surface Temperature Derived from MSG/SEVIRI with in Situ Measurements at Gobabeb, Namibia. *Int. J. Remote Sens.* **2013**, *34*, 3069–3083. [CrossRef]
119. Martin, M.A.; Ghent, D.; Pires, A.C.; Göttsche, F.-M.; Cermak, J.; Remedios, J.J. Comprehensive In Situ Validation of Five Satellite Land Surface Temperature Data Sets over Multiple Stations and Years. *Remote Sens.* **2019**, *11*, 479. [CrossRef]
120. Srivastava, P.K.; Majumdar, T.J.; Bhattacharya, A.K. Surface Temperature Estimation in Singhbhum Shear Zone of India Using Landsat-7 ETM+ Thermal Infrared Data. *Adv. Space Res.* **2009**, *43*, 1563–1574. [CrossRef]
121. Feng, H.; Zhao, X.; Chen, F.; Wu, L. Using Land Use Change Trajectories to Quantify the Effects of Urbanization on Urban Heat Island. *Adv. Space Res.* **2014**, *53*, 463–473. [CrossRef]
122. Anding, D.; Kauth, R. Estimation of Sea Surface Temperature from Space. *Remote Sens. Environ.* **1970**, *1*, 217–220. [CrossRef]
123. Price, J.C. Estimating Surface Temperatures from Satellite Thermal Infrared Data—A Simple Formulation for the Atmospheric Effect. *Remote Sens. Environ.* **1983**, *13*, 353–361. [CrossRef]

124. Price, J.C. Land Surface Temperature Measurements from the Split Window Channels of the NOAA 7 Advanced Very High Resolution Radiometer. *J. Geophys. Res. Atmos.* **1984**, *89*, 7231–7237. [CrossRef]
125. Qin, Z.; Karnieli, A.; Berliner, P. A Mono-Window Algorithm for Retrieving Land Surface Temperature from Landsat TM Data and Its Application to the Israel-Egypt Border Region. *Int. J. Remote Sens.* **2001**, *22*, 3719–3746. [CrossRef]
126. Guha, S.; Govil, H.; Mukherjee, S. Dynamic Analysis and Ecological Evaluation of Urban Heat Islands in Raipur City, India. *J. Appl. Remote Sens.* **2017**, *11*, 036020. [CrossRef]
127. Saha, P.; Bandopadhyay, S.; Kumar, C.; Mitra, C. Multi-Approach Synergic Investigation between Land Surface Temperature and Land-Use Land-Cover. *J. Earth Syst. Sci.* **2020**, *129*, 74. [CrossRef]
128. Jiménez-Muñoz, J.C.; Sobrino, J.A. A Generalized Single-Channel Method for Retrieving Land Surface Temperature from Remote Sensing Data. *J. Geophys. Res. Atmos.* **2003**, *108*. [CrossRef]
129. Li, L.; Tan, Y.; Ying, S.; Yu, Z.; Li, Z.; Lan, H. Impact of Land Cover and Population Density on Land Surface Temperature: Case Study in Wuhan, China. *J. Appl. Remote Sens.* **2014**, *8*, 084993. [CrossRef]
130. Eresanya, E.O.; Daramola, M.T.; Durowoju, O.S.; Awoyele, P. Investigation of the Changing Patterns of the Land Use Land Cover over Osogbo and Its Environs. *R. Soc. Open Sci.* **2019**, *6*, 191021. [CrossRef]
131. Rizvi, S.H.; Fatima, H.; Iqbal, M.J.; Alam, K. The Effect of Urbanization on the Intensification of SUHIs: Analysis by LULC on Karachi. *J. Atmos. Sol. Terr. Phys.* **2020**, *207*, 105374. [CrossRef]
132. Oke, T.R. Towards Better Scientific Communication in Urban Climate. *Theor. Appl. Climatol.* **2006**, *84*, 179–190. [CrossRef]
133. Deilami, K.; Kamruzzaman, M.; Hayes, J.F. Correlation or Causality between Land Cover Patterns and the Urban Heat Island Effect? Evidence from Brisbane, Australia. *Remote Sens.* **2016**, *8*, 716. [CrossRef]
134. Siqi, J.; Yuhong, W. Effects of Land Use and Land Cover Pattern on Urban Temperature Variations: A Case Study in Hong Kong. *Urban Clim.* **2020**, *34*, 100693. [CrossRef]
135. Chakraborti, S.; Banerjee, A.; Sannigrahi, S.; Pramanik, S.; Maiti, A.; Jha, S. Assessing the Dynamic Relationship among Land Use Pattern and Land Surface Temperature: A Spatial Regression Approach. *Asian Geogr.* **2019**, *36*, 93–116. [CrossRef]
136. Kamali Maskooni, E.; Hashemi, H.; Berndtsson, R.; Daneshkar Arasteh, P.; Kazemi, M. Impact of Spatiotemporal Land-Use and Land-Cover Changes on Surface Urban Heat Islands in a Semiarid Region Using Landsat Data. *Int. J. Digit. Earth* **2021**, 14, 250–270. [CrossRef]
137. Wang, R.; Hou, H.; Murayama, Y.; Derdouri, A. Spatiotemporal Analysis of Land Use/Cover Patterns and Their Relationship with Land Surface Temperature in Nanjing, China. *Remote Sens.* **2020**, *12*, 440. [CrossRef]
138. Dutta, I.; Das, A. Exploring the Spatio-Temporal Pattern of Regional Heat Island (RHI) in an Urban Agglomeration of Secondary Cities in Eastern India. *Urban Clim.* **2020**, *34*, 100679. [CrossRef]
139. Huang, Q.; Huang, J.; Yang, X.; Fang, C.; Liang, Y. Quantifying the Seasonal Contribution of Coupling Urban Land Use Types on Urban Heat Island Using Land Contribution Index: A Case Study in Wuhan, China. *Sustain. Cities Soc.* **2019**, *44*, 666–675. [CrossRef]
140. Xiong, Y.; Peng, F.; Zou, B. Spatiotemporal Influences of Land Use/Cover Changes on the Heat Island Effect in Rapid Urbanization Area. *Front. Earth Sci.* **2019**, *13*, 614–627. [CrossRef]
141. Amiri, R.; Weng, Q.; Alimohammadi, A.; Alavipanah, S.K. Spatial-Temporal Dynamics of Land Surface Temperature in Relation to Fractional Vegetation Cover and Land Use/Cover in the Tabriz Urban Area, Iran. *Remote Sens. Environ.* **2009**, *113*, 2606–2617. [CrossRef]
142. Amanollahi, J.; Tzanis, C.; Ramli, M.F.; Abdullah, A.M. Urban Heat Evolution in a Tropical Area Utilizing Landsat Imagery. *Atmos. Res.* **2016**, *167*, 175–182. [CrossRef]
143. El-Hattab, M.; Amany, S.M.; Lamia, G.E. Monitoring and Assessment of Urban Heat Islands over the Southern Region of Cairo Governorate, Egypt. *Egypt. J. Remote Sens. Space Sci.* **2018**, *21*, 311–323. [CrossRef]
144. Hu, Y.; Jia, G. Influence of Land Use Change on Urban Heat Island Derivedfrom Multi-Sensor Data. *Int. J. Climatol.* **2010**, *30*, 1382–1395. [CrossRef]
145. Ranagalage, M.; Estoque, R.C.; Murayama, Y. An Urban Heat Island Study of the Colombo Metropolitan Area, Sri Lanka, Based on Landsat Data (1997–2017). *ISPRS Int. J. Geo Inf.* **2017**, *6*, 189. [CrossRef]
146. Rani, M.; Kumar, P.; Pandey, P.C.; Srivastava, P.K.; Chaudhary, B.S.; Tomar, V.; Mandal, V.P. Multi-Temporal NDVI and Surface Temperature Analysis for Urban Heat Island Inbuilt Surrounding of Sub-Humid Region: A Case Study of Two Geographical Regions. *Remote Sens. Appl. Soc. Environ.* **2018**, *10*, 163–172. [CrossRef]
147. Saleem, M.S.; Ahmad, S.R.; Shafiq-Ur-Rehman; Javed, M. A. Impact Assessment of Urban Development Patterns on Land Surface Temperature by Using Remote Sensing Techniques: A Case Study of Lahore, Faisalabad and Multan District. *Environ. Sci. Pollut. Res.* **2020**, *27*, 39865–39878. [CrossRef]
148. Sannigrahi, S.; Rahmat, S.; Chakraborti, S.; Bhatt, S.; Jha, S. Changing Dynamics of Urban Biophysical Composition and Its Impact on Urban Heat Island Intensity and Thermal Characteristics: The Case of Hyderabad City, India. *Model. Earth Syst. Environ.* **2017**, *3*, 647–667. [CrossRef]
149. Sultana, S.; Satyanarayana, A.N.V. Assessment of Urbanisation and Urban Heat Island Intensities Using Landsat Imageries during 2000–2018 over a Sub-Tropical Indian City. *Sustain. Cities Soc.* **2020**, *52*, 101846. [CrossRef]
150. Rotem-Mindali, O.; Michael, Y.; Helman, D.; Lensky, I.M. The Role of Local Land-Use on the Urban Heat Island Effect of Tel Aviv as Assessed from Satellite Remote Sensing. *Appl. Geogr.* **2015**, *56*, 145–153. [CrossRef]

151. Karakuş, C.B. The Impact of Land Use/Land Cover (LULC) Changes on Land Surface Temperature in Sivas City Center and Its Surroundings and Assessment of Urban Heat Island. *Asia Pac. J. Atmos. Sci.* **2019**, *55*, 669–684. [CrossRef]
152. Lin, Y.; Jim, C.Y.; Deng, J.; Wang, Z. Urbanization Effect on Spatiotemporal Thermal Patterns and Changes in Hangzhou (China). *Build. Environ.* **2018**, *145*, 166–176. [CrossRef]
153. Makinde, E.O.; Agbor, C.F. Geoinformatic Assessment of Urban Heat Island and Land Use/Cover Processes: A Case Study from Akure. *Environ. Earth Sci.* **2019**, *78*, 483. [CrossRef]
154. Sahana, M.; Dutta, S.; Sajjad, H. Assessing Land Transformation and Its Relation with Land Surface Temperature in Mumbai City, India Using Geospatial Techniques. *Int. J. Urban Sci.* **2019**, *23*, 205–225. [CrossRef]
155. Song, Y.; Song, X.; Shao, G. Effects of Green Space Patterns on Urban Thermal Environment at Multiple Spatial-Temporal Scales. *Sustainability* **2020**, *12*, 6850. [CrossRef]
156. Turner, M.G. Landscape Ecology: What Is the State of the Science? *Annu. Rev. Ecol. Evol. Syst.* **2005**, *36*, 319–344. [CrossRef]
157. Dobrovolný, P. The Surface Urban Heat Island in the City of Brno (Czech Republic) Derived from Land Surface Temperatures and Selected Reasons for Its Spatial Variability. *Theor. Appl. Climatol.* **2013**, *112*, 89–98. [CrossRef]
158. Li, Y.-Y.; Liu, Y.; Ranagalage, M.; Zhang, H.; Zhou, R. Examining Land Use/Land Cover Change and the Summertime Surface Urban Heat Island Effect in Fast-Growing Greater Hefei, China: Implications for Sustainable Land Development. *ISPRS Int. J. Geo Inf.* **2020**, *9*, 568. [CrossRef]
159. Chen, Z.; Gong, C.; Wu, J.; Yu, S. The Influence of Socioeconomic and Topographic Factors on Nocturnal Urban Heat Islands: A Case Study in Shenzhen, China. *Int. J. Remote Sens.* **2012**, *33*, 3834–3849. [CrossRef]
160. Kleerekoper, L.; van Esch, M.; Salcedo, T.B. How to Make a City Climate-Proof, Addressing the Urban Heat Island Effect. *Resour. Conserv. Recycl.* **2012**, *64*, 30–38. [CrossRef]
161. Wong, N.H.; Tan, C.L.; Kolokotsa, D.D.; Takebayashi, H. Greenery as a Mitigation and Adaptation Strategy to Urban Heat. *Nat. Rev. Earth Environ.* **2021**, *2*, 166–181. [CrossRef]
162. Zhang, R. Cooling Effect and Control Factors of Common Shrubs on the Urban Heat Island Effect in a Southern City in China. *Sci. Rep.* **2020**, *10*, 17317. [CrossRef]
163. Zhou, G.; Wang, H.; Chen, W.; Zhang, G.; Luo, Q.; Jia, B. Impacts of Urban Land Surface Temperature on Tract Landscape Pattern, Physical and Social Variables. *Int. J. Remote Sens.* **2020**, *41*, 683–703. [CrossRef]
164. Ghosh, S.; Das, A. Modelling Urban Cooling Island Impact of Green Space and Water Bodies on Surface Urban Heat Island in a Continuously Developing Urban Area. *Model. Earth Syst. Environ.* **2018**, *4*, 501–515. [CrossRef]
165. Khamchiangta, D.; Dhakal, S. Time Series Analysis of Land Use and Land Cover Changes Related to Urban Heat Island Intensity: Case of Bangkok Metropolitan Area in Thailand. *J. Urban Manag.* **2020**, *9*, 383–395. [CrossRef]
166. Swain, D.; Roberts, G.J.; Dash, J.; Lekshmi, K.; Vinoj, V.; Tripathy, S. Impact of Rapid Urbanization on the City of Bhubaneswar, India. *Proc. Natl. Acad. Sci. India Sect. A Phys. Sci.* **2017**, *87*, 845–853. [CrossRef]
167. Wan, Z. New Refinements and Validation of the MODIS Land-Surface Temperature/Emissivity Products. *Remote Sens. Environ.* **2008**, *112*, 59–74. [CrossRef]
168. Fu, P.; Weng, Q. A Time Series Analysis of Urbanization Induced Land Use and Land Cover Change and Its Impact on Land Surface Temperature with Landsat Imagery. *Remote Sens. Environ.* **2016**, *175*, 205–214. [CrossRef]
169. Peres, L.D.F.; Lucena, A.J.D.; Rotunno Filho, O.C.; França, J.R.D.A. The Urban Heat Island in Rio de Janeiro, Brazil, in the Last 30 Years Using Remote Sensing Data. *Int. J. Appl. Earth Obs. Geoinf.* **2018**, *64*, 104–116. [CrossRef]
170. Athukorala, D.; Murayama, Y. Spatial Variation of Land Use/Cover Composition and Impact on Surface Urban Heat Island in a Tropical Sub-Saharan City of Accra, Ghana. *Sustainability* **2020**, *12*, 7953. [CrossRef]
171. Apolonio Callejas, I.J.; de Oliveira, A.S.; de Moura Santos, F.M.; Durante, L.C.; de Jesus Albuquerque Nogueira, M.C.; Zeilhofer, P. Relationship between Land Use/Cover and Surface Temperatures in the Urban Agglomeration of Cuiaba-Varzea Grande, Central Brazil. *J. Appl. Remote Sens.* **2011**, *5*, 053569. [CrossRef]
172. Silva, J.S.; da Silva, R.M.; Guimaraes Santos, C.A. Spatiotemporal Impact of Land Use/Land Cover Changes on Urban Heat Islands: A Case Study of Paco Do Lumiar, Brazil. *Build. Environ.* **2018**, *136*, 279–292. [CrossRef]
173. Imran, M.; Mehmood, A. Analysis and Mapping of Present and Future Drivers of Local Urban Climate Using Remote Sensing: A Case of Lahore, Pakistan. *Arab. J. Geosci.* **2020**, *13*, 278. [CrossRef]
174. Tucker, C.J. Red and Photographic Infrared Linear Combinations for Monitoring Vegetation. *Remote Sens. Environ.* **1979**, *8*, 127–150. [CrossRef]
175. Gillies, R.R.; Kustas, W.P.; Humes, K.S. A Verification of the "triangle" Method for Obtaining Surface Soil Water Content and Energy Fluxes from Remote Measurements of the Normalized Difference Vegetation Index (NDVI) and Surface e. *Int. J. Remote Sens.* **1997**, *18*, 3145–3166. [CrossRef]
176. Huete, A.R. A Soil-Adjusted Vegetation Index (SAVI). *Remote Sens. Environ.* **1988**, *25*, 295–309. [CrossRef]
177. Bannari, A.; Asalhi, H.; Teillet, P.M. Transformed Difference Vegetation Index (TDVI) for Vegetation Cover Mapping. In Proceedings of the IEEE International Geoscience and Remote Sensing Symposium, Toronto, ON, Canada, 24–28 June 2002; Volume 5, pp. 3053–3055.
178. Huete, A.; Didan, K.; Miura, T.; Rodriguez, E.P.; Gao, X.; Ferreira, L.G. Overview of the Radiometric and Biophysical Performance of the MODIS Vegetation Indices. *Remote Sens. Environ.* **2002**, *83*, 195–213. [CrossRef]

179. Zha, Y.; Gao, J.; Ni, S. Use of Normalized Difference Built-up Index in Automatically Mapping Urban Areas from TM Imagery. *Int. J. Remote Sens.* **2003**, *24*, 583–594. [CrossRef]
180. Xu, H. A New Index for Delineating Built-up Land Features in Satellite Imagery. *Int. J. Remote Sens.* **2008**, *29*, 4269–4276 [CrossRef]
181. As-syakur, A.R.; Adnyana, I.W.S.; Arthana, I.W.; Nuarsa, I.W. Enhanced Built-Up and Bareness Index (EBBI) for Mapping Built-Up and Bare Land in an Urban Area. *Remote Sens.* **2012**, *4*, 2957–2970. [CrossRef]
182. Xu, H. Analysis of Impervious Surface and Its Impact on Urban Heat Environment Using the Normalized Difference Impervious Surface Index (NDISI). *Photogramm. Eng. Remote Sens.* **2010**, *76*, 557–565. [CrossRef]
183. Rasul, A.; Balzter, H.; Ibrahim, G.R.F.; Hameed, H.M.; Wheeler, J.; Adamu, B.; Ibrahim, S.; Najmaddin, P.M. Applying Built-Up and Bare-Soil Indices from Landsat 8 to Cities in Dry Climates. *Land* **2018**, *7*, 81. [CrossRef]
184. Gao, B. NDWI—A Normalized Difference Water Index for Remote Sensing of Vegetation Liquid Water from Space. *Remote Sens. Environ.* **1996**, *58*, 257–266. [CrossRef]
185. McFeeters, S.K. The Use of the Normalized Difference Water Index (NDWI) in the Delineation of Open Water Features. *Int. J. Remote Sens.* **1996**, *17*, 1425–1432. [CrossRef]
186. Xu, H. Modification of Normalised Difference Water Index (NDWI) to Enhance Open Water Features in Remotely Sensed Imagery. *Int. J. Remote Sens.* **2006**, *27*, 3025–3033. [CrossRef]
187. Xiao, X.; Shen, Z.; Qin, X. Assessing the Potential of VEGETATION Sensor Data for Mapping Snow and Ice Cover: A Normalized Difference Snow and Ice Index. *Int. J. Remote Sens.* **2001**, *22*, 2479–2487. [CrossRef]
188. Zhao, H.; Chen, X. Use of Normalized Difference Bareness Index in Quickly Mapping Bare Areas from TM/ETM+. In Proceedings of the 2005 IEEE International Geoscience and Remote Sensing Symposium, 2005. IGARSS '05, Seoul, Korea, 29 July 2005; Volume 3, pp. 1666–1668.
189. McGarigal, K.; Cushman, S.A.; Neel, M.C.; Ene, E. *FRAGSTATS: Spatial Pattern Analysis Program for Categorical Maps*; The University of Massachusetts: Amherst, MA, USA, 2002.

remote sensing

MDPI

Article

The Influence of Green Space Patterns on Land Surface Temperature in Different Seasons: A Case Study of Fuzhou City, China

Liuqing Yang [1,2], Kunyong Yu [2,3], Jingwen Ai [1,2], Yanfen Liu [1,2], Lili Lin [1,2], Lingchen Lin [2,3] and Jian Liu [1,2,3,*]

1 College of Landscape Architecture, Fujian Agriculture and Forestry University, Fuzhou 350000, China; 2181775004@fafu.edu.cn (L.Y.); 2191775006@fafu.edu.cn (J.A.); 1191775023@fafu.edu.cn (Y.L.); lilil@g.clemson.edu (L.L.)

2 University Key Lab for Geomatics Technology and Optimize Resource Utilization in Fujian Province, Fujian Agriculture and Forestry University, Fuzhou 350000, China; yuyky@126.com (K.Y.); lingchen_lin@fafu.edu.cn (L.L.)

3 College of Forestry, Fujian Agriculture and Forestry University, Fuzhou 350000, China

* Correspondence: fjliujian@fafu.edu.cn; Tel.: +86-136-0598-5226

Abstract: Background: Urban green space (UGS) has been shown to play an important role in mitigating urban heat island (UHI) effects. In the context of accelerating urbanization, a better understanding of the landscape pattern mechanisms affecting the thermal environment is important for the improvement of the urban ecological environment. Methods: In this study, the relationship between land surface temperature (LST) and the spatial patterns of green space was analyzed using a bivariate spatial autocorrelation and spatial autoregression model in three seasons (summer, transition season (spring), and winter) with different grid scales in Fuzhou city. Results: Our results indicated that the LST in Fuzhou City has a significant spatial autocorrelation. The percentage of landscape and patch density area were negatively correlated with surface temperature. The results of our indicators differed according to the season, with population density and distance to the water indicators not being significant in the winter. The coefficient of determination was higher at the 510 m grid scale on this study's scale. Conclusion: This study extends our understanding on the influence of UHI effects after accounting for different seasonal and spatial scale factors. It also provides a reference for urban planners to mitigate heat islands in the future.

Keywords: land surface temperature; greenspace spatial patterns; landscape metrics; spatial autoregressive model; seasonal variation

Citation: Yang, L.; Yu, K.; Ai, J.; Liu, Y.; Lin, L.; Lin, L.; Liu, J. The Influence of Green Space Patterns on Land Surface Temperature in Different Seasons: A Case Study of Fuzhou City, China. *Remote Sens.* **2021**, *13*, 5114. https://doi.org/10.3390/rs13245114

Academic Editors: Yuji Murayama and Ruci Wang

Received: 24 October 2021
Accepted: 13 December 2021
Published: 16 December 2021

1. Introduction

China is currently experiencing rapid urbanization, with the urbanization rate expected to reach 70% by 2030. As urbanization accelerates, the subsequent expansion of roads and buildings has led to the reduction of green space, resulting in a significant increase in the temperature of the city relative to the peripheral suburbs, a phenomenon also known as the urban heat island (UHI) effect [1]. Not only does the UHI effect affect the local and regional climate [2], it also affects water resources, air quality [3,4], the growth of plants [5], biodiversity, and the ecosystem in general [6]. It also has an adverse impact on the health of urban residents [7,8]. Developing effective adaptation measures and mitigation strategies for UHI effect is thus a common challenge that urban planners and climatologists are currently facing.

Rapid urbanization has significantly affected land cover characteristics. This change affects the characteristics of surface radiation, heat, and water in urban areas, and increases the urban surface temperature. On the one hand, remote sensing collected by satellites or airborne platforms has been proven effective in evaluating large-scale and local-scale land surface temperature (LST) [9]. By monitoring the relationship between the spatial change of LST and UHI combined with satellite images, it is found that impervious surfaces and urban green space (UGS) are highly related to LST. The average LST of impervious surfaces is 4–6 °C higher than that of green space, which indicates that urban expansion is the fundamental reason behind the UHI effect [10–15]. At the same time, urbanization has led to land use/land cover (LULC) change on the surface, while the transformation of forest land for commercial land use intensified the change in surface temperature and humidity [16,17].

UGS has been proven to be an important measure for alleviating the UHI effect [12], given that increasing vegetation coverage in urban areas has been proven to effectively reduce LST [18–20]. As such, many scholars have carried out research on the shape [21], area, structure, composition, and configuration of green space [22]. It has been found that the connectivity [23], complexity [24], and fragmentation of greenfield patches are all strongly correlated with LST [25–27]. It is confirmed that the structure and configuration of green space can effectively reduce urban temperature by rational optimization [28].

As a fast-growing city in China, Fuzhou's rapid urban expansion has led to a large reduction in wetlands and woodlands, which has earned Fuzhou the title of a "burning stove" in recent years [18]. Therefore, studying the spatial distribution and composition of UGS in Fuzhou City and analyzing the changes in its landscape patterns can guide us toward alleviating the urban heat island effect. However, the cooling effect of the UGS spatial pattern differs regionally [29]. Thus, further research is needed to optimize the effect of UGS on UHI and clarify the effect of UGS on UHI mitigation in different cities. Considering the changes in UGS, the relationship between UGS landscape patterns and LST is mostly analyzed via correlation or linear regression [20,30], which ignores the spatial heterogeneity of the two. Moreover, due to the difference in landscape patterns, the relationship between landscape patterns of different geographical environments and LST is not consistent [12,31]. Therefore, while considering the spatial patterns of green space by combining spatial autocorrelation and spatial regression models, this paper analyzes the spatial relationship between green space landscape patterns and LST in different seasons and scales in Fuzhou, while adding further discussion on the impact mechanism of green space on the thermal environment.

2. Methods

2.1. Study Area and Data Source

Fuzhou is the capital of Fujian Province in China, which is located at 25°15′–26°39′ N, 118°08′–120°31′ E, and has a subtropical monsoon climate. The urbanization process of Fuzhou has been particularly obvious in recent years. According to data from the Fuzhou Bureau of Statistics, by the end of 2017, the population of Fuzhou was 7.21 million, with 5.105 million urban residents; the urbanization rate of Fuzhou had already exceeded 70 percent (70.8 percent). The conversion of water bodies and UGS into major built-up areas amounted to 13.3 and 20.2 km^2. From 1985 to 2015, the area of built-up areas doubled. Large UGS losses during urbanization are the main impact of the urban thermal environment [18,32]. The study area covers a total of 1448.479 km^2 (Figure 1).

Fuzhou has a subtropical monsoon climate, which is marked by longer summers and shorter winters. Spring starts in March and ends in June, while autumn starts in October and ends in November. After analyzing the temperature changes in Fuzhou from 2000–2020 (Figure 2), we found that the temperature distributions in spring and autumn were relatively similar, and that there was little difference in temperature between April and May in spring, and October and November in autumn. Therefore, the study divided the seasons in Fuzhou into three types: summer, transitional seasons, and winter [33–35]. Landsat8 image data (11 December 2019, 16 March 2020, and 22 July 2020) for three different periods were selected from the USGS (https://espa.cr.usgs.gov/index/ accessed on 18 March 2021) and radiometric and geometric corrections were applied [36]. Based on the atmospheric correction method, LST was retrieved from the thermal band of Landsat8 images to characterize the LST of different seasons.

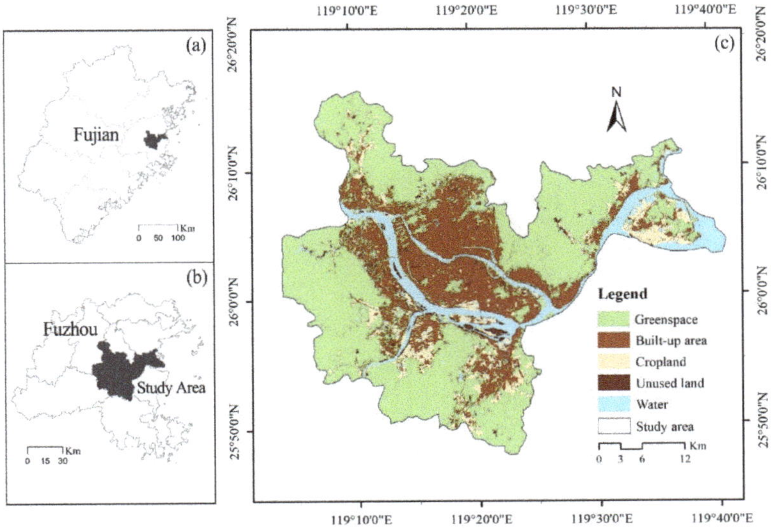

Figure 1. Study area: (**a**) Fujian Province, China; (**b**) Fuzhou City, Fujian Province; (**c**) the land classification of Fuzhou City in this study area.

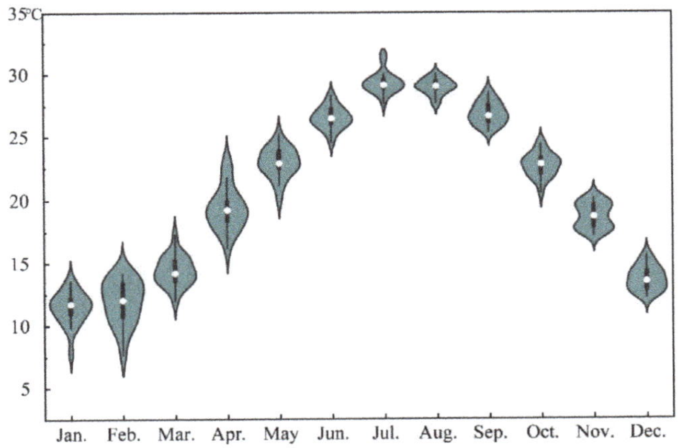

Figure 2. Violin plot of the monthly average air temperature in Fuzhou City during 2000–2020.

Land use data were obtained using the 2018 Gaofen-1 satellite (GF-1) image, which has a panchromatic resolution of 2 m and a multispectral resolution of 8 m [37]. Using eCognition software (Definiens Imaging, Inc., Munich, Germany), five main landscape types were extracted based on an object-oriented classification approach—green space, built-up area, water body, cropland, and unused land—and combined with high-resolution historical images from Google Earth™, they were used as a reference layer to assess classification accuracy for all land types. The classification accuracy assessment was calculated based on a confusion matrix of 100 randomly selected points, with an overall classification accuracy of 88.19%. At the same time, the results of land use classification were further used to analyze the landscape metrics of UGS.

2.2. Retrieving Land-Surface Temperature (LST)

This research is based on the atmospheric correction method, and the Landsat8 TIRS was used to retrieve the surface temperature with a data accuracy of 30 m × 30 m.

$$L_\lambda = [\varepsilon B(T_s) + (1-\varepsilon)L_d]\tau + L_\mu \quad (1)$$

In Equation (1), L_μ is the upwelling radiance, L_d is the downwelling radiance, ε L_λ is the top-of-atmosphere (TOA) radiance, where ε is the surface emissivity, $B(T_s)$ is the black body heat radiation brightness, and τ is the transmittance of the atmosphere in the thermal infrared band. Thus, the radiance of a black body with a temperature of T in the thermal infrared band $B(T_s)$ is:

$$B(T_s) = \frac{L_\lambda - L_\mu - \tau(1-\varepsilon)L_d}{\tau\varepsilon} \quad (2)$$

T_s is the true surface temperature (LST) (°C), which can be obtained by the function of Planck's formula:

$$T_s = \frac{K_2}{\ln(K_1/B(T_s)+1)} \quad (3)$$

For Landstat8, K_1 = 774.89 W/(m^2·μm·sr), K_2 = 1321.08 K.

2.3. Influencing Factors Selection

The landscape metrics can effectively quantify the structural composition and spatial configuration characteristics of the patch [38], given that studies have proven landscape composition and configuration to have a significant impact on LST [39]. Based on the results of previous studies, six commonly used landscape indicators were selected as explanatory variables for LST changes [30]. To account for the correlation between the various landscape metrics, to fully reflect the landscape pattern of green space, and to avoid potential correlations caused by redundant metrics, we selected percentage of landscape (PLAND), patch density (PD), edge density (ED), aggregation index (AI), mean patch size (AREA_MN), and mean shape index (SHAPE_MN) as our key indicators [40] to analyze the relationship between green space landscape pattern and LST. These indicators reflect the main characteristics of green space patterns. Generally, PLAND and AREA_MN represent the patch area, PD and ED are indicator density and patch boundary, while SHAPE_MN is the patch shape index and AI represents the spread of landscape patches.

We used the grid analysis method to analyze the green landscape pattern metrics using the FRAGSTATS 4.2 software [31]. However, the appropriate grid scale is still argued in the literature, and previous studies have proved that the green space cooling effect is scale-dependent [41], while the optimal scale varies depending on study area [12]. To explore the suitable grid scale in Fuzhou, we applied four grid scales: 360 m × 360 m, 510 m × 510 m, 720 m × 720 m, and a 960 m × 960 m grid scale (all grid scales are integer multiples of 30 m accuracy of landsat8 images) [25,42–44]. By extracting the green landscape metrics of each grid, a grid image of each green landscape metric were obtained. We then used ArcGIS 10.2 (ESRI, Redlands, CA, USA) software to calculate the average LST and landscape metrics for each grid. We normalize the indicator results before analysis. Considering the impact of the water bodies and population density on LST [11,45–47], we also added closest distance to a water (water_distance) [17] and population density as research variables (https://www.worldpop.org/ accessed on 30 June 2021). The accuracy of the population density map was 100 m × 100 m (Table 1). We applied the Pearson correlation to check the strength of the association between LST and each variable indicator.

Table 1. Descriptions of the variables used in this study [48].

Variables	Description	Formula	Unit
Percentage of landscape (PLAND)	The proportional abundance of each patch type in the landscape within an analysis unit.	$PLAND = p_i = \frac{\sum_{j=1}^{n} a_{ij}}{A} * 100$	Percent
Patch density (PD)	Density of landscape patches within an analysis unit, reflects the degree of fragmentation and spatial heterogeneity of landscape patches.	$PD = \frac{N_i}{A}(10000) * (100)$	n/km^2
Edge density (ED)	The total perimeter of landscape patches per ha within an analysis unit, reflects the degree of fragmentation of the patch.	$ED = \sum_{k=1}^{m} e_{ik} \times \frac{10000}{A}$	m/ha
Aggregation Index (AI)	Aggregating degree of the corresponding patches within an analysis unit.	$AI = \left\lvert \frac{g_{ii}}{\max - g_{ii}} \right\rvert * 100$	Percent
Mean patch area (AREA_MN)	The average area of landscape patches within an analysis unit.	$\text{AREA\_MN} = \frac{1}{10000 \times n} \times \sum_{i=1}^{n} a_{ij}$	Hectares
Mean patch shape index (SHAPE_MN)	The average shape index of landscape patches within an analysis unit, for reflecting the complexity of individual patch shapes.	$\text{SHAPE\_MN} = \frac{1}{n} \times \frac{0.25 p_{ij}}{\sqrt{a_{ij}}}$	unitless
Water_distance	Analyze the Euclidean distance of each unit to the water bodies.		Km
Population density	The population density within an analysis unit.		People/km^2

A, total landscape area (m^2); a_{ij}, area (m^2) of patch ij; N_i, number of patches in the landscape of patch type (class); n, number of patches p_i, proportion of the landscape occupied by patch type (class) i; e_{ik}, total length (m) of edge in landscape involving patch type (class) i; p_{ij}, perimeter (m) of patch ij; g_{ii}, number of like adjacencies (joins) between pixels of patch type (class) i based on the single-count method; max-g_{ii}, maximum number of like adjacencies (joins) between pixels of patch type (class) i based on the single-count method.

2.4. Spatial Autocorrelation and Spatial Autoregressive Model

Spatial autocorrelation analysis reveals whether the variables are spatially correlated, and the degree of autocorrelation [49]. Spatial autocorrelation includes global spatial autocorrelation and local spatial autocorrelation, and here we use global Moran's I (the spatial correlation of the whole study area) and local indicators of spatial association (LISA) (the spatial correlation of each spatial unit and its neighboring units for a certain attribute) to describe the above results [50,51]. Moran's I values range from −1 to 1, with negative values indicating negative spatial autocorrelation, and positive values indicating positive spatial autocorrelation. LISA plots depict the results of five scenarios: clusters of high values (high-high), clusters of low values (low-low), outliers where high values are surrounded by low values (high-low), and outliers where low values are surrounded by high values (low-high) and have no significant local autocorrelation (not-significant). It determines the degree of spatial clustering present in the data (p-value). Additionally, the bivariate spatial autocorrelation analysis was used to analyze the spatial correlation of different variables with LST [52]. The GeoDa (v.1.14.0; Luc Anselin, AZ, USA) software was used for the above-mentioned spatial correlation analysis.

The formula for global spatial autocorrelation can be expressed as:

$$I = \frac{n}{S_0} \frac{\sum_{i=1}^{n} \sum_{j=1}^{n} w_{i,j} z_i z_j}{\sum_{i=1}^{n} z_i^2} \quad (4)$$

where n is the total number of spatial units, z_i is the attribute of element i with its mean $(x_i\text{-}\overline{x})$, $w_{i,j}$ is the spatial weight value between i,j, and S_0 is the aggregation of all spatial weights.

$$S_0 = \sum_{i=1}^{n} \sum_{j=1}^{n} w_{ij}, \quad (5)$$

The z_x score for the statistic is given by:

$$z_x = \frac{I - E[I]}{\sqrt{V[I]}} \quad (6)$$

where

$$E[I] = \frac{-1}{(n-1)} \quad (7)$$

$$V[I] = E\left[I^2\right] - E[I]^2 \quad (8)$$

A traditional ordinary least squares (OLS) model was employed to examine the effect of each variable on the change in LST.

$$y = X\beta + u \quad (9)$$

where X is the matrix of explanatory variables, β denotes a vector of slopes, and u represents a vector of random error terms.

Spatial lag model (SLM):

$$y = \rho Wy + X\beta + u \quad (10)$$

where ρ is a spatial autocorrelation parameter, W is the spatial weight matrix, Wy is the spatial lag operator, and β is the unknown coefficient vector.

Spatial error model (SEM):

$$y = X\beta + \gamma W\varepsilon + \delta \quad (11)$$

where γ is the spatial autocorrelation parameter, $W\varepsilon$ denotes the spatial weight matrix, and δ represents a vector of the error terms.

The spatial error model and the spatial lag model were determined using Lagrangian multiplier diagnostics (LM) [36,44]. The LM and robust LM values of the SEM model are larger than that of SLM, indicating that SEM performed better than SLM. Results show that the spatial error model was suitable for the data in this study (Table A1).

3. Results

3.1. Spatial Characteristics of LST

In general, the spatial distribution characteristics of LST in Fuzhou during different seasons tend to be consistent. High-temperature areas were mainly concentrated in the city center (Figure 3). As Fuzhou is surrounded by mountains on three sides (Gushan Mountain to the east, Wuhu Mountain to the south, and Qishan Mountain to the west), the vegetation coverage is high and the forests are relatively abundant. Thus, low-temperature areas mainly surround the urban areas. In summer, the temperature in Fuzhou can reach above 50 °C (with an average of 43 °C) (Table 2), which is quite different from the other two seasons and can seriously affect daily travel [53].

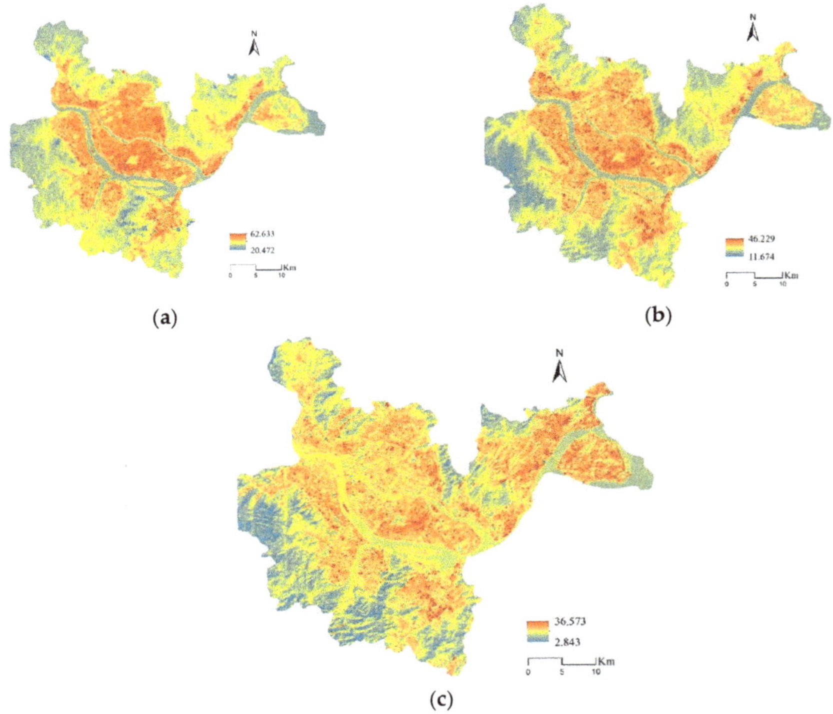

Figure 3. The three images represent the spatial distribution of land surface temperature (LST, °C) in (**a**) summer, (**b**) transitional season, and (**c**) winter, respectively.

Table 2. Temperature in different seasons.

Season	Data Date	Local Time	Minimum	Maximum	Average	Std. Dev.
Summer	22 July 2020	02:32:25	20.473	62.633	43.417	5.556
Transition season	16 March 2020	02:32:22	11.674	46.229	22.305	3.168
Winter	11 December 2019	02:32:49	2.843	36.573	18.225	1.901

Through the analysis of the study area, the LST distribution of global Moran's I is positively autocorrelated (Table 3). The overall distribution of LST and the spatial autocorrelation of LST were also significant at the 0.01 level, indicating that the spatial distribution of LST exhibited aggregation. With an increase in the grid scale, the larger the scale, the lower the correlation to a certain extent. At the same time, the global Moran's I of LST in summer and the transition season was greater than that in winter. It can be seen from (Figure A1) that the spatial aggregation distribution trends of all scales in the study area are basically the same. The low-low cluster (L-L) is distributed in the forest areas to the west, east, and north, while the proportion of high-high clusters (H-H) is relatively large and distributed mostly in the urban center and a little to the south. With a decrease in temperature, the L-L distribution in the urban area decreased. The distribution proportions of the high-low cluster (H-L) and low-high cluster (L-H) were small and sporadic. In winter, H-H accumulation in the urban center decreased significantly, but increased in the east. The distribution of L-L increased with a decrease in temperature, especially in the south.

Table 3. Global Moran's I (Equation (4)) of land surface temperature (LST).

Scale (m)	Global Moran's *I*			
	Summer	Transition Season	Winter	*p*-Value
360 m	0.766	0.773	0.653	<0.01
510 m	0764	0772	0668	<0.01
720 m	0.664	0.673	0.546	<0.01
960 m	0.620	0.644	0.548	<0.01

3.2. LST and UGS Spatial Pattern Analysis

3.2.1. Analysis of the Spatial Pattern of Green Space

By comparing the cities around Fuzhou [54], the patch density (PD) and edge density (ED) in Fuzhou were higher than the surrounding cities, and the mean patch area (AREA_MN) value was lower. The higher the patch density, the higher the fragmentation of the landscape; the higher the edge density, the lower the aggregation of patches at the edge of the landscape, while the small average patch area indicates that there are many small green areas with a certain degree of fragmentation in the Fuzhou UGS. Meanwhile, the global Moran's I is also greater than 0.7 (Table 4), thus indicating that the green space has obvious spatial aggregation.

Table 4. Green space landscape metrics and global spatial autocorrelation analysis.

Landscape Metrics						Moran's *I*
PLAND (%)	PD (n/km^2)	ED (m/ha)	AI (%)	AREA_MN (ha)	SHAPE_MN	
58.379	33.869	65.510	82.582	10.222	1.197	0.869

3.2.2. Bivariate Analysis of Green Space Landscape Pattern Index and LST

Pearson's correlation coefficient and bivariate spatial autocorrelation analysis showed that the relationship between each indicator and LST was more significant in summer and transition seasons than in winter (Figure 4). The correlations of PLAND, AI, AREA_MN, and water_distance were all negative. Among the negative indicators, PLAND and AREA_MN were more strongly correlated with LST, indicating that the increase in the proportion of green patches can have a cooling effect. ED, SHAP_MN, and population density were positive numbers. This indicates that the increase of patch edge density leads to the fragmentation of green areas. Under different grid scales, the correlation between the positive indexes SHAPE_MN and ED increases with an increase in the grid scale, while the negative index AI is the opposite. The larger the grid, the lower the correlation. There was no strong correlation between PD and spatial autocorrelation (see Figure 4b).

3.2.3. Spatial Autoregressive Analysis

Spatial autoregressive analysis was carried out using LST as the dependent variable. At the same time, before performing spatial regression analysis, all indicators passed the collinearity test with variance inflation factor (VIF) values less than 10, indicating that there was no problem with collinearity among the indicators.

From the results of the OLS analysis (Table 5), the correlation of the overall indicators is the highest in the transition season and lowest in winter. The contribution of PLAND and PD to LST cooling is larger (especially PLAND), while ED, AI, and population density are obviously positive. At the same time, the correlation of PLAND decreased from summer to winter, while PD remained relatively stable. In terms of positive correlation indicators, the LST interpretation rates of ED, AI, and population density increased with an increase in temperature. Water distance is positively correlated in summer and transition seasons, while negatively correlated in winter. AREA_MN loses explanatory significance as the temperature decreases. On the grid scale, as the grid scale increases, the fit of each index (R2) in different seasons grows higher under the OLS model.

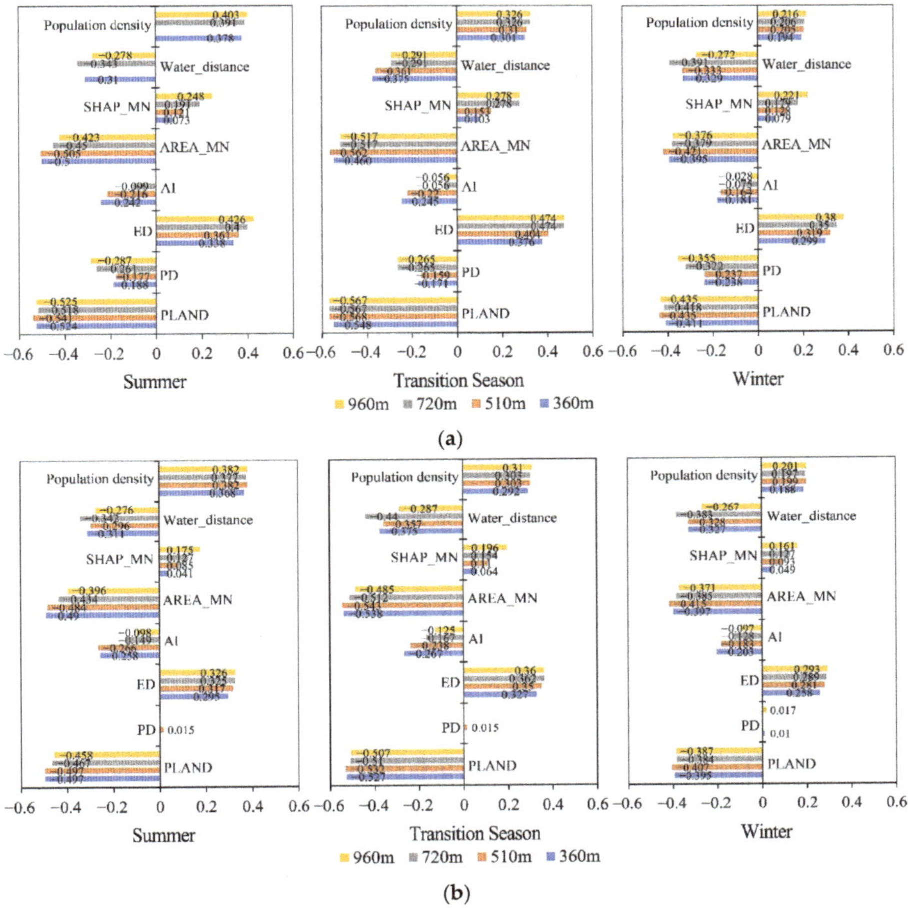

Figure 4. Pearson correlation analysis at different scales and bivariate spatial autocorrelation. Values without significant correlation were excluded from the graph. (**a**) Pearson correlation analysis; (**b**) bivariate spatial autocorrelation.

However, it was found through the results that the residuals of the OLS model Moran's I had significant spatial clustering (Table A1), and the traditional OLS regression model may have ignored the spatial correlation of the indicators, so the SEM model was used for further analysis (Table 6).

Specifically, the coefficients of ED, AI were significantly positive. The coefficients for PLAND and PD were significantly negative. SHAPE_MN was not statistically significant. The correlation of overall indicators was better in the summer and transition season than in the winter. Under the standardized coefficient, the indexes of PLAND and PD make a great contribution to the reduction of LST; with the increase in grid scale, the effect of index on LST is enhanced. At the same time, with the increase in scale, the correlation between AREA_MN, water_distance, population density, and LST disappeared. From the fitting results, the fitting effect (R^2) of the 510 m scale is higher than that of the other scales in this study. Compared with the OLS model, the interpretation degree of the SEM index for seasonal changes is lower, but the overall fitting effect of the index is better than that of OLS.

Table 5. Results from the ordinary least squares (OLS) multiple linear regressions.

Season	Scale (m)	PLAND	PD	ED	AI	AREA_MN	SHAPE_MN	Water_Distance	Population Density	AIC	R^2
Summer	360 m	−0.081 *** (−0.610)	−0.002 *** (−0.181)	0.014 *** (0.250)	0.025 *** (0.155)	0.075 *** (0.077)	−0.179 (−0.019)	0.000 *** (0.058)	0.022 *** (0.215)	67,392.8	0.424
	510 m	−0.076 ** (−0.585)	−0.003 *** (−0.221)	0.017 *** (0.289)	0.024 *** (0.143)	0.028 ** (0.057)	−0.220 (−0.022)	0.000 *** (0.077)	0.022 *** (0.217)	34,568.3	0.459
	720 m	−0.077 ** (−0.562)	−0.001 *** (−0.226)	0.017 *** (0.256)	0.039 *** (0.193)	0.016 * (0.061)	−0.022 0.000 **	0.000 ** (0.051)	0.024 *** (0.219)	17,156.5	0.474
	960 m	−0.073 ** (−0.533)	−0.001 *** (−0.241)	0.019 *** (0.264)	0.056 *** (0.238)	0.008 * (0.056)	−0.177 (−0.015)	0.000 (0.013)	0.024 *** (0.223)	9729.38	0.551
Transition Season	360 m	−0.040 *** (−0.540)	−0.001 *** (−0.169)	0.007 *** (0.235)	0.012 *** (0.131)	−0.008 (−0.014)	0.102 (0.019)	0.000 * (−0.018)	0.007 *** (0.117)	53,706	0.432
	510 m	−0.038 *** (−0.526)	−0.002 *** (−0.207)	0.009 *** (0.277)	0.012 *** (0.123)	−0.011 * (0.040)	0.040 (0.007)	0.000 (0.008)	0.006 *** (0.112)	27,337.2	0.472
	720 m	−0.036 *** (−0.474)	−0.001 *** (−0.212)	0.009 *** (0.236)	0.019 *** (0.167)	−0.006 (0.041)	0.197 (0.032)	0.000 *** (0.070)	0.007 *** (0.114)	13,485	0.495
	960 m	−0.040 *** (−0.533)	−0.001 *** (−0.226)	0.010 *** (0.258)	0.030 *** (0.231)	−0.006 * (−0.068)	−0.007 (−0.001)	0.000 (0.019)	0.007 *** (0.113)	7571.76	0.538
Winter	360 m	−0.019 *** (−0.425)	−0.001 *** (−0.231)	0.004 *** (0.218)	0.004 *** (0.081)	0.022 ** (0.065)	0.034 (0.010)	0.000 *** (−0.076)	0.002 *** (0.049)	45,041.6	0.290
	510 m	−0.018 *** (−0.407)	−0.002 *** (−0.274)	0.005 *** (0.248)	0.004 ** (0.071)	0.004 (0.023)	0.070 (0.021)	0.000 *** (−0.064)	0.001 *** (0.044)	22,555	0.334
	720 m	−0.016 *** (−0.333)	−0.001 *** (−0.299)	0.005 *** (0.202)	0.006 ** (0.089)	0.003 (0.039)	0.193 (0.050)	0.000 *** (−0.151)	0.002 ** (0.048)	11,323.9	0.362
	960 m	−0.018 *** (−0.392)	−0.001 *** (−0.325)	0.005 *** (0.225)	0.012 ** (0.150)	−0.001 (0.012)	0.084 (0.020)	0.000 * (−0.064)	0.002 * (0.044)	6371.76	0.399

Note: ***, **, and * represent significance at the 0.001, 0.01, and 0.05 levels respectively. We included both unstandardized regression coefficients and standardized coefficients (included in parentheses). The standardized coefficient estimates help determine the relative importance of the explanatory variables in different grids. The larger the absolute value of the standardized coefficient estimate, the more important the variable. AIC is Akaike information criterion.

Table 6. The results of spatial error models (SEM).

Season	Scale(m)	PLAND	PD	ED	AI	AREA_MN	SHAPE_MN	Water_Distance	Population Density	AIC	R^2
Summer	360 m	−0.035 *** (−0.269)	−0.001 *** (−0.174)	0.005 *** (0.089)	0.008 *** (0.049)	0.0422 *** (0.043)	−0.009 (0.000)	0.000 *** (0.154)	0.007 *** (0.064)	55,629.1	0.827
	510 m	−0.049 *** (−0.380)	−0.003 *** (−0.205)	0.008 *** (0.146)	0.006 ** (0.033)	0.022 ** (0.044)	−0.083 (−0.009)	0.000 *** (0.183)	0.003 * (0.034)	28,148.5	0.845
	720 m	−0.054 *** (−0.394)	−0.001 *** (−0.236)	0.008 *** (0.134)	0.016 ** (0.081)	0.009 (0.035)	−0.022 (0.002)	0.000 *** (0.177)	0.006 * (0.059)	15,234.9	0.763
	960 m	−0.059 *** (−0.437)	−0.002 *** (−0.281)	0.012 *** (0.175)	0.018 ** (0.077)	0.003 (0.024)	0.155 (0.013)	0.000 * (0.162)	0.006 * (0.063)	8734	0.763
Transition Season	360 m	−0.016 *** (−0.218)	−0.000 *** (−0.162)	0.003 *** (0.093)	0.005 *** (0.054)	0.018 ** (0.033)	0.024 (0.005)	0.000 ** (0.099)	0.003 *** (0.059)	41,941.6	0.830
	510 m	−0.023 *** (−0.318)	−0.002 *** (−0.186)	0.005 *** (0.154)	0.0038 ** (0.040)	0.011 ** (0.040)	−0.032 (−0.006)	0.000 *** (0.119)	0.0012 * (0.033)	21,081.8	0.844
	720 m	0.026 *** (−0.338)	−0.000 *** (−0.214)	0.005 *** (0.144)	0.008 *** (0.076)	0.003 (0.024)	0.077 (0.012)	0.000 (0.039)	0.003 * (0.050)	11,714.4	0.759
	960 m	−0.029 *** (−0.397)	−0.000 *** (−0.239)	0.007 *** (0.188)	0.014 *** (0.106)	−0.000 (−0.005)	0.057 (0.009)	0.000 (0.096)	0.003 (0.048)	6637	0.767
Winter	360 m	−0.009 *** (−0.207)	−0.000 *** (−0.217)	0.002 *** (0.093)	0.003 *** (0.057)	0.014 ** (0.043)	−0.012 (−0.004)	−0.000 *** (−0.139)	0.002 ** (0.046)	36,329	0.711
	510 m	−0.014 *** (−0.326)	−0.001 *** (−0.257)	0.003 *** (0.163)	0.003 ** (0.054)	0.010 *** (0.062)	−0.077 (−0.023)	0.000 (−0.016)	0.000 (0.028)	17,549.6	0.749
	720 m	−0.013 *** (−0.284)	−0.000 *** (−0.289)	0.002 *** (0.128)	0.006 ** (0.079)	0.005 * (0.053)	−0.028 (−0.007)	−0.000 *** (−0.153)	0.001 (0.042)	9942.35	0.639
	960 m	−0.016 *** (−0.336)	−0.000 *** (−0.319)	0.004 *** (0.166)	0.007 *** (0.091)	0.001 (0.031)	−0.029 (−0.007)	−0.000 (−0.025)	0.001 (0.047)	5439.88	0.693

Note: ***, **, and * represent significance at the 0.001, 0.01, and 0.05 levels, respectively. We included both unstandardized regression coefficients and standardized coefficients (included in parentheses). AIC is Akaike information criterion.

4. Discussion

4.1. Spatial Variation of LST

Urban LST decreases with the change from summer to winter. Owing to the characteristics of urban development in Fuzhou, the suburbs are basically surrounded by mountains and less developed, which is why the distribution characteristics of the high-temperature areas across the whole study area are obvious and H-H accumulation is mainly distributed in the urban center where the population is concentrated. Urban expansion leads to the continuous reduction of green space and the continuous expansion of construction land, man-made emissions, etc., resulting in high LST. L-L aggregation is mainly distributed around the suburbs. With the decrease in LST, the aggregation degree of H-H decreased, while the L-L aggregation in the suburbs increased significantly.

4.2. Differences in the Impact of UGS on UHI Mitigation

Our OLS analysis showed that the indexes which contributed greatly to the cooling of LST under the standardized coefficient are PLAND and PD. This supports the claim that PLAND has great explanatory significance for cooling [55]. ED, AI, and population density also had significant effects on LST. At the same time, the effect of model fitting increased with an increase in the grid scale.

The results of the index analysis by the SEM model were slightly different from those of the OLS model. The coefficient of determination (R^2) for each variable at the 510 m grid scale is higher than the other three scales of this study. With an increase in scale, the value of each index coefficient increases. The landscape metrics PLAND and PD of green space showed an obvious negative correlation with LST, which is consistent with the research results of [40,55]. This shows that an increase in green space area, a larger patch area, and more patch quantity can reduce LST, given that green space can produce cold island effects through evapotranspiration, as well as shade to prevent the surface from being directly heated by the sun [56]. At the same time, the more aggregated the patch area, the better the connectivity and interaction between vegetation, and the better the cooling effect on the area [57]. However, the results of ED contradicted some previous studies. LST was found to decrease with an increase in the complexity of plant shape and edge density, probably because the shape complexity of edge density can increase the area of shadow provided by plants and reduce LST [40]. Although the patch edge density in Fuzhou was high, there were many small patches in Fuzhou and the green space distribution was relatively scattered, which weakened the cooling capacity of the green space and reduced the cooling effect [58–60]. Several studies have pointed out that population density is an important reason for the rise in LST. The more densely populated an area is, the more anthropogenic heat emissions there are, and the higher the UHI [34,61]. The heat island effect is more pronounced closer to the urban center. In the summer and transition seasons, the effect of population density on LST was positively correlated, but the correlation was not obvious in winter.

At the same time, different seasons have a significant impact on indicators. The nearest distance to the water was positively correlated with LST in the summer and transition seasons, but not in winter. It may be that rivers are scattered, mainly distributed around roads and urban areas, and are seriously disturbed by human factors. This shows that the distance and area of the water from the urban area are different, the degree of impact of human activities is different, and the impact on LST is different in different seasons. Because of the low temperature in winter, the temperature difference between the water body and the surface decreases, so the factors affecting LST will also decrease and the cooling effect will not be so obvious [33,62]. Overall, the research shows that the impact of the index in summer and transition season is higher than that in winter; in summer, the heat brought by the reduction of thermal radiation and evaporation is greater than that in winter, the temperature in winter is low, the impact of vegetation on LST is reduced [62], and the amount of solar thermal radiation is small [36,62,63].

4.3. UHI Mitigation Implications by Urban Greening

Our findings help planners to implement differentiated UGS management strategies for LST alleviation in specific regions. From the analysis, it is clear that increasing the UGS coverage is the most effective way to promote cooling effects. However, with the acceleration of urbanization in Fuzhou, green space resources are gradually being reduced, while the spread of impervious surfaces has also reduced the aggregation of green space patches and dispersed them. Under limited circumstances, it is difficult to conduct greening in large areas, thus optimizing space allocation and improving space utilization, such as increasing roof greening and pocket parks [20] in undeveloped urban suburbs, is necessary. During the construction process, the patch areas of green space can be increased, and a certain scale of parks and green spaces can be built to improve the evapotranspiration of plants and reduce the development intensity of the city and the aggregation of buildings. With the urbanization of Fuzhou in recent years, a large number of UGS have been reduced, which have resulted in many scattered patches [18]. The edge density of these patches is large, which results in a low cooling effect. Therefore, improving the connectivity between patches and reducing the fragmentation of patches can improve the mitigation effect of green spaces on UHIs. At the same time, Fuzhou has a rich network of urban waterways since it has a significant effect on reducing LST. With the optimization and rational use of the water system and the protection of existing wetlands from damage, we can alleviate the negative impact of the thermal environment [32,64].

4.4. Limitations and Future Research

The changes in urban ecosystems and land use/land cover caused by accelerated urbanization have largely exacerbated the UHI effect [20]. Combining land use data from multiple years can better identify the changing patterns of LST; this study only used one year's worth of data to analyze the spatial characteristics of green space and LST in Fuzhou. Meanwhile, there are many factors affecting LST, and the study failed to comprehensively analyze the interactions between multiple LST influencing elements. By analyzing the LST temperature changes across different years using the long-time series and multi-cycle data analysis in combination with different indicators in the future, the UHI effect can be analyzed more comprehensively. Incorporating it with the above analysis in future studies can also help us study and alleviate the heat island effects in Fuzhou more comprehensively.

5. Conclusions

Taking Fuzhou as the study area, this paper used a bivariate spatial autocorrelation and spatial autoregression model to analyze the spatial relationship between different green landscape elements and LST in different seasons from multiple grid scales. The results showed that the percentage of landscape and patch density were negatively correlated with LST, while edge density and aggregation index were positively correlated with LST. Mean patch shape index had no correlation. At the same time, the dominant factors affecting LST differed according to the season. Distance to the water body and population density had positive impact on LST during the summer, but the correlation weakened in the transition season and winter. With the decrease of seasonal temperature, the explanatory effect of each index on LST decreased accordingly. Because the factors affecting LST in the winter differed from the other seasons, this indicates that more factors need to be considered in the analysis of LST, and mitigation measures need to be taken for different seasons. At the same time, it was found that the coefficient of the spatial error model was lower than that of the OLS regression model, indicating that SEM considers the spatial autocorrelation of each index. The fitting effect of SEM is better than that of OLS, which proves it can better explain the spatial relationship between the greenspace landscape pattern and LST. This study also analyzed the impact of landscape indicators on LST in different seasons at different grid scales to identify seasonal changes that can promote development and guide urban planning, and ultimately reduce the UHI effect.

Author Contributions: Conceptualization, J.L. and K.Y.; methodology, L.Y.; software, L.Y.; validation, L.Y., L.L. (Lili Lin) and L.L. (Lingchen Lin); formal analysis, L.Y. and L.L. (Lingchen Lin); investigation, L.Y.; resources, L.Y.; data curation, Y.L. and J.A.; writing—original draft preparation, L.Y.; writing—review and editing, L.Y.; visualization, L.Y.; supervision, K.Y.; project administration, J.L.; funding acquisition, J.L. and K.Y. All authors have read and agreed to the published version of the manuscript.

Funding: This research was funded by Fujian Province Key Laboratory of Geographic Information Technology and Resource Optimization Construction Project, grant number (PTJH17014).

Institutional Review Board Statement: Not applicable.

Informed Consent Statement: Not applicable.

Data Availability Statement: The data that support the findings of this study are available from the author upon reasonable request.

Acknowledgments: We sincerely thank the editor and anonymous reviewers for their valuable comments and suggestions to improve the quality of this paper.

Conflicts of Interest: The authors declare no conflict of interest.

Appendix A

Local indicators of spatial association (LISA) at different grid scales in the study area.

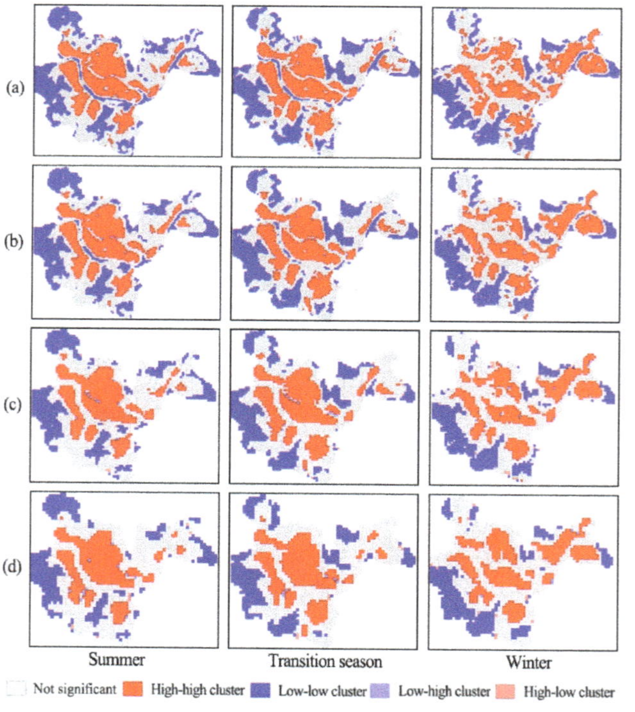

Figure A1. (**a**) 360 m × 360 m; (**b**) 510 m × 510 m; (**c**) 720 m × 720 m; (**d**) 960 m × 960 m. (1) High- high means spatial clustering of similarly high values, (2) low-low means spatial clustering of similarly low values, (3) low-high cluster means low values surrounded by high neighboring values, and (4) high-low cluster means high values surrounded by low values. (5) no significant value.

Appendix A

Table A1. Lagrange multiplier (LM) diagnostics for spatial dependence.

		Value			p-Value		
Scale (m)		**Summer**	**Transition Season**	**Winter**	**Summer**	**Transition Season**	**Winter**
360 m	LM (SLM)	17,669.722	17,554.956	14,749.657	0.000	0.000	0.000
	Robust LM (SLM)	751.682	817.425	335.424	0.000	0.000	0.000
	LM (SEM)	18,289.119	17,995.990	15,079.633	0.000	0.000	0.000
	Robust LM (SEM)	1371.078	1258.459	665.401	0.000	0.000	0.000
	Moran's I (error)	0.638	0.633	0.576	0.000	0.000	0.000
510 m	LM (SLM)	8980.604	8785.118	8081.876	0.000	0.000	0.000
	Robust LM (SLM)	315.210	375.142	134.112	0.000	0.000	0.000
	LM (SEM)	9733.919	9383.335	8577.889	0.000	0.000	0.000
	Robust LM (SEM)	1068.525	973.335	630125	0.000	0.000	0.000
	Moran's I (error)	0.646	0.634	0.606	0.000	0.000	0.000
720 m	LM (SLM)	2777.740	2600.889	2244.589	0.000	0.000	0.000
	Robust LM (SLM)	164.529	165.802	62.053	0.000	0.000	0.000
	LM (SEM)	2892.608	2658.432	2314.228	0.000	0.000	0.000
	Robust LM (SEM)	279.137	223.345	131.692	0.000	0.000	0.000
	Moran's I (error)	0.506	0.485	0.453	0.000	0.000	0.000
960 m	LM (SLM)	1229.250	1196.510	1335.168	0.000	0.000	0.000
	Robust LM (SLM)	52.147	77.669	38.198	0.000	0.000	0.000
	LM (SEM)	1393.353	1308.945	1443.972	0.000	0.000	0.000
	Robust LM (SEM)	216.250	190.104	147.002	0.000	0.000	0.000
	Moran's I (error)	0.466	0.451	0.474	0.000	0.000	0.000

Note: p-value is the probability of observing an event at least as extreme as the test statistic.

References

1. Oke, T.R. The energetic basis of the urban heat island. *Q. J. R. Meteorol. Soc.* **1982**, *108*, 1–24. [CrossRef]
2. Morris, K.I.; Chan, A.; Morris, K.J.K.; Ooi, M.C.G.; Oozeer, M.Y.; Abakr, Y.A.; Nadzir, M.S.M.; Mohammed, I.Y.; Al-Qrimli, H.F. Impact of urbanization level on the interactions of urban area, the urban climate, and human thermal comfort. *Appl. Geogr.* **2017**, *79*, 50–72. [CrossRef]
3. Li, H.; Meier, F.; Lee, X.; Chakraborty, T.; Liu, J.; Schaap, M.; Sodoudi, S. Interaction between urban heat island and urban pollution island during summer in Berlin. *Sci. Total Environ.* **2018**, *636*, 818–828. [PubMed]
4. Jochner, S.; Menzel, A. Urban phenological studies—Past, present, future. *Environ. Pollut.* **2015**, *203*, 250–261. [CrossRef]
5. Zhou, D.; Zhao, S.; Zhang, L.; Liu, S. Remotely sensed assessment of urbanization effects on vegetation phenology in China's 32 major cities. *Remote. Sens. Environ.* **2016**, *176*, 272–281. [CrossRef]
6. Grimm, N.B.; Faeth, S.H.; Golubiewski, N.E.; Redman, C.L.; Wu, J.; Bai, X.; Briggs, J.M. Global change and the ecology of cities. *Science* **2008**, *319*, 756–760. [CrossRef]
7. Ingole, V.; Marí-Dell'Olmo, M.; Deluca, A.; Borrell, C.; Basagaa, X. Spatial Variability of Heat-Related Mortality in Barcelona from 1992-2015: A Case Crossover Study Design. *Int. J. Environ. Res. Public Health* **2020**, *17*, 2553. [CrossRef] [PubMed]
8. Kim, S.W.; Brown, R.D. Urban heat island (UHI) variations within a city boundary: A systematic literature review. *Renew. Sustain. Energy Rev.* **2021**, *148*, 111256. [CrossRef]
9. Xu, H.Q. Quantitative analysis on the relationship of urban impervious surface with other components of the urban ecosystem. *Acta Ecol. Sin.* **2009**, *29*, 2456–2462.
10. Yu, Z.; Yao, Y.; Yang, G.; Wang, X.; Vejre, H. Spatiotemporal patterns and characteristics of remotely sensed region heat islands during the rapid urbanization (1995–2015) of Southern China. *Sci. Total Environ.* **2019**, *674*, 242–254. [CrossRef]
11. Shukla, A.; Jain, K. Analyzing the impact of changing landscape pattern and dynamics on land surface temperature in Lucknow city, India. *Urban For. Urban Green.* **2021**, *58*, 126877. [CrossRef]
12. Estoque, R.C.; Murayama, Y.; Myint, S.W. Effects of landscape composition and pattern on land surface temperature: An urban heat island study in the megacities of Southeast Asia. *Sci. Total Environ.* **2017**, *577*, 349–359. [CrossRef]
13. Chrysoulakis, N.; Feigenwinter, C.; Triantakonstantis, D.; Penyevskiy, I.; Tal, A.; Parlow, E.; Fleishman, G.; Düzgün, S.; Esch, T.; Marconcini, M. A Conceptual List of Indicators for Urban Planning and Management Based on Earth Observation. *ISPRS Int. J. Geo-Inf.* **2014**, *3*, 980–1002. [CrossRef]

14. Mitraka, Z.; Chrysoulakis, N.; Doxani, G.; Del Frate, F.; Berger, M. Urban Surface Temperature Time Series Estimation at the Local Scale by Spatial-Spectral Unmixing of Satellite Observations. *Remote Sens.* **2015**, *7*, 4139–4156. [CrossRef]
15. Fidani, S.; Daliakopoulos, I.; Manios, T.; Grillakis, M.; Charalampopoulou, V.; Nektarios, P. Evaluating the Effect of Green Infrastructure in Mitigating the Urban Heat Island Effect Using Remote Sensing. In Proceedings of the EGU General Assembly 2021, online, 19–30 April 2021. [CrossRef]
16. Jiang, Y.; Fu, P.; Weng, Q. Assessing the Impacts of Urbanization-Associated Land Use/Cover Change on Land Surface Temperature and Surface Moisture: A Case Study in the Midwestern United States. *Remote Sens.* **2015**, *7*, 4880–4898. [CrossRef]
17. Xie, Q.; Sun, Q. Monitoring thermal environment deterioration and its dynamic response to urban expansion in Wuhan, China. *Urban Clim.* **2021**, *39*, 100932. [CrossRef]
18. Cai, Y.; Chen, Y.; Tong, C. Spatiotemporal evolution of urban green space and its impact on the urban thermal environment based on remote sensing data: A case study of Fuzhou City, China. *Urban For. Urban Green.* **2019**, *41*, 333–343. [CrossRef]
19. Oliveira, S.; Andrade, H.; Vaz, T. The cooling effect of green spaces as a contribution to the mitigation of urban heat: A case study in Lisbon. *Build. Environ.* **2011**, *46*, 2186–2194. [CrossRef]
20. Zhou, W.; Huang, G.; Cadenasso, M.L. Does spatial configuration matter? Understanding the effects of land cover pattern on land surface temperature in urban landscapes. *Landsc. Urban Plan.* **2011**, *102*, 54–63.
21. Liu, G.; Zhang, Q.; Li, G.; Doronzo, D.M. Response of land cover types to land surface temperature derived from Landsat-5 TM in Nanjing Metropolitan Region, China. *Environ. Earth Sci.* **2016**, *75*, 1381–1386. [CrossRef]
22. Asgarian, A.; Amiri, B.J.; Sakieh, Y. Assessing the effect of green cover spatial patterns on urban land surface temperature using landscape metrics approach. *Urban Ecosyst.* **2015**, *18*, 209–222. [CrossRef]
23. Peng, J.; Hu, Y.; Dong, J.; Liu, Q.; Liu, Y. Quantifying spatial morphology and connectivity of urban heat islands in a megacity: A radius approach. *Sci. Total Environ.* **2020**, *714*, 136792. [CrossRef]
24. Lv, H.; Yang, Y.; Zhang, D.; Du, H.; Zhang, J.; Wang, W.; He, X. Perimeter-area ratio effects of urbanization intensity on forest characteristics, landscape patterns and their associations in Harbin City, Northeast China. *Urban Ecosyst.* **2019**, *22*, 631–642. [CrossRef]
25. Maimaitiyiming, M.; Ghulam, A.; Tiyip, T.; Pla, F.; Latorre-Carmona, P.; Halik, Ü.; Sawut, M.; Caetano, M. Effects of green space spatial pattern on land surface temperature: Implications for sustainable urban planning and climate change adaptation. *ISPRS J. Photogramm.* **2014**, *89*, 59–66. [CrossRef]
26. Richards, D.R.; Fung, T.K.; Belcher, R.N.; Edwards, P.J. Differential air temperature cooling performance of urban vegetation types in the tropics. *Urban For. Urban Green.* **2020**, *50*, 126651. [CrossRef]
27. Rakoto, P.Y.; Deilami, K.; Hurley, J.; Amati, M.; Sun, Q.C. Revisiting the cooling effects of urban greening: Planning implications of vegetation types and spatial configuration. *Urban For. Urban Green.* **2021**, *64*, 127266. [CrossRef]
28. Li, J.; Song, C.; Cao, L.; Zhu, F.; Meng, X.; Wu, J. Impacts of landscape structure on surface urban heat islands: A case study of Shanghai, China. *Remote Sens. Environ.* **2011**, *115*, 3249–3263. [CrossRef]
29. Ke, X.; Men, H.; Zhou, T.; Li, Z.; Zhu, F. Variance of the impact of urban green space on the urban heat island effect among different urban functional zones: A case study in Wuhan. *Urban For. Urban Green.* **2021**, *62*, 127159. [CrossRef]
30. Li, X.; Zhou, W.; Ouyang, Z. Relationship between land surface temperature and spatial pattern of greenspace: What are the effects of spatial resolution? *Landsc. Urban Plan.* **2013**, *114*, 1–8. [CrossRef]
31. Hou, H.; Estoque, R.C. Detecting Cooling Effect of Landscape from Composition and Configuration: An Urban Heat Island Study on Hangzhou. *Urban For. Urban Green.* **2020**, *53*, 126719. [CrossRef]
32. Du, J.; Xiang, X.; Zhao, B.; Zhou, H. Impact of urban expansion on land surface temperature in Fuzhou, China using Landsat imagery. *Sustain. Cities Soc.* **2020**, *61*, 102346. [CrossRef]
33. Peng, J.; Jia, J.; Liu, Y.; Li, H.; Wu, J. Seasonal contrast of the dominant factors for spatial distribution of land surface temperature in urban areas. *Remote Sens. Environ.* **2018**, *215*, 255–267. [CrossRef]
34. Geng, S.; Yang, L.; Sun, Z.; Wang, Z.; Qian, J.; Jiang, C.; Wen, M. Spatiotemporal patterns and driving forces of remotely sensed urban agglomeration heat islands in South China. *Sci. Total Environ.* **2021**, *800*, 149499. [CrossRef] [PubMed]
35. Feng, L.; Zhao, M.; Zhou, Y.; Zhu, L.; Tian, H. The seasonal and annual impacts of landscape patterns on the urban thermal comfort using Landsat. *Ecol. Indic.* **2020**, *110*, 105798. [CrossRef]
36. Lu, Y.; Yue, W.; Liu, Y.; Huang, Y. Investigating the spatiotemporal non-stationary relationships between urban spatial form and land surface temperature: A case study of Wuhan, China. *Sustain. Cities Soc.* **2021**, *72*, 103070. [CrossRef]
37. Fu, B.; Wang, Y.; Campbell, A.; Li, Y.; Zhang, B.; Yin, S.; Xing, Z.; Jin, X. Comparison of object-based and pixel-based Random Forest algorithm for wetland vegetation mapping using high spatial resolution GF-1 and SAR data. *Ecol. Indic.* **2017**, *73*, 105–117. [CrossRef]
38. Peng, J.; Wang, Y.; Zhang, Y.; Wu, J.; Li, W.; Li, Y. Evaluating the effectiveness of landscape metrics in quantifying spatial patterns. *Ecol. Indic.* **2010**, *10*, 217–223. [CrossRef]
39. Zhang, Y.; Balzter, H.; Zou, C.; Xu, H.; Tang, F. Characterizing bi-temporal patterns of land surface temperature using landscape metrics based on sub-pixel classifications from Landsat TM/ETM+. *Int. J. Appl. Earth Obs.* **2015**, *42*, 87–96. [CrossRef]
40. Li, X.; Zhou, W.; Ouyang, Z.; Xu, W.; Zheng, H. Spatial pattern of greenspace affects land surface temperature: Evidence from the heavily urbanized Beijing metropolitan area, China. *Landsc. Ecol.* **2012**, *27*, 887–898. [CrossRef]

41. Liu, H.; Weng, Q. Scaling Effect on the Relationship between Landscape Pattern and Land Surface Temperature. *Photogramm. Eng. Remote. Sens.* **2009**, *75*, 291–304. [CrossRef]
42. Guo, G.; Zhou, X.; Wu, Z.; Xiao, R.; Chen, Y. Characterizing the impact of urban morphology heterogeneity on land surface temperature in Guangzhou, China. *Environ. Model. Softw.* **2016**, *84*, 427–439. [CrossRef]
43. Yang, C.; He, X.; Wang, R.; Yan, F.; Lingxue, Y.; Bu, K.; Yang, J.; Chang, L.; Zhang, S. The Effect of Urban Green Spaces on the Urban Thermal Environment and Its Seasonal Variations. *Forests* **2017**, *8*, 153. [CrossRef]
44. Song, J.; Du, S.; Feng, X.; Guo, L. The relationships between landscape compositions and land surface temperature: Quantifying their resolution sensitivity with spatial regression models. *Landsc. Urban Plan.* **2014**, *123*, 145–157. [CrossRef]
45. Tan, M.; Li, X. Quantifying the effects of settlement size on urban heat islands in fairly uniform geographic areas. *Habitat Int.* **2015**, *49*, 100–106. [CrossRef]
46. Yao, L.; Sun, S.; Song, C.; Li, J.; Xu, W.; Xu, Y. Understanding the spatiotemporal pattern of the urban heat island footprint in the context of urbanization, a case study in Beijing, China. *Appl. Geogr.* **2021**, *133*, 102496. [CrossRef]
47. Liang, Z.; Huang, J.; Wang, Y.; Wei, F.; Wu, S.; Jiang, H.; Zhang, X.; Li, S. The mediating effect of air pollution in the impacts of urban form on nighttime urban heat island intensity. *Sustain. Cities Soc.* **2021**, *74*, 102985. [CrossRef]
48. Mcgarigal, K.S.; Cushman, S.A.; Neel, M.C.; Ene, E. FRAGSTATS: Spatial Pattern Analysis Program for Categorical Maps. 2002. Available online: www.umass.edu/landeco/research/fragstats/fragstats (accessed on 15 September 2020).
49. Getis, A. Spatial Autocorrelation: A Primer. *Econ. Geogr.* **1988**, *64*, 88. [CrossRef]
50. Chi, G.; Zhu, J. Spatial Regression Models for Demographic Analysis. *Popul. Res. Policy Rev.* **2008**, *27*, 17–42. [CrossRef]
51. Scott, L.M.; Janikas, M.V. *Handbook of Applied Spatial Analysis*; Springer: Berlin/Heidelberg, Germany, 2010.
52. Anselin, L. Local Indicators of Spatial Association—ISA. *Geogr. Anal.* **2010**, *27*, 93–115. [CrossRef]
53. Tian, X.; Deng, Y.; Wargocki, P.; Liu, W. Effects of increased activity level on physiological and subjective responses at different high temperatures. *Build. Environ.* **2021**, *201*, 108011. [CrossRef]
54. Shen, Z.; Zeng, J.; Liang, C. Spatial relationship of greenspace landscape pattern with land surface temperature in three cities of southern Fujian. *Chin. J. Ecol.* **2020**, *39*, 245–253.
55. Chen, A.; Yao, L.; Sun, R.; Chen, L. How many metrics are required to identify the effects of the landscape pattern on land surface temperature? *Ecol. Indic.* **2014**, *45*, 424–433. [CrossRef]
56. Zhou, W.; Wang, J.; Cadenasso, M.L. Effects of the spatial configuration of trees on urban heat mitigation: A comparative study. *Remote Sens. Environ.* **2017**, *195*, 1–12. [CrossRef]
57. Yang, J.; Sun, J.; Ge, Q.; Li, X. Assessing the impacts of urbanization-associated green space on urban land surface temperature: A case study of Dalian, China. *Urban For. Urban Green.* **2017**, *22*, 1–10. [CrossRef]
58. Kong, F.; Yin, H.; James, P.; Hutyra, L.R.; He, H.S. Effects of spatial pattern of greenspace on urban cooling in a large metropolitan area of eastern China. *Landsc. Urban Plan.* **2014**, *128*, 35–47. [CrossRef]
59. Guo, G.; Wu, Z.; Chen, Y. Complex mechanisms linking land surface temperature to greenspace spatial patterns: Evidence from four southeastern Chinese cities. *Sci. Total Environ.* **2019**, *674*, 77–87. [CrossRef] [PubMed]
60. Zhang, X.; Zhong, T.; Feng, X.; Wang, K. Estimation of the relationship between vegetation patches and urban land surface temperature with remote sensing. *Int. J. Remote Sens.* **2009**, *30*, 2105–2118. [CrossRef]
61. Zhou, D.; Zhao, S.; Liu, S.; Zhang, L.; Zhu, C. Surface urban heat island in China's 32 major cities: Spatial patterns and drivers. *Remote Sens. Environ.* **2014**, *152*, 51–61. [CrossRef]
62. Wu, W.; Li, L.; Li, C. Seasonal variation in the effects of urban environmental factors on land surface temperature in a winter city. *J. Clean. Prod.* **2021**, *299*, 126897. [CrossRef]
63. Xiang, Y.; Huang, C.; Huang, X.; Zhou, Z.; Wang, X. Seasonal variations of the dominant factors for spatial heterogeneity and time inconsistency of land surface temperature in an urban agglomeration of central China. *Sustain. Cities Soc.* **2021**, *75*, 103285. [CrossRef]
64. Cai, Z.; Han, G.; Chen, M. Do water bodies play an important role in the relationship between urban form and land surface temperature? *Sustain. Cities Soc.* **2018**, *39*, 487–498. [CrossRef]

remote sensing

MDPI

Article

Heterogeneous Urban Thermal Contribution of Functional Construction Land Zones: A Case Study in Shenzhen, China

Han Wang [1,2], Bingxin Li [3], Tengyun Yi [1] and Jiansheng Wu [1,2,*]

1 Key Laboratory for Urban Habitat Environmental Science and Technology, School of Urban Planning and Design, Peking University, Shenzhen 518055, China; han.wang@stu.pku.edu.cn (H.W.); tengyun_yi@stu.pku.edu.cn (T.Y.)

2 Key Laboratory for Earth Surface Processes, Ministry of Education, College of Urban and Environmental Sciences, Peking University, Beijing 100871, China

3 School of Resource and Environmental Sciences, Wuhan University, Wuhan 430079, China; whulibingxin@whu.edu.cn

* Correspondence: wujs@pkusz.edu.cn; Tel.: +86-755-2603-5508

Abstract: Anthropogenic interferences through various intensive social-economic activities within construction land have induced and strengthened the Urban Heat Island (UHI) effects in global cities. Focused on the relative heat effect produced by different social-economic functions, this study established a general framework for functional construction land zones (FCLZs) mapping and investigated their heterogeneous contribution to the urban thermal environment, and then the thermal responses in FCLZs with 12 environmental indicators were analyzed. Taking Shenzhen as an example city, the results show that the total contribution and thermal effects within FCLZs are significantly different. Specifically, the FCLZs contribution to UHI regions highly exceeds the corresponding proportions of their area. The median warming capacity order of FCLZs is: Manufacture function (3.99 °C) > Warehousing and logistics function (3.69 °C) > Street and transportation function (3.61 °C) > Business services function (3.06 °C) > Administration and public services function (2.54 °C) > Green spaces and squares function (2.40 °C) > Residential function (2.21 °C). Both difference and consistency coexist in the responses of differential surface temperature (DST) to environmental indicators in FCLZs. The thermal responses of DST to biophysical and building indicators in groups of FCLZs are approximately consistent linear relationships with different intercepts, while the saturation effects shown in location and social-economic indicators indicate that distance and social-economic development control UHI effects in a non-linear way. This study could extend the understanding of urban thermal warming mechanisms and help to scientifically adjust environmental indicators in urban planning.

Keywords: functional construction land zones; urban thermal environment; differential surface temperature; environmental indicators; Shenzhen

Citation: Wang, H.; Li, B.; Yi, T.; Wu, J. Heterogeneous Urban Thermal Contribution of Functional Construction Land Zones: A Case Study in Shenzhen, China. *Remote Sens.* **2022**, *14*, 1851. https://doi.org/10.3390/rs14081851

Academic Editors: Yuji Murayama and Ruci Wang

Received: 27 February 2022
Accepted: 9 April 2022
Published: 12 April 2022

Publisher's Note: MDPI stays neutral with regard to jurisdictional claims in published maps and institutional affiliations.

1. Introduction

Intensive social-economic activities in urban areas have produced severe and irreversible interferences on natural landscape patterns and climate conditions in the urbanization processes [1–3]. According to [4], more than two-thirds of the world population will live in cities and modern towns by 2050. Over the decades, Urban Heat Islands (UHI) has already received critical attention [5,6]. Due to the induced heat threats and health risks to urban residents [7,8], UHI has become one of the key issues in the realization of sustainable urban development goals (SDG11) [9,10].

As a proxy in the social-ecological process of human activities on the natural environment, land use and land cover change (LUCC) have been considered the main drivers of UHI effects. The existing literature on UHI has revealed that urban land use and the resulting changes in surface biophysical properties of the landscape have led to the energy unbalance of the thermal environment [11–13], and the thermal difference between urban and rural areas in local regions have been the main object in UHI studies. To date, various studies have investigated the relationship between UHI and urban climate-related properties from different perspectives from urban land cover and land use differences. For example, the local climate zones (LCZ) framework [14] provides a basic framework to map land surface into 17 types of urban climate zones according to urban physical structure parameters [15–17] and evaluate UHI effects [18–20]. Urban Function Zones (UFZ) [21] is another LUCC scheme in UHI studies that divides the whole urban area into sub-zones in consideration of similar social-economic activities in the local blocks. General findings that the causes and characteristics of UHI are closely related to the landscape patterns and environmental parameters of the land surface have been demonstrated in many academic publications [22]. It has been argued that the spatial heterogeneity of land surface temperature (LST) across LCZ or UFZ is influenced by scale [5], composition [23], and patterns [24,25] in categories of the surface landscape. Conclusions about differentiated thermal effects between impervious surfaces [26] and green-blue landscapes [27] have been applied to provide empirical planning strategies to alleviate UHI effects.

However, there are still limitations in UHI planning practice and research. Landscape with natural cooling effects such as the waterbody [28], vegetation [29], and artificial green infrastructures (GI) [13,27,30,31] could help to alleviate urban heat island effects with natural-based solutions, while the fixed location and scale of areas of forests, rivers, and lakes mean that these methods are only locally effective and show a lack of extensibility in the whole area. The effects and causes of UHI have already been investigated at the block level in recent studies, however, blocks divided by road networks are frequently used as the basic research units in LCZ and UFZ frameworks, researchers may lose the delicate insight into the construction land of different functions in a single block caused by different sizes of the blocks in the real world. In addition, the quantitative distinction between anthropogenic and environmental heat in UHI studies remains unclear. The difference between anthropogenic and natural heat still needs further study. Therefore, it is crucial to measure the thermal contribution in urban construction land.

Understanding human activities in urban land and the resulting quantitative thermal effects would help to regulate human behaviors to control artificial heating effects on the urban environment. Nowadays, crowdsourced data like points of interest (POIs) are gradually used to promote urban mapping due to its directivity to specific social and economic activities and the lower cost and more convenient access than field surveys [32,33]. Compared to the visual features in remote sensing images [17,34–36], categories or tags of POIs, which are collected in the searching requirements of daily life and travel activities, could better quantify the patterns of human activities in the region [37,38] and provide a good option for anthropogenic heat measurement. However, they are established in different classification systems by commercial firms, e.g., AMap [39], Baidu Map [40], and Google Maps [41]. They are also usually inconsistent with the land use standards in the urban planning field. Moreover, the absolute difference in quantity across different categories of POIs and different research objectives make it a problem to determine the function in the research unit just by the quantities or densities [42]. Thus, we still need to integrate the advantages of POIs for the mapping of social-economic functions of construction land for urban planning and management in a more spatial-continuous way. When relocating and resizing the natural cooling area is diseconomy and impossible for the whole city [43], reasonable allocation of the function of construction land, after understanding its diversified thermal capacity and contribution to the urban thermal environment, may provide new scientific pathways for heat island effect reduction.

After more than 40 years of rapid development under the Reform and Opening policy, the urbanization rate of Shenzhen has continued to take the lead in China in the recent decade [44]. Thus, Shenzhen has become a megacity with high risks of the UHI effects [45] and limited natural cooling areas [46,47]. Taking Shenzhen as an example city, three questions are focused on: (1) Are there differential thermal effects within urban construction land among various social-economic functions? (2) What are the contributions of construction land with different social-economic functions to the UHI effect? (3) What are the response processes of thermal effects to the environmental factors in construction land? In this study, we proposed an extensible framework to map urban construction land for social-economic functions at the grid level into functional construction land zones (FCLZs) using POIs and then evaluated the relative thermal effects based on differential surface temperature (DST) among urban areas to the surrounding areas. This study aims to provide implications for the regulation of the anthropogenic influences in UHI reduction in both the natural and social solutions for urban social-ecological problems.

2. Materials and Methods

2.1. Study Area

Shenzhen is a hilly megacity (21°25′N~24°30′N, 111°12′E~115°35′E) with a population of 17.56 million. Situated on the south coastline of China's mainland and adjutant to the Peral River in the west, Shenzhen has a subtropical monsoon climate with a rainy and hot summer for 8 months and dry winter of less than 1 month during average recent years. The highly utilized construction land with a wide range of functional types is now widely distributed around woody hills and water bodies in the city (Figure 1), even if under the control of the urban master plan. Therefore, Shenzhen is one of the first cities to face not only insufficient space for urban development but also many severe urban ecological problems, e.g., an urban heat island.

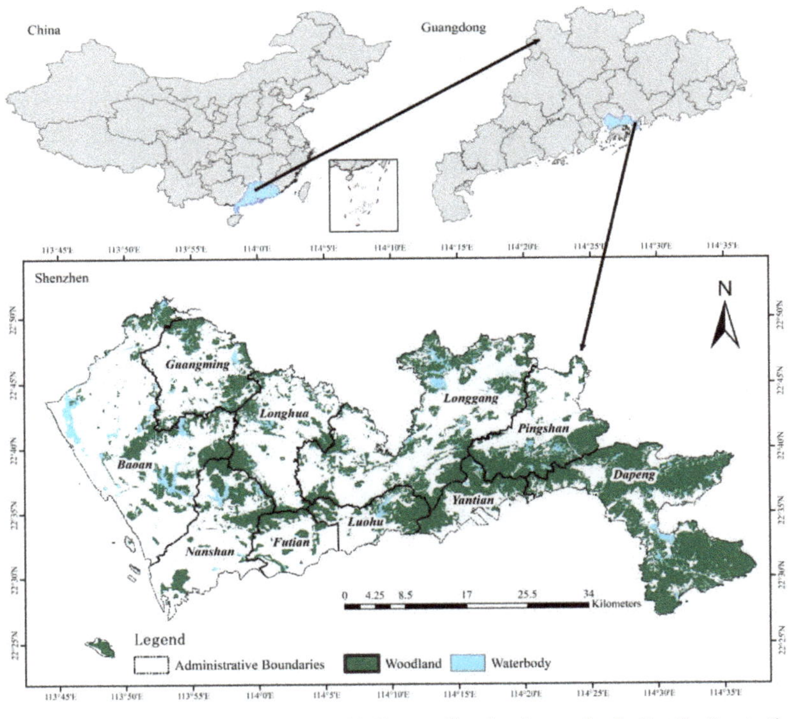

Figure 1. Location of Shenzhen: limited hilly woodland and waterbody distribution in the city.

2.2. Data Sources

In this study, we used points of interest (POIs) data collected from AMap [48] in June 2019 to recognize FCLZ types. The Landsat 8 Collection 2 Level 2 Science Product (L2SP) data processed based on Landsat 8 OLI/TIRS images (Table 1) were downloaded from the

USGS website [49] due to similar higher air temperature and less cloud cover (<5%). To obtain the natural land surface temperature in non-construction areas, we extracted the waterbody and woodland cover from GlobeLand30 (2020) dataset [50]. Building survey data with shape and floor were provided by the Planning and Natural Resources Bureau of Shenzhen [51]. DEM data from ASTER GDEMV2 [52], night light data from Visible Infrared Imaging Radiometer [53], and population data from Landscan [54] were also employed to take into account the influences of building, topography, urban economic development, and demographic factors to the urban thermal environment, respectively.

Table 1. Information about the Landsat images used in this study.

Scene ID	Path/Row	Air Temperature of the Day (°C)	Average Air Temperature for Ten Days (°C)	Satellite Transit Time (UTC + 8)
LC81220442021035LGN00	122/44	16/24	18	10:45:33.11 a.m.
LC81210442019071LGN00	121/44	19/24	20	10:52:14.37 a.m.

2.3. Methods

As shown in Figure 2, based on social sourced and remote sensing data, the research framework of this study consisted of three main steps.

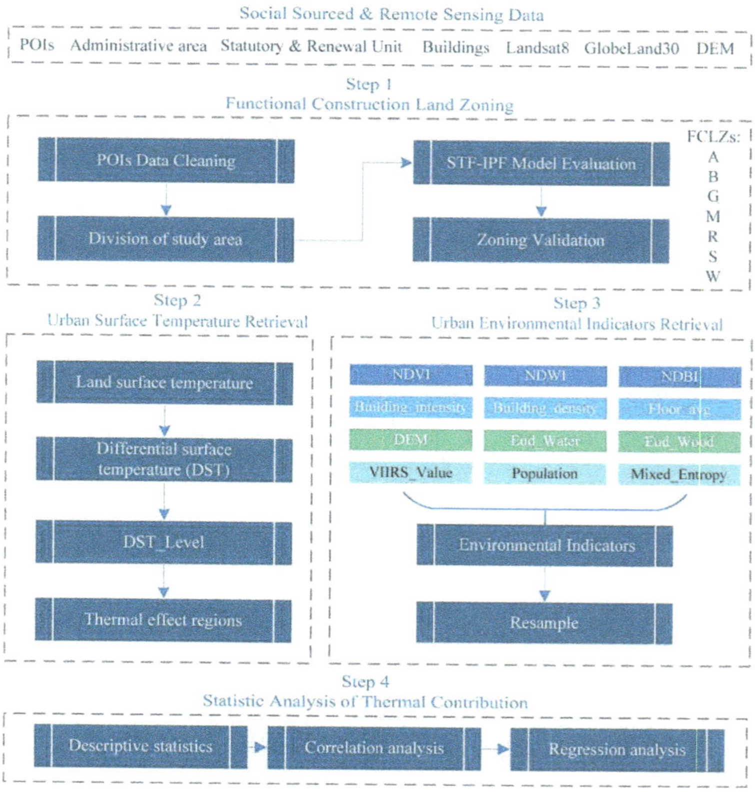

Figure 2. The research framework of heterogeneous thermal contribution analysis in urban construction areas.

First, functional construction land zoning (FCLZ) was implemented through the process of data cleaning, division of study area, spatial term frequency-inversed patches frequency (STF-IPF) model evaluation, and a results validation. Then, urban surface indicators including land surface temperature and four groups of environmental parameters, i.e., biophysical indicators, building indicators, location, and social-economic indicators, were retrieved and reprocessed across FCLZS using domain-based approaches. Finally, statistical analysis methods were used to explore the relationship between DST and environmental parameters in FCLZs. Next, those key processes are described in detail.

2.3.1. Functional Construction Land Zoning

1. Division of study area

To depict the actual functional construction land in a more precise way, the first step was to divide the whole urban space into continuous grids at a certain scale to break the traditional boundary divided by road networks (Figure 3). An appropriate grid scale should be able to make the recognition results of functional constructional land as close as possible to reflect the spatial distribution and the focused problems. It was considered that 500 m might be enough distance to capture neighboring POIs in cities [55,56]. Therefore, we divided Shenzhen into 8575 grids at a scale of 500 m.

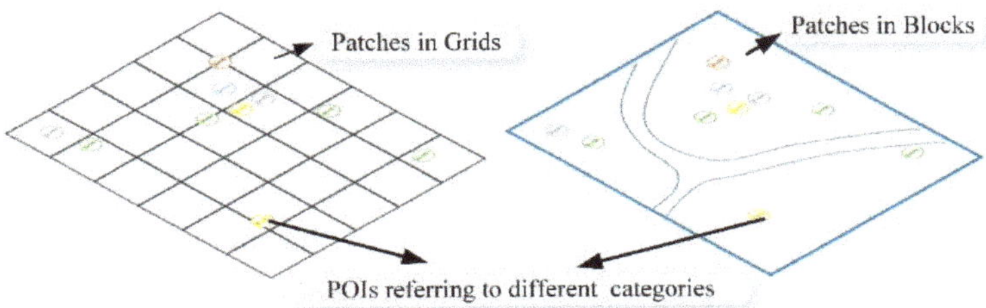

Figure 3. Patches diagrams in continuous grids (**left**) and road level blocks (**right**).

2. Matching POIs attributions to functional construction land zones

To facilitate the FCLZs to guide urban planning and construction land use, we reclassified the POIs collected from AMap according to [57] into 7 types of construction land use by their defined categories and tags (Table A1), i.e., Residential function (R), Administration and public services function (A), Business services function (B), Green spaces and squares function (G), Street and transportation function (S), Manufacture function (M), and Warehousing and logistics function (W).

3. Calculation of POIs representativeness in patches

Different types of POIs usually correspond to various degrees of public awareness although they have differences in quantity. For instance, the commercial service facilities such as stores, supermarkets, catering, and so on, are always around some places in large numbers while the more concerned POIs in urban planning, e.g., railway stations, logistics parks, or large-scale residential communities may have only one or several POIs. As a consequence, density and frequency methods [42,58] (Equation (1)) often mistakenly identify the functional construction land zone type based on biased quantity observation caused by differences in the number of collected points.

$$POI\_Density_{i,j} = \frac{Count\ of\ POI\ matched\ to\ function\ j\ in\ Patch\ i}{Total\ Count\ of\ POI\ in\ Patch\ i} \quad (1)$$

Inspired by the Term Frequency-Inversed Document Frequency (TF-IDF) model (Equation (2)) [59] in the field of Natural Language Process (NLP), we proposed a Spatial Term Frequency-Inversed Patches Frequency (STF-IPF) model in consideration of both the amount and the global significance to recognize representative functional construction land types (Equation (3)).

$$TF\_IDF = \frac{Count\ of\ word\ occurances}{Total\ words\ in\ document} \times log \frac{Total\ number\ of\ docs}{Number\ of\ docs\ word\ is\ in} \tag{2}$$

$$STF\_IPF = \frac{Count\ of\ POI\ matched\ to\ specific\ function}{Total\ number\ of\ POI\ in\ patches} \times log \frac{Total\ number\ of\ patches}{Number\ of\ patches\ where\ specific\ POI\ are\ in} \tag{3}$$

Given a study area divided into patches with a total number of m and a set of POIs matched to m kinds of functional construction land, then the vector UNump_j and the matrix NU could be defined as the following Equations (4) and (5), where $Nu_{i,j}$ is the number of POIs in the patch i matched functional construction land type j.

$$\text{UNump}_j = \left(Nu_{1,j}, Nu_{2,j}, \ldots, Nu_{m-1,j}, Nu_{m,j}\right) \tag{4}$$

$$NU = \left(\text{UNump}_1{}^T, \text{UNump}_1{}^T, \ldots, \text{UNump}_{n-1}{}^T, \text{UNump}_n{}^T\right) \tag{5}$$

Then the normalized STF-IPF value for function j in patch i could be calculated according to Equation (6), which takes both numbers and repetitiveness into account by combining Equations (1) and (3).

$$STF\_IPF_{i,j} = \frac{Nu_{i,j} * \frac{Nu_{i,j}}{\sum_{j=1}^{n} Nu_{i,j}} * \log \frac{\text{m}}{COUNT(\text{UNump}_j, Nu_{m,j} > 0)}}{\sum_{i=1}^{m} Nu^s{}_{i,j} * \frac{Nu_{i,j}}{\sum_{j=1}^{n} Nu_{i,j}} * \log \frac{\text{m}}{COUNT(\text{UNump}_j, Nu_{m,j} > 0)}} \tag{6}$$

where $COUNT(\text{UNump}_j, Nu_{m,j} > 0)$ calculates the counts of patches where the functional construction land zone type j is not null. Higher normalized $STF\_IPF_{i,j}$ means that functional construction land zone type j in patch i is more dominant and distinct from other types.

4. Recognition vectors evaluation

As shown in Equation (7), an FCLZ recognition vector for patch i could be constructed after sorting the value of normalized STF_IPF values in the descending order with the corresponding functional zone type index j'.

$$Rec\_vec_i = \left(STF\_IPF^{max}{}_{i,j}, STF\_IPF^{secondary}{}_{i,j'}, \ldots, STF\_IPF^{min}{}_{i,j''}\right) \tag{7}$$

The corresponding functions in the vector whose cumulative percentage is greater than a threshold value ε (0~100%) are defined as the functions of the patches, and the function with the maximal value is the main function in the patch. If there is no POI in the patch, the type is set to a non-construction area (N). According to suggestions on compatible types and proportions of urban construction land [57], ε was set to 70%. Therefore, we could infer the FCLZs types in each patch by evaluating the recognition vectors, and then describe the mixed status of FCLZs by calculating the mixed entropy in each patch by Equation (8).

$$Mixed\_entropy = \sum_{j=1}^{n} (STF\_IPF_j * \ln STF\_IPF_j) \tag{8}$$

In this study, we implemented the above processes in python scripts.

2.3.2. Urban Surface Temperature Retrieval

The land surface temperature (LST) data were derived from Landsat 8 Collection 2 Level 2 Science Product (L2SP) [49]. This product provides thermal infrared bands using the Radiative Transfer Equation method [60] after radiometric calibration and atmospheric correction. Digital numbers (DN) were converted to land surface temperature (LST) using Equation (9).

$$LST = DN \times SF + AF - 273.15 \tag{9}$$

where DN is the digital number for a given pixel. SF and AF are the multiplicative and additive scale factors of ST products, and their values are 0.00341802 and 149, respectively. The item −273.15 is the additive factor that converts LST value from Kelvin to Celsius.

As there is no rural area in this concept but many forests in Shenzhen, we used the average LST in the woodland ($\overline{LST}_{wood}$) to reflect the natural surface land temperature because of the low impact of human activities. Then DST could be calculated using Equation (10), which is a more reprehensive and comparable indicator to quantify the urban thermal environment change induced by human activities.

$$DST = LST - \overline{LST}_{wood} \tag{10}$$

2.3.3. Urban Environmental Indicators Retrieval

5. Biophysical indicators

L2SP also provides surface reflectance (SR) data in 9 bands processed by the LaSRC algorithm after calibration and atmospheric correction routines [61]. Similar to Equation (9), the SR data was converted from DN value to reflectance using Equation (11).

$$SR_i = DN \times SF + AF \tag{11}$$

where SR_i is the digital number for a given pixel in band i. $SF = 0.0000275$, $AF = -0.2$. The valid value range for each band except band 6 (1~65,535) is from 7272 to 43,636.

Then based on the previous findings, the normalized difference vegetation index (NDVI), normalized difference water index (NDWI), and normalized building index (NDBI) were calculated as indicators to quantify the surface bio-physical characters considering vegetation cover [26,62], moisture content [28,63], and building materials [64,65] in the land surface.

6. Building indicators

Buildings are still the most important feature in the construction area [25,66,67], thus we designed four building indicators (Table 2) using building survey data to quantify the building characters in the land grid.

Table 2. Definition of building indicators in construction land grid.

Indicator	Definition	Diagram Explanation
Floor_avg	The ratio of the sum of the total area of buildings to the sum of the base area of buildings	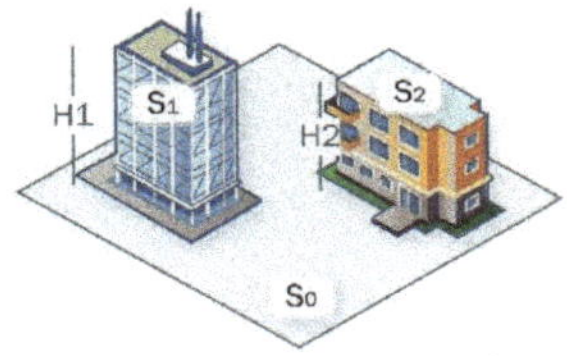
Building_density	The ratio of the sum of the base area of buildings to the area of the grid.	
Building_intensity	The ratio of the sum of the total building area of buildings to the area of the grid.	$Floor\_avg = \frac{H_1 \times S_1 + H_2 \times S_2}{S_1 + S_2}$ $Building\_density = \frac{S_1 + S_2}{S_0}$ $Building\_intensity = \frac{H_1 \times S_1 + H_2 \times S_2}{S_0}$

Note: S_0 means the area of the land grid. S_1,S_2, … means the projected area of the buildings, while H_1,H_2, … means the average height of the corresponding building.

7. Location and social-economic indicators

DEM, Euclidean distance from the waterbody (*Eud_Water*) and wood (*Eud_Wood*) were used as location and neighborhood factors that would influence the urban thermal environment [19,68,69]. Night light data (*VIIRS_Value*) and Landscan data (*Population*) were indirect but explicit indicators to reflect the intensity of the economic development density and social activities intensity [70,71].

They were all extracted or calculated, resampled, and matched to the grids under the ArcGIS 10.2 platform.

2.3.4. Statistical Analysis

The Shapiro-Wilk normality test [72] was used to explore the distributions of DST among FCLZs, and the Kruskal-Wallis H test [73] was applied to find if there was a significant thermal difference between FCLZs. The Dunn test with Holm-Bofferoni correction [74] was used as a post hoc method in multiple comparisons to verify the specific difference in each pair of FCLZs. Then Spearman coefficients [75] between DST and environmental indicators among various types of FCLZs were calculated to figure out factors influencing the difference in thermal contribution capacity. Finally, multiple simple regression analyses were conducted to quantify the response processes of DST on environmental indicators.

3. Results

3.1. Mapping of FCLZs

According to the main functional type in each grid, grids with functional construction land of the 7 types are unevenly distributed in the 10 districts in Shenzhen (Figure 4).

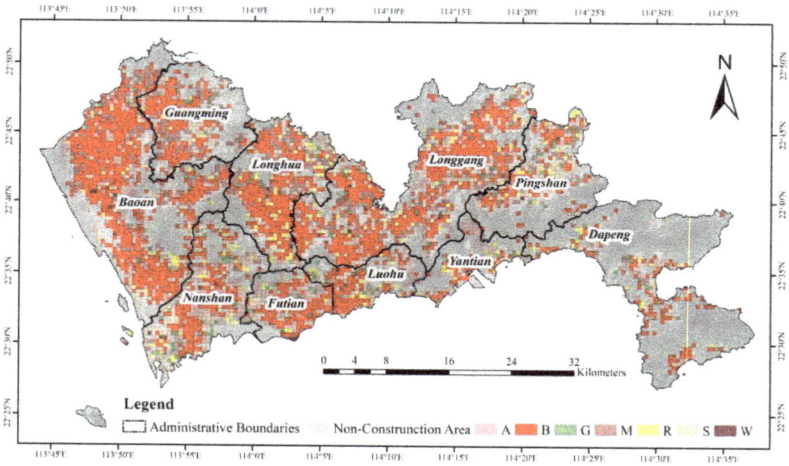

Figure 4. The main type of functional construction land zone recognition results at a 500 m grid scale. A: Administration and public services function; B: Business services function; G: Green spaces and squares function; S: Street and transportation function; M: Manufacture function; R: Residential function; W: Warehousing and logistics function.

To be more specific, the function of the most widely distributed grids which are agglomerated and distributed in clusters and strips is the Business services function (B), accounting for 41.62% of the total area of construction land. Grids of Administration and public services function (A), Street and transportation function (S), and Manufacture function (M) are clustered in patches of different sizes and show the aggregated distribution in each district, which takes up 19.77%, 14.90%, and 11.98% in the construction area, respectively. Residential function (R) and Green spaces and squares function (G) grids hold lesser area proportions of 6.94% and 4.16%, while they both show scattered distribution across districts. Warehousing and logistics function (W) grids with an area proportion of only 0.64% are significantly lower than others, which are distributed mainly at the edge and the center of every district.

As Figure 5 shows, the mixed entropy (ME) of functional construction land in each grid was calculated and showed obvious spatial aggregation. The natural breaks (Jenks) method [76] was used to divide the mixing entropy values into three groups (0–0.62, 0.62–0.98, >0.98), representing the mixed degree of low, medium, and high, respectively. Most grids with a high mixed degree are concentrated in Futian, Luohu, Nanshan, and southern Baoan, which shows a similar distribution as grids of Business services function (B) and Street and transportation function (S).

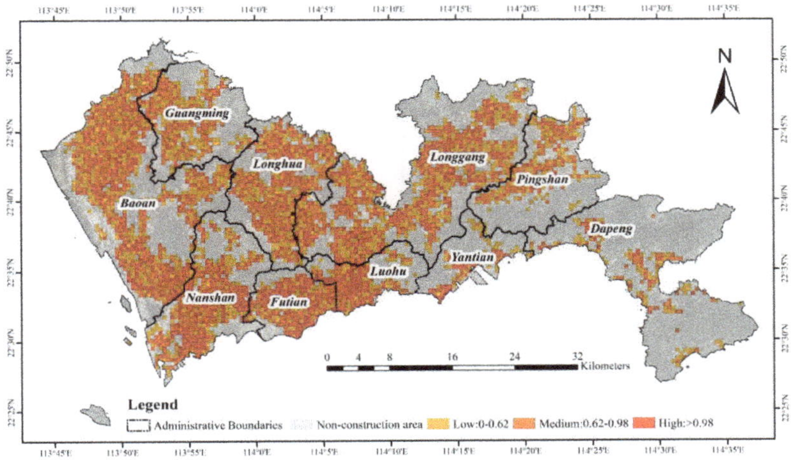

Figure 5. Mixed entropy distribution of functional construction land zones at 500 m grid scale.

In addition, the grids along the center of other districts are also with a high-level mixed degree, as it can be inferred that the grids with a high mixed degree are usually the regions with mature commercial activities and convenient traffic lines. Meanwhile, the grids with a medium mixed degree are mainly distributed in northern Baoan, Longhua, Pingshan, Longgang, and Yantian, where there is Manufacture function (M) and Residential function (R) grids are mainly distributed. Besides, the grids with a low mixed degree are mainly concentrated in mountainous areas, e.g., northern Yantian, Dapeng, and northern Nanshan, where there is less construction land but more ecological land (woodland and waterbody) distributed.

3.2. *Differential Surface Temperature in FCLZs*

The LSTP images cannot cover the study area in a single scene due to the satellite orbits across both the two paths in different periods (Figure 6a,b), thus we scattered the retrieved LST values in the overlapping areas to find the relationship between them. The fitting curve and regression coefficient (Figure 6c) showed a high degree of linear consistency between the two images. Therefore, we finally obtained the DST data (Figure 6d) covering the whole area of Shenzhen computed based on mosaicking the two LST images after linear adjustment.

To reflect the DST difference in more detail, the mean-standard deviation method [77,78] was used to classify the DST data into five urban thermal effect levels, which correspond to the significant cooling effect region (SCR), the unnoticeable thermal difference region (UTR), the weak heat effect region (WHR), the moderate heat effect region (MHR), and the significant heat effect region (SHR) in Table 3 and Figure 7. It could be seen from the comparison between

Figures 6 and 7 that the regions with different levels of heat effect (WHR, MHR, and SHR) show a spatial distribution of high similarity to that of FCLZs. While SCR presents a spatial conjugate relationship to heat effect regions, UTR is mainly distributed in the mountainous woodland area and places adjacent to the waterbody areas.

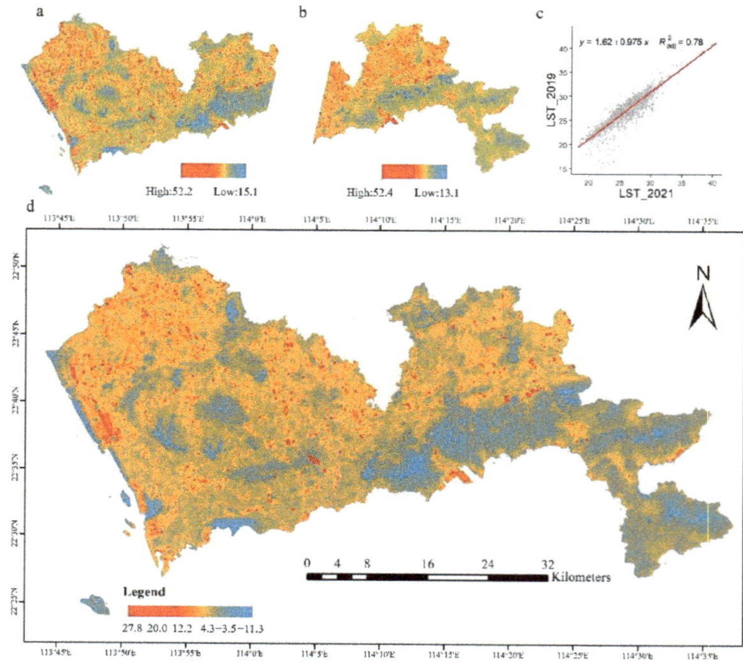

Figure 6. Urban surface temperature results in Shenzhen. (**a**) land surface temperature retrieved from scene LC81220442021035LGN00; (**b**) land surface temperature retrieved from scene LC81210442019071LGN00; (**c**) scatter plot and regression between (land surface temperature) LST of the overlap region in 2019 and 2021. (**d**) Distribution of differential surface temperature (DST) (°C).

Table 3. Area Proportion (%) of functional construction land zones (FCLZs) in each thermal effect level region.

Thermal Effect Region	DST Range (°C)	Region Area	Functional Construction Land Zones							Non-Construction Areas
			A	B	G	M	R	S	W	
SCR	$DST < -2.96$	5.21	5.29	2.66	0.00	0.00	0.24	0.29	0.00	91.53
UTR	$-2.96 \leq DST < -0.23$	17.87	6.35	5.80	1.33	0.14	1.46	1.07	0.00	83.85
WHR	$-0.23 \leq DST < 2.51$	33.40	11.11	27.64	2.77	3.86	3.38	6.16	0.17	44.91
MHR	$2.51 \leq DST < 5.25$	34.75	13.84	43.96	3.17	11.46	3.66	8.79	0.57	14.54
SHR	$5.25 \leq DST$	8.77	10.69	32.63	1.98	20.01	2.26	13.49	1.57	17.37
Total	/	100	10.87	28.54	2.44	7.05	2.87	6.50	0.39	41.33

Note: SCR: significant cooling effect region; UTR: unnoticeable thermal difference region; WHR: weak heat effect region; MFR: moderate heat effect region; SHR: significant heat effect region.

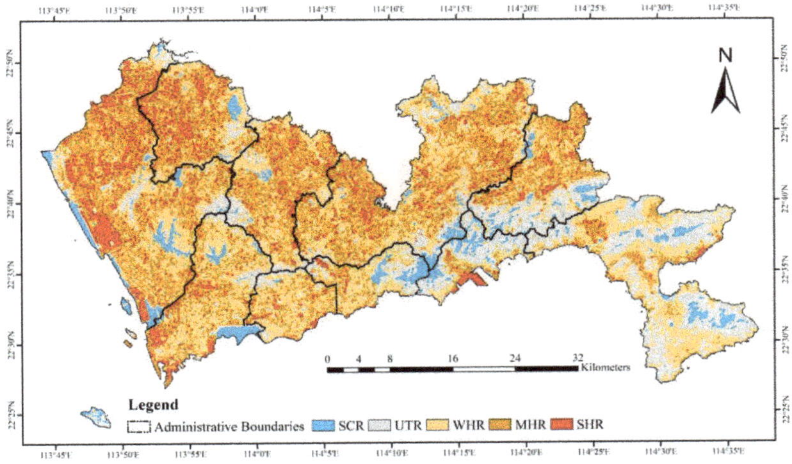

Figure 7. Thermal effect levels distribution in Shenzhen. SCR: significant cooling effect region; UTR: unnoticeable thermal difference region; WHR: weak heat effect region; MFR: moderate heat effect region; SHR: significant heat effect region.

Meanwhile, we calculated the area proportion of heat effect regions in both construction land and non-construction areas to explore the contribution of various types of FCLZs to the urban thermal environment. Results (Table 3) show that although it is clear that construction land contributes more to heat effect regions than non-construction areas, both the different degree of contribution between various FCLZs and the contribution from non-construction areas (at least >14.54%) cannot be ignored.

The Shapiro-Wilk test results (Table 4) show all DSTs in each functional construction land zone and non-construction areas were not statistically normal distributions except for functions G and R. The results of the Kruskal-Wallis H test and the multiple comparisons (Table 5) prove significant differences of DST among different FCLZs.

Table 4. The statistics and Shapiro-Wilk test of DST in urban land (°C).

Statistics	Functional Construction Land Zone							Non-Construction Areas
	A	B	G	M	R	S	W	
Avg	2.33	2.98	2.27	3.97	2.22	3.42	4.00	0.18
Std	2.27	1.91	2.10	1.77	1.95	2.14	2.44	2.73
Med	2.54	3.06	2.40	3.99	2.21	3.61	3.69	−0.06
p-value	0.000 ***	0.000 **	0.5816	0.000 ***	0.1345	0.000 **	0.0229 *	0.000 ***

Note: *** indicates $p < 0.001$; ** indicates $p < 0.01$; * indicates $p < 0.05$.

Table 5. Dunn test of DST with Holm-Bofferoni correction among FCLZs.

Functional Land Type	A	B	G	M	R	S	W
A	/	/	/	/	/	/	/
B	0.000 ***	/	/	/	/	/	/
G	0.876	0.000	/	/	/	/	/
M	0.000 ***	0.000	0.000	/	/	/	/
R	0.354	0.000	1.000	0.000	/	/	/
S	0.000 ***	0.000	0.000	0.000	0.000	/	/
W	0.006 **	0.334	0.004	1.000	0.001	0.568	/

Note: *** indicates $p < 0.001$; ** indicates $p < 0.01$.

Although there were no significant differences in DST between functional land G and R, either W or M; it was significantly different for most pairs of construction land types. The above results indicate that the DST distribution in functional land G and R is similar to that in functional A, while the functional land W showed similar characteristics in DST of functional B, M, and S.

3.3. DST Relationships with Surface Environmental Indictors

Figure 8 shows that the coefficients in functions A, G, and R are more consistent with non-construction areas while the building factors in function types B, M, S, and W present opposite effects to those in non-construction areas at a significant level of 0.05, which also supports the results in Table 5 from another quantitative perspective. Similar coefficients suggest similar relationships between DST to environmental indicators, while the different sizes of the coefficients indicate there are diversified differences in thermal capacity in construction land. Furthermore, according to the results of statistical differences (Table 5) and coefficients (Figure 8), we divided 7 types of construction land functional areas into 2 groups (group A, G, and R; group B, M, S, and W).

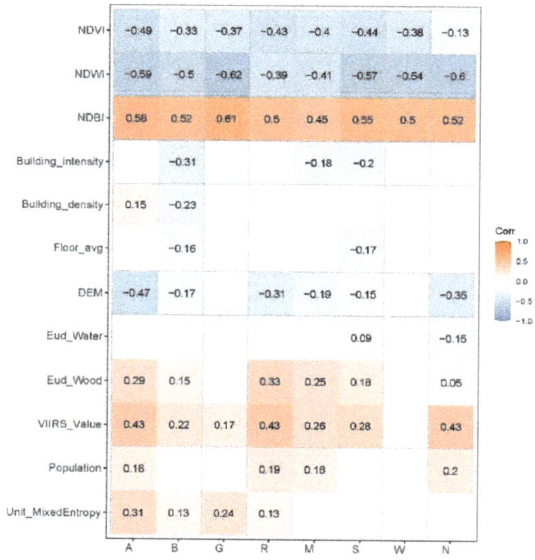

Figure 8. Spearman correlation coefficients of DST with 12 biophysical and social-economic indicators in seven types of functional construction land zones and non-construction areas.

With different intercepts in the linear and approximate linear relationships between thermal responses of the three biophysical indicators to DST, it is found that the main differences are in the degree rather than the mode.

DST in two groups of FCLZs shows a downward trend on the whole (Figure 9a,b), even if an insignificant upward trend is shown when the NDVI value is small (<0.2). In the non-construction areas, the first-increase and then-decreasing trends are more obvious (Figure 9c). As for NDWI in both groups FCLZs and non-construction areas, a consistent linear decline appears simultaneously. Although the increase of NDWI reduces the size of DST in group B, M, S, and W, it is consistently found that the increase of NDWI cannot offset the warming effect (Figure 9e). While the cooling effect (DST < 0 °C) of NDWI is found both in the group for functions of A, G, and R and in non-construction areas when the values are larger than 0.15 and 0.1, respectively (Figure 9d,f). Similar findings are observed in the warming effect of NDBI. DST presents similar linear growing trends in both groups of FCLZs and the non-construction areas (Figure 9g–i) with the increase of NDBI. However, the same value of NDBI has different warming effects in the two groups, while in non-construction areas, the warming effect (DST > 0 °C) is coexisting with the cooling effect (DST > 0 °C) with a boundary value ~−0.25.

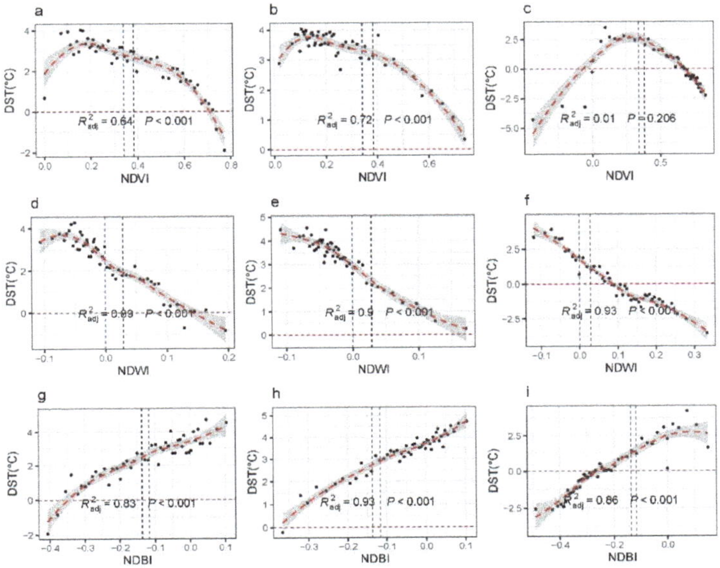

Figure 9. Single-factor scatter plots and fitting curves of DST to three surface biophysical indicators. Subgraphs (**a**,**d**,**g**) are for the group of functions for A, G and R; (**b**,**e**,**h**) are for the group of functions for B, M, S and W; (**c**,**f**,**i**) are for Non-construction areas. The vertical blue dotted line is the median value and the black one is the mean value of the corresponding independent variables; the horizontal red dotted line is the zero value of DST.

Figure 10 shows that the responses of DST to building indicators present different trends between the group of functions for A, G, and R and the group of functions for B, M, S, and W. The increases in Building_density and Building_intensity finally display stable warming effects with DST above 2.0 °C, however, the trends of curves indicate that intensifying and heightening buildings weaken the warm effects in the group of functions for B, M, S, and W (Figure 10b,e) but makes little significant contribution to the DST change in the group of function for A, G, and R (Figure 10a,d). A slight reduction of warming effects is also shown with the increase of Floor_avg in the group of functions for B, M, S, and W while in the group of functions for A, G, and R, the increase of Floor_avg leads to the DST increase first and then tends to be stable.

As is shown in Figure 11c, DST in the two groups both presents a short rising stage with the increase of elevation (DEM) before the altitude is less than average values (~25 m) and the cooling effect appears in the non-construction areas with the increase of DEM value. When the elevation is larger than median values (~30 m), the DST in the group of functions for A, G, and R shows a declining trend of warming effect and finally presents cooling effects when the elevation is larger than ~150 m. However, a stable state of warming effect at 2.75 °C after a rapid decline is also shown in the group of functions for B, M, S, and W.

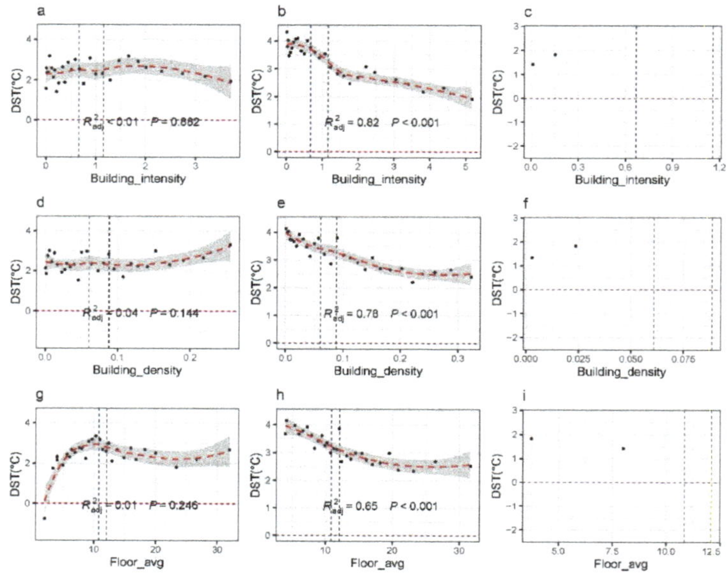

Figure 10. Single-factor scatter plots and fitting curves of DST to three surface building indicators. Subgraphs (**a**,**d**,**g**) are for the group of functions for A, G and R; (**b**,**e**,**h**) are for the group of functions for B, M, S and W; (**c**,**f**,**i**) are for Non-construction areas. The vertical blue dotted line is the median value and the black one is the mean value of the corresponding independent variables; the horizontal red dotted line is the zero value of DST.

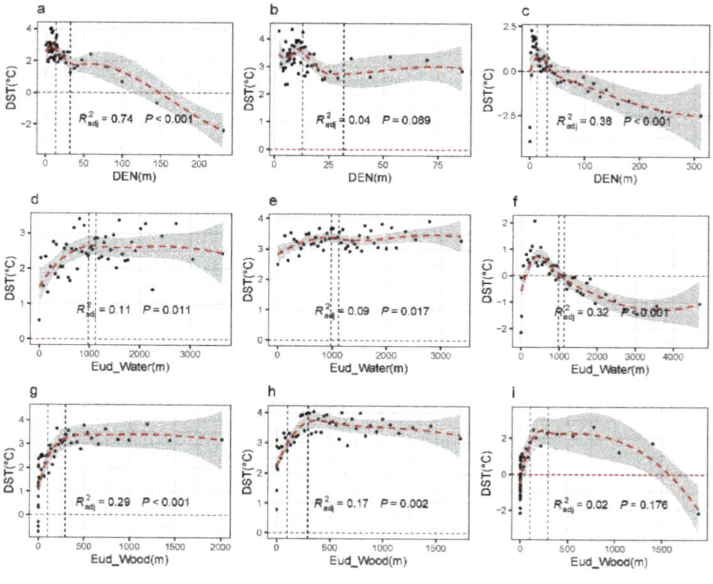

Figure 11. Single-factor scatter plots and fitting curves of DST to three surface location indicators. Subgraphs (**a**,**d**,**g**) are for the group of functions for A, G and R; (**b**,**e**,**h**) are for the group of functions for B, M, S and W; (**c**,**f**,**i**) are for Non-construction areas. The vertical blue dotted line is the median value and the black one is the mean value of the corresponding independent variables; the horizontal red dotted line is the zero value of DST.

The comparison of the other two location indicators (Eud_Water and Eud_Wood) represents the responses of DST to the distances from these two types of cooling areas. Different from the natural distance effects in non-construction areas (Figure 11f,i), DST increases then tend to be stable after different values as the increase of distance from cooling areas in different groups of FCLZs, showing that the limited spatial ranges of cooling effect are diversified (Figure 11d–h). Besides, the stable points of Eud_Water and Eud_Wood to the group functions for A, G, and R and B, M, S, and W are close to the median distances with values of ~1000 m and ~300 m, respectively.

Further analysis revealed there is a saturation effect with similar trends in social-economic indicators like that with location indicators and the thresholds for saturation likewise varied between the groups of FCLZs (Figure 12). As with the increases of VIIRS_Value, DST increases at first and then tends to be stable below 4.0 °C for all FCLZs (Figure 12a,b) and in the non-construction areas, there is a linear increase trend (Figure 12c). The stable points are ~30 in the group of functions for B, M, S, and W, and ~60 in the group of functions for A, G, R. The increase in population (Figure 12d–f) would cause the warming effect to reach saturation points (3~4 °C for construction land and 2.5 °C for non-construction areas) at about 3000~5000 people per grid (0.25 square kilometers). Curves in Figure 12g,h indicate that even the same level of the mixed utilization of urban construction land (Unit_MixedEntropy) could lead to a stronger warming effect in FCLZs.

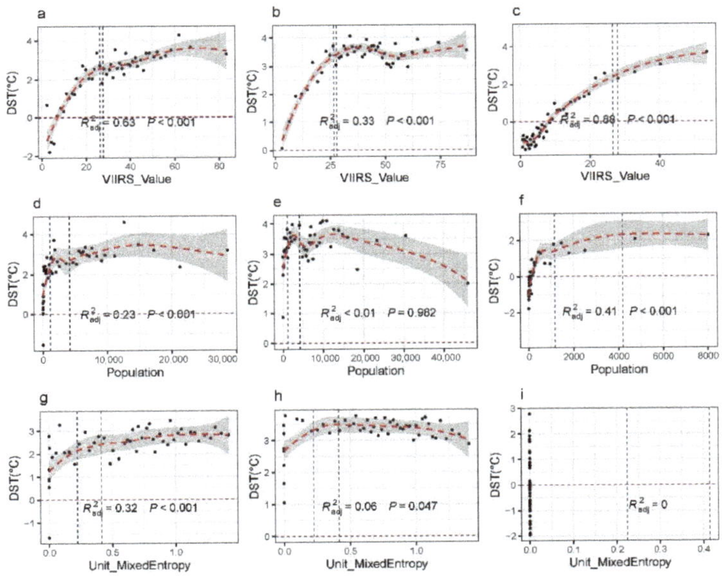

Figure 12. Single-factor scatter plots and fitting curves of DST to three surface social-economic indicators. Subgraphs (**a**,**d**,**g**) are for the group of functions for A, G and R; (**b**,**e**,**h**) are for the group of functions for B, M, S and W; (**c**,**f**,**i**) are for Non-construction areas. The vertical blue dotted line is the median value and the black one is the mean value of the corresponding independent variables; the horizontal red dotted line is the zero value of DST.

4. Discussion

4.1. Consistency Analysis of Recognized FCLZs

In this study, 4755 grids were identified as functional construction land zones and 3820 are non-construction areas (Figure 4). Comparing the total amount of construction land and manual discrimination of random samples could evaluate the effectiveness of recognized FCLZs quantitatively and qualitatively. Calculated with the functional proportion in each grid, the total area of functional construction land zones is 1171.1 square kilometers, of which the difference is 16.4% from the construction land area (1005.9 square kilometers) in the official report by the Planning and Natural Resources Bureau and Statistics Bureau of Shenzhen [51]. In addition, considering that the area of new construction land in 2019 was likely to increase according to the government's land supply plan and development demands, the actual smaller difference suggests the method of FCLZs could measure the construction land total amount with relative quantity accuracy at a low cost and in a fast way. Moreover, the degree of mixed functional construction land zones in Shenzhen is closely related to the development status of the region, i.e., the higher the mixed degree of the grid indicates the more diverse human activities and land use patterns (Figure 5). Through random sampling and manual interpretation (Figure A1), we verified that different social-economic functions and the characteristic of mixed-use of construction land within the indistinguishable image features could be identified and reflected through FCLZs.

FCLZs map the attributes of POIs to the corresponding social function on a particular unit of space, which is different from land use and land cover types because the classification of construction land is more research-goal oriented due to the diversity and versatility of human activities in urban construction land. POIs or any other data with location information referring to functional land could be used and patches could be any spatial unit with valid meaning and shape besides grids or blocks. However, the grid size has a great influence on the final recognition result of urban functional land because the functional types of each grid unit are determined through the spatial distribution and quantities of POIs contained in each grid. Due to the limited number of POIs in real life, if the grid scale is too small or lower than the precision of the coordinates of POIs, the functional types inferred will not be meaningful due to collection errors. At the same time, too large a grid scale will make it difficult to show the transition and differences of regional changes in different types of functional construction land zones.

4.2. Differential Thermal Contribution in FCLZs

The non-parametric methods were used in the difference test, as the DST distributions in FCLZs except G and R followed left-skew distribution rather than a normal distribution [79]. FCLZs were divided into two groups according to the Dunn test (Table 5) and environmental indicators correlation validation (Figure 8), which indicates similarity and difference coexist in the thermal effects of urban construction land among various functional types. The results proved that the UHI is not only caused by construction land [66,80] but also has significant inner differences due to functions [70,79]. Although LCZ could divide the urban land through buildings and image characteristics [16], it lacks the detailed reflection of social-economic functions and is usually applied to study the problems caused by the physical environment in the heat island issues [14,15].

The thermal contribution proportions of different FCLZs are usually affected by the area size in terms of the total contribution, however, the thermal contribution at urban thermal effect levels is not completely consistent with the area. We found that the Business services function (B) contributes most with proportions of 32.63~43.96% to the considerable heat effect regions (MHR and SHR), which should be given priority in UHI control and mitigation in Shenzhen. Meanwhile, the non-construction areas (N) cannot be ignored, of which the contribution to WHR is 44.91%. All of the contribution proportions to heat effect regions for Street and transportation function (S), Manufacture function (M), and Warehousing and logistics function (W) increase rapidly from weak thermal effect to significant thermal effect.

DST between the FCLZs shows significantly different thermal capacity in construction land. The thermal difference in degrees among various FCLZs (Table 4) caused by social functions was measured. We found that the median thermal effects ranked from strong to weak are as follows: M (3.99 °C)> W (3.69 °C) > S (3.61 °C) > B (3.06 °C) > A (2.54 °C) > G (2.40 °C) > R (2.21 °C) > N (−0.06 °C), in which the ranks are similar to a case study conducted in Beijing [36] and thus it shows consistency in the degree of the thermal effect of social functions between cities in different climate zones [79]. In addition, it is noticeable that the minimum warming effect (R) is about 36.8 times that of non-construction areas in degrees, which indicates that regardless of the intensity and the way, the relative change of urban surface temperature induced by human activities is huge enough to increase hot extremes [81].

4.3. Differential Responses of DST to Environmental Indicators

Response processes of DST to environmental indicators have been key points to understand and regulate urban thermal effects. The indicators including the fraction of impervious surface have been discussed in previous studies [34,63,82], but the similarities and differences within construction land have not been clarified in detail. Our research investigated and refined the difference and consistency of various environmental indicators on DST in FCLZs. To further figure out the impacts of indicators on DST, we used the median resampled values to reduce random errors rather than the mean values because of the non-normality of DST data [79]. Equivalent repeated observation experiments were constructed by reorganizing the original grid-level data to ordered DST and corresponding indicator pairs. Scatter plots and single factor regression analyses were carried out on each pair of resampled values in the two groups of FCLZs with non-construction areas as the controlled group, to preliminarily explore the effect of each indicator on DST. On the whole, most of the indicators show similar trends but distinguished degrees of the effects on DST in the two groups, which proves that these indicators have similar effect modes [65] across the scale from a single city to global cities [26]. Non-linear relationships suggest the saturation effects exhibit in both distance and corresponding values of social-economic levels, for location indicators and social-economic indicators, respectively.

4.4. Potential Implication and Future Directions

This study investigated the thermal effects in construction land through a continuous division of social-economic functions at a 500-m grid-level using the FCLZs framework, which could extend the UHI analyses routine based on LUCC with more details about human-induced thermal contribution to the urban environment [10]. Our results show that urban heat island is not only mostly caused by the amount of different land use or land cover types, but there are also still significant differences within construction land induced by their social functions. Both similarities and differences exist in the responses to environmental indicators among FCLZs. Therefore, when the cooling capacities of GITs and [47,68,69] are insufficient or there are limits in maintaining and appropriately increasing green space and water bodies [27,68], except for controlling the expansion of urban areas [44], the relocation of the construction land with different social-economic functions and configuration with cooling and warming indicators at appropriate thresholds may be helpful ways to alleviate urban heat island problems.

The limitation of this work is that FCLZs do not have clear boundaries in reality because of fuzzy positions of POIs, and it is hard to get an absolute ground truth map to calculate a confusion matrix for absolute accuracy comparison. Random sampling for verification and comparison with official statics is the currently limited method to compare relative accuracy. Due to the different response processes among FCLZs, the tradeoff and synergies between environment indicators in construction should be the focus of future research.

5. Conclusions

This research established a general framework for functional construction land zones (FCLZs) mapping based on multi-sourced data to investigate their different contribution to the urban thermal environment in Shenzhen, China. The thermal environment is characterized by DST extracted from land surface temperature data. Based on the difference test of DST, FCLZs were then divided into groups of functions for A, G, and R and groups of functions for B, M, S, and W to analyze the thermal response with 12 environmental indicators considering surface biophysics, buildings, location, and social-economic development.

We found that: (1) The thermal effect and total contribution FCLZs are significantly different. Although construction land leads to an obvious warming effect with a median value of 2.94 °C in the urban environment, thermal contribution in FCLZs is significantly different. FCLZs contribute to MHR and SHR (moderate and significant heat regions) with a proportion of 82.63~85.46%, which highly exceeds the corresponding proportion in non-construction areas. As for weaker warming effect regions, the contribution of non-construction areas cannot be ignored because of the size (44.91%). The median thermal effects of various FCLZs are as follows: M (3.99 °C)> W (3.69 °C) > S (3.61 °C) > B (3.06 °C) > A (2.54 °C) > G (2.40 °C) > R (2.21 °C), and the minimum thermal warming effect of R is about 36.8 times than unnoticeable thermal effect (−0.06 °C) in the non-construction areas. (2) Difference and consistency coexist in responses of DST to various environmental indicators in FCLZs. Different intercepts in the consistent linear and approximate linear relationships indicated the differences between thermal responses of biophysical indicators (NDBI, NDVI, and NDWI) in FCLZs were mainly in degree rather than mode. Buildings indicators (Building_density, Building_intensity, and Floor_avg) showed weak inversed relationships with DST in the two groups. The saturation effects shown in response of DST to location (DEM, Eud_Water, and Eud_Wood) and social-economic indicators (Unit_MixedEnropy, VIIRS_Value, and Population) proved that distance and social-economic development contribute to the nonlinear change of urban thermal environment. The stable points for the two groups are ~1000 m and ~300 m, respectively, both of which are almost double the distances from the cooling region than the truing points in non-construction areas. Social-economic indicators would have no more impact on the thermal environment when reaching stable points.

It is an attempt to measure the differences in thermal environment in view of functional construction land zones. The findings of this research could extend the understanding of urban thermal warming mechanisms from the different social-economic activities reflected by the agent of FLCZs and provide new macroscopic perspectives on reducing the negative impacts of urban heat islands by combining scientific adjustment of environmental indicators according to their responses processes with the allocation of construction land in urban planning.

Author Contributions: Conceptualization, methodology, writing—original draft, visualization, writing—review and editing, H.W.; methodology, validation, writing—review and editing, B.L.; validation, writing—review and editing, T.Y.; formal analysis, supervision, funding acquisition, J.W. All authors have read and agreed to the published version of the manuscript.

Funding: This research was funded by Shenzhen Fundamental Research Program, grant number GXWD20201231165807007-20200816003026001.

Conflicts of Interest: The authors declare no conflict of interest.

Appendix A

Table A1. Reclassification system of POIs to functional construction land zones.

Functional Class	Sub-Functional Class	Class Code	Tag of POIs
Residential Function	—	R	Residential community, villas, community centers
Administration and public services Function	Administration	A1	Government agency, Industrial and commercial bureau, public security bureau, procuratorates, courts, democratic Parties, social organization, public institutions
	Cultural facilities	A2	Public library, museum, science, and technology museum, art gallery, archives center, exhibition center, convention center

Table A1. *Cont.*

Functional Class	Sub-Functional Class	Class Code	Tag of POIs
Administration and public services Function	Education and research development	A3	Colleges and universities, technical secondary school, high school, middle school, primary school, research, and development institution
	Sports	A4	Gymnasium, court, sports training sites
	Medical Treatment and Public Health	A5	Health care services, general hospital, specialized hospital, clinic, emergency center, disease prevention agency
	Public welfare	A6	Welfare house, nursing home, orphanage
	Conservation of historic landmarks and sites	A7	Scenic spots and historical sites, tourist attractions, revolutionary site
	Religious facilities	A9	Church, mosque, temple
Business Services Function (B)	Commercial Facilities	B1	Retail business (shopping malls, supermarkets, shops, etc.)
			Wholesale market
			Catering services (restaurant, bar, tea house, cake shop, cafe, cold drink, and dessert shop)
			Accommodation services (hotels, guest houses, and resorts)
	Business Facilities	B2	Financial insurance (banking and insurance company, ATM, securities company, financial and insurance service organization)
			Art Media (Media organizations such as music, fine arts, film, television, advertising, network media, art groups)
			Other business facilities companies
	Recreation facilities	B3	Entertainment facilities (theatre, concert hall, cinema, song, dance hall, Internet cafe, amusement park)
			Recreation and Sports facilities (Golf Driving Range Racecourse Skating Rink Skydiving Range Motorcycle Range Shooting Range)
	Public utilities	B4	Refueling and filling stations (refueling and filling stations and other energy stations)
			Public facilities business outlets (telecommunications, postal service, water supply, gas supply, heat supply, etc.)
	Others	B9	Scientific, educational and cultural services (training institutions) medical and health services (clinics, medical and health sales shops, animal medical places) automobile services life services funeral services
Green spaces and squares (G)	park green space	G1	Park, zoo, botanical garden
	street and square green area	G3	City square

Table A1. *Cont.*

Functional Class	Sub-Functional Class	Class Code	Tag of POIs
Street and transport function (S)	Transport hub	S3	Railway station, long distance bus station, port and pier
	Transport stations	S4	Transport facilities (car parks, bus stops, MTR stations)
	Others	S9	Car training ground
Manufacture Function (M)	–	–	Industrial park, factory
Warehousing and logistics Function (W)	–	–	Logistics warehouse

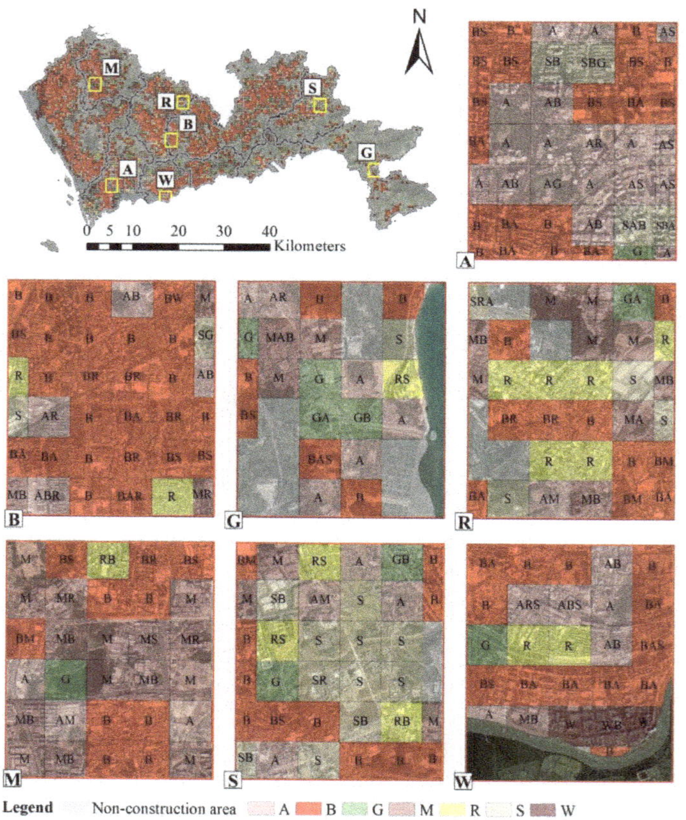

Figure A1. Random sampling grids for functional construction land zones manual verification.

References

1. Kalnay, E.; Cai, M. Impact of urbanization and land-use change on climate. *Nature* **2003**, *423*, 528–531. [CrossRef] [PubMed]
2. Deng, J.S.; Wang, K.; Hong, Y.; Qi, J.G. Spatio-temporal dynamics and evolution of land use change and landscape pattern in response to rapid urbanization. *Landsc. Urban Plan.* **2009**, *92*, 187–198. [CrossRef]
3. Sun, L.; Chen, J.; Li, Q.; Huang, D. Dramatic uneven urbanization of large cities throughout the world in recent decades. *Nat. Commun.* **2020**, *11*, 5366. [CrossRef] [PubMed]
4. Ritchie, H.; Roser, M. Urbanization. Available online: https://ourworldindata.org/urbanization (accessed on 25 June 2021).

5. Peng, S.; Piao, S.; Ciais, P.; Friedlingstein, P.; Ottle, C.; Bréon, F.M.; Nan, H.; Zhou, L.; Myneni, R.B. Surface urban heat island across 419 global big cities. *Environ. Sci. Technol.* **2012**, *46*, 696–703. [CrossRef] [PubMed]
6. Manoli, G.; Fatichi, S.; Schläpfer, M.; Yu, K.; Crowther, T.W.; Meili, N.; Burlando, P.; Katul, G.G.; Bou-Zeid, E. Magnitude of urban heat islands largely explained by climate and population. *Nature* **2019**, *573*, 55–60. [CrossRef] [PubMed]
7. Tomlinson, C.J.; Chapman, L.; Thornes, J.E.; Baker, C.J. Including the urban heat island in spatial heat health risk assessment strategies: A case study for Birmingham, UK. *Int. J. Health Geogr.* **2011**, *10*, 42. [CrossRef] [PubMed]
8. Estoque, R.C.; Ooba, M.; Seposo, X.T.; Togawa, T.; Hijioka, Y.; Takahashi, K.; Nakamura, S. Heat health risk assessment in Philippine cities using remotely sensed data and social-ecological indicators. *Nat. Commun.* **2020**, *11*, 1581. [CrossRef]
9. Ravanelli, R.; Nascetti, A.; Cirigliano, R.V.; Di Rico, C.; Leuzzi, G.; Monti, P.; Crespi, M. Monitoring the impact of land cover change on surface urban heat island through Google Earth Engine: Proposal of a global methodology, first applications and problems. *Remote Sens.* **2018**, *10*, 1488. [CrossRef]
10. Dong, J.; Peng, J.; He, X.; Corcoran, J.; Qiu, S.; Wang, X. Heatwave-induced human health risk assessment in megacities based on heat stress-social vulnerability-human exposure framework. *Landsc. Urban Plan.* **2020**, *203*, 103907. [CrossRef]
11. Coseo, P.; Larsen, L. How factors of land use/land cover, building configuration, and adjacent heat sources and sinks explain Urban Heat Islands in Chicago. *Landsc. Urban Plan.* **2014**, *125*, 117–129. [CrossRef]
12. Zhou, D.; Zhao, S.; Liu, S.; Zhang, L.; Zhu, C. Surface urban heat island in China's 32 major cities: Spatial patterns and drivers. *Remote Sens. Environ.* **2014**, *152*, 51–61. [CrossRef]
13. Lin, P.; Lau, S.S.Y.; Qin, H.; Gou, Z. Effects of urban planning indicators on urban heat island: A case study of pocket parks in high-rise high-density environment. *Landsc. Urban Plan.* **2017**, *168*, 48–60. [CrossRef]
14. Stewart, I.D.; Oke, T.R. Local climate zones for urban temperature studies. *Bull. Am. Meteorol. Soc.* **2012**, *93*, 1879–1900. [CrossRef]
15. Bechtel, B.; Alexander, P.J.; Böhner, J.; Ching, J.; Conrad, O.; Feddema, J.; Mills, G.; See, L.; Stewart, I. Mapping local climate zones for a worldwide database of the form and function of cities. *ISPRS Int. J. Geo-Inf.* **2015**, *4*, 199–219. [CrossRef]
16. Quan, J. Multi-temporal effects of urban forms and functions on urban heat islands based on local climate zone classification. *Int. J. Environ. Res. Public Health* **2019**, *16*, 2140. [CrossRef]
17. Liu, S.; Shi, Q. Local climate zone mapping as remote sensing scene classification using deep learning: A case study of metropolitan China. *ISPRS J. Photogramm. Remote Sens.* **2020**, *164*, 229–242. [CrossRef]
18. Zhao, C.; Jensen, J.L.R.; Weng, Q.; Currit, N.; Weaver, R. Use of Local Climate Zones to investigate surface urban heat islands in Texas. *GISci. Remote Sens.* **2020**, *57*, 1083–1101. [CrossRef]
19. Zhou, X.; Okaze, T.; Ren, C.; Cai, M.; Ishida, Y.; Watanabe, H.; Mochida, A. Evaluation of urban heat islands using local climate zones and the influence of sea-land breeze. *Sustain. Cities Soc.* **2020**, *55*, 102060. [CrossRef]
20. Yu, Z.; Jing, Y.; Yang, G.; Sun, R. A new urban functional zone-based climate zoning system for urban temperature study. *Remote Sens.* **2021**, *13*, 1–17. [CrossRef]
21. Sun, R.H.; Lü, Y.; Chen, L.D.; Yang, L.; Chen, A.L.; Lu, Y.H.; Chen, L.D.; Yang, L.; Chen, A.L. Assessing the stability of annual temperatures for different urban functional zones. *Build. Environ.* **2013**, *65*, 90–98. [CrossRef]
22. Liu, Y.; Peng, J.; Wang, Y. Relationship between urban heat island and landscape patterns: From city size and landscape composition to spatial configuration. *Acta Ecol. Sin.* **2017**, *37*, 7769–7780. [CrossRef]
23. Yan, J.; Zhou, W.; Jenerette, G.D. Testing an energy exchange and microclimate cooling hypothesis for the effect of vegetation configuration on urban heat. *Agric. For. Meteorol.* **2019**, *279*, 107666. [CrossRef]
24. Wang, J.; Meng, B.; Fu, D.; Pei, T.; Xu, C. Mapping spatiotemporal patterns and multi-perspective analysis of the surface urban heat islands across 32 major cities in China. *ISPRS Int. J. Geo-Inf.* **2018**, *7*, 207. [CrossRef]
25. Sun, F.; Liu, M.; Wang, Y.; Wang, H.; Che, Y. The effects of 3D architectural patterns on the urban surface temperature at a neighborhood scale: Relative contributions and marginal effects. *J. Clean. Prod.* **2020**, *258*, 120706. [CrossRef]
26. Yang, Q.; Huang, X.; Yang, J.; Liu, Y. The relationship between land surface temperature and artificial impervious surface fraction in 682 global cities: Spatiotemporal variations and drivers. *Environ. Res. Lett.* **2021**, *16*, 24032. [CrossRef]
27. Bartesaghi-Koc, C.; Osmond, P.; Peters, A. Quantifying the seasonal cooling capacity of 'green infrastructure types' (GITs): An approach to assess and mitigate surface urban heat island in Sydney, Australia. *Landsc. Urban Plan.* **2020**, *203*, 103893. [CrossRef]
28. Nakayama, T.; Fujita, T. Cooling effect of water-holding pavements made of new materials on water and heat budgets in urban areas. *Landsc. Urban Plan.* **2010**, *96*, 57–67. [CrossRef]
29. Kotharkar, R.; Bagade, A. Evaluating urban heat island in the critical local climate zones of an Indian city. *Landsc. Urban Plan.* **2018**, *169*, 92–104. [CrossRef]
30. Georgescu, M.; Morefield, P.E.; Bierwagen, B.G.; Weaver, C.P. Urban adaptation can roll back warming of emerging megapolitan regions. *Proc. Natl. Acad. Sci. USA* **2014**, *111*, 2909–2914. [CrossRef]
31. Zhang, Y.; Murray, A.T.; Turner, B.L. Optimizing green space locations to reduce daytime and nighttime urban heat island effects in Phoenix, Arizona. *Landsc. Urban Plan.* **2017**, *165*, 162–171. [CrossRef]
32. Andrade, R.; Alves, A.; Bento, C. Exploring different combinations of data and methods for urban land use analysis: A survey. In Proceedings of the 2019 Joint Poster and Workshop Sessions of AmI, AmI 2019 and 2019 European Conference on Ambient Intelligence, Rome, Italy, 13–15 November 2019; Strinati, E.C., Charitos, D., Chatzigiannakis, I., Ciampolini, P., Cuomo, F., Di Lorenzo, P., Gavalas, D., Hanke, S., Komninos, A., Mylonas, G., Eds.; CEUR-WS: Aachen, Germany, 2019; Volume 2492, pp. 55–65.

33. Niu, H.; Silva, E.A. Crowdsourced Data Mining for Urban Activity: Review of Data Sources, Applications, and Methods. *J. Urban Plan. Dev.* **2020**, *146*, 04020007. [CrossRef]
34. Feng, Y.; Du, S.; Myint, S.W.; Shu, M. Do urban functional zones affect land surface temperature differently? A case study of Beijing, China. *Remote Sens.* **2019**, *11*, 1802. [CrossRef]
35. Huang, X.; Wang, Y. Investigating the effects of 3D urban morphology on the surface urban heat island effect in urban functional zones by using high-resolution remote sensing data: A case study of Wuhan, Central China. *ISPRS J. Photogramm. Remote Sens.* **2019**, *152*, 119–131. [CrossRef]
36. Li, T.; Cao, J.F.; Xu, M.X.; Wu, Q.Y.; Yao, L. The influence of urban spatial pattern on land surface temperature for different functional zones. *Landsc. Ecol. Eng.* **2020**, *16*, 249–262. [CrossRef]
37. Liu, X.; Long, Y. Automated identification and characterization of parcels with OpenStreetMap and points of interest. *Environ. Plan. B Plan. Des.* **2016**, *43*, 341–360. [CrossRef]
38. Zhang, Y.; Li, Q.; Tu, W.; Mai, K.; Yao, Y.; Chen, Y. Functional urban land use recognition integrating multi-source geospatial data and cross-correlations. *Comput. Environ. Urban Syst.* **2019**, *78*, 101374. [CrossRef]
39. Amap. Web Service API Related Downloads. POI Classification Code. Available online: https://a.amap.com/lbs/static/amap_3dmap_lite/amap_poicode.zip (accessed on 15 December 2021).
40. Baidu. LBS. Cloud Service. POITags. Available online: https://lbsyun.baidu.com/index.php?title=lbscloud/poitags (accessed on 15 December 2021).
41. Google Maps Platform. Places API. Place Types. Available online: https://developers.google.com/maps/documentation/places/web-service/supported_types (accessed on 15 December 2021).
42. Liu, H.; Xu, Y.; Tang, J.; Deng, M.; Huang, J.; Yang, W.; Wu, F. Recognizing urban functional zones by a hierarchical fusion method considering landscape features and human activities. *Trans. GIS* **2020**, *24*, 1359–1381. [CrossRef]
43. Wu, J.; Li, S.; Shen, N.; Zhao, Y.; Cui, H. Construction of cooling corridors with multiscenarios on urban scale: A case study of Shenzhen. *Sustainability* **2020**, *12*, 5903. [CrossRef]
44. Qian, J.; Peng, Y.; Luo, C.; Wu, C.; Du, Q. Urban land expansion and sustainable land use policy in Shenzhen: A case study of China's rapid urbanization. *Sustainability* **2016**, *8*, 16. [CrossRef]
45. Cao, J.; Zhou, W.; Zheng, Z.; Ren, T.; Wang, W. Within-city spatial and temporal heterogeneity of air temperature and its relationship with land surface temperature. *Landsc. Urban Plan.* **2021**, *206*, 103979. [CrossRef]
46. Wu, J.; Li, C.; Zhang, X.; Zhao, Y.; Liang, J.; Wang, Z. Seasonal variations and main influencing factors of the water cooling islands effect in Shenzhen. *Ecol. Indic.* **2020**, *117*, 106699. [CrossRef]
47. Peng, J.; Dan, Y.; Qiao, R.; Liu, Y.; Dong, J.; Wu, J. How to quantify the cooling effect of urban parks? Linking maximum and accumulation perspectives. *Remote Sens. Environ.* **2021**, *252*, 112135. [CrossRef]
48. Amap. Guides for Developers: API Documents for Serarching POI. Available online: https://lbs.amap.com/api/webservice/guide/api/search/ (accessed on 15 March 2019).
49. Kristi, S. Landsat 8 Collection 2 (C2) Level 2 Science Product (L2SP) Guide. Available online: https://www.usgs.gov/media/files/landsat-8-collection-2-level-2-science-product-guide (accessed on 8 October 2021).
50. Ministry of Natural Resources. Globeland30: Global Geo-Information Public Product. Available online: http://www.globallandcover.com/ (accessed on 13 May 2021).
51. Shenzhen Municipal Bureau of Planning and Natural Resources; Shenzhen Municipal Bureau of Statistics. Report of the Main Data Results of Shenzhen Land Change Survey in 2018. Available online: http://pnr.sz.gov.cn/xxgk/sjfb/tjsj/content/post_7058772.html (accessed on 20 April 2021).
52. Geospatial Data Cloud Digital Elevation Data of GDEMV2 30M. Available online: https://www.gscloud.cn/sources/accessdata/421?pid=302 (accessed on 14 May 2021).
53. Earth Observation Group EOG Nighttime Light. Available online: https://eogdata.mines.edu/nighttime_light/annual/v20/ (accessed on 27 May 2021).
54. Rose, A.N.; McKee, J.J.; Sims, K.M.; Bright, E.A.; Reith, A.E.; Urban, M.L. LandScan. 2019. Available online: https://landscan.ornl.gov/ (accessed on 27 May 2021).
55. Liu, X.; Andris, C.; Rahimi, S. Place niche and its regional variability: Measuring spatial context patterns for points of interest with representation learning. *Comput. Environ. Urban Syst.* **2019**, *75*, 146–160. [CrossRef]
56. Liu, K.; Yin, L.; Lu, F.; Mou, N. Visualizing and exploring POI configurations of urban regions on POI-type semantic space. *Cities* **2020**, *99*, 102610. [CrossRef]
57. Ministry of Housing and Urban-Rural Development (MOHURD). *Code for Classification of Urban Land Use and Planning Standards of Development Land*; China Architecture & Building Press: Beijing, China, 2011; pp. 1–60.
58. Kang, Y.; Wang, Y.; Xia, Z.; Chi, J.; Jiao, L.; Wei, Z. Identification and classification of Wuhan urban districts based on POI. *J. Geomat.* **2018**, *43*, 81–85. [CrossRef]
59. Ramos, J. Using TF-IDF to Determine Word Relevance in Document Queries. In Proceedings of the first instructional conference on machine learning, Piscataway, NJ, USA, 3–8 December 2003; Volume 242, pp. 29–48.
60. Malakar, N.K.; Hulley, G.C.; Hook, S.J.; Laraby, K.; Cook, M.; Schott, J.R. An Operational Land Surface Temperature Product for Landsat Thermal Data: Methodology and Validation. *IEEE Trans. Geosci. Remote Sens.* **2018**, *56*, 5717–5735. [CrossRef]

61. Cook, M.; Schott, J.R.; Mandel, J.; Raqueno, N. Development of an operational calibration methodology for the Landsat thermal data archive and initial testing of the atmospheric compensation component of a land surface temperature (LST) product from the archive. *Remote Sens.* **2014**, *6*, 11244–11266. [CrossRef]
62. Pettorelli, N.; Vik, J.O.; Mysterud, A.; Gaillard, J.M.; Tucker, C.J.; Stenseth, N.C. Using the satellite-derived NDVI to assess ecological responses to environmental change. *Trends Ecol. Evol.* **2005**, *20*, 503–510. [CrossRef]
63. Wang, Y.; Yi, G.; Zhou, X.; Zhang, T.; Bie, X.; Li, J.; Ji, B. Spatial distribution and influencing factors on urban land surface temperature of twelve megacities in China from 2000 to 2017. *Ecol. Indic.* **2021**, *125*, 107533. [CrossRef]
64. Hu, Y.; Dai, Z.; Guldmann, J.M. Modeling the impact of 2D/3D urban indicators on the urban heat island over different seasons: A boosted regression tree approach. *J. Environ. Manag.* **2020**, *266*, 110424. [CrossRef] [PubMed]
65. Silva, A.G.L.; Torres, M.C.A. Proposing an effective and inexpensive tool to detect urban surface temperature changes associated with urbanization processes in small cities. *Build. Environ.* **2021**, *192*, 107634. [CrossRef]
66. Li, Y.; Schubert, S.; Kropp, J.P.; Rybski, D. On the influence of density and morphology on the Urban Heat Island intensity. *Nat. Commun.* **2020**, *11*, 2647. [CrossRef] [PubMed]
67. Yang, J.; Menenti, M.; Wu, Z.; Wong, M.S.; Abbas, S.; Xu, Y.; Shi, Q. Assessing the impact of urban geometry on surface urban heat island using complete and nadir temperatures. *Int. J. Climatol.* **2021**, *41*, E3219–E3238. [CrossRef]
68. Li, X.; Zhou, W. Optimizing urban greenspace spatial pattern to mitigate urban heat island effects: Extending understanding from local to the city scale. *Urban For. Urban Green.* **2019**, *41*, 255–263. [CrossRef]
69. Peng, J.; Liu, Q.; Xu, Z.; Lyu, D.; Du, Y.; Qiao, R.; Wu, J. How to effectively mitigate urban heat island effect? A perspective of waterbody patch size threshold. *Landsc. Urban Plan.* **2020**, *202*, 103873. [CrossRef]
70. Liang, Z.; Wu, S.; Wang, Y.; Wei, F.; Huang, J.; Shen, J.; Li, S. The relationship between urban form and heat island intensity along the urban development gradients. *Sci. Total Environ.* **2020**, *708*, 135011. [CrossRef]
71. Zhang, J.; Yuan, X.D.; Lin, H. The Extraction of Urban Built-Up Areas by Integrating Night-Time Light and POI Data-A Case Study of Kunming, China. *IEEE Access* **2021**, *9*, 22417–22429. [CrossRef]
72. Royston, P. Approximating the Shapiro-Wilk W-test for non-normality. *Stat. Comput.* **1992**, *2*, 117–119. [CrossRef]
73. Vargha, A.; Delaney, H.D. The Kruskal-Wallis Test and Stochastic Homogeneity. *J. Educ. Behav. Stat.* **1998**, *23*, 170–192. [CrossRef]
74. Sture Holm A Simple Sequentially Rejective Multiple Test Procedure. *Scand. J. Stat.* **1979**, *6*, 65–70.
75. De Winter, J.C.F.; Gosling, S.D.; Potter, J. Comparing the pearson and spearman correlation coefficients across distributions and sample sizes: A tutorial using simulations and empirical data. *Psychol. Methods* **2016**, *21*, 273–290. [CrossRef]
76. Chen, J.; Yang, S.; Li, H.; Zhang, B.; Lv, J. Research on geographical environment unit division based on the method of natural breaks (Jenks). *Int. Arch. Photogramm. Remote Sens. Spat. Inf. Sci. ISPRS Arch.* **2013**, *40*, 47–50. [CrossRef]
77. Chen, S.; Wang, T. Comparison Analyses of Equal Interval Method and Mean-standard Deviation Method Used to Delimitate Urban Heat Island. *Geo-Inf. Sci.* **2009**, *11*, 145–150. [CrossRef]
78. Qiao, Z.; Sun, Z.; Sun, X.; Xu, X.; Yang, J. Prediction and analysis of urban thermal environment risk and its spatio- temporal pattern. *Shengtai Xuebao/Acta Ecol. Sin.* **2019**, *39*, 649–659. [CrossRef]
79. Li, N.; Yang, J.; Qiao, Z.; Wang, Y.; Miao, S. Urban thermal characteristics of local climate zones and their mitigation measures across cities in different climate zones of China. *Remote Sens.* **2021**, *13*, 1468. [CrossRef]
80. Huang, Q.; Huang, J.; Yang, X.; Fang, C.; Liang, Y. Quantifying the seasonal contribution of coupling urban land use types on Urban Heat Island using Land Contribution Index: A case study in Wuhan, China. *Sustain. Cities Soc.* **2019**, *44*, 666–675. [CrossRef]
81. Liao, W.; Li, D.; Malyshev, S.; Shevliakova, E.; Zhang, H.; Liu, X. Amplified Increases of Compound Hot Extremes Over Urban Land in China. *Geophys. Res. Lett.* **2021**, *48*, e2020GL091252. [CrossRef]
82. Shi, Y.; Liu, S.; Yan, W.; Zhao, S.; Ning, Y.; Peng, X.; Chen, W.; Chen, L.; Hu, X.; Fu, B.; et al. Influence of landscape features on urban land surface temperature: Scale and neighborhood effects. *Sci. Total Environ.* **2021**, *771*, 145381. [CrossRef]

remote sensing

MDPI

Article

Quantifying the Cooling Effect and Scale of Large Inner-City Lakes Based on Landscape Patterns: A Case Study of Hangzhou and Nanjing

Yaoyao Zheng [1], Yao Li [2], Hao Hou [1,*], Yuji Murayama [3], Ruci Wang [3] and Tangao Hu [1]

1 Institute of Remote Sensing and Earth Sciences, College of Science, Hangzhou Normal University, Hangzhou 311121, China; zhengyaoyao@stu.hznu.edu.cn (Y.Z.); hutangao@hznu.edu.cn (T.H.)
2 Faculty of Geo-Information Science and Earth Observation (ITC), University of Twente, 7500AE Enschede, The Netherlands; yao.li@utwente.nl
3 Faculty of Life and Environmental Sciences, University of Tsukuba, 1-1-1 Tennodai, Tsukuba City, Ibaraki 305-8572, Japan; mura@geoenv.tsukuba.ac.jp (Y.M.); wang.ruci.fw@u.tsukuba.ac.jp (R.W.)
* Correspondence: houhao@hznu.edu.cn; Tel.: +86-150-6889-0483

Abstract: The rapid urbanization worldwide has brought various environmental problems. The urban heat island (UHI) phenomenon is one of the most concerning issues because of its strong relation with daily lives. Water bodies are generally considered a vital resource to relieve the UHI. In this context, it is critical to develop a method for measuring the cooling effect and scale of water bodies in urban areas. In this study, West Lake and Xuanwu Lake, two famous natural inner-city lakes, are selected as the measuring targets. The scatter plot and multiple linear regression model were employed to detect the relationship between the distance to the lake and land surface temperature based on Landsat 8 Operational Land Imager/Thermal Infrared Sensor (OLI/TIRS) and Sentinel-2 data. The results show that West Lake and Xuanwu Lake massively reduced the land surface temperature within a few hundred meters (471 m for West Lake and 336 m for Xuanwu Lake) and have potential cooling effects within thousands of meters (2900 m for West Lake and 3700 m for Xuanwu Lake). The results provide insights for urban planners to manage tradeoffs between the large lake design in urban areas and the cooling effect demands.

Keywords: cooling effect; distance analysis; landscape pattern; urban heat island; urban lake

Citation: Zheng, Y.; Li, Y.; Hou, H.; Murayama, Y.; Wang, R.; Hu, T. Quantifying the Cooling Effect and Scale of Large Inner-City Lakes Based on Landscape Patterns: A Case Study of Hangzhou and Nanjing. *Remote Sens.* **2021**, *13*, 1526. https://doi.org/10.3390/rs13081526

Academic Editor: Zina Mitraka

Received: 16 March 2021
Accepted: 13 April 2021
Published: 15 April 2021

1. Introduction

Due to the rapid urbanization and high-density population, various environmental changes occurred in urban areas [1]. The urban heat island (UHI) is one of the most significant climate changes caused by human activities [2]. UHI, firstly mentioned by Howard [3], refers to the common phenomenon in which the temperatures in urban areas are higher than that in the surrounding non-urbanized areas [4]. The high temperatures caused by the UHI phenomenon not only change the local climate environment resulting in extreme weather conditions [5], increased energy and water consumption [6,7], but also raise the risk of human health issues [8,9]. Thus, UHI mitigation strategies should be studied and incorporated into future city design and planning to reduce the adverse effects.

UHI represents the temperature difference between urban and suburban areas [10]. There are mainly three ways to measure the temperature for UHI studies, including meteorological station observation records [11,12], thermometers mounted on vehicles [13], and remote sensing observation of the surface temperature [14–16]. Among them, the meteorological stations provide detailed records but fail to show the spatial distribution characteristics of the temperature on a large scale. The vehicle's temperature records are limited in space and biased since the readings are made on the spot where the equipment is situated [17]. The land surface temperature (LST) derived from remote sensing observations provides easy access and large-scale temperature data and gradually becomes the primary way to assess the UHI effect [18,19]. For example, Khorchani et al. explored the temporal and spatial distributing characteristics of LST in peninsular Spain based on advanced very high resolution radiometer (AVHRR) data [20]. Clinton and Gong used the moderate resolution imaging spectroradiometer (MODIS) sensor to obtain global surface temperature and surface greenness at 1 km resolution [19]. Chen et al. utilized the Landsat thematic mapper (TM) and enhanced thematic mapper plus (ETM+) thermal infrared (TIR) data with 120 m and 60 m spatial resolutions, respectively, for local-scale studies of UHI [21]. Satellite-derived LST measurements have been conducted primarily by MODIS [18,19,22] and AVHRR data for large-scale regional (including national, continental, and even global) LST studies [23]. On the other hand, Landsat data provide better resolution and less frequent LST observations, and have been widely used in city-level studies [17].

Urbanization changes the surface temperature in cities by modifying the characteristics of the natural surface [24]. Landscape composition and configuration are considered two main factors in LST variation [25–28]. Of these impact factors, large water bodies in urban areas (e.g., lakes, rivers, and streams) are regarded as an effective resource to reduce the UHI [29]. The specific heat capacity of water bodies is more remarkable than other materials [30]. Thus, they have a lower warming-up speed than the surrounding areas, resulting in "cool islands" during the daytime [31]. Urban lakes play a significant role in regional climate regulation, maintenance of ecosystem balance, and diversification of urban derivatives [32]. Thus, it is crucial to quantitatively evaluate the cooling effects of the urban lakes and determine their cooling scale. In previous studies, the distances-LST scatter diagram has been widely used to qualitatively describe the cooling effects of lakes [25,33]. However, the cooling effects of the lake can be affected by surrounding land covers (e.g., trees, grasses, buildings and pavements) and the urban landscape patterns [11]. It is still a lack of explicit discussion on the impact of surrounding urban landscape patterns on the urban lake cooling effect. To bridge this knowledge gap, this article proposed a scientific method to assess the urban lake cooling effects and the potential cooling effect scale based on multiple linear regression methods by considering the urban landscape patterns surrounding the lakes with multi-source remote sensing images.

Here, two famous Lakes (West Lake in Hangzhou, Zhejiang province; Xuanwu Lake in Nanjing, Jiangsu province) were selected as the targets to quantitatively evaluate the cooling effect of the large natural inner-city lakes by considering their surrounding landscape patterns. West Lake, located in Hangzhou, is among the 55 United Nations Educational, Scientific and Cultural Organization (UNESCO) World Heritage sites in China [34]. Xuanwu Lake, the largest imperial garden lake in China, is located in Nanjing. As two of the highest gross domestic product (GDP) capitals of provinces, Hangzhou and Nanjing suffer from high temperatures during summer. Both of them are listed in the top 10 hottest cities in China (http://www.cma.gov.cn/) (accessed on 26 January 2021). As the landmarks of the two cities, West Lake and Xuanwu Lake are regarded as the typical inner-city lakes to detect the cooling scale of water bodies in cities.

This study aims to quantify the cooling effect and scale of urban lakes based on landscape patterns and provide important insights on landscape design and urban planning in the perspective of relieving UHI intensity. To achieve the primary purpose, the study has three sub-objectives, including (1) detect the variables affecting the LST in the surrounding areas of the urban lake; (2) build the regression models for LST in different scales; and (3) determine the cooling scale and the corresponding importance of lake for relieving the LST in the neighborhoods.

2. Materials and Methods

2.1. Study Area

Hangzhou, the capital of Zhejiang province, is situated in the southern wing of the Yangtze River Delta, with a latitude of 29°11′ N to 30°34′ N and a longitude of 118°20′ E to 120°37′ E) (Figure 1). Hangzhou has a subtropical monsoon climate with clearly divided four seasons. Hangzhou owns the natural environment that integrates rivers, lakes, and hills. As one of the three core cities (Shanghai, Nanjing, and Hangzhou) in the Yangtze River Delta urban agglomerations, Hangzhou boasts a population of 10.36 million and steady economic growth of a local GDP of RMB 1.54 trillion in 2019 [35]. The world-famous scenic spot, West Lake, lies in Hangzhou city's main urban area, 1.4 km away from the city center (Wulin Square). It is an oval-shaped lake covering a water area of 6.38 km^2, with its three sides surrounded by the mountains, one side by the urban area. The bottom of West Lake is relatively flat, with an average water depth of 2.27 m.

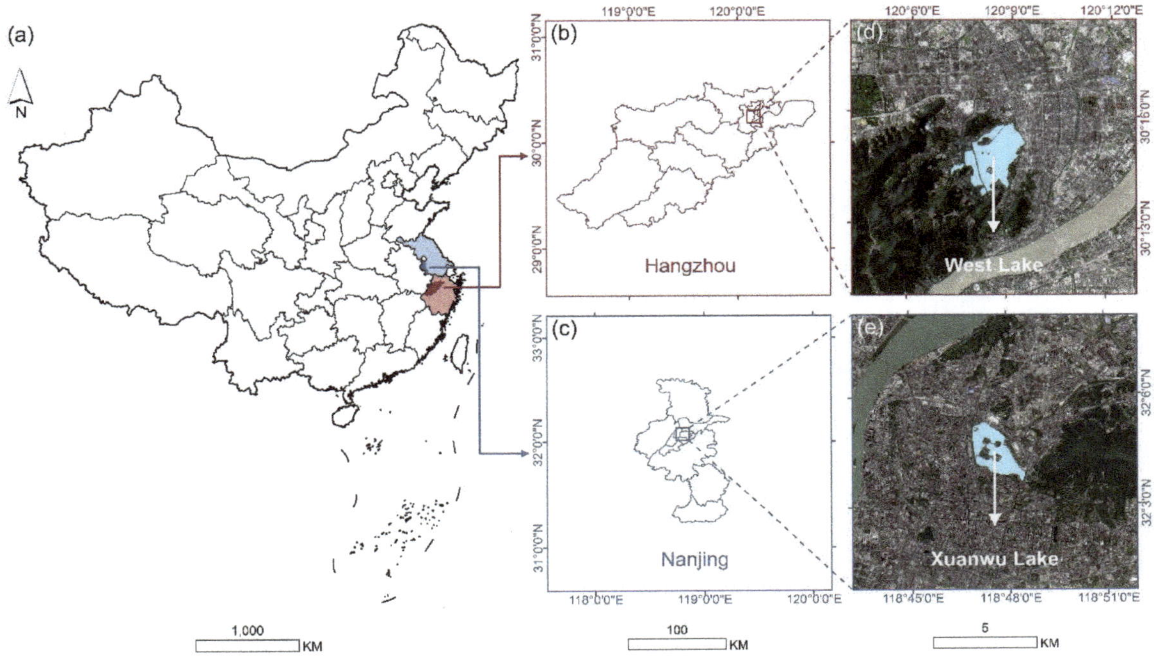

Figure 1. Location of the study area. (**a**) Map of China; (**b**) map of Hangzhou; (**c**) map of Nanjing; (**d**) Sentinel-2 image of West Lake and its neighboring areas; and (**e**) Sentinel-2 image of Xuanwu Lake and its neighboring areas. The sentinel-2 images were displayed in true color composite (Red—Band 4; Green—Band 3; and Bule—Band 2).

As the capital of Jiangsu province, Nanjing is located at the lower reaches of the Yangtze River (31°14′ N–32°37′ N, 118°22′ E–119°14′ E) (Figure 1). Nanjing is in the monsoon climate area of the north subtropical zone, with four distinct seasons. In terms of topographic conditions, Nanjing is connected with the vast Jianghuai (the Yellow River and Huai River) Plain to the north and the prosperous Yangtze River Delta to the east. Owing to its superior natural condition, Nanjing is one of the fastest urbanized cities in China. At the end of 2019, Nanjing has a population of 8.5 million and a GDP of RMB 1.40 trillion [36]. Xuanwu Lake is 2.7 km from the city center of Nanjing (Xinjiekou), with Zijin Mountain in the east. Xuanwu Lake is diamond-shaped, with a water area of 3.78 km^2 and an average water depth of 1.14 m.

Hangzhou and Nanjing are suffering from high temperatures during summer. In 2019, Hangzhou suffered 38 days of high temperature (>35 °C), with the highest temperature reaching 39.4 °C. In contrast, there were 18 days of high temperature (>35 °C) in Nanjing, with the highest temperature reaching 38.3 °C. The two cities are all in great need of reducing the temperature in summer.

2.2. Data Source and Pre-Processing

The Landsat 8 OLI/TIRS images provided by the United States Geological Survey (USGS) were employed to estimate the LST. Since the impact of the high-temperature weather mainly occurs in summer, one cloud-free image in the summer season was selected to capture the LST dynamics for each study area [37]. The acquisition dates of the images covering Hangzhou and Nanjing were the 22nd of July 2020 (path/row 119/39) and the 13th of September 2019 (path/row 120/38), respectively. Within the environment for visualizing images (ENVI), radiometric calibration and atmospheric correction were applied to the multispectral bands. Moreover, the thermal bands were pre-processed into an analysis-ready radiance format using radiometric calibration.

Two Sentinel-2 Level 1C products were downloaded from the Copernicus Open Access Hub (https://scihub.copernicus.eu/) (accessed on 20 January 2021) shared by the European Satellite Agency (ESA). The date of the selected products was close to that of the Landsat 8 OLI/TIRS images with no clouds cover the study area (Table 1). The publicly available ESA command-line program Sen2Cor was used to convert the top-of-atmosphere (TOA) Level-1C tiles to the bottom-of-atmosphere (BOA) Level-2A tiles [38]. The spatial resolution of the 13 spectral bands of Sentinel 2 varies from 10 m to 60 m. Herein, four bands (bands 2, 3, 4, and 8) with 10 m resolution and six bands (bands 5, 6, 7, 8a, 11, and 12) with 20 m resolution were used. The 20 m bands were up-sampled to 10 m resolution using nearest-neighbor interpolation [39]. Bands 1, 9, and 10 at 60 m spatial resolution, dedicated to atmospheric correction and cirrus detection, were discarded [40,41].

Table 1. Information of the remote sensing data used in this study.

Study Area	Platform	Resolution	Acquisition Data	Local Time
West Lake	Landsat 8	30 m	22 July 2020	10:31:30
	Sentinel-2	10 m	22 July 2020	10:35:51
Xuanwu Lake	Landsat 8	30 m	13 September 2019	10:37:38
	Sentinel-2	10 m	19 September 2019	10:45:51

2.3. Overall Workflow

This study attempted to detect the cooling effects and scales of urban lakes using Landsat 8 OLI/TIRS and Sentinel-2 data. Herein, to accomplish this objective, the overall workflow was designed as follows (Figure 2), with four main procedures including: (1) LST inversion based on Landsat 8 OLI/TIRS images; (2) supervised maximum likelihood classification based on Sentinel-2 images; (3) landscape metrics measurement; and (4) multiple regression modeling of LST based on Euclidean distance maps, landscape composition and configuration indices, and digital elevation model (DEM).

2.4. LST Retrieval

The retrieval of LST followed the radiative transfer equation (RTE) method. The equation used to compute the thermal infrared radiance received by the sensor (L_λ) is mentioned as given [42,43]:

$$L_\lambda = [\varepsilon B(T_S) + (1-\varepsilon)L_{atm}\downarrow]\tau + L_{atm}\uparrow \quad (1)$$

where ε is the land surface emissivity, $B(T_S)$ is the blackbody radiance (W/(m^2·sr·μm)), T_S is the LST (K), τ is atmospheric transmittance and $L_{atm}\uparrow$ and $L_{atm}\downarrow$ are upwelling and downwelling atmospheric radiance (W/(m^2·sr·μm)), respectively. τ, $L_{atm}\uparrow$ and $L_{atm}\downarrow$ can be calculated on the Atmospheric Correction Parameter Calculator.

$B(T_S)$ is computed according to Equation (1). Subsequently, the LST is derived by Equation (2) [44]:

$$T_S = \frac{K_2}{\ln\left(\frac{K_1}{B(T_S)} + 1\right)} \quad (2)$$

where K_1 is conversion constant 774.8853 W/(m^2·sr·µm) and K_2 is conversion constant 1321.0789 K.

The land surface emissivity of water is 0.995, and the values of buildings and soil are obtained by the following equations [45]:

$$\varepsilon_{building} = 0.9589 + 0.0860P_v - 0.0671P_v^2 \quad (3)$$

$$\varepsilon_{soil} = 0.9625 + 0.0614P_v - 0.0461P_v^2 \quad (4)$$

where P_V is the vegetation proportion calculated by Equation (5) [46]:

$$P_V = \frac{NDVI - NDVI_{Soil}}{NDVI_{Veg} - NDVI_{Soil}} \quad (5)$$

where NDVI is the normalized difference vegetation index, $NDVI_{Soil}$ is the NDVI value of bare soil, $NDVI_{Veg}$ is the NDVI value of the area completely covered by vegetation. The equation of NDVI is [47]:

$$NDVI = \frac{\rho_{NIR} - \rho_{Red}}{\rho_{NIR} + \rho_{Red}} \quad (6)$$

where ρ_{NIR} refers to the surface reflectance of NIR band (band 5, 0.845–0.885 µm), ρ_{Red} refers to the surface reflectance of Red band (band 4, 0.630–0.680 µm).

In addition, the LST data was up-sampled to 10 m resolution by nearest-neighbor interpolation to match the spatial resolution of Sentinel-2 data.

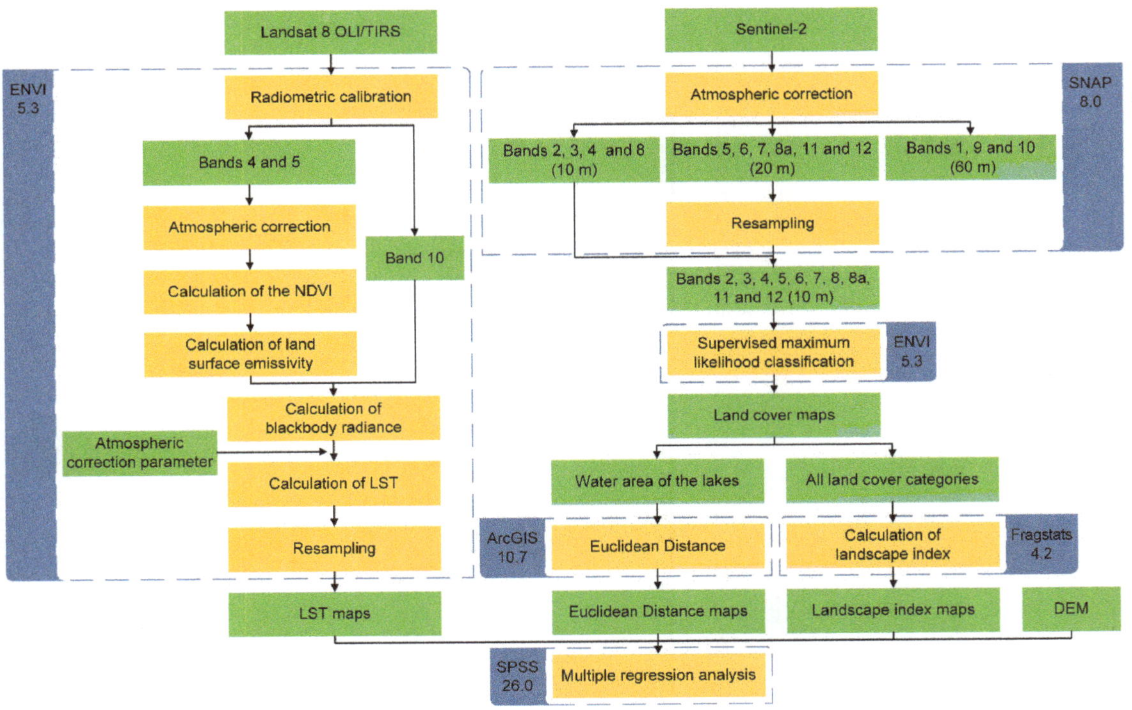

Figure 2. Overall workflow of this study. Green, yellow, and blue represent data, operations, and software, respectively.

2.5. Land Cover Classification

The land cover classification was derived from the pre-processed Sentinel-2 images using the maximum likelihood supervised classification method. The land cover classification system developed by Cadenasso et al. [48] was adopted to classify the land cover into five categories, namely coarse-textured vegetation (CV), fine-textured vegetation (FV), impervious surface (IS), bare lands (BL), and water. The detailed information of these categories is shown in Table 2. The accuracy of classification was evaluated using 500 random points located in each classified land-cover map with reference to the ground-truth data in Google Earth. The results showed that the overall accuracies [49] of the classifications were 91.20% for Hangzhou and 91.67% for Nanjing. The kappa coefficients were 0.878 for Hangzhou and 0.883 for Nanjing.

Table 2. Land cover categories and their detailed information.

Category	Description
CV	Coarse-textured vegetation which includes forest, woodland and shrub land
FV	Fine-textured vegetation which includes cropland and grassland
IS	Impervious surface which includes buildings and pavements
BL	Bare lands which include bare soil and bare rock
Water	Water bodies which include natural-flowing river and lake as well as artificial pond and reservoir

2.6. Landscape Metrics-Based Analysis

Numerous landscape metrics have arisen as a method to quantify landscape patterns [50]. Herein, five commonly used class-level landscape metrics [51–54] were employed to relate the spatial variability of LST with the landscape patterns, including one composition metric: percentage of landscape (PLAND), and four configuration metrics: (1) Largest patch index (LPI); (2) Mean shape index (SHAPE_MN); (3) Aggregation index (AI); and (4) Patch density (PD) (Table 3). These metrics were selected according to the following principles [55,56]: (1) importance in both theory and practice; (2) simplicity in both calculation and interpretation; and (3) minimal redundancy. Fragstats Version 4.2.1 was employed to calculate the selected metrics. The uniform tiles method was adopted to subdivide the landscape into square tiles representing sub-landscapes. Bartesaghi-Koc et al. [57] proposed that a 50 m × 50 m grid size is proper for local-scale studies if very-high resolution data are available. In comparison, studies employing data with coarser spatial resolutions (>10 m) may require larger grid resolutions (>100 m). Masoudi et al. [58] and Masoudi and Tan [59] recommended an optimal grid size of 240 × 240 m to explore the relationship between LST and the spatial pattern of urban green spaces. Herein, considering that the resolutions of LST and land cover maps are 10 m, the side length was chosen as integer multiples of 100 m (10 × 10 pixels).

Table 3. Landscape metrics used in this study [60].

Metrics (Abbreviation)	Equation (Unit)	Description
Percentage of landscape (PLAND)	$PLAND = P_i = \frac{\sum_{j=1}^{n} a_{ij}}{A} \times 100$	The percentage of the landscape consisting of the corresponding patches.
Largest patch index (LPI)	$LPI = \frac{max(a_{ij})}{A} \times 100$	The percentage of the landscape comprised by the largest patch.
Mean shape index (SHAPE_MN)	$SHAPE\_MN = \sum_{j=1}^{n} \frac{0.25 p_{ij}}{\sqrt{a_{ij}}}$	Mean shape index of the corresponding patches within an analysis unit.
Aggregation index (AI)	$AI = \left[\frac{g_{ii}}{max(g_{ii})}\right] \times 100$	The degree of the corresponding patches' aggregation within an analysis unit.
Patch density (PD)	$PD = \frac{n_i}{A} \times 10^6$	The ratio of the corresponding patches' number to the total landscape area within an analysis unit.

P_i = proportion of the landscape occupied by patch type (class) i. a_{ij} = area (m^2) of patch ij. A = total landscape area (m^2). p_{ij} = perimeter (m) of patch ij. g_{ii} = number of like adjacencies (joins) between pixels of patch type (class) i based on the single-count method. $max(g_{ii})$ = maximum number of like adjacencies (joins) between pixels of patch type (class) i based on the single-count method. n_i = number of patches in the landscape of patch type (class) i.

2.7. Statistical Analysis

The scatter diagram has been widely employed for qualitatively discussing the relationships between LST and the distance to the water bodies in previous studies [33,61]. In addition, the affecting scale was acquired by seeking the stationary point of the fitting function of the scatter diagram [4].

Relationships between LST and multiple influencing factors (i.e., the distance to the lakes, landscape metrics, and elevation) were quantitatively examined through the multiple linear regression (MLR) (Equation (7)) [62].

$$Y = \beta_0 + \sum_{i=1}^{n} \beta_i X_i + \varepsilon \tag{7}$$

where n is the number of the independent variable, β_0 is the intercept, β_i is the regression coefficient for X_i, and ε is the error term of the model.

To reduce the redundancy, a stepwise regression approach was adopted to identify closely linked variables with LST from the MLR models and provide their predictive importance for LST [6]. The significance levels at 0.05 and 0.1 were adopted as thresholds for adding and removing predictor variables, respectively. Additionally, in order to get more stable predictions, the variance inflation factor (VIF) was employed to detect multicollinearity. A VIF value equal to or larger than ten means near collinearity [63], indicating that the corresponding variables should be excluded. The remaining variables were analyzed again by the above-mentioned regression analysis until the absence of multicollinearity amongst selected variables.

We took 100 m as the step length and gradually added samples according to the distance to the lake, which was defined as the nearest distance to the lake shoreline. In other words, as the distance to the lake increased by 100 m, the samples covered by a certain distance were added to the existing samples. The process continued until the distance to the lake was not considered an independent variable of the stepwise regression model, indicating the influence of the target lakes on LST can be neglected. Each variable was pre-processed using the normalization method (Equation (8)) before the regression analysis [64].

$$\hat{X}_i = \frac{X_i - X_{min}}{X_{max} - X_{min}} \tag{8}$$

where X_i, X_{min}, and X_{max} are the original, minimum, maximum value of a variable, respectively. The regression analyses were carried out with the Statistical Package for the Social Sciences (SPSS) Version 26.

3. Results

3.1. Spatial Distributions of Land Cover and LST

The spatial distribution of LST in Hangzhou and Nanjing were shown in Figure 3. The LST of Hangzhou ranged from 15.37 to 57.96 °C, with an average of 37.21 °C. The LST of Nanjing was between 28.28 and 50.72 °C, with an average of 36.45 °C. Temperature variation over impervious surface tended to be more variable, while the water-covered surface showed the opposite trend due to its high heat capacity. In Hangzhou, the north and east side of West Lake, mainly occupied by built-up areas, owned the highest LST. On the other hand, the south and west parts, dominated by CV, had lower LST. In addition, the Qiantang River passed through the southeast of the study area showing the lowest LST. For the case of Nanjing, Xuanwu Lake and the adjacent Zijin Mountain were the low-value areas of LST. The Yangtze River in the northwest of the study area owned the lowest LST, while the urban areas dominated by buildings had the highest LST.

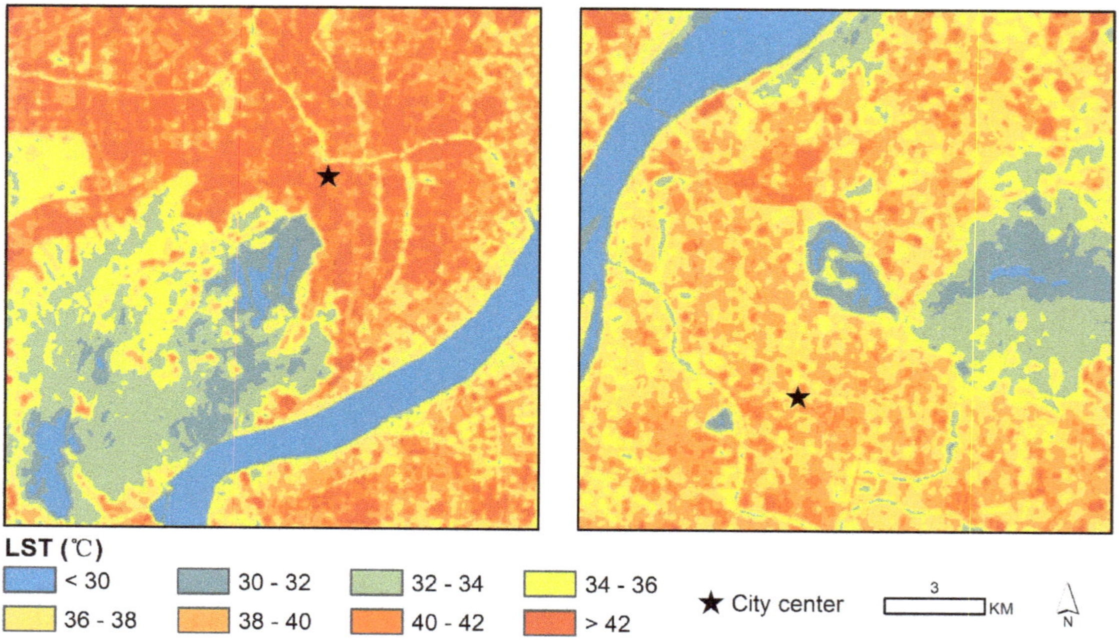

(a) (b)

Figure 3. The land surface temperature (LST) maps of (**a**) West Lake and its neighboring areas in Hangzhou, and (**b**) Xuanwu Lake and its neighboring areas in Nanjing.

Similar LST patterns were observed in Hangzhou and Nanjing despite the different temperature conditions of the cities (Figure 4). The study areas were mainly covered by built-up areas, coarse-textured vegetation, and water body. The high-temperature region corresponding to the area was dominated by built-up, and the low-temperature region corresponding to the area were covered with water and coarse-textured vegetation. Compared with the average temperature (37.21 °C for Hangzhou and 36.45 °C for Nanjing), impervious surface (40.25 °C for Hangzhou and 38.30 °C for Nanjing) and bare land (38.45 °C for Hangzhou and 38.16 °C for Nanjing) had higher LST (Figure 5). In contrast, the LST of coarse-textured vegetation (35.62 °C for Hangzhou and 35.51 °C for Nanjing), fine-textured vegetation (35.42 °C for Hangzhou and 35.76 °C for Nanjing), and water (28.32 °C for Hangzhou and 30.27 °C for Nanjing) fell below the average line (Figure 5).

3.2. Drivers of LST Variations in the Lakes and Their Surrounding Areas

The results of the scatter diagrams showed an increasing trend of LST in response to the increase in distance to the lake within a certain scale (741 m for West Lake and 336 m for Xuanwu Lake) (Figure 6). The maximum cooling effects for West Lake and Xuanwu Lake were 4.86 °C and 4.78 °C, respectively. Additionally, the correlation analysis showed significant positive correlations between mean LST and the distance to the lake inside a certain scale. The Pearson correlation coefficients were 0.637 ($p < 0.01$) for West Lake, and 0.841 ($p < 0.01$) for Xuanwu Lake, respectively (Figure 6).

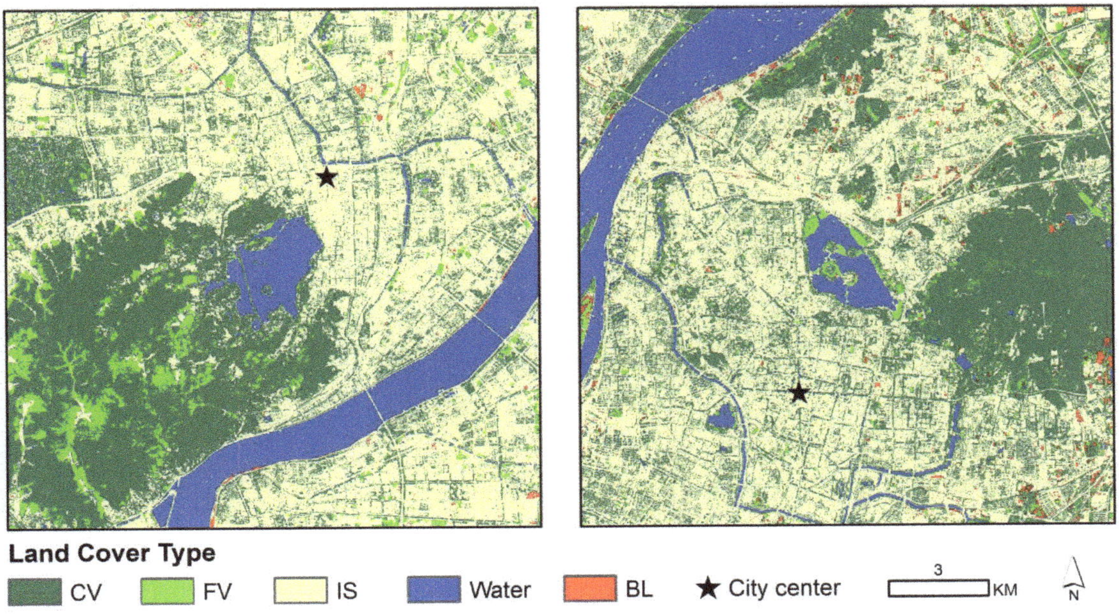

(a) (b)

Figure 4. The land cover maps of (**a**) West Lake and its neighboring areas in Hangzhou, and (**b**) Xuanwu Lake and its neighboring areas in Nanjing (CV: coarse-textured vegetation; FV: fine-textured vegetation; IS: impervious surface; BL: bare lands; Water: water).

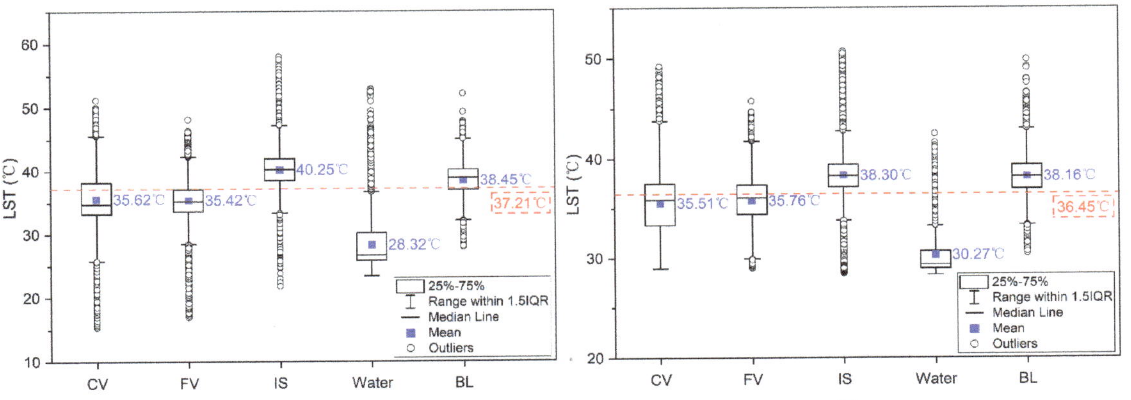

(a) (b)

Figure 5. LST of land cover categories in (**a**) West Lake and its neighboring areas in Hangzhou, and (**b**) Xuanwu Lake and its neighboring areas in Nanjing (CV: coarse-textured vegetation; FV: fine-textured vegetation; IS: impervious surface; BL: bare lands; Water: water).

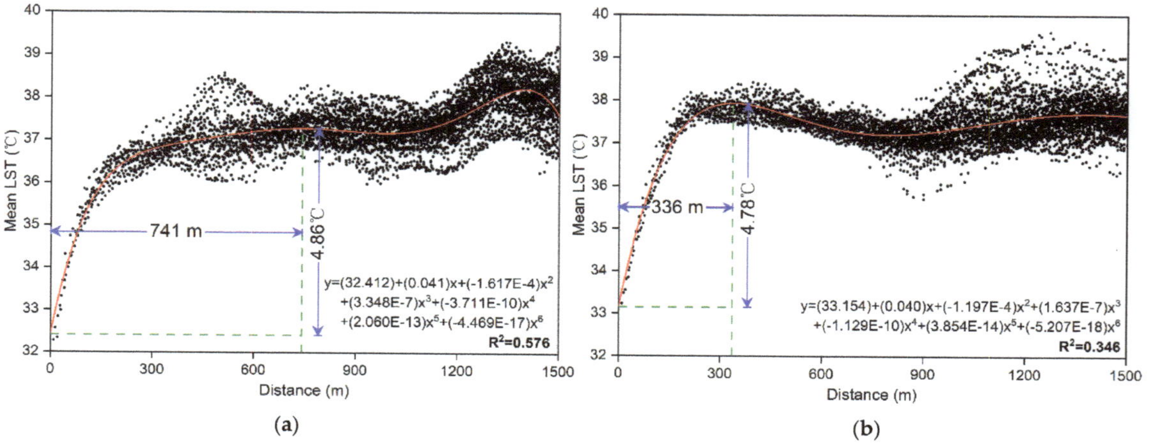

Figure 6. Mean LST at different distances to (**a**) West Lake in Hangzhou and (**b**) Xuanwu Lake in Nanjing.

The scatter diagrams showed the relationships between LST and the distance to the water bodies intuitively but not quantitatively. Based on the above-mentioned affecting scale, multiple linear regression models were established to present the relationships quantitatively, and the significance tests were listed in Tables 4 and 5. The models for Hangzhou ($R^2 > 0.81$) and Nanjing ($R^2 > 0.65$) were all of desirable goodness-of-fit.

Table 4. Significance tests of the multiple linear regressions in Hangzhou.

Model	R^2	ΔR^2	AIC	F	Model	R^2	ΔR^2	AIC	F
700 m	0.817	0.815	−3642.549	475.977 ***	1900 m	0.854	0.853	−12,057.291	1468.306 ***
800 m	0.820	0.819	−4197.190	425.007 ***	2000 m	0.850	0.849	−12,793.736	1519.813 ***
900 m	0.828	0.826	−4821.972	561.296 ***	2100 m	0.843	0.843	−13,502.545	1546.701 ***
1000 m	0.835	0.834	−5454.464	670.202 ***	2200 m	0.840	0.839	−14,323.481	1607.948 ***
1100 m	0.844	0.842	−6161.756	677.878 ***	2300 m	0.833	0.833	−15,061.922	1510.684 ***
1200 m	0.848	0.847	−6838.319	780.868 ***	2400 m	0.829	0.828	−15,915.979	1555.192 ***
1300 m	0.851	0.850	−7501.743	816.920 ***	2500 m	0.825	0.825	−16,762.030	1728.390 ***
1400 m	0.856	0.855	−8228.796	933.408 ***	2600 m	0.823	0.822	−17,648.922	1792.588 ***
1500 m	0.858	0.857	−8975.791	1318.902 ***	2700 m	0.820	0.820	−18,556.259	1858.858 ***
1600 m	0.858	0.857	−9772.715	1128.923 ***	2800 m	0.815	0.815	−19,416.599	1894.329 ***
1700 m	0.857	0.856	−10,546.830	1301.174 ***	2900 m	0.807	0.806	−20,228.078	1751.485 ***
1800 m	0.856	0.855	−11,318.390	1393.030 ***					

*** $p < 0.001$. R^2 = determinant coefficient; ΔR^2 = adjusted R^2; AIC = Akaike information criterion; F = F-test.

The multiple linear regressions showed that the distance to West Lake had significantly positive effects on LST in the range of 2900 m at the 5% significance level (Table 4 and Figure 7a). The distance to the Xuanwu Lake was a significant influencing factor in the range of 3700 m at the 5% significance level (Table 5 and Figure 7b).

Table 5. Significance tests of the multiple linear regressions in Nanjing.

Model	R^2	ΔR^2	AIC	F	Model	R^2	ΔR^2	AIC	F
300 m	0.664	0.661	−864.991	182.782 ***	2100 m	0.714	0.713	−9074.115	568.348 ***
400 m	0.662	0.658	−1203.617	151.728 ***	2200 m	0.718	0.717	−9736.086	620.828 ***
500 m	0.654	0.651	−1586.683	191.253 ***	2300 m	0.722	0.721	−10,431.697	629.963 ***
600 m	0.658	0.655	−1985.280	241.875 ***	2400 m	0.724	0.723	−11,123.377	725.120 ***
700 m	0.681	0.676	−2398.239	159.392 ***	2500 m	0.726	0.725	−11,801.854	778.349 ***
800 m	0.678	0.675	−2793.519	233.355 ***	2600 m	0.727	0.726	−12,514.335	829.313 ***
900 m	0.690	0.687	−3208.182	207.705 ***	2700 m	0.729	0.728	−13,217.408	884.023 ***
1000 m	0.700	0.697	−3624.970	248.397 ***	2800 m	0.732	0.731	−13,957.272	949.271 ***
1100 m	0.708	0.706	−4062.281	321.778 ***	2900 m	0.737	0.737	−14,753.843	1030.338 ***
1200 m	0.718	0.715	−4504.243	269.532 ***	3000 m	0.744	0.743	−15,557.309	1120.447 ***
1300 m	0.721	0.719	−4969.837	354.215 ***	3100 m	0.748	0.747	−16,436.357	1206.056 ***
1400 m	0.726	0.724	−5491.625	369.638 ***	3200 m	0.750	0.750	−17,272.656	1281.379 ***
1500 m	0.723	0.721	−5968.423	399.837 ***	3300 m	0.752	0.752	−18,154.546	1358.456 ***
1600 m	0.717	0.715	−6443.835	424.270 ***	3400 m	0.754	0.754	−19,048.449	1438.198 ***
1700 m	0.710	0.708	−6883.467	482.297 ***	3500 m	0.757	0.757	−19,956.757	1526.397 ***
1800 m	0.704	0.703	−7324.604	468.843 ***	3600 m	0.763	0.762	−20,979.320	1646.591 ***
1900 m	0.699	0.698	−7808.448	495.792 ***	3700 m	0.766	0.766	−21,943.199	1749.128 ***
2000 m	0.707	0.706	−8408.943	551.379 ***					

*** $p < 0.001$. R^2 = determinant coefficient; ΔR^2 = adjusted R^2; AIC = Akaike information criterion; F = F-test.

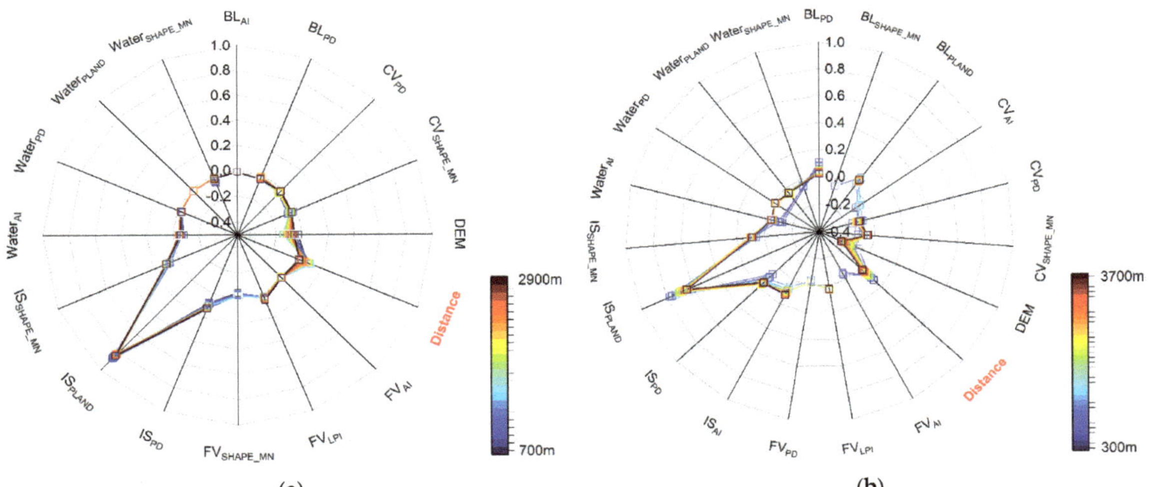

Figure 7. Radar diagrams showing the standardized coefficients of the multiple linear regressions for (**a**) West Lake and its neighboring areas, and (**b**) Xuanwu Lake and its neighboring areas. The significance levels of the coefficients are all less than 0.05 except for DEM (sig. = 0.058) from the 700 m model in Hangzhou, FV_AI (sig. = 0.067) from the 700 m model in Nanjing, FV_AI (sig. = 0.063) from the 900 m model in Nanjing, FV_PD (sig. = 0.062) from the 1200 m model in Nanjing and BL_PD (sig. = 0.050) from the 1400 m model in Nanjing.

As expected, the distance to the lake in both cities was positively related to LST regardless of the scale (Figure 7, Tables S1 and S2). Moreover, it is quite observant that as we examined LSTs at increasingly larger spatial extents, the standardized coefficients of distance generally had a decreasing trend, indicating that the explanatory power of distance parameters declined. The coefficients fluctuated between 0.078 and 0.110 in the first 1400 m around West Lake and kept decreasing in the extent of 1400 to 2900 m. For the Xuanwu Lake, the coefficients varied from 0.064 to 0.132 in the first 1200 m from the shoreline and showed a decreasing trend when the distance from the lake was more than 1200 m.

Although results from the MLRs showed that the distance to the lakes is an essential predictor of LST, the spatial configuration of impervious surface has a more significant effect on LST. The coefficients of the PLAND of IS were maintained at a high level both for West Lake (range from 0.811 to 0.867) and Xuanwu Lake (range from 0.639 to 0.779). In addition, with the decrease of the coefficients of distance, the coefficients of SHAPE_MN of IS showed higher values and tended to be stable after a certain distance (1800 m for West Lake and 1400 m for Xuanwu Lake). The coefficient values of PD of IS were steadily higher than that of distance after a certain distance (1500 m for West Lake and 1200 m for Xuanwu Lake). After 1500 m from Xuanwu Lake, the AI of IS also showed the same trend. The CV's coefficients of SHAPE_MN (range from −0.077 to −0.036 for West Lake and −0.121 to −0.048 for Xuanwu Lake), PD (range from −0.051 to −0.027 for West Lake and −0.153 to −0.076 for Xuanwu Lake), and AI in Xuanwu Lake (range from −0.086 to −0.047) were negative regardless of the scale. DEM was negatively related to LST, and its coefficients varied greatly in the model with different scales. With the expansion of the spatial extent, the coefficients of DEM decreased first and then increased, and bottomed out at −0.150 in the model with a spatial extent of 1500 m for West Lake. On the scale of 600 m around Xuanwu Lake, DEM was not selected as the models' variable, indicating that topography was not the main factor affecting LST. The first five models (600 m to 1100 m) that considered DEM as a dependent variable saw a dramatic fall in the coefficients of DEM, to a low of −0.204 in the model of 1100 m. Moreover, the coefficients of DEM dwindled to −0.231, after the rise to −0.152 in the extent of 1100 m to 3700 m.

4. Discussion

West Lake and Xuanwu Lake, located in the central areas of Hangzhou and Nanjing, showed a significant reduction of the LST in the surrounding areas up to a certain distance (741 m for West Lake and 336 m for Xuanwu Lake), demonstrating the ability of urban lakes in cooling the environment. Within the scale, the maximum cooling effects for West Lake and Xuanwu Lake were 4.86 and 4.78 °C, respectively. The cooling effect occurs when the thermal energy absorbed by the water is converted from sensible heating to latent heating with the production of water vapor [61]. Moreover, due to the high thermal capacity, the lakes own a lower temperature than the impervious surface during the daytime, which provides a higher pressure gradient for convective heat transfer [65]. Horizontal cooler air is generated above the lake and transported to the neighboring environment by the wind. The intensity of the processes decreases with the increasing distance to the lake, leading to the limited influence of the lakes on LST.

The ability of water bodies to adjust surrounding temperatures is determined both by the distance and its interactions with the surrounding environment [31]. However, most of the studies one-sidedly considered the distance to the water bodies, with little discussion on the integrated dynamics between the two features. Moyer and Hawkins [12] assessed the cooling effect of a fairly large river using urban temperature sensors deployed near the river. They reported that the UHI decreased by 0.6 °C to 0.3 °C for every 1000 m increase in distance from the river, depending on the season. Wu and Zhang [25] revealed that the horizontal cooling distance of Suzhou Bay could reach 800 m, and the maximum cooling effect was 3.02 °C. Cheval et al. [66] detected that the temperature regularly increases with the distance from the lake shoreline. Compared with these studies, our work underscored the necessity of considering the influence of its surrounding landscape patterns when quantifying the effects of the distance to the lakes on LST. In fact, most of the studies only focused on the scatter diagrams of LST to determine the cooling scale [4,33,61]. In contrast, after confirming the turning point from scatter diagrams, we continuously conducted a step-by-step multiple linear regression to detect the cooling potential of the two lakes on a much larger scale. The affecting scale obtained by seeking the stationary point of the fitting function of the scatter diagram was considered to be the region where the urban lakes massively lower the LST. The scale obtained by the multiple linear regression considering the influence of landscape patterns was deemed as the region where the urban lakes showed a significant capacity in cooling the environment. Although this study is conducted for West Lake and Xuanwu Lake, the general analysis procedure is flexible and can be extended to the case studies of urban lakes in other cities.

The regression results showed that although Xuanwu Lake was smaller in size, the maximum cooling scale could reach 3700 m, which was much larger than West Lake (2900 m). The cooling effect difference results from the surrounding landscape patterns, topography, and wind conditions (including wind directions and wind speed). It is demonstrated in Figure 8 that the PD of CV and the SHAPE_MN of IS around Xuanwu Lake were obviously higher than those of West Lake. High-density of trees and shrubs around the lake was conducive to forming a solid local circulation. The small and scattered architectural composition was instrumental in the formation of effective urban ventilation corridors [67], resulting in the expansion of the cooling range. In addition, large and dense buildings lead to significant heat effects [68,69]. To maximize the cooling potential of urban lakes, urban planners are suggested to arrange more trees and shrubs in the area around the lake and control the concentration of buildings. Meanwhile, West Lake is neighbored by mountains and is strongly impacted by a mountain valley breeze. Hence, the valley breeze in the northeast direction prevails in West Lake during the day [70], leading to the limited cooling effect on the urban area on the east side. Xuanwu Lake is mainly affected by east and southeast winds [71], which helps transport cooler air from the lake surface to the urban area.

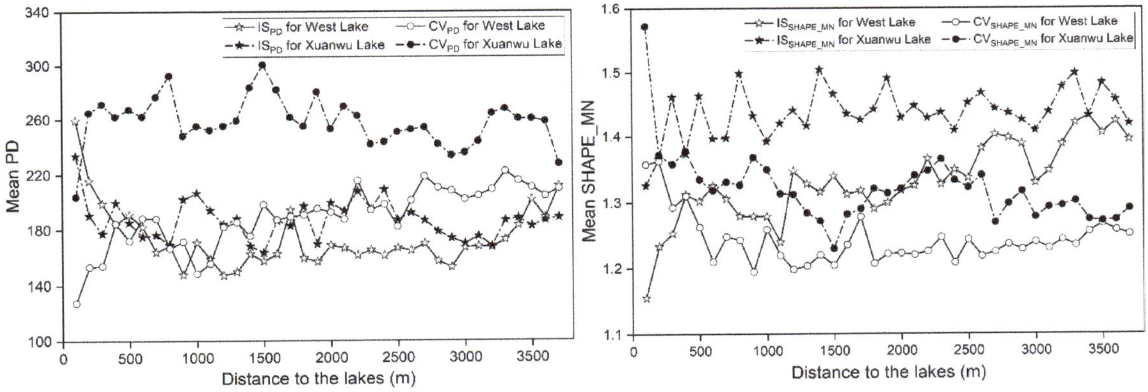

Figure 8. Mean (**a**) patch density (PD), and (**b**) SHAPE_MN at different distances to the lakes.

Furthermore, as a famous tourist spot, West Lake, which ranked No.1 in the 2018 China National Scenic Spot Popularity Index Ranking, received 833,700 tourists only one day on 3 October 2018 [72]. The overloaded West Lake tourism was also considered a contribution to the emergence of heat islands in West Lake and its neighboring areas. Previous studies have shown the significant cooling effects of urban greenspace on LST [73–75]. Nonetheless, the PLAND of CV, which corresponds closely to the LST [76–78], was not considered as a predictor variable in the linear regression models. This is because a large number of samples in the study were composed of residential buildings and the greening around the buildings, resulting in a significant negative correlation between the PLAND of IS and CV ($r = -0.738$ and -0.670 in Hangzhou and Nanjing, respectively, $p < 0.01$). The VIF values of partial variables were much more significant than 10 when the PLAND of IS and CV were both considered as independent variables, indicating the existence of multicollinearity.

In order to get satisfactory results in model accuracy and applicability, the problem of multicollinearity was eliminated by excluding the factor, the PLAND of CV.

Several limitations of this work should be addressed for future research avenues. First, the influence of the inherent properties of lakes (e.g., the size, shape, and water volume) on their cooling effect is not considered due to the limited number of lakes. Moreover, it is found that the relationship between landscape metrics and LST varies with the change of grid size. In this context, multi-scale analysis is needed in further studies to explore the cooling effects of urban lakes fully. Finally, this study was conducted in two cities with similar climatic conditions. The results may differ for other cities with a different climate. Thus, future studies involving other urban lakes with different climatic conditions are expected.

5. Conclusions

This study applied a distance—LST scatter diagram and multiple linear regression method to detect the cooling effect and scale of urban lakes based on Landsat 8 OLI/TIRS and Sentinel-2 data. Taking two famous city inner lakes, West Lake and Xuanwu Lake, as the study sites, the cooling effect and the potential cooling scale of large inner-city lakes were quantitatively analyzed. Meanwhile, the landscape patterns in the surrounding areas of the urban lake were considered as independent variables to detect the relationship in different scales. The whole research flow provides a complete procedure to detect the maximum cooling distance of a large cooling resource and can be applied not only to lakes but also to other specific complete cooling resources such as a park, a river, and a wetland. The results show that West Lake and Xuanwu Lake massively reduced the land surface temperature of surrounding areas within a certain distance (471 m for West Lake and 336 m for Xuanwu Lake) and had potential cooling effects on a larger scale (2900 m for West Lake and 3700 m for Xuanwu Lake). The results proved that the turning point in the temperature diagram could not reflect the cooling scale, and the exact cooling scale could be much larger. In addition, from the comparative study between Hangzhou and Nanjing, we detect the phenomenon that the surrounding landscape composition and configuration could strongly affect the maximum cooling scale. In detail, the high density of trees and shrubs, together with small and scattered buildings, could contribute to the extension of the cooling scale of inner lakes. However, to completely show the influencing factors for the cooling intensity and scale, future studies are suggested to consider more samples and consider the size and shape of lakes. In general, this research expands our scientific understanding of urban lakes' cooling effects, especially the potential cooling scale. These findings provide insights for urban planners to arrange the landscape of surrounding areas of large inner-city lakes to pursuit better environmental outcomes.

Supplementary Materials: The following are available online at https://www.mdpi.com/article/10.3390/rs13081526/s1, Table S1: The coefficients of the multiple linear stepwise regression models at different scales for the West Lake. The bold and italic rows are standardized coefficients, Table S2: The coefficients of the multiple linear stepwise regression models at different scales for Xuanwu Lake. The bold and italic rows are standardized coefficients.

Author Contributions: Conceptualization, Y.Z., Y.L., and H.H.; methodology, Y.Z. and H.H.; software, Y.Z.; formal analysis, Y.Z. and H.H.; investigation, Y.Z., Y.L., and H.H.; resources, Y.Z.; data curation, Y.Z.; writing—original draft preparation, Y.Z. and Y.L.; writing—review and editing, H.H., Y.M., R.W., and T.H.; visualization, Y.Z.; supervision, H.H.; project administration, H.H.; funding acquisition, H.H and R.W. All authors have read and agreed to the published version of the manuscript.

Funding: This research was supported by the Natural Science Foundation of Zhejiang Province: LQ20D010008 (Representative: Hao Hou), and the Japan Society for the Promotion of Science (JSPS) Doctoral Fellowship: 19J11409 (Representative: Ruci Wang).

Institutional Review Board Statement: Not applicable.

Informed Consent Statement: Not applicable.

Data Availability Statement: The data that support the findings of this study are available from the author upon reasonable request.

Acknowledgments: The authors appreciate the editors and anonymous reviewers for their valuable comments and suggestions.

Conflicts of Interest: The authors declare no conflict of interest.

References

1. Grimm, N.B.; Faeth, S.H.; Golubiewski, N.E.; Redman, C.L.; Wu, J.; Bai, X.; Briggs, J.M. Global change and the ecology of cities. *Science* **2008**, *319*, 756–760. [CrossRef] [PubMed]
2. Pascual, J.I.; Lorente, N.; Song, Z.; Conrad, H.; Rust, H.P. Selectivity in vibrationally mediated single-molecule chemistry. *Nature* **2003**, *423*, 525–528. [CrossRef]
3. Howard, L. *The Climate of London*; London Harvey and Dorton: London, UK, 1833; Volume 2, pp. 1818–1820.
4. Hou, H.; Estoque, R.C. Detecting Cooling Effect of Landscape from Composition and Configuration: An Urban Heat Island Study on Hangzhou. *Urban For. Urban Green.* **2020**, *53*, 126719. [CrossRef]
5. Ganeshan, M.; Murtugudde, R.; Imhoff, M.L. A multi-city analysis of the UHI-influence on warm season rainfall. *Urban Clim.* **2013**, *6*, 1–23. [CrossRef]
6. Zhou, W.; Huang, G.; Cadenasso, M.L. Does spatial configuration matter? Understanding the effects of land cover pattern on land surface temperature in urban landscapes. *Landsc. Urban Plan.* **2011**, *102*, 54–63. [CrossRef]
7. Santamouris, M.; Cartalis, C.; Synnefa, A.; Kolokotsa, D. On the impact of urban heat island and global warming on the power demand and electricity consumption of buildings—A review. *Energy Build.* **2015**, *98*, 119–124. [CrossRef]
8. Rocklöv, J.; Forsberg, B.; Ebi, K.; Bellander, T. Susceptibility to mortality related to temperature and heat and cold wave duration in the population of Stockholm County, Sweden. *Glob. Health Action* **2014**, *7*, 22737. [CrossRef]
9. Wong, L.P.; Alias, H.; Aghamohammadi, N.; Aghazadeh, S.; Sulaiman, N.M.N. Urban heat island experience, control measures and health impact: A survey among working community in the city of Kuala Lumpur. *Sustain. Cities Soc.* **2017**, *35*, 660–668. [CrossRef]
10. Wu, Y.; Zhang, M.; Hong, L.; Xian, X.; Yang, K. Temporal and Spatial Variation of Urban Heat Island Effect in Plateau Lake Region Based on RS—A 30-Year Case Study in Dianchi Basin. In Proceedings of the 2018 26th International Conference on Geoinformatics, Kunming, China, 28–30 June 2018; pp. 1–6.
11. Hathway, E.A.; Sharples, S. The interaction of rivers and urban form in mitigating the Urban Heat Island effect: A UK case study. *Build. Environ.* **2012**, *58*, 14–22. [CrossRef]
12. Moyer, A.N.; Hawkins, T.W. River effects on the heat island of a small urban area. *Urban Clim.* **2017**, *21*, 262–277. [CrossRef]
13. Voogt, J.A.; Oke, T.R. Thermal remote sensing of urban climates. *Remote Sens. Environ.* **2003**, *86*, 370–384. [CrossRef]
14. Hu, L.; Sun, Y.; Collins, G.; Fu, P. Improved estimates of monthly land surface temperature from MODIS using a diurnal temperature cycle (DTC) model. *ISPRS J. Photogramm. Remote Sens.* **2020**, *168*, 131–140. [CrossRef]
15. Chen, L.; Wang, X.; Cai, X.; Yang, C.; Lu, X. Seasonal Variations of Daytime Land Surface Temperature and Their Underlying Drivers over Wuhan, China. *Remote Sens.* **2021**, *13*, 323. [CrossRef]
16. Cao, H.; Gao, B.; Gong, T.; Wang, B. Analyzing Changes in Frozen Soil in the Source Region of the Yellow River Using the MODIS Land Surface Temperature Products. *Remote Sens.* **2021**, *13*, 180. [CrossRef]
17. Ngie, A.; Abutaleb, K.; Ahmed, F.; Darwish, A.; Ahmed, M. Assessment of urban heat island using satellite remotely sensed imagery: A review. *S. Afr. Geogr. J.* **2014**, *96*, 198–214. [CrossRef]
18. Zhou, D.; Zhao, S.; Liu, S.; Zhang, L.; Zhu, C. Surface urban heat island in China's 32 major cities: Spatial patterns and drivers. *Remote Sens. Environ.* **2014**, *152*, 51–61. [CrossRef]
19. Clinton, N.; Gong, P. MODIS detected surface urban heat islands and sinks: Global locations and controls. *Remote Sens. Environ.* **2013**, *134*, 294–304. [CrossRef]
20. Khorchani, M.; Vicente-Serrano, S.M.; Azorin-Molina, C.; Garcia, M.; Martin-Hernandez, N.; Peña-Gallardo, M.; Kenawy, A.E.; Domínguez-Castro, F. Trends in LST over the peninsular Spain as derived from the AVHRR imagery data. *Global Planet. Chang.* **2018**, *166*, 75–93. [CrossRef]
21. Chen, X.; Zhao, H.; Li, P.; Yin, Z. Remote sensing image-based analysis of the relationship between urban heat island and land use/cover changes. *Remote Sens. Environ.* **2006**, *104*, 133–146. [CrossRef]
22. Zhao, L.; Lee, X.; Smith, R.B.; Oleson, K. Strong contributions of local background climate to urban heat islands. *Nature* **2014**, *511*, 216–219. [CrossRef]
23. Streutker, D.R. Satellite-measured growth of the urban heat island of Houston, Texas. *Remote Sens. Environ.* **2003**, *85*, 282–289. [CrossRef]
24. Wu, C.; Li, J.; Wang, C.; Song, C.; Chen, Y.; Finka, M.; Rosa, D.L. Understanding the relationship between urban blue infrastructure and land surface temperature. *Sci. Total. Environ.* **2019**, *694*, 133742. [CrossRef]
25. Wu, Z.; Zhang, Y. Water Bodies' Cooling Effects on Urban Land Daytime Surface Temperature: Ecosystem Service Reducing Heat Island Effect. *Sustainability* **2019**, *11*, 787. [CrossRef]

26. Deilami, K.; Kamruzzaman, M.; Liu, Y. Urban heat island effect: A systematic review of spatio-temporal factors, data, methods, and mitigation measures. *Int. J. Appl. Earth Obs. Geoinf.* **2018**, *67*, 30–42. [CrossRef]
27. Arnfield, A.J. Two decades of urban climate research: A review of turbulence, exchanges of energy and water, and the urban heat island. *Int. J. Climatol.* **2003**, *23*, 1–26. [CrossRef]
28. Schwarz, N.; Manceur, A.M. Analyzing the Influence of Urban Forms on Surface Urban Heat Islands in Europe. *J. Urban Plan. Dev.* **2015**, *141*, A4014003. [CrossRef]
29. Theeuwes, N.E.; Solcerová, A.; Steeneveld, G.J. Modeling the influence of open water surfaces on the summertime temperature and thermal comfort in the city. *J. Geophys. Res.: Atmos.* **2013**, *118*, 8881–8896. [CrossRef]
30. Spronken-Smith, R.A.; Oke, T.R.; Lowry, W.P. Advection and the surface energy balance across an irrigated urban park. *Int. J. Climatol.* **2000**, *20*, 1033–1047. [CrossRef]
31. Gunawardena, K.R.; Wells, M.J.; Kershaw, T. Utilising green and bluespace to mitigate urban heat island intensity. *Sci. Total Environ.* **2017**, *584–585*, 1040–1055. [CrossRef]
32. Song, W.; Xu, Q.; Fu, X.; Zhang, P.; Pang, Y.; Song, D. Research on the Relationship between Water Diversion and Water Quality of Xuanwu Lake, China. *Int. J. Environ. Res. Public Health* **2018**, *15*, 1262. [CrossRef]
33. Cai, Z.; Han, G.; Chen, M. Do water bodies play an important role in the relationship between urban form and land surface temperature? *Sustain. Cities Soc.* **2018**, *39*, 487–498. [CrossRef]
34. Xu, S.; Liu, Y.; Qian, Y.; Wang, Q. The significance of the West Lake pattern and its heuristic implications for creating China's heritage tourism economics. *Tour. Manag.* **2017**, *58*, 286–292. [CrossRef]
35. Hangzhou Municipal Bureau of Statistics. An Overview of Hangzhou. Available online: http://tjj.hangzhou.gov.cn/ (accessed on 26 January 2021).
36. Nanjing Municipal Bureau of Statistics. Nanjing Statistical Yearbook 2020. Available online: http://tjj.nanjing.gov.cn/material/njnj_2020/ (accessed on 26 January 2021).
37. Wang, L.; Hou, H.; Weng, J. Ordinary least squares modelling of urban heat island intensity based on landscape composition and configuration: A comparative study among three megacities along the Yangtze River. *Sustain. Cities Soc.* **2020**, *62*, 102381. [CrossRef]
38. Gascon, F.; Bouzinac, C.; Thépaut, O.; Jung, M.; Francesconi, B.; Louis, J.; Lonjou, V.; Lafrance, B.; Massera, S.; Gaudel-Vacaresse, A.; et al. Copernicus Sentinel-2A Calibration and Products Validation Status. *Remote Sens.* **2017**, *9*, 584. [CrossRef]
39. Shao, Z.; Cai, J.; Fu, P.; Hu, L.; Liu, T. Deep learning-based fusion of Landsat-8 and Sentinel-2 images for a harmonized surface reflectance product. *Remote Sens. Environ.* **2019**, *235*, 111425. [CrossRef]
40. Fernandez-Beltran, R.; Baidar, T.; Kang, J.; Pla, F. Rice-Yield Prediction with Multi-Temporal Sentinel-2 Data and 3D CNN: A Case Study in Nepal. *Remote Sens.* **2021**, *13*, 1391. [CrossRef]
41. Persson, H.J.; Jonzén, J.; Nilsson, M. Combining TanDEM-X and Sentinel-2 for large-area species-wise prediction of forest biomass and volume. *Int. J. Appl. Earth Obs. Geoinf.* **2021**, *96*, 102275. [CrossRef]
42. Sobrino, J.A.; Jiménez-Muñoz, J.C.; Paolini, L. Land surface temperature retrieval from LANDSAT TM 5. *Remote Sens. Environ.* **2004**, *90*, 434–440. [CrossRef]
43. Windahl, E.; Beurs, K.d. An intercomparison of Landsat land surface temperature retrieval methods under variable atmospheric conditions using in situ skin temperature. *Int. J. Appl. Earth Obs. Geoinf.* **2016**, *51*, 11–27. [CrossRef]
44. Chander, G.; Markham, B.L.; Helder, D.L. Summary of current radiometric calibration coefficients for Landsat MSS, TM, ETM+, and EO-1 ALI sensors. *Remote Sens. Environ.* **2009**, *113*, 893–903. [CrossRef]
45. Qin, Z.; Li, W.; Xu, B.; Chen, Z.; Liu, J. The estimation of land surface emissivity for Landsat TM6. *Remote Sens. Land Resour.* **2004**, *16*, 28–32,36,41. [CrossRef]
46. Carlson, T.N.; Ripley, D.A. On the relation between NDVI, fractional vegetation cover, and leaf area index. *Remote Sens. Environ.* **1997**, *62*, 241–252. [CrossRef]
47. Becker, F.; Choudhury, B.J. Relative sensitivity of normalized difference vegetation Index (NDVI) and microwave polarization difference Index (MPDI) for vegetation and desertification monitoring. *Remote Sens. Environ.* **1988**, *24*, 297–311. [CrossRef]
48. Cadenasso, M.L.; Pickett, S.T.A.; Schwarz, K. Spatial heterogeneity in urban ecosystems: Reconceptualizing land cover and a framework for classification. *Front. Ecol. Environ.* **2007**, *5*, 80–88. [CrossRef]
49. Congalton, R.G. A review of assessing the accuracy of classifications of remotely sensed data. *Remote Sens. Environ.* **1991**, *37*, 35–46. [CrossRef]
50. O'Neill, R.V.; Krummel, J.R.; Gardner, R.H.; Sugihara, G.; Jackson, B.; Deangelis, D.L.; Milne, B.T.; Turner, M.G.; Zygmunt, B.; Christensen, S.W.; et al. Indices of landscape pattern. *Landsc. Ecol.* **1988**, *1*, 153–162. [CrossRef]
51. Chen, A.; Yao, L.; Sun, R.; Chen, L. How many metrics are required to identify the effects of the landscape pattern on land surface temperature? *Ecol. Indic.* **2014**, *45*, 424–433. [CrossRef]
52. Estoque, R.C.; Murayama, Y.; Myint, S.W. Effects of landscape composition and pattern on land surface temperature: An urban heat island study in the megacities of Southeast Asia. *Sci. Total Environ.* **2017**, *577*, 349–359. [CrossRef]
53. Yao, L.; Li, T.; Xu, M.; Xu, Y. How the landscape features of urban green space impact seasonal land surface temperatures at a city-block-scale: An urban heat island study in Beijing, China. *Urban For. Urban Green.* **2020**, *52*, 126704. [CrossRef]
54. Zhou, W.; Cao, F. Effects of changing spatial extent on the relationship between urban forest patterns and land surface temperature. *Ecol. Indic.* **2020**, *109*, 105778. [CrossRef]

55. Maimaitiyiming, M.; Ghulam, A.; Tiyip, T.; Pla, F.; Latorre-Carmona, P.; Halik, Ü.; Sawut, M.; Caetano, M. Effects of green space spatial pattern on land surface temperature: Implications for sustainable urban planning and climate change adaptation. *ISPRS J. Photogramm. Remote Sens.* **2014**, *89*, 59–66. [CrossRef]
56. Zhou, W.; Wang, J.; Cadenasso, M.L. Effects of the spatial configuration of trees on urban heat mitigation: A comparative study. *Remote Sens. Environ.* **2017**, *195*, 1–12. [CrossRef]
57. Bartesaghi-Koc, C.; Osmond, P.; Peters, A. Spatio-temporal patterns in green infrastructure as driver of land surface temperature variability: The case of Sydney. *Int. J. Appl. Earth Obs. Geoinf.* **2019**, *83*, 101903. [CrossRef]
58. Masoudi, M.; Tan, P.Y.; Fadaei, M. The effects of land use on spatial pattern of urban green spaces and their cooling ability. *Urban Clim.* **2021**, *35*, 100743. [CrossRef]
59. Masoudi, M.; Tan, P.Y. Multi-year comparison of the effects of spatial pattern of urban green spaces on urban land surface temperature. *Landsc. Urban Plan.* **2019**, *184*, 44–58. [CrossRef]
60. McGarigal, K.; Cushman, S.A.; Neel, M.C.; Ene, E. FRAGSTATS: Spatial Pattern Analysis Program for Categorical Maps. Available online: http://www.umass.edu/landeco/research/fragstats/fragstats.html (accessed on 26 January 2021).
61. Gupta, N.; Mathew, A.; Khandelwal, S. Analysis of cooling effect of water bodies on land surface temperature in nearby region: A case study of Ahmedabad and Chandigarh cities in India. *Egypt. J. Remote Sens. Space Sci.* **2019**, *22*, 81–93. [CrossRef]
62. Pardoe, I. Multiple Linear Regression. In *Applied Regression Modeling*, 3rd ed.; Pardoe, I., Ed.; John Wiley & Sons, Inc: Hoboken, NJ, USA, 2020; Volume 3, pp. 96–100.
63. Franklin, J.B.; Sathish, T.; Vinithkumar, N.V.; Kirubagaran, R. A novel approach to predict chlorophyll-a in coastal-marine ecosystems using multiple linear regression and principal component scores. *Mar. Pollut. Bull.* **2020**, *152*, 110902. [CrossRef]
64. Yuan, D.; Elvidge, C.D. Comparison of relative radiometric normalization techniques. *ISPRS J. Photogramm. Remote Sens.* **1996**, *51*, 117–126. [CrossRef]
65. Sun, R.; Chen, L. How can urban water bodies be designed for climate adaptation? *Lands. Urban Plan.* **2012**, *105*, 27–33. [CrossRef]
66. Cheval, S.; Popa, A.-M.; Șandric, I.; Iojă, I.-C. Exploratory analysis of cooling effect of urban lakes on land surface temperature in Bucharest (Romania) using Landsat imagery. *Urban. Clim.* **2020**, *34*, 100696. [CrossRef]
67. Zhang, Q.; Wen, Y.; Wu, Z.; Chen, Y. Seasonal Variations of the Cooling Effect of Water Landscape in High-density Urban Built-up Area: A Case Study of the Center Urban District of Guangzhou. *Ecol. Environ. Sci.* **2018**, *27*, 1323–1334. [CrossRef]
68. Athukorala, D.; Murayama, Y. Spatial Variation of Land Use/Cover Composition and Impact on Surface Urban Heat Island in a Tropical Sub-Saharan City of Accra, Ghana. *Sustainability* **2020**, *12*, 7953. [CrossRef]
69. Liu, F.; Hou, H.; Murayama, Y. Spatial Interconnections of Land Surface Temperatures with Land Cover/Use: A Case Study of Tokyo. *Remote Sens.* **2021**, *13*, 610. [CrossRef]
70. Yu, B.; He, X.; Wei, L.; Chen, L.; Zhou, W. Primary exploration for construction of urban multilevel ventilation corridors system in Hangzhou. *J. Meteorol. Sci.* **2018**, *38*, 625–636. [CrossRef]
71. Zhang, Y.; Rui, D.; Zhou, L. Nanjing Zhongshan Mountain Scenic's Influence on the Air Quality in Main Urban Area. *Environ. Monit. Forewarning* **2010**, *2*, 47–50. [CrossRef]
72. Fei, J.; Chen, W.; Bao, Q.; Zhang, W. Evolution Characteristics and Multi-Value of Scenic Byway of the West Lake Mountain Forest in Hangzhou. *J. Chin. Urban For.* **2020**, *18*, 98–102. [CrossRef]
73. Jenerette, G.D.; Harlan, S.L.; Brazel, A.; Jones, N.; Larsen, L.; Stefanov, W.L. Regional relationships between surface temperature, vegetation, and human settlement in a rapidly urbanizing ecosystem. *Lands. Ecol.* **2007**, *22*, 353–365. [CrossRef]
74. Soydan, O. Effects of landscape composition and patterns on land surface temperature: Urban heat island case study for Nigde, Turkey. *Urban. Clim.* **2020**, *34*, 100688. [CrossRef]
75. Liu, Y.; Huang, X.; Yang, Q.; Cao, Y. The turning point between urban vegetation and artificial surfaces for their competitive effect on land surface temperature. *J.Clean. Prod.* **2021**, *292*, 126034. [CrossRef]
76. Kong, F.; Yin, H.; James, P.; Hutyra, L.R.; He, H.S. Effects of spatial pattern of greenspace on urban cooling in a large metropolitan area of eastern China. *Lands. Urban Plan.* **2014**, *128*, 35–47. [CrossRef]
77. Fan, H.; Yu, Z.; Yang, G.; Liu, T.Y.; Liu, T.Y.; Hung, C.H.; Vejre, H. How to cool hot-humid (Asian) cities with urban trees? An optimal landscape size perspective. *Agric. For. Meteorol.* **2019**, *265*, 338–348. [CrossRef]
78. Wang, R.; Hou, H.; Murayama, Y.; Derdouri, A. Spatiotemporal Analysis of Land Use/Cover Patterns and Their Relationship with Land Surface Temperature in Nanjing, China. *Remote Sens.* **2020**, *12*, 440. [CrossRef]

remote sensing

Article

Exploring the Spatio-Temporal Characteristics of Urban Thermal Environment during Hot Summer Days: A Case Study of Wuhan, China

Weifang Shi, Jiaqi Hou, Xiaoqian Shen and Rongbiao Xiang *

College of Resources and Environment, Huazhong Agricultural University, Wuhan 430070, China
* Correspondence: xiangrb@mail.hzau.edu.cn; Tel.: +86-27-87282027

Abstract: An urban thermal environment is an area receiving special attention. In order to effectively explore its spatio-temporal characteristics during hot summer days, this study introduced the standard deviational ellipse (SDE) to construct an urban heat island index to describe the general spatial character of an urban thermal environment, and then used local Moran's I to identify its local spatial cluster characteristics. Finally, the regressions of ordinary least squares (OLS) and spatial lag model (SLM) were adopted to explore the effect of woodland, water body and impervious surface on the thermal environment. Taking the city of Wuhan as a study area and using the air temperature on seven consecutive days, from 17 July to 23 July in 2018, from the China Meteorological Administration Land Data Assimilation System (CLDAS-V2.0), the results show that the urban heat island index can effectively represent the general characteristics of the thermal environment. The general trends of heat island intensity decrease first and then increase from 00:00 to 24:00. The heat island intensity is at its minimum from 10:00 to 16:00, and at its maximum from 22:00 to 4:00 the next day. Local Moran's I values indicate that the clusters of high air temperature at 06:00 and at 22:00 are associated with the impervious surface and the water body. This is further illustrated by the regression analysis of OLS, which can explain 50–60% of the spatial variation of the air temperature. Then, the fitness of the SLM is greatly improved; the coefficients of determination at 06:00 and at 22:00 are all not less than 0.97. However, the explanation of the local land uses accounting for the spatial variation of the air temperature becomes lower. The regression analysis also shows that the woodland always has the effect of decreasing air temperature at 06:00, 14:00 and 22:00, implying that increasing the vegetation may be the most effective way to mitigate the adverse circumstance of the urban thermal environment.

Keywords: thermal environment; heat island intensity; spatio-temporal characteristics; local spatial pattern; land use; regression analysis

Citation: Shi, W.; Hou, J.; Shen, X.; Xiang, R. Exploring the Spatio-Temporal Characteristics of Urban Thermal Environment during Hot Summer Days: A Case Study of Wuhan, China. *Remote Sens.* **2022**, *14*, 6084. https://doi.org/10.3390/rs14236084

Academic Editors: Yuji Murayama and Ruci Wang

Received: 31 October 2022
Accepted: 26 November 2022
Published: 30 November 2022

1. Introduction

Human activities, especially urbanization, have gradually changed the Earth's surface landforms, and then caused changes in the urban thermal environments. Urban heat island (UHI) effect, a phenomenon according to which urban areas have higher air temperatures than the surrounding rural areas, is a prominent problem of the urban thermal environment. The effect of UHI on heat wave intensifies during the period of prolonged extreme heat [1], and the synergy [2] between the two even causes an increased heat stress that is higher than the combined effect of the background UHI and heat waves [3], which may increase the mortality related to high temperatures in urban areas [4] and have a profound impact on the human living environments [5].

Land cover change through urban sprawl and cultivated land degradation has a dramatic effect on UHI intensity [6], as the city with a higher urban land cover usually has the higher UHI intensity [7]. Moreover, the urban radiant heat is varied and depends largely on the type of land cover and urban surface [8]. Different land uses have different effects on

UHI [9]. The lack of vegetation may contribute to higher temperatures in industrial parks, while the temperatures go up by up to 2 °C in commercial districts on weekends [10]. It is found that the impervious surface area (ISA) is the most significant factor causing the increasing temperature [11], while the land surface temperature (LST) generally increases in parallel with the ISA [12]. Then, the urban green space and its spatial configurations, such as larger patches, simpler shapes, being more connected and less fragmented, have a significant effect on the LST [13]; in particular, they contribute to a lower LST [14,15].

Effective water resources management is also of great significance to reduce the heat island effect [16]. Studies have found that the intensity of the UHI effect decreases by 11.33% with the increase in water-body coverage by 10% during the daytime [17]. However, the water will increase the UHI effect after the evening, for its high heat capacity inhibits its circulation and makes it remain at relatively high temperatures [18].

There are many types of UHIs [19], of which the most commonly used are surface UHI (SUHI), based on LST, and canopy UHI (CUHI), based on air temperature within the urban canopy layer. These two types of UHI depict the thermal environment characteristics in different vertical layers of the urban structure. The approaches to measure the two types of UHI both have advantages and disadvantages [20]. Most of the studies focus on SUHIs because the LST data are easy to obtain from remote sensing, but the uncertainty of the LST is too great to obtain reliable urban thermal trends [21]. Moreover, the CUHIs described by using air temperature, which is obtained through meteorological field measurements, is more relevant to public health [22] and is a key parameter for studying the UHI effect [23].

Although there are a large number of studies focusing on UHI, some problems with respect to it are still ambiguous. The UHI intensity is conventionally defined as the difference in temperature between an urban and rural area [24]. The traditional detection of intensity is conducted at two fixed in situ stations in urban and rural regions [25]. Similarly, the calculation of the surface UHI is conducted over selected pixels located in the urban and rural regions based on remote sensing data [26]. The estimations both depend on the determination of urban and rural stations or pixels [23,27]. However, the air temperatures of the urban area are heterogeneous in space, and the choice of a rural area also affects the quantification. There are some modified methods to improve the estimation [28–32], such as calculating the average temperature of the corresponding areas, but the spatial characteristics of the urban thermal environment are still not good enough to present. In addition, the studies on the evolution of the spatio-temporal characteristics of the urban thermal environment during a day are scarce.

Thus, this study focuses on the air temperature and introduces the standard deviational ellipse (SDE) to explore the general spatial distribution of an urban thermal environment. We then further construct an UHI index to characterize UHI intensity. This index is not just determined by the air temperatures of two points but with the purpose of effectively reflecting the general spatial character of the urban thermal environment. Taking the city of Wuhan as a study area, and using the air temperature data of seven consecutive days with sunny and cloudy weather conditions in the hot summer of 2018, we first analyzed the general spatial thermal character and calculated the UHI index every two hours during the seven days. Then, we chose three time points in each day and used local Moran's I to explore the local spatial pattern of the thermal environment in the study area. Furthermore, the regression analysis including ordinary least squares (OLS) and the spatial lag model (SLM) are adopted to explore the land uses that contribute to the spatial variation of air temperature. The efforts of this work are aimed at trying to provide a method to characterize the spatial distribution of the thermal environment and the UHI effect more effectively as well as to illustrate their evolutions with the high temporal resolution during a day.

2. Materials and Methods

2.1. Study Area and Data Source

Wuhan, the capital of Hubei Province, is the city with the largest population in Central China, between the latitudes 29°58′19.04″N and 31°21′44.07″N and between the longitudes 113°41′32.86″E and 115°4′55.70″E, covering a total area of approximately 8624.69 km^2 (Figure 1a), and with an average elevation of about 37 m above sea level. The city possesses the distinctive character of having a large number of water bodies (Figure 1b), which is why it was once called "the city of a thousand lakes". According to Köppen climate classification, the study area has a Cfa climate with abundant rainfall, cool winters and hot summers. As one of the Four Furnaces of China, Wuhan is always suffering heat waves and high air temperatures in the summertime.

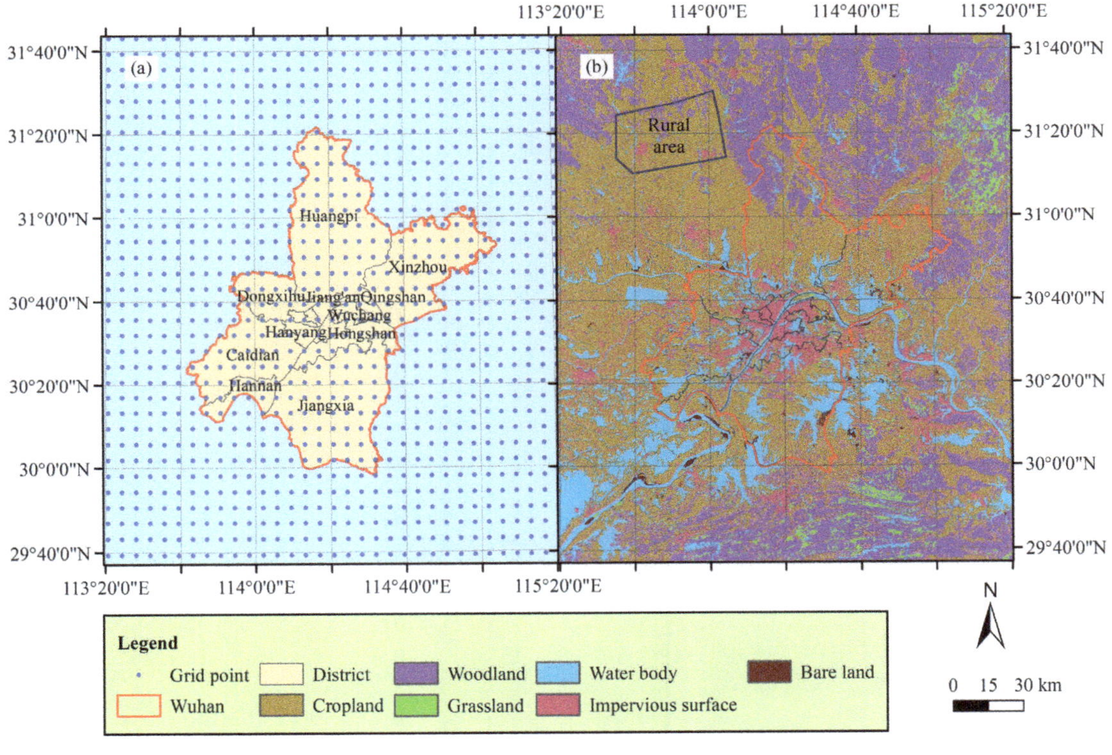

Figure 1. Study area. (**a**) Grid points of the air temperature; (**b**) Land uses.

In this study, we focused on the days that are the hottest in the summer. Additionally, both sunny and cloudy days are included in our analysis to compare the differences in thermal environmental characteristics between the two weather conditions. Thus, we chose the days from 17 July to 23 July in 2018 as the study period. The weather of the middle three days in the study area was sunny, while for the four other days it was cloudy, which is illustrated in Table 1. The air temperature (2 m above the land surface) of these seven days was obtained from the China Meteorological Administration (CMA) Land Data Assimilation System (CLDAS) version 2.0 dataset, called "CLDAS-V2.0" dataset, in China Meteorological Data Service Centre. The "CLDAS-V2.0" dataset is a grid fusion analysis dataset with the resolution of 0.0625° × 0.0625° and 1 h. There are 207 grid points in the city of Wuhan; the distance between two adjacent grid points in the east–west direction is about 6.0 km and that in the north–south direction is about 6.9 km (Figure 1a). The dataset is developed by using a multiple-grid variational assimilation called Space and Time Mesoscale Analysis System (STMAS), along with physical inversion, terrain correction based on the ground, satellite observations from various sources and CLM3.5, CoLM, Noah-MP land surface model. Evaluated by 2400 in situ national automatic stations of the China Meteorological Administration (CMA), the RMSE of the 2 m air temperature of the dataset is 0.88 °C; the bias is −0.13 °C; and the correlation coefficient is 0.97 (http://data.cma.cn/en/?r=data/detail&dataCode=NAFP_CLDAS2.0_NRT, accessed on 16 November 2020). Studies evaluating the near-surface air temperature of CLDAS also show that it has a high reliability in China and is significantly correlated with the observations [33,34]. The dataset has been widely used in various studies [35–38].

Table 1. Weather conditions of the study area in the study period.

Date	Weather Condition
17 July	Cloudy
18 July	Cloudy
19 July	Sunny
20 July	Sunny
21 July	Sunny
22 July	Cloudy
23 July	Cloudy

The land use data of the study area are from Resource and Environment Science and Data Center, Chinese Academy of Sciences (https://www.resdc.cn, accessed on 17 October 2021). The data include 6 major categories of cropland, woodland, water body, impervious surface and bare land as well as 25 secondary classes, with a resolution of 30 m by 30 m (Figure 1b). The overall accuracy of the data is about 95% for 25 secondary land use classes [39].

2.2. Standard Deviational Ellipse and Urban Heat Island Index

The standard deviational ellipse (SDE) is a technique to explore spatial distributions. It was first introduced in study of geographic density in sociology, such as population distribution [40], and is now widely used to estimate spatial distribution evolutions and trends [41,42]. We employed this technique in this study. Moreover, we further developed a new index based on it.

The analysis of the SDE includes weighted SDE and non-weighted SDE. The steps of calculating the weighted SDE are as follows. The first is to calculate the weighted mean center of the point set, which is given by:

$$\overline{x}_w = \frac{\sum_{i=1}^{n} w_i x_i}{\sum_{i=1}^{n} w_i}, \overline{y}_w = \frac{\sum_{i=1}^{n} w_i y_i}{\sum_{i=1}^{n} w_i}, \quad (1)$$

where x_i and y_i are the coordinates of the ith point; n is the total number of the points; w_i is the weight of the ith point. Thereafter, the transformed coordinates, which are the different metric spaces between the ith point and the weighted mean center, are given as follows:

$$x'_i = x_i - \overline{x}_w, y'_i = y_i - \overline{y}_w. \quad (2)$$

Then, the angle of rotation, measured clockwise from North to the long axis of the ellipse, can be calculated by:

$$\tan\theta = \frac{\left(\sum_{i=1}^{n} w_i^2 x_i'^2 - \sum_{i=1}^{n} w_i^2 y_i'^2\right) + \sqrt{\left(\sum_{i=1}^{n} w_i^2 x_i'^2 - \sum_{i=1}^{n} w_i^2 y_i'^2\right)^2 + 4\left(\sum_{i=1}^{n} w_i^2 x'_i y'_i\right)^2}}{2\sum_{i=1}^{n} w_i^2 x'_i y'_i}. \quad (3)$$

Finally, the standard distances of the ellipse on x and y directions can be produced by:

$$\sigma_x = \sqrt{\frac{\sum_{i=1}^{n} \left(w_i x_i' \cos\theta - w_i y_i' \cos\theta\right)^2}{\sum_{i=1}^{n} w_i^2}}, \quad (4)$$

$$\sigma_y = \sqrt{\frac{\sum_{i=1}^{n} \left(w_i x_i' \cos\theta + w_i y_i' \cos\theta\right)^2}{\sum_{i=1}^{n} w_i^2}}. \quad (5)$$

When $w_i = 1$ in the above equations, the results of the weighted SDE are the same as the non-weighted SDE. In this study, we use the weighted SDE to explore the spatial distribution of air temperature and combine its results with those of the non-weighted SDE for further analysis. Evidently, the weight of a point relates to the air's temperature. However, if the weight is set equal to the air temperature of the point, the weight discrimination is very small for the minor difference between the air temperature of the maximum and the minimum in the study area, which is about 5 °C or even less, whereas the air temperature of the point is generally around 30 °C. Therefore, we introduce the air temperature of an area identified as rural and use it combined with the air temperature of the grid point to construct the weight. Since the weight cannot be negative, the functional form of the weight is constructed as:

$$w_i = f(Ta_i, Ta_r) = \exp(Ta_i - Ta_r), \quad (6)$$

where Ta_i is the air temperature (°C) of the ith point and Ta_r indicates the air temperature (°C) of the rural area. Therefore, what an area is considered as rural, distinguished from the urban area, is a key point. We first analyzed the air temperature isotherm and the air temperature profile of an extended area covering the city of Wuhan. Then, based on combined analysis of air temperature distribution and land use configuration, we identified a typical area as rural. Thereafter, we calculated the mean value of the air temperature with respect to the grid points in the typical rural area and used this mean value as the Ta_r. Since a higher w_i is obtained with a higher Ta_i, it is evident that the higher the Ta_i near the center of the study area, the smaller the standard distances of the ellipse in Equations (4) and (5). This suggests that the smaller the area of the weighted SDE, the more intensive the heat island effect. As for the non-weighted SDE, it is equivalent to considering all the grid points to have equal weights; that is to say, all the grid points with the same air temperature, representing the state of having no heat island effect in the study area. Therefore, we construct an UHI index that reflect the heat island intensity, as follows:

$$I = \frac{S_{SDE-nonW}}{S_{SDE-W}} = \frac{\pi\sigma_{x-nonW}\sigma_{y-nonW}}{\pi\sigma_{x-W}\sigma_{y-W}} = \frac{\sigma_{x-nonW}\sigma_{y-nonW}}{\sigma_{x-W}\sigma_{y-W}}, \quad (7)$$

where $S_{SDE-nonW}$, σ_{x-nonW} and σ_{y-nonW} are the area and standard deviations on x and y directions of the non-weighted SDE, respectively; S_{SDE-W}, σ_{x-W} and σ_{y-W} are those of the weighted SDE. When the value of this index is larger than 1, this indicates that the corresponding area presents the state of the heat island effect. Moreover, the weaker the heat island effect of the study area, the closer the value of this index is to 1, and the more intensive the heat island effect, the larger the value of the index. Conversely, a value of the UHI index less than 1 suggests that the weights of the grid points near the center are lesser than those far from the center. Since the w_i monotonically increases with the Ta_i base on Equation (6), the air temperature has the same spatial distribution trend as the weights, therefore representing the state of cool island.

2.3. Spatial Pattern Analysis

Spatial autocorrelation refers to the potential interdependence of observed data of some variables within the same distribution area. Moran's I is a commonly used spatial autocorrelation statistic to describe the concentration or dispersion of variables in space. Moran's I can be divided into global Moran's I and local Moran's I. Global Moran's I shows whether there exists aggregation or outliers in a space, and is then used for exploring the global correlation between regions. The local Moran's I [43], a statistic to measure the local spatial autocorrelation index, is used to reflect the characteristics of agglomeration and differentiation of related variables in a small-scale spatial structure, which is used in this study to analyze the spatial pattern of the air temperature of the study area. It can be expressed as follows:

$$I_i = Z_i W_{Z_i}, \tag{8}$$

where

$$Z_i = \frac{Ta_i - \overline{Ta}}{S}, \tag{9}$$

$$W_{Z_i} = \sum_{j=1, j \neq i}^{n} w_{ij}^* Z_j = \sum_{j=1, j \neq i}^{n} w_{ij}^* \left(\frac{Ta_j - \overline{Ta}}{S} \right), \tag{10}$$

where Ta_i is the air temperature of grid point I; $\overline{Ta}$ and S are the mean and the standard deviation of the air temperatures of all grid points in the study area, respectively; n is the total number of the grid points; and $w_{ij}{}^*$ is a normalized spatial weight of point j with respect to point i, derived from a non-normalized spatial weight. In this study, the non-normalized spatial weight is defined as the inverse of the distance squared, which is given by:

$$w_{ij} = \begin{cases} 1/d_{ij}^2, & d_{ij} \leq b \\ 0, & d_{ij} > b \end{cases}, \tag{11}$$

where d_{ij} is the distance between grid point i and j; b is a distance bandwidth. In this study, b is specified with the value of 11 km, which implies that only the grid points surrounding the grid point i have weights larger than 0, with other grid points given the weights of 0. Usually, there are 8 points surrounding a grid point, but less than 8 points for each of the points at the edge of the study area. Thereafter, the normalized spatial weight is given by:

$$w_{ij}^* = w_{ij} / \sum_{j=1, j \neq i}^{n} w_{ij}. \tag{12}$$

In Equations (8)–(10), Z_i describes the air temperature of point i in deviations from the mean of air temperature of the study area and W_{Zi} represents the weighted summation of air temperature deviations of the grid points surrounding point i from the mean of air temperature. Through the analysis of local Moran's I, the patterns of spatial clusters of grid points and the outliers with a statistically significance can be obtained. If $Z_i > 0$ and $W_{Zi} > 0$, this means the grid point and the outliers have high values (relative to the mean value), called the cluster of high values (HH). If $Z_i < 0$ and $W_{Zi} < 0$, this means the grid point and the outliers have low values (relative to the mean value), called the cluster of low values (LL). The other two patterns are the grid point having a high value and surrounded by grid points with low values (HL), with $Z_i > 0$ and $W_{Zi} < 0$, and the grid point having a low value and surrounded by grid points with high values (LH), with $Z_i < 0$ and $W_{Zi} > 0$.

2.4. Regression Analysis

Linear regression analysis is a common method to analyze the relationship of a dependent variable driven by other factors. In this study, the dependent variable is the air temperature of the grid point, and the independent variables are the proportions of the land uses in the area with respect to the Thiessen polygon of each grid point. The ordinary least squares (OLS) model, involving a global multiple linear regression (MLR) in this study, is formed as:

$$\boldsymbol{Ta} = \boldsymbol{X\beta} + \boldsymbol{\varepsilon}, \tag{13}$$

where $\boldsymbol{Ta}$ is a vector with respect to the air temperature of the grid points of the study area; $\boldsymbol{X}$ is a matrix containing intercept (a vector of ones with the length equal to the total number of the grid points) and values of land use fractions of the Thiessen polygon corresponding to each grid point; $\boldsymbol{\beta}$ is a vector of the regression coefficients; and $\boldsymbol{\varepsilon}$ is a vector of random errors. Additionally, stepwise regression technology is used to select the proper explanatory variables from the proportions of the six land use types illustrated in Figure 1b.

Usually, there are spatial autocorrelations with geographical variables. Thus, spatial regression models may be more effective in explaining the relationship of the variables compared with the OLS model. Spatial lag model (SLM) and spatial error model (SEM) are two common types of spatial regression model [44]. Since SLM is mostly suitable to deal with the problem associated with spatial diffusion, a characteristic of heat, we adopt SLM to further analyze air temperature with driving factors. The SLM adds an autocorrelation term of the air temperature of the grid points to the right of Equation (13). Additionally, the driving factors that we use for the model are the final selected explanatory variables based on the stepwise regression analysis. The model is described as follows:

$$\boldsymbol{Ta} = \rho \boldsymbol{WTa} + \boldsymbol{X\beta} + \boldsymbol{\varepsilon}, \tag{14}$$

where ρ is the parameter characterizing the contribution of spatial autocorrelation; $\boldsymbol{W}$ is a row-sum standardized spatial weight matrix with the elements of $w_{ii}{}^* = 0$ and $w_{ij}{}^*$ calculated by Equation (12); and $\boldsymbol{WTa}$ represents the spatially lagged response of the air temperature of the grid points.

3. Results

3.1. General Spatio-Temporal Distribution of the Thermal Environment

We calculated the mean air temperature of the grid points in an extended area including the city of Wuhan during the nighttime (20:31 the day before to 06:30) and the daytime (06:31 to 20:30) from 17 July to 23 July. We then plotted the air temperature isotherms, which are illustrated in Figure 2. It can be seen that the islands of air temperature protrude above the surrounding air temperature field both during nighttime and daytime, revealing that heat islands are present in the city of Wuhan. The center of the heat island is slightly southwest of the city center during the nighttime, and slightly west of the city center during the daytime. The northern region is relatively cool compared to the entire city. Furthermore, we analyzed the air temperature profile along the line of AB in Figure 2. The profiles of the air temperature are shown in Figure 3. Based on the combined analysis of air temperature distribution and land use configuration, we chose a typical rural area located to the northwest of the city of Wuhan, illustrated in Figure 2. The air temperature of the designated rural area is flat, which means that the change rate is small, as seen from the profiles (Figure 3), and the air temperature gradient on the designated rural area boundary toward the city center is uphill and increased. On the other hand, as seen in Figure 1b, the designated rural area is dominated by the cropland and contains some woodland and a small amount of impervious surface. All these temperature and land use characteristics are typical of rural areas. Therefore, we used the mean air temperature of the grid points in this area to represent the rural air temperature.

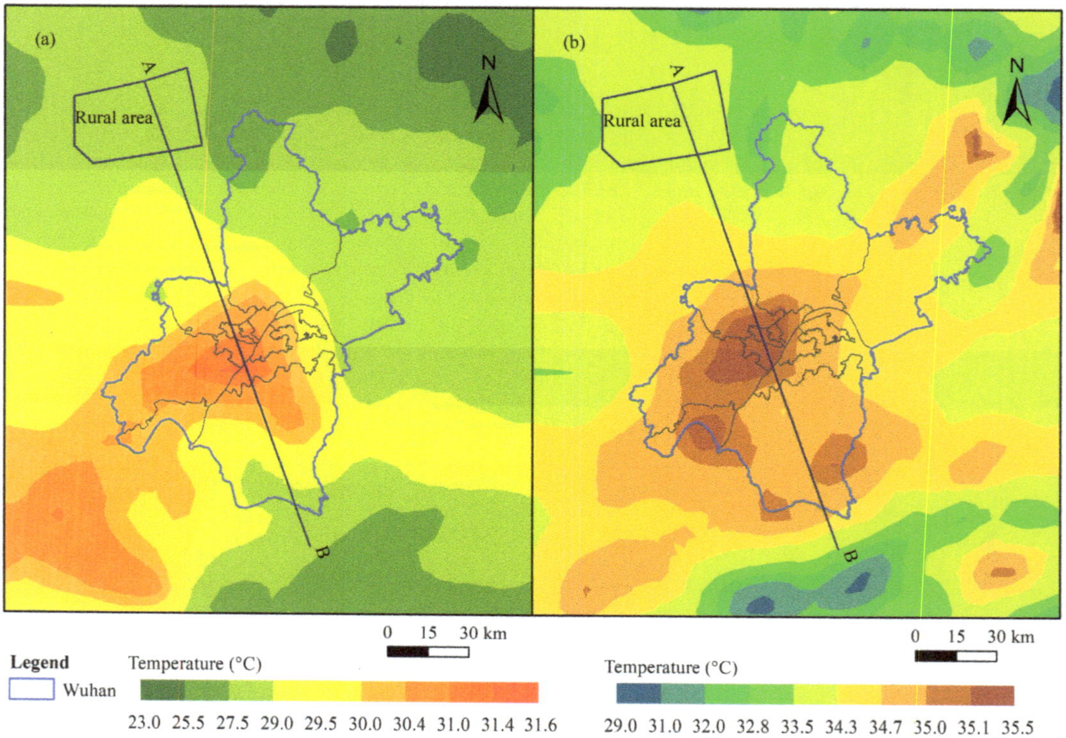

Figure 2. The mean air temperature isotherm of an extended area including the city of Wuhan from 17 July to 23 July. (**a**) Nighttime; (**b**) daytime.

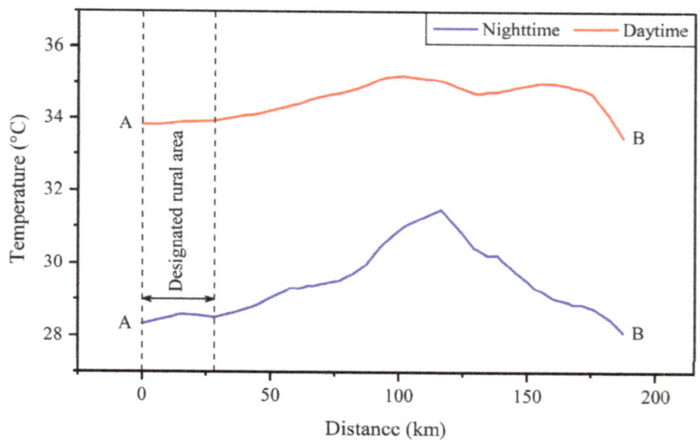

Figure 3. The mean air temperature profiles along line AB.

After determining the rural air temperature, the weights of the grid points were obtained according to Equation (6). Then, we took the 207 grid points in the city of Wuhan for the SDE analysis. Furthermore, based on Equation (7), the values of the UHI index for every two hours from 00:00 on 17 July to 00:00 on 24 July were derived, which are shown in Figure 4. The trends of the UHI index within each day are very similar, generally decreasing from 04:00 to 10:00 and increasing from 16:00 to 22:00. The UHI index minimum of each day occurs between 10:00 and 16:00, and the maximum occurs between 22:00 and 04:00 the next day. Additionally, the difference between the UHI index minimum of these seven days is very small. However, the peak values of the UHI index curve from the evening of 19 July to the early morning of 22 July are obviously higher than those of the other days. The weather from 19 July to 21 July is sunny, while the other days it is cloudy, which indicates that sunny days enhance the heat island effect at night compared to cloudy days, but the heat island effect with respect to the two weather conditions during the day is not much different.

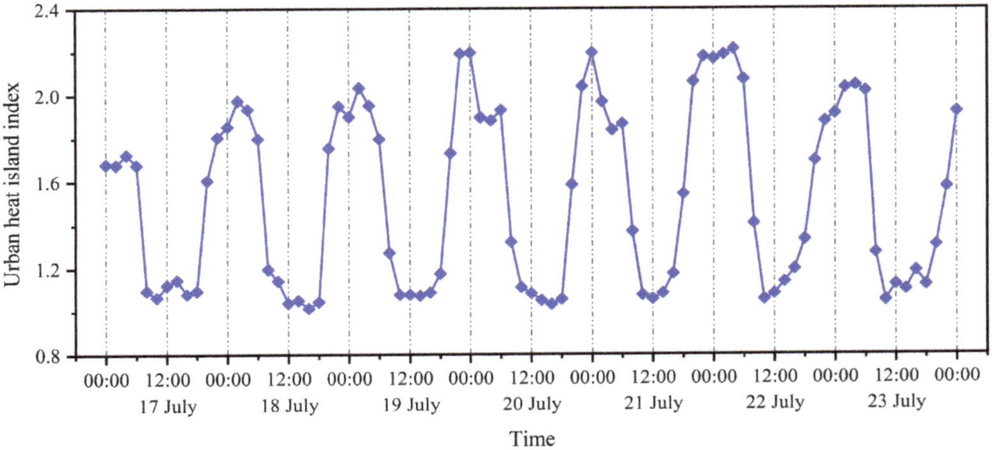

Figure 4. UHI index of the city of Wuhan in the summer of 2018.

The non-weighted SDEs and the weighted SDEs at 06:00, 14:00 and 22:00 of the seven days are illustrated in Figure 5. The center of the non-weighted SDEs represents the geographical center of the study area, and that of the weighted SDEs represents the center of the thermal field. It is evident that the centers of the thermal fields in the morning and at night are southwest of the geographical center of the city. Moreover, the shape of the weighted SDEs can reflect the dispersion of air temperature. Similar to the spatially normal distribution, the ellipse of the one standard deviation encompasses approximately 68% of the cumulate air temperature weights (w_i in Equation (6)) in the area. Except for 21 and 23 July, the weighted SDEs at 06:00 and 22:00 each day are close in shape, position and size. This illustrates that the spatial distributions of the thermal characteristics at 06:00 and at 22:00 are very similar. Likewise, the weighted SDE at 14:00 and the non-weighted SDEs are close in shape, position and size except for 17 and 22 July. Since the non-weighted SDE is considered as having no heat island effect, the results illustrate that the heat islands at 14:00 are weak. On the contrary, as shown in Figure 5, the distinction of the weighted SDEs at 06:00 and at 22:00 from the non-weighted SDEs are broad, especially in size and position, indicating that the heat island effect is intense at these two time points.

3.2. Local Spatial Pattern of the Thermal Environment

A local spatial analysis of the air temperature distribution is performed at 06:00, 14:00 and 22:00 on the seven days. Identified by calculating local Moran's I values, the spatial clusters and the outliers of the grid points at the three time points are, respectively, shown in Figures 6–8. Only the HH and LL clusters are of statistical significance in the study area at the three time points on all the seven days. The areas, except for the HH and LL clusters, are of no significance. These results indicate that the air temperature distribution has similarly high or low values in neighboring zones; there is no grid point with high air temperature surrounded by grid points with low air temperature nor grid point with low air temperature surrounded by grid points with high air temperature. This also implies that the air temperature of a grid point is affected by the surrounding points.

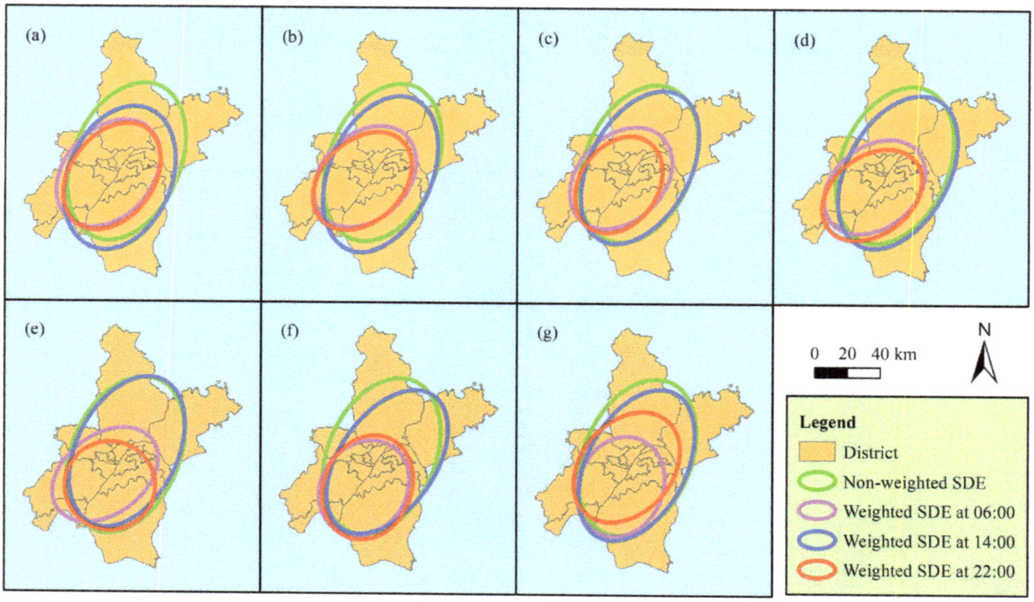

Figure 5. The SDEs of the city of Wuhan. (**a**) 17 July; (**b**) 18 July; (**c**) 19 July; (**d**) 20 July; (**e**) 21 July; (**f**) 22 July; (**g**) 23 July.

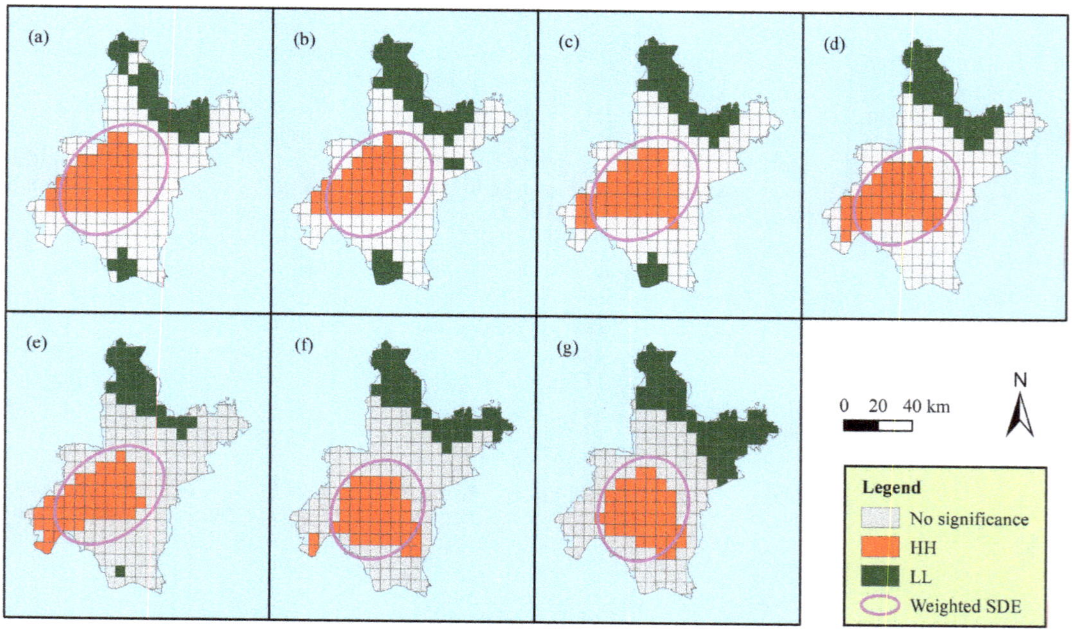

Figure 6. Local spatial pattern of the air temperature at 06:00 (significant level at 0.05). (**a**) 17 July; (**b**) 18 July; (**c**) 19 July; (**d**) 20 July; (**e**) 21 July; (**f**) 22 July; (**g**) 23 July.

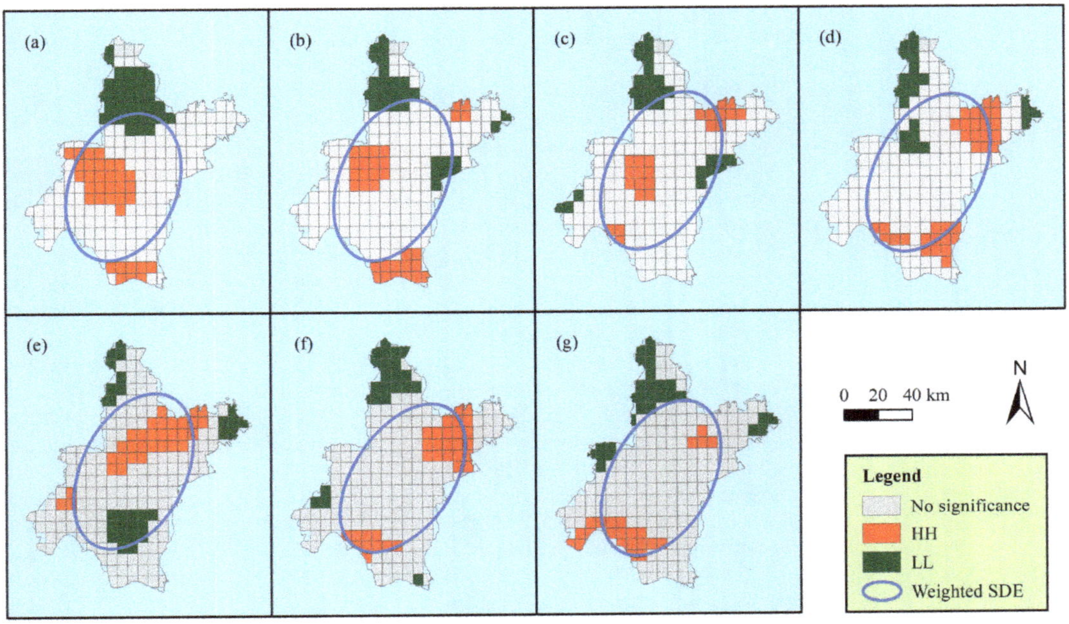

Figure 7. Local spatial pattern of the air temperature at 14:00 (significant level at 0.05). (**a**) 17 July; (**b**) 18 July; (**c**) 19 July; (**d**) 20 July; (**e**) 21 July; (**f**) 22 July; (**g**) 23 July.

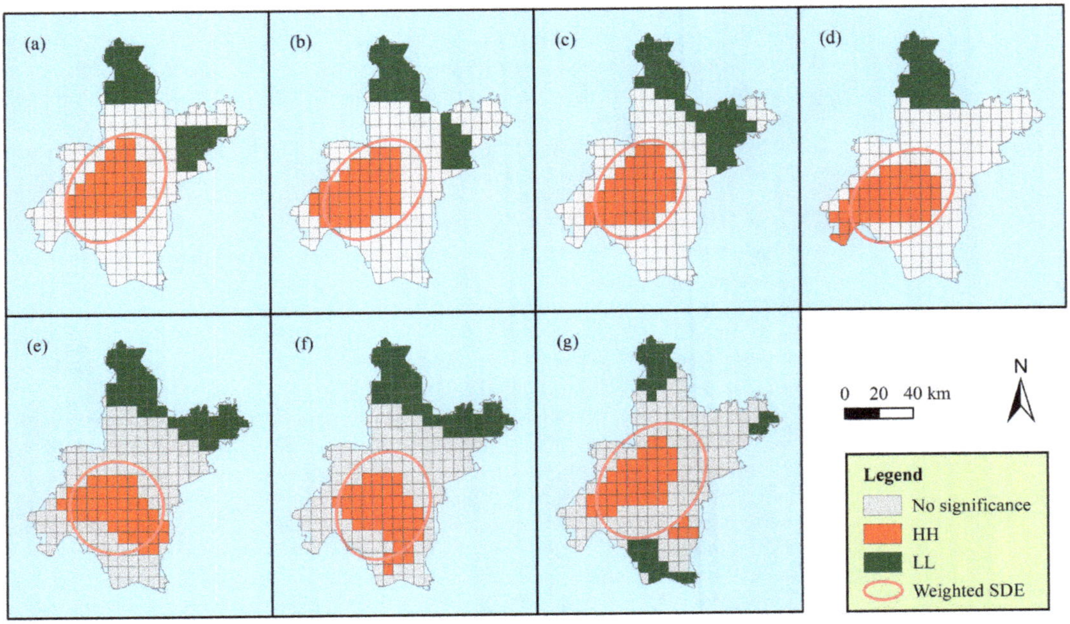

Figure 8. Local spatial pattern of the air temperature at 22:00 (significant level at 0.05). (**a**) 17 July; (**b**) 18 July; (**c**) 19 July; (**d**) 20 July; (**e**) 21 July; (**f**) 22 July; (**g**) 23 July.

As seen from Figure 6, the local spatial patterns of the thermal characteristic at 06:00 on seven days mostly overlap, especially on 18 and 19 July. As for 22:00 (Figure 8), the local spatial patterns from 17 July to 19 July mostly overlap and those on 21 and 22 July mostly overlap. Additionally, the overlap proportion of the HH clusters at 06:00 and 22:00 is large in general. However, the HH clusters at 14:00 are substantially different from those at 06:00 and 22:00. Additionally, the corresponding weighted SDEs are displayed in Figures 6–8. It can be seen that the results of the SDE analysis are consistent with the local spatial pattern analysis. The shape, position, and size of the weighted SDEs at 06:00 and 22:00 are dominantly determined by the distribution of the HH cluster.

A further analysis regarding the land uses is illustrated in Figure 9, which shows the proportions of the land uses within the HH clusters, LL clusters and no significance clusters. In the HH clusters (Figure 9a), there are four time points at which the proportion of impervious surface is the largest (22:00 on 17 July; 06:00 on 18 July; 06:00 and 22:00 on 19 July), whereas the proportion of the cropland is the largest at the other time points. However, the proportion of cropland is always the largest in the LL clusters (Figure 9b) and no significance clusters (Figure 9c). This probably suggests that cropland is not associated with either high air temperature or low air temperature. The large proportion of cropland in the HH clusters and LL clusters may be due to the fact that the air temperature over cropland is easily affected by neighboring with other land uses. When excluding cropland, as in the HH clusters at 06:00 and 22:00, the proportions of the impervious surface and water bodies are the two largest; additionally, the proportion of impervious surface is generally larger than that of water bodies, except for 22:00 on 20 July and 06:00 on 21 July, which are approximately equal. This may suggest that the HH clusters at 06:00 and 22:00 are associated with the impervious surface and water bodies. Since the proportion of water bodies in the no significance clusters is not very small, this may indicate that the association of the water bodies with the HH clusters at 06:00 and 22:00 is weak. As for the HH clusters at 14:00, the proportion of water bodies is larger than that of impervious surface on five out of seven days; however, these proportions of water bodies is not fairly larger and even sometimes less than those in the no significance clusters. Thus, the water bodies may even be more weakly associated with the HH clusters at 14:00. When considering the land uses in the LL clusters after excluding cropland, the proportion of woodland is always the largest; moreover, the proportion of woodland in the HH clusters and no significance clusters is always very small. This suggests that the LL clusters are strongly associated with woodland (mainly distributed in the northern part of the city).

3.3. Land Uses Contributing to Air Temperature Variation

We used a regression analysis to analyze the contribution of land uses to the special distribution of air temperature. We carried out the ordinary least squares (OLS) regression analysis of the air temperature of the grid points with the proportions of land uses with respect to the Thiessen polygons of the grid points. Additionally, stepwise regression technology was used to select the proper variables and eliminate the variables without significance (at the significant level of 0.05).

Among the proportions of the six land use types illustrated in Figure 1b, the main variables contributing to the variation of air temperature are the proportion of woodland area (P_{WDA}), the proportion of water body area (P_{WBA}) and the proportion of impervious surface area (P_{ISA}); more details of the OLS regression results are shown in Table 2. The stepwise regression technology can mitigate the collinearity of the predictor variables in the model to some extent. Moreover, the variance inflation factors (VIFs) for the predictors in the final regression model were evaluated and the maximum VIF is presented (Table 2). All the values of the maximum VIF are less than 2, indicating essentially no collinearity. Therefore, the predictor variables could independently predict the value of the dependent variable. The coefficients of P_{WDA} are all significant and all negative at 06:00, 14:00 and 22:00 from 17 July to 23 July, which means that by increasing the proportion of woodland area, the air temperature will decrease. As for P_{WBA} and P_{ISA}, both coefficients are significant and positive at 06:00 and 22:00 on the seven days; however, at 14:00, the coefficients of P_{WBA} are only significant on two days and those of P_{ISA} are only significant on three days. This means that in the morning and evening, increasing the proportions of water body area and impervious surface area will increase the air temperature. However, at noon, the effects of the water body and the impervious surface may require further consideration, for the coefficients of P_{WBA} and P_{ISA} are not always significant.

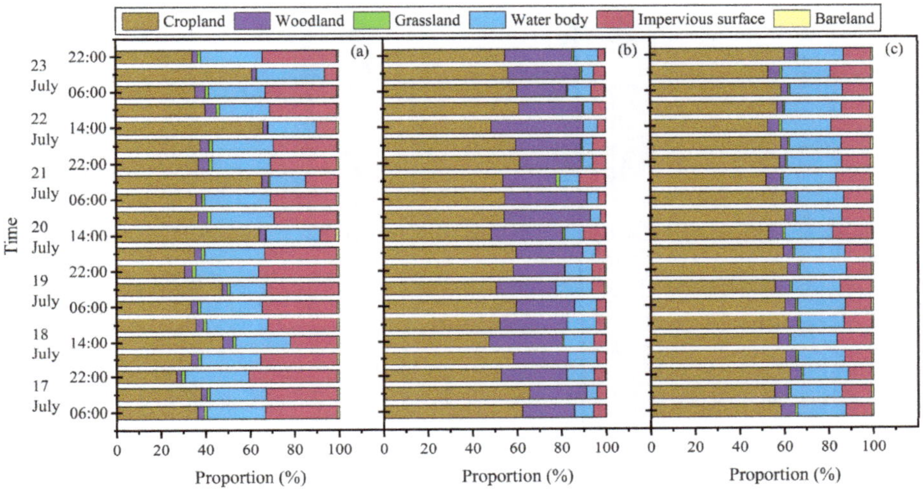

Figure 9. Proportions of land uses in the local spatial patterns. (**a**) HH cluster; (**b**) LL cluster; (**c**) no significance cluster.

When considering the fitness of the OLS model varying in time (R^2 in Table 2), it can be seen that the R^2 value at 22:00 is larger than that at 06:00 each day, and that the R^2 value at 06:00 is larger than that at 14:00. With the exception of 22 July, the OLS model can explain more than 50% of the variation in the spatial distribution of air temperature at 22:00; at 06:00, there are three days on which more than 50% of the variation can be explained by the model. However, at 14:00, the model can explain not more than 40% of the variation, with the exception of 17 and 22 July, and even less than 20% on 21 July. This suggests that there are some other factors affecting the spatial distribution of air temperature. Moreover, we used the spatial lag model (SLM) with the same independent variables than for the regression analysis. The results are shown in Table 3. The R^2 of the model is very high at all the time points, much larger than that of the OLS model. None of the R^2 values at 06:00 and 22:00 are lesser than 0.97; only the R^2 values at 14:00 on 19 and 23 July are slightly lesser than 0.90, but also greater than 0.89. This means that the explanation of the SLM is improved. However, the absolute values of the coefficients of the explanatory variables become smaller. This is because some variation of the dependent variable explained by the independent variable is substituted by the spatially lagged term (***WTa*** in Equation (14)). When the absolute value becomes very small and close to zero, it may not even be statistically significantly different from the zero value. As seen from Table 3, all the coefficients with absolute values of less than 0.1 are not significant. Two of the coefficients of P_{ISA} that are significant in the OLS model become not significant. Even the coefficients of P_{WBA} all become not significant. Conversely, the Lag coefficients of the SLM are very high (higher than 0.95) and extremely significant at all the time points, which indicates that the contribution of spatial autocorrelation to the spatial variation of air temperature is very high and that the air temperature of one grid point is greatly affected by its neighboring points. This implies that the effect of spatial diffusion is significant in the spatial distribution of air temperature.

Table 2. Coefficients of the OLS regression with respect to air temperature.

Date	Time	Constant	P_{WDA}	P_{WBA}	P_{ISA}	Max VIF	R^2	Adjusted R^2
17 July	06:00	26.77 ***	−1.202 ***	1.223 ***	2.594 ***	1.207	0.430	0.422
	14:00	35.67 ***	−0.963 ***	1.026 ***	1.145 ***	1.207	0.408	0.399
	22:00	30.33 ***	−2.247 ***	1.175 ***	2.756 ***	1.207	0.614	0.608
18 July	06:00	27.12 ***	−1.590 ***	1.279 ***	3.009 ***	1.207	0.499	0.492
	14:00	37.02 ***	−1.517 ***	-	-	-	0.329	0.326
	22:00	30.82 ***	−2.860 ***	1.287 ***	2.877 ***	1.207	0.555	0.548
19 July	06:00	27.19 ***	−1.749 ***	1.488 ***	2.995 ***	1.207	0.522	0.515
	14:00	37.45 ***	−1.061 ***	-	0.327 **	1.068	0.335	0.328
	22:00	30.75 ***	−2.710 ***	1.749 ***	3.367 ***	1.207	0.523	0.516
20 July	06:00	27.10 ***	−2.375 ***	1.742 ***	3.160 ***	1.207	0.504	0.497
	14:00	37.68 ***	−1.568 ***	-	−0.634 ***	1.068	0.373	0.367
	22:00	30.47 ***	−3.102 ***	2.048 ***	2.960 ***	1.207	0.546	0.539
21 July	06:00	27.34 ***	−2.751 ***	1.630 ***	2.902 ***	1.207	0.591	0.585
	14:00	38.34 ***	−0.895 ***	-	-	-	0.169	0.165
	22:00	30.99 ***	−3.622 ***	2.528 ***	3.152 ***	1.207	0.526	0.519
22 July	06:00	26.84 ***	−2.669 ***	2.176 ***	2.658 ***	1.207	0.487	0.479
	14:00	36.63 ***	−1.862 ***	-	-	-	0.473	0.471
	22:00	30.44 ***	−2.325 ***	1.862 ***	2.431 ***	1.207	0.495	0.487
23 July	06:00	26.97 ***	−2.218 ***	1.847 ***	2.715 ***	1.207	0.463	0.455
	14:00	36.72 ***	−0.923 ***	0.362 ***	-	1.113	0.355	0.349
	22:00	30.15 ***	−2.300 ***	0.882 ***	2.147 ***	1.207	0.535	0.528

*** Significant at $p < 0.001$, ** significant at $p < 0.01$.

Table 3. Coefficients of the SLM regression with respect to the air temperature.

Date	Time	Lag Coefficient	Constant	P_{WDA}	P_{WBA}	P_{ISA}	R^2
17 July	06:00	0.986 ***	0.358 *	−0.264 ***	0.068	0.154 *	0.977
	14:00	0.962 ***	1.368 ***	−0.289 ***	0.071	0.068	0.962
	22:00	0.965 ***	1.057 ***	−0.373 ***	0.040	0.221 ***	0.986
18 July	06:00	0.984 ***	0.429 *	−0.299 ***	0.051	0.199 **	0.981
	14:00	0.964 ***	1.356 ***	−0.395 ***	-	-	0.934
	22:00	0.972 ***	0.862 ***	−0.388 ***	0.016	0.202 ***	0.990
19 July	06:00	0.982 ***	0.476 **	−0.329 ***	0.049	0.184 **	0.982
	14:00	0.956 ***	1.682 ***	−0.388 ***	-	−0.009	0.892
	22:00	0.980 ***	0.611 **	−0.348 ***	0.029	0.235 ***	0.990
20 July	06:00	0.981 ***	0.497 **	−0.341 ***	0.041	0.214 **	0.985
	14:00	0.962 ***	1.456 ***	−0.420 ***	-	−0.119 **	0.934
	22:00	0.970 ***	0.894 **	−0.301 ***	0.099	0.254 ***	0.989
21 July	06:00	0.963 ***	1.005 ***	−0.404 ***	0.077	0.215 ***	0.987
	14:00	0.973 ***	1.074 **	−0.331 ***	-	-	0.900
	22:00	0.979 ***	0.643 **	−0.329 ***	0.041	0.207 ***	0.993
22 July	06:00	0.980 ***	0.514 **	−0.322 ***	0.099	0.188 **	0.989
	14:00	0.941 ***	2.167 ***	−0.457 ***	-	-	0.941
	22:00	0.974 ***	0.793 **	−0.340 ***	−0.003	0.213 ***	0.984
23 July	06:00	0.980 ***	0.531**	−0.320 ***	−0.004	0.191 ***	0.990
	14:00	0.947 ***	1.952 ***	−0.334 ***	0.011	-	0.891
	22:00	0.969 ***	0.927 **	−0.353 ***	0.035	0.134 *	0.970

*** Significant at $p < 0.001$, ** significant at $p < 0.01$, * significant at $p < 0.05$.

4. Discussion

4.1. Advantage of the UHI Index

Figure 10 displays the standard deviation of air temperature (*Ta*-SD) of the grid points, its trend generally corresponding to that of the UHI index (Figure 4). It may be supposed that the UHI index is simply determined by the variation of air temperature with respect to the grid points. However, this is not exactly true. In order to illustrate this issue, we first calculated the difference in the UHI index and the difference in the *Ta*-SD in 2 h steps. We then calculated the quotient of the differences, which is shown in Figure 10. As it can be seen, the difference quotient is positive most of the time, but there are still some times when the difference quotient is negative. This means that at some time points, not very few, the change trend of the UHI index is opposite to that of the *Ta*-SD. As the UHI index is derived from an SDE analysis, it considers not only the value distribution of air temperature, but also the spatial distribution. With the same *Ta*-SD, the spatial distribution could be different, so the values of the UHI index may be distinct. In specific situations, with the same *Ta*-SD, an area may present the state of heat island or the state of cool island. However, these situations can be distinguished by the UHI index, for its value of the former is greater than 1 and that of the latter less than 1. Thus, the UHI index has an advantage in characterizing the spatial distribution of the thermal environment and the UHI effect.

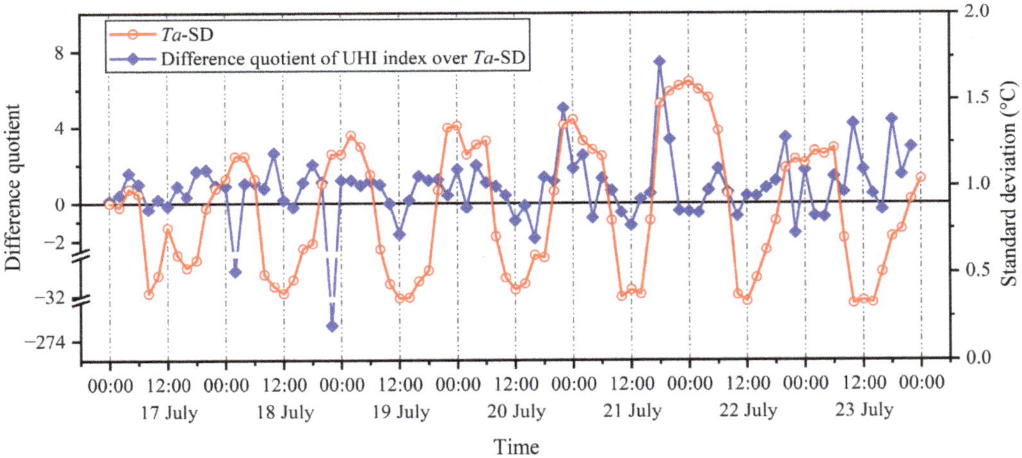

Figure 10. Comparison of UHI index and the standard deviation of the air temperature (*Ta*-SD) in the study area.

4.2. Thermal Environment in Relation to Weather Condition

The weather conditions of the seven days on which we analyzed the thermal environmental characteristics are cloudy and sunny. As it can be seen, the UHI index from 10:00 to 16:00 does not vary much on these seven days. However, the values of the UHI index at 22:00 and at 24:00 on the three sunny days are all greater than those on the four cloudy days. These results are in accordance with the other literature. There is wide consensus that the UHI effect is strong during the nighttime but weak or even disappears during the daytime [45–48]. Thus, the influence of weather conditions on the UHI effect is slight during the daytime. However, during the nighttime, previous studies have demonstrated that that the UHI intensity in clear conditions is evidently greater than in overcast conditions during the nighttime [47,49,50].

Conventionally, UHI intensity is usually the difference between urban peak air temperatures and rural air temperatures. In this study, the curve of the UHI index reflects that the air temperature difference between urban center and the outskirts of the city during the nighttime on sunny days is generally greater than that on cloudy days, implying that the spatial heterogeneity of thermal environmental characteristics increases in sunny days.

4.3. Limitations of the Regression Models

In this study, we attribute the spatial variation of the air temperature to local land uses by using a regression analysis. In fact, wind is also an important factor affecting air temperature, which is not considered in the regression model. This may be a reason why the coefficients of the determination with respect to the OLS model are not very satisfactory. Yet, the magnitude of the variation in air temperature explained by the land use regression model is also not very satisfactory in previous studies [51–54]. However, it is difficult to consider the contribution of wind in a global regression model, for the effect of wind on local air temperature is complex. Even if the wind remains in a constant direction, it may decrease the air temperature in one local area but may increase the air temperature in another local area, not to mention that the wind's direction frequently changes. Thus, it may not be suitable to consider the contribution of wind to local air temperature in space when using the OLS model. When the SLM is adopted, the coefficients of the determination greatly improve. Since the SLM can reflect the effect of spatial diffusion, it can be supposed that the model takes into account the effect of the wind to some extent. However, this comes at the cost of losing the explanation of the local independent variables, namely the proportion of land uses, for the absolute values of the regression coefficients become smaller and some of them are even without significance.

5. Conclusions

In the light of the ambiguous measurements in UHI intensity and the scarce studies on the spatial characteristics' evolution of the urban thermal environment with the high temporal resolution during a day, this study introduces the SDE method to depict the general spatial character of the urban thermal environment and constructs an UHI index to evaluate the UHI intensity based on the SDE. Taking the city of Wuhan as a study area, the results illustrate that the UHI index can effectively represent the general characteristics of the thermal environment in the study area. The trends of the UHI intensity generally decrease from midnight to midday, and generally increase from midday to evening. The local spatial pattern analysis through local Moran's I shows that there are only high–high clusters and low–low clusters of air temperature with statistical significance in the study area. The former clusters are mainly associated with the imperious surface, and are also associated with water bodies in the morning and the evening. The latter clusters are mainly associated with woodland. This can be further demonstrated by a regression analysis. The regression analysis of the SLM also indicates that the effect of spatial diffusion has a significant influence on the distribution of air temperature in space.

Furthermore, the regression analysis indicates that the effects of water body on air temperature are mostly not statistically significant at noon according to the OLS model and are even all insignificant in the morning, noon and evening according to the SLM model. Nevertheless, woodland always has a cooling effect in the morning, at noon and in the evening, implying that increasing the vegetation may be the most effective strategy to mitigate the urban thermal environment. The regression results also show that the OLS model could not fit the spatial distribution of air temperature well enough, while the fitness of the SLM is greatly improved. However, the explanation of the proportion of the local land uses accounting for the spatial variation of air temperature becomes lower. An improved model is still required to explain the spatial distribution of the urban air temperature.

Author Contributions: Design of the research, W.S. and R.X.; methodology and data processing, W.S.; writing of the article, W.S. and J.H.; verification of the results, X.S. All authors have read and agreed to the published version of the manuscript.

Funding: This research was funded by National Natural Science Foundation of China (NSFC), grant number 41701179.

Data Availability Statement: Data is contained within the article.

Conflicts of Interest: The authors declare no conflict of interest.

References

1. Jedlovec, G.; Crane, D.; Quattrochi, D. Urban heat wave hazard and risk assessment. *Results Phys.* **2017**, *7*, 4294–4295. [CrossRef]
2. He, B.J.; Wang, J.S.; Liu, H.M.; Ulpiani, G. Localized synergies between heat waves and urban heat islands: Implications on human thermal comfort and urban heat management. *Environ. Res.* **2021**, *193*, 18. [CrossRef]
3. Li, D.; Bou-Zeid, E. Synergistic Interactions between Urban Heat Islands and Heat Waves: The Impact in Cities Is Larger than the Sum of Its Parts. *J. Appl. Meteorol. Climatol.* **2013**, *52*, 2051–2064. [CrossRef]
4. Tan, J.G.; Zheng, Y.F.; Tang, X.; Guo, C.Y.; Li, L.P.; Song, G.X.; Zhen, X.R.; Yuan, D.; Kalkstein, A.J.; Li, F.R.; et al. The urban heat island and its impact on heat waves and human health in Shanghai. *Int. J. Biometeorol.* **2010**, *54*, 75–84. [CrossRef] [PubMed]
5. Grimm, N.B.; Faeth, S.H.; Golubiewski, N.E.; Redman, C.L.; Wu, J.G.; Bai, X.M.; Briggs, J.M. Global change and the ecology of cities. *Science* **2008**, *319*, 756–760. [CrossRef]
6. Chen, X.L.; Zhao, H.M.; Li, P.X.; Yin, Z.Y. Remote sensing image-based analysis of the relationship between urban heat island and land use/cover changes. *Remote Sens. Environ.* **2006**, *104*, 133–146. [CrossRef]
7. Xian, G.; Shi, H.; Auch, R.; Gallo, K.; Zhou, Q.; Wu, Z.T.; Kolian, M. The effects of urban land cover dynamics on urban heat Island intensity and temporal trends. *GISci. Remote Sens.* **2021**, *58*, 501–515. [CrossRef]
8. Radhi, H.; Fikry, F.; Sharples, S. Impacts of urbanisation on the thermal behaviour of new built up environments: A scoping study of the urban heat island in Bahrain. *Landsc. Urban Plan.* **2013**, *113*, 47–61. [CrossRef]
9. Connors, J.P.; Galletti, C.S.; Chow, W.T.L. Landscape configuration and urban heat island effects: Assessing the relationship between landscape characteristics and land surface temperature in Phoenix, Arizona. *Landscape Ecol.* **2013**, *28*, 271–283. [CrossRef]
10. Hart, M.; Sailor, D. Quantifying the influence of land-use and surface characteristics on spatial variability in the urban heat island. *Theor. Appl. Climatol.* **2009**, *95*, 397–406. [CrossRef]
11. Morini, E.; Touchaei, A.G.; Castellani, B.; Rossi, F.; Cotana, F. The Impact of Albedo Increase to Mitigate the Urban Heat Island in Terni (Italy) Using the WRF Model. *Sustainability* **2016**, *8*, 999. [CrossRef]
12. Zhang, Y.S.; Wang, X.Q.; Balzter, H.; Qiu, B.W.; Cheng, J.Y. Directional and Zonal Analysis of Urban Thermal Environmental Change in Fuzhou as an Indicator of Urban Landscape Transformation. *Remote Sens.* **2019**, *11*, 2810. [CrossRef]
13. Maimaitiyiming, M.; Ghulam, A.; Tiyip, T.; Pla, F.; Latorre-Carmona, P.; Halik, U.; Sawut, M.; Caetano, M. Effects of green space spatial pattern on land surface temperature: Implications for sustainable urban planning and climate change adaptation. *ISPRS-J. Photogramm. Remote Sens.* **2014**, *89*, 59–66. [CrossRef]
14. Kong, F.H.; Yin, H.W.; James, P.; Hutyra, L.R.; He, H.S. Effects of spatial pattern of greenspace on urban cooling in a large metropolitan area of eastern China. *Landsc. Urban Plan.* **2014**, *128*, 35–47. [CrossRef]
15. Masoudi, M.; Tan, P.Y. Multi-year comparison of the effects of spatial pattern of urban green spaces on urban land surface temperature. *Landsc. Urban Plan.* **2019**, *184*, 44–58. [CrossRef]
16. Nakayama, T.; Hashimoto, S. Analysis of the ability of water resources to reduce the urban heat island in the Tokyo megalopolis. *Environ. Pollut.* **2011**, *159*, 2164–2173. [CrossRef]
17. Lin, Y.; Wang, Z.F.; Jim, C.Y.; Li, J.B.; Deng, J.S.; Liu, J.G. Water as an urban heat sink: Blue infrastructure alleviates urban heat island effect in mega-city agglomeration. *J. Clean. Prod.* **2020**, *262*, 8. [CrossRef]
18. Steeneveld, G.J.; Koopmans, S.; Heusinkveld, B.G.; Theeuwes, N.E. Refreshing the role of open water surfaces on mitigating the maximum urban heat island effect. *Landsc. Urban Plan.* **2014**, *121*, 92–96. [CrossRef]
19. Oke, T.R.; Mills, G.; Christen, A.; Voogt, J.A. *Urban Climates*, 1st ed.; Cambridge University Press: Cambridge, UK, 2017; pp. 197–200.
20. Schwarz, N.; Schlink, U.; Franck, U.; Grossmann, K. Relationship of land surface and air temperatures and its implications for quantifying urban heat island indicators-An application for the city of Leipzig (Germany). *Ecol. Indic.* **2012**, *18*, 693–704. [CrossRef]
21. Sun, T.; Sun, R.H.; Chen, L.D. The Trend Inconsistency between Land Surface Temperature and Near Surface Air Temperature in Assessing Urban Heat Island Effects. *Remote Sens.* **2020**, *12*, 1271. [CrossRef]
22. Venter, Z.S.; Chakraborty, T.; Lee, X.H. Crowdsourced air temperatures contrast satellite measures of the urban heat island and its mechanisms. *Sci. Adv.* **2021**, *7*, 9. [CrossRef] [PubMed]
23. Azevedo, J.A.; Chapman, L.; Muller, C.L. Quantifying the Daytime and Night-Time Urban Heat Island in Birmingham, UK: A Comparison of Satellite Derived Land Surface Temperature and High Resolution Air Temperature Observations. *Remote Sens.* **2016**, *8*, 153. [CrossRef]
24. Oke, T.R. City size and urban heat island. *Atmos. Environ.* **1973**, *7*, 769–779. [CrossRef]
25. Earl, N.; Simmonds, I.; Tapper, N. Weekly cycles in peak time temperatures and urban heat island intensity. *Environ. Res. Lett.* **2016**, *11*, 10. [CrossRef]
26. Stewart, I.D. A systematic review and scientific critique of methodology in modern urban heat island literature. *Int. J. Climatol.* **2011**, *31*, 200–217. [CrossRef]

27. Du, H.Y.; Wang, D.D.; Wang, Y.Y.; Zhao, X.L.; Qin, F.; Jiang, H.; Cai, Y.L. Influences of land cover types, meteorological conditions, anthropogenic heat and urban area on surface urban heat island in the Yangtze River Delta Urban Agglomeration. *Sci. Total Environ.* **2016**, *571*, 461–470. [CrossRef]
28. Cui, Y.P.; Xu, X.L.; Dong, J.W.; Qin, Y.C. Influence of Urbanization Factors on Surface Urban Heat Island Intensity: A Comparison of Countries at Different Developmental Phases. *Sustainability* **2016**, *8*, 706. [CrossRef]
29. Deilami, K.; Kamruzzaman, M.; Hayes, J.F. Correlation or Causality between Land Cover Patterns and the Urban Heat Island Effect? Evidence from Brisbane, Australia. *Remote Sens.* **2016**, *8*, 716. [CrossRef]
30. Estoque, R.C.; Murayama, Y.; Myint, S.W. Effects of landscape composition and pattern on land surface temperature: An urban heat island study in the megacities of Southeast Asia. *Sci. Total Environ.* **2017**, *577*, 349–359. [CrossRef]
31. Yao, R.; Wang, L.C.; Huang, X.; Zhang, W.W.; Li, J.L.; Niu, Z.G. Interannual variations in surface urban heat island intensity and associated drivers in China. *J. Environ. Manag.* **2018**, *222*, 86–94. [CrossRef]
32. Yao, R.; Wang, L.C.; Huang, X.; Gong, W.; Xia, X.G. Greening in Rural Areas Increases the Surface Urban Heat Island Intensity. *Geophys. Res. Lett.* **2019**, *46*, 2204–2212. [CrossRef]
33. Han, S.; Liu, B.C.; Shi, C.X.; Liu, Y.; Qiu, M.J.; Sun, S. Evaluation of CLDAS and GLDAS Datasets for Near-Surface Air Temperature over Major Land Areas of China. *Sustainability* **2020**, *12*, 4311. [CrossRef]
34. Huang, X.L.; Han, S.; Shi, C.X. Evaluation of Three Air Temperature Reanalysis Datasets in the Alpine Region of the Qinghai-Tibet Plateau. *Remote Sens.* **2022**, *14*, 4447. [CrossRef]
35. Leng, P.; Li, Z.L.; Duan, S.B.; Gao, M.F.; Huo, H.Y. A practical approach for deriving all-weather soil moisture content using combined satellite and meteorological data. *ISPRS-J. Photogramm. Remote Sens.* **2017**, *131*, 40–51. [CrossRef]
36. Zhou, W.; Peng, B.; Shi, J.C.; Wang, T.X.; Dhital, Y.P.; Yao, R.Z.; Yu, Y.C.; Lei, Z.T.; Zhao, R. Estimating High Resolution Daily Air Temperature Based on Remote Sensing Products and Climate Reanalysis Datasets over Glacierized Basins: A Case Study in the Langtang Valley, Nepal. *Remote Sens.* **2017**, *9*, 959. [CrossRef]
37. Rao, Y.H.; Liang, S.L.; Wang, D.D.; Yu, Y.Y.; Song, Z.; Zhou, Y.; Shen, M.G.; Xu, B.Q. Estimating daily average surface air temperature using satellite land surface temperature and top-of-atmosphere radiation products over the Tibetan Plateau. *Remote Sens. Environ.* **2019**, *234*, 14. [CrossRef]
38. Chen, Y.; Liang, S.L.; Ma, H.; Li, B.; He, T.; Wang, Q. An all-sky 1 km daily land surface air temperature product over mainland China for 2003–2019 from MODIS and ancillary data. *Earth Syst. Sci. Data* **2021**, *13*, 4241–4261. [CrossRef]
39. Ran, Y.H.; Li, X.; Lu, L. Evaluation of four remote sensing based land cover products over China. *Int. J. Remote Sens.* **2010**, *31*, 391–401. [CrossRef]
40. Lefever, D.W. Measuring geographic concentration by means of the standard deviational ellipse. *Am. J. Sociol.* **1926**, *32*, 88–94. [CrossRef]
41. Johnson, D.P.; Wilson, J.S. The socio-spatial dynamics of extreme urban heat events: The case of heat-related deaths in Philadelphia. *Appl. Geogr.* **2009**, *29*, 419–434. [CrossRef]
42. Xu, J.H.; Zhao, Y.; Zhong, K.W.; Zhang, F.F.; Liu, X.L.; Sun, C.G. Measuring spatio-temporal dynamics of impervious surface in Guangzhou, China, from 1988 to 2015, using time-series Landsat imagery. *Sci. Total Environ.* **2018**, *627*, 264–281. [CrossRef] [PubMed]
43. Anselin, L. Local indicators of spatial association—LISA. *Geogr. Anal.* **1995**, *27*, 93–115. [CrossRef]
44. Anselin, L.; Rey, S. Properties of tests for spatial dependence in linear-regression models. *Geogr. Anal.* **1991**, *23*, 112–131. [CrossRef]
45. Oke, T.R. The energetic basis of the urban heat-island. *Q. J. R. Meteorol. Soc.* **1982**, *108*, 1–24. [CrossRef]
46. Basara, J.B.; Hall, P.K.; Schroeder, A.J.; Illston, B.G.; Nemunaitis, K.L. Diurnal cycle of the Oklahoma City urban heat island. *J. Geophys. Res.-Atmos.* **2008**, *113*, 16. [CrossRef]
47. Wolters, D.; Brandsma, T. Estimating the Urban Heat Island in Residential Areas in the Netherlands Using Observations by Weather Amateurs. *J. Appl. Meteorol. Climatol.* **2012**, *51*, 711–721. [CrossRef]
48. Hardin, A.W.; Liu, Y.; Cao, G.; Vanos, J.K. Urban heat island intensity and spatial variability by synoptic weather type in the northeast U.S. *Urban Clim.* **2018**, *24*, 747–762. [CrossRef]
49. Morris, C.J.G.; Simmonds, I.; Plummer, N. Quantification of the influences of wind and cloud on the nocturnal urban heat island of a large city. *J. Appl. Meteorol.* **2001**, *40*, 169–182. [CrossRef]
50. Ngarambe, J.; Oh, J.W.; Su, M.A.; Santamouris, M.; Yun, G.Y. Influences of wind speed, sky conditions, land use and land cover characteristics on the magnitude of the urban heat island in Seoul: An exploratory analysis. *Sust. Cities Soc.* **2021**, *71*, 22. [CrossRef]
51. Myint, S.W.; Brazel, A.; Okin, G.; Buyantuyev, A. Combined Effects of Impervious Surface and Vegetation Cover on Air Temperature Variations in a Rapidly Expanding Desert City. *GISci. Remote Sens.* **2010**, *47*, 301–320. [CrossRef]
52. Schatz, J.; Kucharik, C.J. Urban climate effects on extreme temperatures in Madison, Wisconsin, USA. *Environ. Res. Lett.* **2015**, *10*, 13. [CrossRef]
53. Coseo, P.; Larsen, L. Accurate Characterization of Land Cover in Urban Environments: Determining the Importance of Including Obscured Impervious Surfaces in Urban Heat Island Models. *Atmosphere* **2019**, *10*, 347. [CrossRef]
54. Tsin, P.K.; Knudby, A.; Krayenhoff, E.S.; Brauer, M.; Henderson, S.B. Land use regression modeling of microscale urban air temperatures in greater Vancouver, Canada. *Urban Clim.* **2020**, *32*, 12. [CrossRef]

remote sensing MDPI

Article

Temporal and Spatial Variation of Land Surface Temperature and Its Driving Factors in Zhengzhou City in China from 2005 to 2020

Shenghui Zhou [1,2], Dandan Liu [1,2], Mengyao Zhu [1,2], Weichao Tang [1,2], Qian Chi [1,2], Siyu Ye [1,2], Siqi Xu [1,2] and Yaoping Cui [1,*]

1 Key Laboratory of Geospatial Technology for the Middle and Lower Yellow River Regions, Ministry of Education, College of Environment and Planning, Henan University, Kaifeng 475004, China
2 Key Laboratory of Integrative Prevention of Air Pollution and Ecological Security of Henan Province, Kaifeng 475004, China
* Correspondence: cuiyp@lreis.ac.cn

Abstract: Rapid urbanization is an important factor leading to the rise in surface temperature. How to effectively reduce the land surface temperature (LST) has become a significant proposition of city planning. For the exploration of LST and the urban heat island (UHI) effect in Zhengzhou, China, the LST was divided into seven grades, and the main driving factors of LST change and their internal relations were discussed by correlation analysis and gray correlation analysis. The results indicated that LST showed an upward trend from 2005 to 2020, and a mutation occurred in 2013. Compared with 2005, the mean value of LST in 2020 increased by 0.92 °C, while the percentage of LST-enhanced areas was 22.77. Furthermore, the spatial pattern of UHI was irregularly distributed, gradually spreading from north to south from 2005 to 2020; it showed a large block distribution in the main city and southeast in 2020, while, in the areas where woodlands were concentrated and in the Yellow River Basin, there was an obvious "cold island" effect. In addition, trend analysis and gray correlation analysis revealed that human factors were positively correlated with LST, which intensified the formation of the UHI effect, and the influence of Albedo on LST showed obvious spatial heterogeneity, while the cooling effect of vegetation water was better than that of topography. The research results can deepen the understanding of the driving mechanism of the UHI effect, as well as provide scientific support for improving the quality of the urban human settlement environment.

Keywords: temporal and spatial variation; land surface temperature; Zhengzhou city; urban heat island

Citation: Zhou, S.; Liu, D.; Zhu, M.; Tang, W.; Chi, Q.; Ye, S.; Xu, S.; Cui, Y. Temporal and Spatial Variation of Land Surface Temperature and Its Driving Factors in Zhengzhou City in China from 2005 to 2020. *Remote Sens.* **2022**, *14*, 4281. https://doi.org/10.3390/rs14174281

Academic Editors: Yuji Murayama and Ruci Wang

Received: 27 July 2022
Accepted: 26 August 2022
Published: 30 August 2022

1. Introduction

The continuous growth of the urban area and population has brought a series of urban environmental problems, such as the urban heat island effect, heat waves, and extreme climate. The thermal environment of cities directly affects people's quality of life, and it is closely related to the urban climate, which has always been a research hotspot [1,2]. Because the increase in land surface temperature (LST) can affect people's life and the environment in multiple ways, the urban heat island (UHI) phenomenon not only affects the local climate, vegetation growth, and air quality, but also affects people's health [3–5]. Therefore, it is a common challenge for climatologists and urban planners to formulate effective mitigation strategies for the UHI effect [6]. Determining the potential driving factors of LST is very important to reduce the urban heat island effect, promote regional sustainable development, and improve the quality of life of city dwellers [7].

According to the method of measuring temperature, the heat island can be divided into canopy layer heat island (UCL), boundary layer heat island (UBL), and surface UHI (SUHI) [8]. Among them, UCL is composed of air between rough elements (such as buildings and tree crowns), and its upper boundary is just below the roof level. UBL is located above UCL, and its lower boundary is affected by the urban surface [9]. Traditionally, the study of heat island usually depends on the on-site measurements of independent stations [10]. For example, LST is usually collected by measuring the air temperature in the urban canopy [11]. This can produce accurate and time-continuous observation results [12]; however, the spatial representation of meteorological stations limits its application, hindering the effective identification of the spatial pattern of LST. The development of remote sensing (RS) technology has provided an effective means to estimate LST from local to global scales [13,14]. Remote sensing inversion using Landsat (30 m–20 m) allows obtaining the fine LST; however, its temporal resolution is approximately every 16 days [15,16]. Moreover, the Moderate Resolution Imaging Spectroradiometer (MODIS) can provide high-resolution global LST data products, which can be directly used for mesoscale surface temperature research [17,18]. Thus, RS data can help researchers study the temporal and spatial changes of LST, and the development of RS technology has greatly promoted the progress of UHI research [19].

The interaction between influencing factors and LST leads to its spatial heterogeneity [7]; hence, exploring this internal driving relationship can provide effective strategies for alleviating the UHI phenomenon [20]. The factors that cause urban temperature change mainly include two types [21]: natural factors, such as topography, vegetation cover, and water body [22], and human factors, including urban construction intensity and socioeconomic activity index [23,24]. Furthermore, researchers have used various models and mathematical analyses to study the relationship between the spatial and temporal changes of the urban thermal environment and various index variables [25]. The main analysis models include Pearson correlation analysis, ordinary least squares regression analysis, principal component analysis, gray correlation analysis, spatial regression model (spatial lag model and spatial error model), and geographic weighted regression model [26–28]. However, existing studies typically focused on one or a few influencing factors, such as building layout, land-use change, and landscape pattern, whereas comprehensive analyses are scarce [29].

According to consensus, an increase in impermeable surfaces in cities reduces vegetation coverage and transpiration, increases the absorption of solar radiation, and leads to changes in the thermal climate and the warming of cities [30,31]. For example, Knight Teri et al. conducted a systematic review on the influence of vegetation on LST. The research showed that the surface temperature of city green space tends to be cooler than city non-green space, and the cooling effect of green space or parks can expand to 1.25 km outside its boundaries [32,33]. Furthermore, taking Shanghai as an example, Yang et al. analyzed the influence of impervious surface (IS) and vegetation cover (VC) on the intensity of the UHI. The results showed that there were obvious differences between urban and rural areas in the gradient distribution of regional land cover and surface temperature, and the heating effect of IS was more obvious than that of VC [34], whereas vegetation and water bodies had obvious cooling effects [35,36]. Liu et al. compared the influence of topography and urban form factors on the urban heat island in Chengdu and Chongqing, indicating that natural factors such as vegetation and water had a similar influence on and contribution to the UHI effect. Nevertheless, the unique topography and urban form played a key role in the difference in UHI between the two cities [37].

Owing to the many factors that can influence the formation of the UHI, a method to qualify the contributions and identify the key factors would help to alleviate the UHI effect and slow down the rising trend, especially for fast-developing cities with high construction intensity and continued growth of non-green spaces [38]. In order to explore the internal relationship between the driving factors of the UHI effect from two aspects of natural and human factors, we chose Zhengzhou as the research area. Zhengzhou is the main economic development center of Henan Province in China; thus, the large number of human activities can easily cause the phenomenon of UHI in this area [39]. However, there has been little research on SUHI in Zhengzhou, while mostly single driving factors were investigated, long timeseries of remote sensing data were not evaluated [40]. Through the analysis of the temporal and spatial variation of land surface temperature and the driving factors of the UHI effect in Zhengzhou, the results of this research can make up for the deficiency of this field, providing a theoretical basis and decision support for the improvement of urban construction and environmental quality of human settlements, as well as providing a reference for future urban planning and design in Zhengzhou.

2. Materials and Methods

2.1. Study Area

The study area was Zhengzhou (112°42′E–114°14′E, 34°16′N–34°58′N), the capital of Henan Province in China, located in the hinterland of China, with the Yellow River in the north, Huanghuai Plain in the southeast, and Songshan Mountain in the west. As shown in Figure 1, the terrain is high in the west and low in the east. D1-D12 are districts, which are Huiji, Zhongyuan, Jinshui, Guancheng Hui, Erqi, Xingyang, Shangjie, Gongyi, Zhongmu, Xinmi, Dengfeng and Xinzheng respectively. Zhengzhou's total population in 2020 reached 1.2601×10^7, representing a densely populated mega city (Statistics Bureau of Zhengzhou, http://tjj.zhengzhou.gov.cn/, accessed on 20 May 2022). In addition, Zhengzhou is located at the intersection of the Beijing–Guangzhou urban development belt and the Longhai urban development belt. It is the central city of the Central Plains urban agglomeration, an important node city in the Zheng–Bian–Luo Industrial Corridor, and one of the most representative cities in urban development in China.

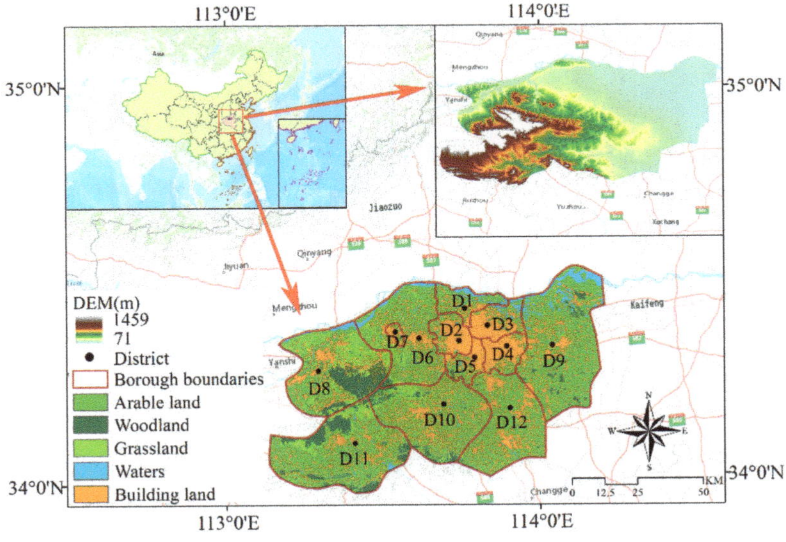

Figure 1. The location map of Zhengzhou.

2.2. Data Resources

The DEM (digital elevation model) data and land-use data in 2020 were obtained from the Resource and Environmental Science Data Center of China, with a spatial resolution of 205 m and 30 m, respectively (http://www.resdc.cn, accessed on 5 March 2022 and 14 March 2022). The Landsat8 OLI_TIRS remote sensing image of 22 May 2020 came from the geospatial data cloud with a cloud content of 6.12% and a spatial resolution of 30 m. The NDISI (normalized difference impervious surface index), NDBBI (normalized difference bareness and built-up index), and MNDWI (modified normalized difference water index) were calculated using the Landsat8 OLI_TIRS remote sensing image (downloaded from http://www.gscloud.cn/#page1/1, accessed on 8 March 2022). The LST, NDVI (normalized difference vegetation index), and Albedo MODIS data products were downloaded from https://modis.gsfc.nasa.gov (accessed on 1 March 2022 and 14 March 2022). The

MODIS data characteristics are shown in Table 1. Zhengzhou POI data in 2020 came from Gaode map (https://ditu.amap.com, accessed on 15 March 2022), including 11 points of interest: catering, living and entertainment, shopping centers, public service facilities, scenic spots, companies and enterprises, government agencies, medical care, commercial housing, accommodation, and sports and leisure, which could be used to represent socioeconomic activities.

Table 1. MODIS data items and descriptions.

Data Items	Spatial Resolution	Time Resolution	Data Resource
LST	1 km	8 days	MYD11A2
NDVI	250 m	16 days	MYD13Q1
Albedo	500 m	Daily	MCD43A3

In this study, the timeseries data mainly included MDOSI LST data from 2005 to 2020, which were used to analyze the trend of surface temperature in Zhengzhou. The cross-sectional data included the Landsat 8 OLI_TIRS remote sensing image, which was used to calculate the NDISI, NDBBI, and MNDWI. In addition, the NDVI, DEM, Albedo, and POI were used to analyze the driving relationship between LST and the influencing factors. In order to study the relationship between LST and the raster data of influencing factors in Zhengzhou, the raster-to-point tool in ArcGIS was used to convert LST raster data into point data, and then the influencing factor data of corresponding points were extracted using a multivalue extraction to point tool.

2.3. *Research Methodology*

2.3.1. Nonparametric Mann–Kendall Trend Test

The nonparametric Mann–Kendall trend test (M–K test) [41,42] was used to test changes in LST over time from 2005 to 2020. The principle is described below [43].

(1) According to the timeseries $X_1, X_2, \ldots, X_n$, construct an ordered sequence as follows:

$$S_K = \sum_{i=1}^{K} R_i,\ R_i = \begin{cases} 1, & X_i > X_j \\ 0, & X_i \le X_j \end{cases},\ (K = 1,\ 2,3,\ldots,n). \quad (1)$$

(2) Calculate the mean and variance of S_K as follows:

$$\mathrm{E}(S_K) = n(n+1)/4. \quad (2)$$

$$\mathrm{Var}(S_K) = n(n-1)(2n+5)/72. \quad (3)$$

(3) Standardize S_K as follows:

$$UF_K = \frac{S_K - E(S_K)}{\sqrt{Var(S_K)}}(K = 1,2,\ldots,n). \quad (4)$$

Here, UF_k is the standard normal distribution, given a significance level α (generally α = 0.05, UF_α = $\pm$1.96); $|UF_k| > U_\alpha$ indicates a significant trend change. $UF_k > 0$ indicates an upward trend, and vice versa. The X-series is inverted to obtain a new timeseries $X_n, X_{n-1}, \ldots, X_1$, and the above process is repeated to obtain $UB_K = -UF_K$, where $UB_1 = 0, K = n, n-1, \ldots, 1$.

2.3.2. Calculation of Surface Information Index

NDISI can be used to represent the impervious surface. Impervious surface refers to artificial ground objects that are impervious to water, and its changing trend can directly or indirectly evaluate the development of a city. The area of impervious surface in a city increases greatly, which can lead to serious urban waterlogging, the UHI effect, and environmental resource pollution [44]. The calculation equation of NDISI is shown below [26,45].

$$NDISI = \frac{TIRS1 - (MNDWI + N + SWIR1)/3}{TIRS1 + (MNDWI + N + SWIR1)/3}. \quad (5)$$

MNDWI is an improved normalized difference water body index, which is used to represent water body information.

$$MNDWI = \frac{G - SWIR1}{G + SWIR1}. \quad (6)$$

NDBBI can be used to obtain the information of urban bare land and built land.

$$NDBBI = \frac{1.5SWIR2 - (N + G)/2}{1.5SWIR2 + (N + G)/2}, \quad (7)$$

where *G*, *N*, *SWIR1*, *SWIR2*, and *TIRS1* correspond to bands 3, 5, 6, 7, and 10, respectively, in Landsat8 OIL-TIR.

Land surface albedo represents the reflective ability of the Earth's surface to solar radiation, and its magnitude is influenced by many factors such as solar altitude angle, land-use type and coverage, and surface roughness. It is an important dynamic dimensionless surface parameter to study the balance of land energy and global climate change [46].

Slope refers to the angle between the actual ground and the horizontal plane, which was calculated using the slope analysis tool in ArcGIS10.5 software.

The socioeconomic activity index can directly reflect the development level of urban areas and indirectly represent the impact of human activities on UHI. In this study, it was expressed using the POI kernel density.

2.3.3. Correlation Analysis and Linear Trend Analysis

In statistics, the Pearson correlation coefficient, also known as the Pearson product-moment correlation coefficient (PPMCC), is used to measure the relationship between two variables X and Y, with values ranging from −1 to +1, and it is widely used in academic research to measure the strength of linear correlation between two variables [47]. The Pearson correlation coefficient between two variables is defined as the quotient of the covariance of these two variables and the product of their standard deviations.

$$\rho_{X,Y} = \frac{cov(X,Y)}{\sigma_X \sigma_Y} = \frac{E[(X - \mu X)(Y - \mu Y)]}{\sigma_X \sigma_Y}. \quad (8)$$

Equation (8) defines the overall correlation coefficient. The Pearson correlation coefficient can be obtained by estimating the covariance and standard deviation of the sample, which is commonly represented by r.

$$\mathrm{r} = \frac{\sum_{i=1}^{n}(X_i - \overline{X})(Y_i - \overline{Y})}{\sqrt{\sum_{i=1}^{n}(X_i - \overline{X})^2}\sqrt{\sum_{i=1}^{n}(Y_i - \overline{Y})^2}}. \quad (9)$$

Furthermore, r can be estimated using the standard score mean of $(X_i,\ Y_i)$ sample points, and the equivalent expression of Equation (9) can be obtained as follows:

$$\mathrm{r} = \frac{1}{n-1}\sum_{i=1}^{n}\left(\frac{X_i - \overline{X}}{\sigma X}\right)\left(\frac{Y_i - \overline{Y}}{\sigma Y}\right), \quad (10)$$

where $\frac{X_i - \overline{X}}{\sigma X}$, $\overline{X}$, and σX are the standardized variable, sample mean, and sample standard deviation, respectively.

In this study, Pearson correlation analysis was used to explore the correlation and interaction between LST and influencing factors, and SPSS software was used to analyze the results.

2.3.4. Gray Relational Analysis Model

Gray correlation analysis [48,49] is a gray process based on the gray system, which compares the timeseries between factors to determine the most influential leading factors. The magnitude of the correlation degree is an external expression of the mutual influence and interaction among factors, and the order of the correlation degree reflects the relative influence of each factor on the reference factor. The principle of gray relational degree analysis is described below [27].

(1) Suppose the original timeseries $X_i = \{X_i(K) | K = 1, 2 \ldots, n; i = 0, 1, 2, \ldots, m-1\}$ is composed of n evaluation samples of m evaluation indicators. First, the original timeseries is averaged to obtain the sequence X_i.

$$X_i(k) = \frac{X_i(k)}{\overline{X_i}}, K = 1, 2 \ldots, n; i = 0, 1, 2, \ldots, m-1, \tag{11}$$

where X_0 is the reference sequence, and the others are comparison sequences; $i \neq 0$ unless otherwise specified.

(2) Calculate the absolute difference between X_0 and X_i at time $\boldsymbol{K}$.

$$\Delta_i(K) = |X_0(K) - X_i(K)|, i = 1, 2, \ldots, m-1. \tag{12}$$

(3) Calculate the correlation coefficient $\xi_i(K)$.

$$\xi_i(K) = \frac{\min_i \min_k \Delta_i(K) + \rho \max_i \max_k \Delta_i(K)}{\Delta_i(K) + \rho \max_i \max_k \Delta_i(K)}, \tag{13}$$

where $\min_i \min_K$ is the minimum difference between two poles, and $\max_i \max_K$ is the maximum difference between two poles. Furthermore, as the resolution coefficient, $\rho \in (0, 1)$, a smaller ρ indicates a greater resolution, which is generally 0.5.

(4) Calculate the gray correlation degree γ_i.

$$\gamma_i = \frac{1}{n} \sum_{k=1}^{n} \xi_i(K). \tag{14}$$

By sorting the gray correlation degree γ_i in order of magnitude, the relative influence degree of various factors on the reference factor can be obtained. The framework of this study is shown in Figure 2.

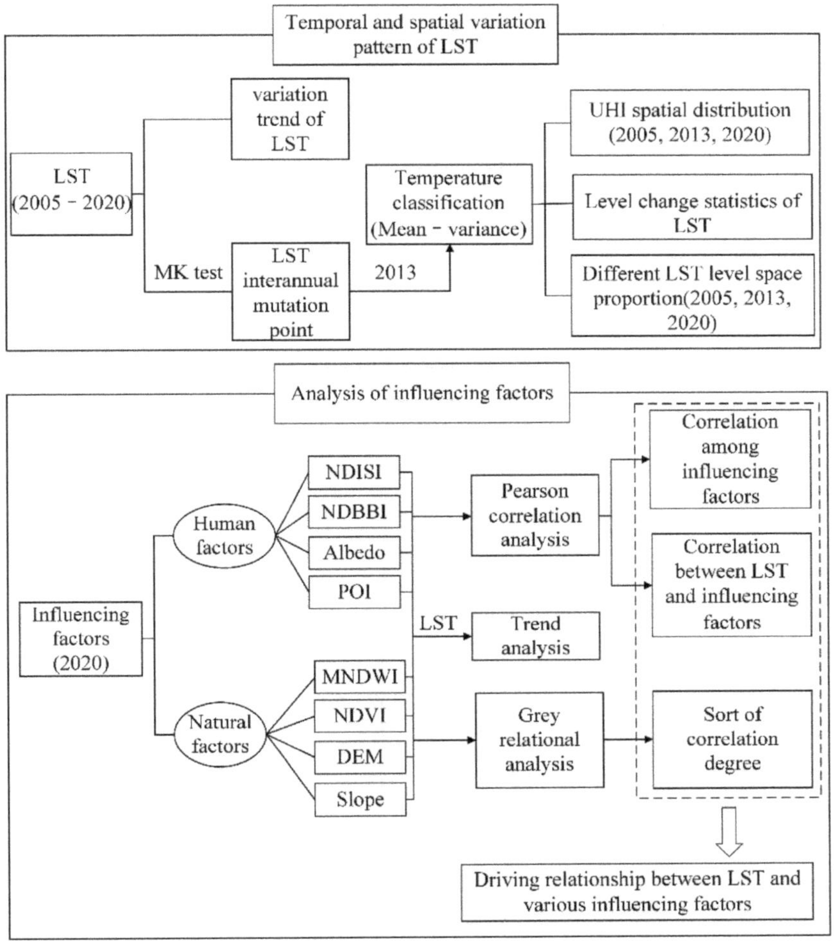

Figure 2. Framework of this study.

3. Results

3.1. Characteristics of Urban Thermal Environment Evolution

3.1.1. Interannual Variation Characteristics of LST

As can be seen from Figure 3A, the interannual variation trend of surface temperature in Zhengzhou from 2005 to 2020 was small, and the variation trends of the mean and maximum values were similar, but they all showed an overall upward trend. From 2005 to 2020, the mean value of LST increased by 0.92 °C, and the maximum value increased by 0.85 °C. Moreover, the mean value and maximum value both reached the maximum in 2019 (26.59 °C and 31.58 °C, respectively).

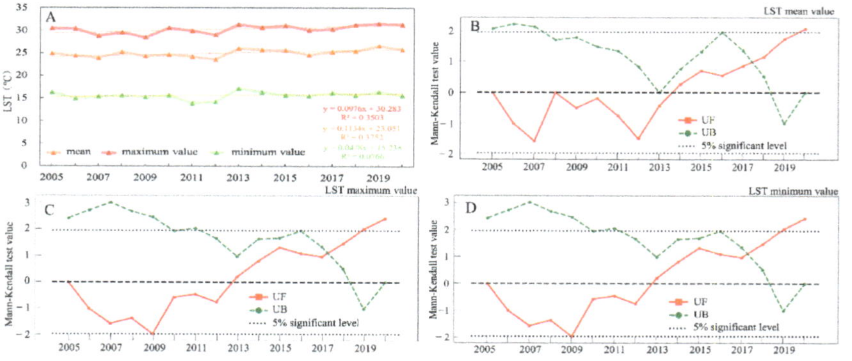

Figure 3. Variation of LST in Zhengzhou from 2005 to 2020 and M–K test values: (**A**) interannual variation; (**B**) M–K mean; (**C**) M–K maximum; (**D**) M–K minimum (yellow: mean, red: maximum, green: minimum).

Figure 3B–D show the M–K test values of the interannual variation of the land surface temperature in Zhengzhou. As can be seen from Figure 3B–D, the UF curve was in a downward trend before 2013, and the trend change was not obvious. In the second half of 2013, it began to show an upward trend, while there was a significant upward trend in the second half of 2019. The UF curve of LST maximum and minimum values began to show an upward trend in 2013, while, after 2019, the upward trend was significant.

3.1.2. Surface Temperature Classification

To identify the reason for the difference in LST in the study area, the mean-variance method was used to classify the LST into seven categories, as shown in Table 2: extremely high temperature, high temperature, relatively high temperature, medium temperature, relatively low temperature, low temperature, and extremely low temperature [26]. The first three categories were considered UHI zones in this study.

Table 2. Classification of LST.

Temperature Rating	Extremely High-Temperature Zone	High-Temperature Zone	Relatively High -Temperature Zone	Medium-Temperature Zone	Relatively Low-Temperature Zone	Low-Temperature Zone	Extremely Low-Temperature Zone
Temperature range	t ≥ u + 2.5 std	u + 1.5 std ≤ t < u +2.5 std	u + 0.5 std ≤ t < u + 1.5 std	u − 0.5 std ≤ t < u + 0.5 std	u − 1.5std ≤ t < u − 0.5 std	u − 2.5 std ≤ t < u − 1.5 std	t < u − 2.5 std

Note: u represents the mean value of LST; std represents the standard deviation of LST.

3.1.3. Spatial Evolution Characteristics of Urban Thermal Environment

Figure 4 shows the spatial evolution of the urban heat island in Zhengzhou in 2005, 2013, and 2020. As shown in Figure 4, the UHI area was irregular and gradually spread from north to south, and it was distributed in the main city and southeast in 2020. High temperature mainly occurred in the main city areas and densely built areas, while low temperature was most concentrated in the areas covered by rivers, grasslands, and woodlands. The spatial distribution characteristics of UHI were similar in the three periods, showing an obvious "cold island" effect in the Yellow River basin and the concentrated distribution area of woodland. The extremely high-temperature zone, relatively high-temperature zone, and high-temperature zone continued to expand along densely populated areas from 2005 to 2020.

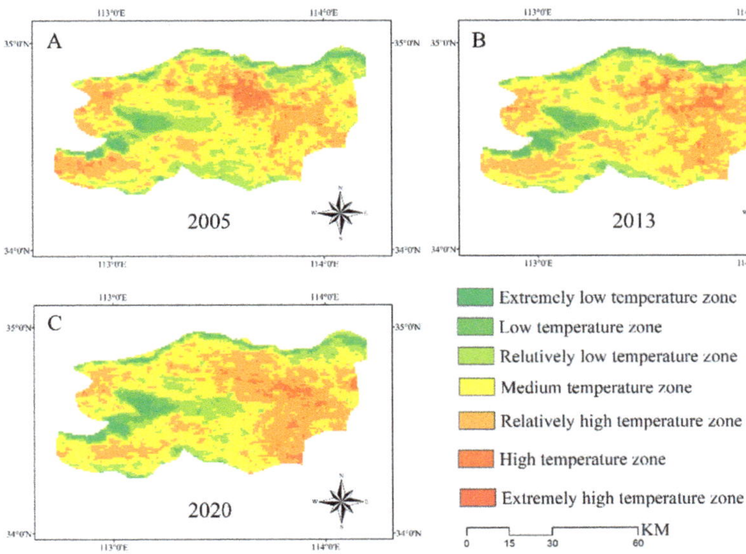

Figure 4. Spatial evolution of UHI in Zhengzhou in 2005, 2013, and 2020: (**A**) 2005, (**B**) 2013, (**C**) 2020.

3.1.4. Temporal Evolution Characteristics of Urban Thermal Environment

As shown in Table 3, the changes in LST grades in 2005, 2013, and 2020 were calculated. Table 3 indicates that the percentage area with weakened LST grades in Zhengzhou from 2005 to 2020 was 23.51%, while unchanged areas accounted for 53.72%, and the enhanced areas accounted for 22.77%. From 2005 to 2013, 20.83% of the area was represented by weakened LST grades, in contrast to 57.48% for unchanged areas and 21.69% for enhanced areas. From 2013 to 2020, the percentage area with weakened LST grades was 21.20%, an increase of 0.37% over the previous period. The percentage area with a constant LST level was 58.43%, which was 0.95% higher compared to the previous period. The percentage area with an enhanced LST level was 20.37%, which was 1.32% lower than that of the previous period. Through the above analysis, it was found that, in the last 15 years, the LST grade in Zhengzhou changed to different degrees. Compared with 2013–2020, the proportion of areas with weakened and unchanged LST grades in 2005–2013 showed an upward trend, while the enhanced areas showed a downward trend.

Table 3. LST grade change detection statistics in 2005–2013, 2013–2020, and 2005–2020.

Category	Range	2005–2013		2013–2020		2005–2020	
		Grade Percentage	Class Percentage	Grade Percentage	Class Percentage	Grade Percentage	Class Percentage
Weaken	−6	0.00%	20.83%	0.00%	21.20%	0.00%	23.51%
	−5	0.00%		0.00%		0.00%	
	−4	0.00%		0.00%		0.00%	
	−3	0.00%		0.01%		0.00%	
	−2	1.24%		0.57%		1.07%	
	−1	19.59%		20.62%		22.44%	
Constant	0	57.48%	57.48%	58.43%	58.43%	53.72%	53.72%
Enhance	1	20.46%	21.69%	19.73%	20.37%	20.95%	22.77%
	2	1.17%		0.63%		1.70%	
	3	0.06%		0.01%		0.12%	
	4	0.00%		0.00%		0.00%	
	5	0.00%		0.00%		0.00%	
	6	0.00%		0.00%		0.00%	

In order to better research the temporal and spatial evolution of the urban heat island in Zhengzhou, the proportions of different temperature grades in 2005, 2013, and 2020 were calculated. As shown in Table 4, in 2005, 2013, and 2020, the proportions of extremely low-temperature area, low-temperature area, and relatively high-temperature area in Zhengzhou showed an upward trend, with the proportion of relatively high-temperature area increasing by 4.03%. On the other hand, extremely high-temperature, high-temperature, and low-temperature areas showed a downward trend, with the high-temperature and relatively low-temperature areas decreasing by 1.72% and 3.22%, respectively, whereby the high-temperature area, extremely high-temperature area, and relatively low-temperature area transformed into relatively high-temperature areas. By 2020, the proportion of the relatively high-temperature zone reached 30.47%, and it was concentrated in the southeast, with a small distribution in the west and southwest. In addition, in 2005, 2013, and 2020, the heat island area accounted for 30.88%, 32.93%, and 32.96%, respectively, showing a 2.08% increase from 2005 to 2020.

Table 4. Proportion of different temperature grades (%) in 2005, 2013, and 2020.

Year	Extremely High	High	Relatively High	Medium	Relatively Low	Low	Extremely Low
2005	0.27	4.17	26.44	43.44	18.02	5.63	2.03
2013	0.12	3.01	29.80	43.98	13.8	6.61	2.67
2020	0.04	2.45	30.47	43.48	14.80	6.00	2.78

3.2. Analysis of Driving Factors of Urban Thermal Environment

3.2.1. Correlation Analysis

According to the analysis of the thermal environment effect in Zhengzhou in 2005, 2013, and 2020, the heat island effect was the most obvious in 2020. Therefore, the influencing factors were analyzed on the basis of the LST data in 2020. In this research, according to the city location and terrain characteristics of Zhengzhou, the relevant human and natural factors were selected for analysis, as shown in Table 5. Among them, the human factors included urban construction intensity (NDISI, NDBBI, and Albedo) and the socioeconomic activity index (POI kernel density), while the natural factors included water body, vegetation, and topographic landforms.

Table 5. Influencing factors of urban thermal environment.

First Level Indicators	Second Level Indicators	Third Level Indicators
Natural factors	Water body	MNDWI
	Vegetation and	NDVI
	Topographic features	Slope
		DEM
Human factors	Intensity of urban construction	NDISI
		NDBBI
		Albedo
	Socioeconomic activities	POI

The LST data and impact factor data of the study area were extracted in GIS using the grid turning point tool before conducting correlation analysis. The Pearson correlation coefficients among indices (Table 6) and the Pearson correlation coefficients between LST and each index (Table 7) were obtained. It can be seen from Tables 6 and 7 that the factors influencing LST in Zhengzhou showed weak and moderate correlation at the 0.01 significance level, and the absolute value range of the correlation coefficient R was between 0.045 and 0.761. Moreover, the eight influencing factors were correlated with LST at the 0.01 significance level, with correlation coefficients $|R| \in [0.027, 0.574]$. Among them, as shown in Table 7, LST had a significant negative correlation with DEM, MNDWI, NDVI, and Slope, indicating that vegetation, water, and high terrain had a cooling effect, which could alleviate the UHI effect. LST had a significant positive correlation with POI, NDBBI, NDISI, and Albedo. The results show that human activities and the intensity of urban construction affected the urban thermal environment, and the heat generated by high-intensity human activities did not dissipate well, causing the urban heat island phenomenon.

Table 6. Pearson correlation among influencing factors.

Factor	NDVI	MNDWI	DEM	Slope	NDISI	NDBBI	Albedo	POI
NDVI	1	−0.379 **	0.368 **	0.330 **	−0.301 **	−0.385 **	−0.659 **	−0.338 **
MNDWI	−0.379 **	1	−0.241 **	−0.197 **	0.687 **	−0.293 **	0.249 **	0.182 **
DEM	0.368 **	−0.241 **	1	0.761 **	−0.208 **	−0.330 **	−0.701 **	−0.136 **
Slope	0.330 **	−0.197 **	0.761 **	1	−0.185 **	−0.349 **	−0.641 **	−0.110 **
NDISI	−0.301 **	0.687 **	−0.208 **	−0.185 **	1	−0.045 **	0.136 **	0.187 **
NDBBI	−0.385 **	−0.293 **	−0.330 **	−0.349 **	−0.045 **	1	0.430 **	0.057 **
Albedo	−0.659 **	0.249 **	−0.701 **	−0.641 **	0.136 **	0.430 **	1	0.180 **
POI	−0.338 **	0.182 **	−0.136 **	−0.110 **	0.187 **	0.057 **	0.180 **	1

Note: ** indicates that the correlation was significant at the level of 0.01 (detection < 0.01).

Table 7. Correlation coefficients between factors and LST.

—	NDVI	MNDWI	DEM	Slope	NDISI	NDBBI	Albedo	POI
LST	−0.301 **	−0.027 **	−0.574 **	−0.568 **	0.141 **	0.457 **	0.527 **	0.195 **

Note: ** indicates that the correlation was significant at the level of 0.01 (detection < 0.01).

Figure 5 reflect the linear relationship between LST and various influencing factors. As shown in Figure 5, LST showed a downward trend under the action of NDVI, MNDWI, DEM, and Slope, with the downward trends of DEM and Slope being more obvious (R^2 = 0.3297 and 0.3222, respectively). Furthermore, LST was increased under the action of NDISI, NDBBI, Albedo, and POI. The trend analysis was consistent with the correlation analysis. Furthermore, the interaction between LST and Albedo showed spatial heterogeneity due to the complex effects of surface cover, elevation, and other factors [50]. Generally speaking, LST and Albedo were on the rise, which was related to the rise in surface temperature, the increase in vegetation and soil water stress, and the decrease in water content, which led to the increase in surface albedo. Moreover, the rise in surface temperature, vegetation growth, and increase in reflectivity in the near-infrared band also led to a significant increase in surface albedo. However, LST and Albedo showed a downward trend in the interval of 0.2–0.3, and these points were mainly distributed in rivers and grasslands. With the increase in Albedo, the light radiation energy absorbed by the surface decreased, resulting in a decrease in LST (Figure 5G).

Figure 6 shows the spatial distribution of each index in Zhengzhou. NDVI was basically consistent with the vegetation distribution in the study area. The improved MNDWI distribution was basically consistent with the water distribution in Zhengzhou. The southwest elevation of Zhengzhou is large, and the slope had a positive correlation with the elevation. The spatial distribution of NDBBI, Albedo, and NDISI was related to the type of urban underlying surface, with high values concentrated in the built-up areas. High POI values were concentrated in the main city area.

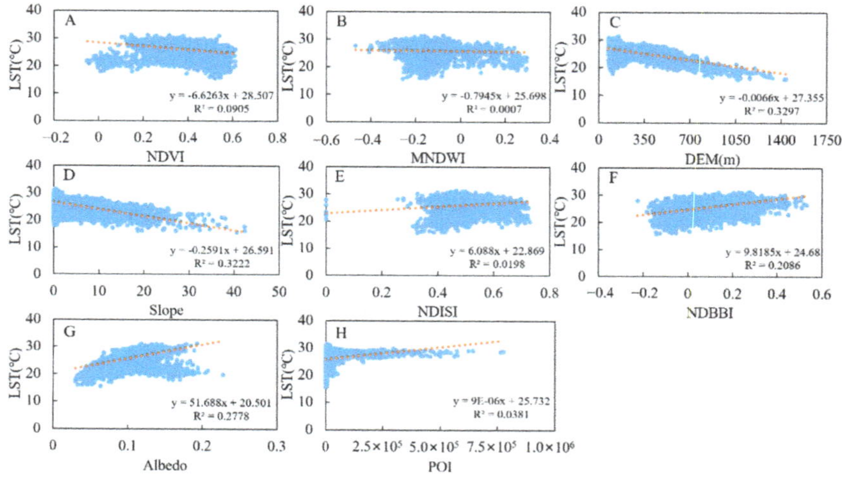

Figure 5. Linear relationship between influencing factors and LST in Zhengzhou: (**A**) NDVI; (**B**) MNDWI; (**C**) DEM; (**D**) Slope; (**E**) NDISI; (**F**) NDBBI; (**G**) Albedo; (**H**) POI.

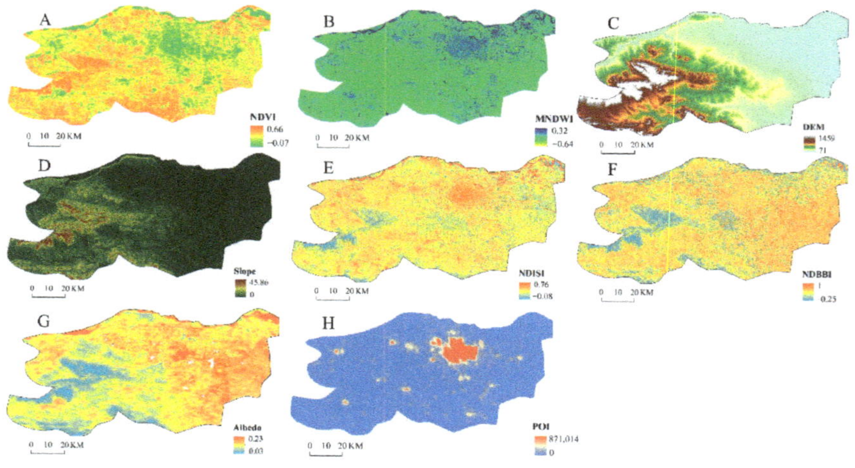

Figure 6. Spatial distribution of indicators in Zhengzhou: (**A**) NDVI; (**B**) MNDWI; (**C**) DEM; (**D**) Slope; (**E**) NDISI; (**F**) NDBBI; (**G**) Albedo; (**H**) POI.

3.2.2. Gray Correlation Analysis

The correlation between the above eight indices and LST was analyzed, but the correlation coefficient does not represent the contribution of each index to the change in LST. Because the correlation coefficient between variables indicates how close they are to each other, when there are many factors, this correlation only reflects their compound relationship, while it does not represent the relative influence degree or effect of each factor on the change in heat island intensity [27,51]. Thus, in order to reveal the contribution of the eight indices to LST, this study took LST as the reference series and the other eight influencing factors as the comparison series, before calculating their gray correlation degree. The results are shown in Table 8.

Table 8. The correlation of LST with each index.

Impact Factor	Correlation	Sort
NDISI	0.99978	1
Albedo	0.99965	2
NDVI	0.99943	3
MNDWI	0.99919	4
DEM	0.99834	5
NDBBI	0.99831	6
Slope	0.99718	7
POI	0.98030	8

As can be seen from Table 8, there were some differences in the order of correlation coefficient between these eight indices and LST. Because the correlation degree in gray theory reflects the correlation degree of independent variables compared with dependent variables, it can explain the dependent variable through the transmission of other independent variables without considering one independent variable [27,52]. In addition, the correlation degrees between the eight indices and LST were all above 0.98, and the difference between the maximum value and minimum value was only 0.01948. This shows that, although the indices were ranked sequentially, they were highly correlated with LST. According to the order of correlation degree, the contribution degree of the urban construction intensity and vegetation water body to LST change was the largest, and the correlation degree was above 0.999. This indicted that these four factors had a high degree of synchronous change with LST in the process of development and change; therefore, they were the most direct factors leading to the change in heat island intensity. However, the correlation of DEM, NDBBI, and Slope with LST was between 0.996 and 0.999, indicating that these three factors contributed extensively to the change in urban LST. In fact, NDBBI was also indirectly reflected in NDISI. Moreover, the correlation degree of social economic activity index POI was the lowest, but its value was 0.98030, indicating that it also had a high impact on the change in LST.

4. Discussion

Urbanization leads to changes in urban atmospheric dynamic characteristics and land use/land cover types, thereby affecting the formation of the urban heat island effect [26]. Furthermore, the spatial heterogeneity of LST may be affected by topography in different areas [36]. The terrain of Zhengzhou is high in the west and low in the east. The results showed that the UHI effect in Zhengzhou had obvious spatial differentiation characteristics. Cold islands mainly occurred in the west, densely forested areas, and Yellow River Basin, whereas heat islands were mainly distributed in southeast plains and built-up areas, where the population was dense, and the proportion impervious water surfaces rapidly increased, thus hindering the dissipation of heat and forming a large area of high temperature. Therefore, LST had the characteristics of "low on the periphery and high in the middle".

A city is a complex dynamic system composed of social connections, human activities, and infrastructure. The UHI is the result of multiple factors of local climate and human activities [26]. The interaction between surface temperature and vegetation dynamics under different land-cover types leads to changes in the spectral radiance and texture of surface temperature, resulting in the spatial pattern of urban heat island [8]. Moreover, impervious surfaces (such as concrete, cement, and asphalt) usually show lower emissivity and higher heat capacity than natural surfaces [53]. In the analysis of LST driving factors in Zhengzhou, the urban construction intensity contributed the most to the formation of the UHI effect, and the expansion of the urban impermeable water surface increased the absorption of solar radiation, resulting in a rapid increase in the land surface temperature [54]. On the other hand, vegetation and water bodies are effective tools to restrain the UHI effect and reduce the LST [55], while controlling the impervious surface percentage of construction land at a low level (e.g., below ~49%) can effectively alleviate the impact of the SUHI [54]. Therefore, this represents a strategic measure to effectively alleviate the UHI effect and build an ecologically livable city by controlling the proportion of impermeable water surface, increasing the urban vegetation coverage area, and rationally utilizing the water distribution, topography, and other features.

This research used multiple sources of data to study the temporal and spatial distribution characteristics and driving factors of Zhengzhou's thermal environment, while correlation analysis, trend analysis, and gray correlation analysis were applied to reveal the correlation between natural and human influencing factors. This study can be helpful for planners to understand the causes and mitigation measures of the UHI effect in Zhengzhou, through reasonably controlling the layout of buildings, effectively utilizing the distribution of vegetation and water, and actively guiding urban ventilation, so as to achieve the purpose of reducing LST and building a livable city. From this perspective, follow-up research can enable planners to more comprehensively understand the driving factors of the UHI effect.

5. Conclusions

Exploring the mitigation strategies of land surface temperature in Zhengzhou is of great significance for sustainable development and environmental quality. In this study, the main conclusions were as follows:

(1) The annual changes in LST in Zhengzhou from 2005 to 2020 were small, with a mutation point in 2013. Furthermore, compared with 2005, in 2020, the mean value of LST increased by 0.92 °C, the percentage of LST-enhanced areas was 22.77%, and the area of the heat island increased by 2.08%.

(2) The spatial pattern of the urban heat island showed an irregular block distribution, gradually spreading from north to south from 2005 to 2020; in 2020, there was a large block distribution in the main city and southeast. In addition, high temperatures mainly occurred in the main urban areas and densely built areas, whereas there was an obvious "cold island" effect in the concentrated distribution areas of forest land and the Yellow River basin.

(3) The results of correlation analysis, trend analysis, and gray correlation analysis showed that human factors (NDISI, NDBBI, Albedo, and POI) were positively correlated with LST, which intensified the formation of the UHI effect, with the influence of Albedo on LST showing obvious spatial heterogeneity. Natural factors (NDVI, MNDWI, DEM, and Slope) were negatively correlated with LST. Among them, the intensity of urban construction had the highest contribution to the formation of the UHI effect, and the cooling effect of vegetation and water was better than that of topography.

Generally, the UHI strength of Zhengzhou City revealed a significant increasing trend from 2005 to 2020. Zhengzhou's altitude is high in the west and low in the east, and there was a negative correlation of DEM and Slope with LST. Therefore, it is possible to reasonably control the layout of urban buildings as a function of the topography, such that mountain wind and cold air can smoothly enter the city and accelerate the air flow. In addition, the cooling effect of vegetation and water was obviously better than that of topography. In urban planning, connecting vegetation, water, and road networks in the urban ventilation corridor represents an effective measure to alleviate the urban heat island effect. As the economic center of Henan Province, Zhengzhou's comprehensive influence of human activities and urban construction intensity on the urban underlying surface is an important factor mediating the UHI. These results can help decision makers and urban planners to make rational and scientific decisions and promote the sustainable development of cities such as Zhengzhou in the future.

Author Contributions: Conceptualization, S.Z., Y.C. and D.L.; methodology, D.L. and S.Z.; software, M.Z.; validation, M.Z., D.L., W.T., S.X. and S.Y.; formal analysis, Q.C. and S.Z.; investigation, S.Z. and D.L.; resources, S.Z.; data curation, D.L.; writing—original draft preparation, D.L. and S.Z.; writing—review and editing, S.Z. and D.L.; visualization, D.L., M.Z., W.T., S.X. and S.Y.; supervision, S.Z. and Y.C.; project administration, Y.C. and S.Z.; funding acquisition, S.Z. The first two authors contributed equally to this work and should be considered co-first authors. All authors have read and agreed to the published version of the manuscript.

Funding: This research was funded by the National Natural Science Foundation of China (grant numbers 42005102 and 42071415), the Qinghai Science and Technology Department Project (grant number 2020-ZJ-711), the Second Tibetan Plateau Comprehensive Scientific Expedition (grant number 2019QZKK0104), the National Key Research and Development Program of China (grant number 2021YFE0106700), and the Outstanding Youth Foundation of Henan Natural Science Foundation (grant number 202300410049).

Data Availability Statement: The land-use data and DEM data in 2020 were obtained from the Resource and Environmental Science Data Center of China (http://www.resdc.cn, accessed on 14 March 2022 and 5 March 2022). The Landsat8 OLI_TIRS remote sensing image of 22 May 2020 was obtained from the geospatial data cloud (http://www.gscloud.cn/#page1/1, accessed on 8 March 2022). LST and NDVI, Albedo MODIS data products were download at https://modis.gsfc.nasa.gov (accessed on 1 March 2022 and 14 March 2022). Zhengzhou POI data in 2020 were obtained from Gaode map (https://ditu.amap.com, accessed on 15 March 2022).

Acknowledgments: The authors thank the reviewers who provided valuable comments to improve the paper.

Conflicts of Interest: The authors declare no conflict of interest.

References

1. Li, N.; Wu, H.; Luan, Q. Land surface temperature downscaling in urban area: A case study of Beijing. *Natl. Remote Sens. Bull.* **2021**, *25*, 1808–1820.
2. Zhu, J.; Zhu, S.; Yu, F.; Zhang, G.; Xu, Y. A downscaling method for ERA5 reanalysis land surface temperature over urban and mountain areas. *Natl. Remote Sens. Bull.* **2021**, *25*, 1778–1791.
3. Stone, B., Jr.; Vargo, J.; Liu, P.; Hu, Y.; Russell, A. Climate change adaptation through urban heat management in Atlanta, Georgia. *Environ. Sci. Technol.* **2013**, *47*, 7780–7786. [CrossRef]
4. Halder, B.; Bandyopadhyay, J.; Banik, P. Evaluation of the Climate Change Impact on Urban Heat Island Based on Land Surface Temperature and Geospatial Indicators. *Int. J. Environ. Res.* **2021**, *15*, 819–835. [CrossRef]
5. Voogt, J.A.; Oke, T.R. Thermal remote sensing of urban climates. *Remote Sens. Environ.* **2003**, *86*, 370–384. [CrossRef]
6. Lu, D.; Weng, Q. Use of impervious surface in urban land-use classification. *Remote Sens. Environ.* **2006**, *102*, 146–160. [CrossRef]
7. Wang, Z.; Sun, D.; Hu, C.; Wang, Y.; Zhang, J. Seasonal Contrast and Interactive Effects of Potential Drivers on Land Surface Temperature in the Sichuan Basin, China. *Remote Sens.* **2022**, *14*, 1292. [CrossRef]
8. Weng, Q.; Lu, D.; Schubring, J. Estimation of Land Surface Temperature–Vegetation Abundance Relationship for Urban Heat Island Studies. *Remote Sens. Environ.* **2004**, *89*, 467–483. [CrossRef]
9. Zhao, C.; Jensen, J.; Weng, Q.; Currit, N.; Weaver, R. Use of Local Climate Zones to investigate surface urban heat islands in Texas. *GIScience Remote Sens.* **2020**, *57*, 1083–1101. [CrossRef]
10. Zhao, C.; Jensen, J.; Weng, Q.; Currit, N.; Weaver, R. Application of airborne remote sensing data on mapping local climate zones: Cases of three metropolitan areas of Texas, US. *Comput. Environ. Urban Syst.* **2019**, *74*, 175–193. [CrossRef]
11. Biggart, M.; Stocker, J.; Doherty, R.M.; Wild, O.; Carruthers, D.; Grimmond, S.; Han, Y.; Fu, P.; Kotthaus, S. Modelling spatiotemporal variations of the canopy layer urban heat island in Beijing at the neighbourhood scale. *Atmos. Chem. Phys.* **2021**, *21*, 13687–13711. [CrossRef]
12. Streutker, D.R. Satellite-measured growth of the urban heat island of Houston, Texas. *Remote Sens. Environ.* **2003**, *85*, 282–289. [CrossRef]
13. He, B.; Zhao, Z.; Shen, L.; Wang, H.; Li, L. An Approach to Examining Performances of Cool/Hot Sources in Mitigating/Enhancing Land Surface Temperature under Different Temperature Backgrounds Based on Landsat 8 Image. *Sustain. Cities Soc.* **2019**, *44*, 416–427. [CrossRef]
14. Yang, J.; Zhan, Y.; Xiao, X.; Xia, J.C.; Sun, W.; Li, X. Investigating the Diversity of Land Surface Temperature Characteristics in Different Scale Cities Based on Local Climate Zones. *Urban Clim.* **2020**, *34*, 100700. [CrossRef]
15. Roy, S.; Pandit, S.; Eva, E.A.; Bagmar, M.S.H.; Papia, M.; Banik, L.; Dube, T.; Rahman, F.; Razi, M.A. Examining the nexus between land surface temperature and urban growth in Chattogram Metropolitan area of Bangladesh using long term Landsat series data. *Urban Clim.* **2020**, *32*, 100593. [CrossRef]

16. Hough, I.; Just, A.C.; Zhou, B.; Dorman, M.; Lepeule, J.; Kloog, I. A multi-resolution air temperature model for France from MODIS and Landsat thermal data. *Environ. Res.* **2020**, *183*, 109244. [CrossRef]
17. Xing, Z.; Li, Z.; Duan, S.; Liu, X.; Zheng, X.; Leng, P.; Gao, M.; Zhang, X.; Shang, G. Estimation of daily mean land surface temperature at global scale using pairs of daytime and nighttime MODIS instantaneous observations. *ISPRS J. Photogramm. Remote Sens.* **2021**, *178*, 51–67. [CrossRef]
18. Wu, W.; Li, L.; Li, C. Seasonal variation in the effects of urban environmental factors on land surface temperature in a winter city. *J. Clean. Prod.* **2021**, *299*, 126897. [CrossRef]
19. Wei, B.; Bao, Y.; Yu, S.; Yin, S.; Zhang, Y. Analysis of land surface temperature variation based on MODIS data a case study of the agricultural pastural ecotone of northern China. *Int. J. Appl. Earth Obs. Geoinf.* **2021**, *100*, 102342. [CrossRef]
20. Li, L.; Zha, Y.; Zhang, J. Spatially Non-Stationary Effect of Underlying Driving Factors on Surface Urban Heat Islands in Global Major Cities. *Int. J. Appl. Earth Obs. Geoinf.* **2020**, *90*, 102131. [CrossRef]
21. Wicht, M.; Wicht, A.; Osińska-Skotak, K. Detection of ventilation corridors using a spatiotemporal approach aided by remote sensing data. *Eur. J. Remote Sens.* **2017**, *50*, 254–267. [CrossRef]
22. Liu, D.; Zhou, S.; Wang, L.; Chi, Q.; Zhu, M.; Tang, W.; Zhao, X.; Xu, S.; Ye, S.; Lee, J.; et al. Research on the Planning of an Urban Ventilation Corridor Based on the Urban Underlying Surface Taking Kaifeng City as an Example. *Land* **2022**, *11*, 206. [CrossRef]
23. Yao, L.; Sun, S.; Song, C.; Li, J.; Xu, W.; Xu, Y. Understanding the spatio-temporal pattern of the urban heat island footprint in the context of urbanization, a case study in Beijing, China. *Appl. Geogr.* **2021**, *133*, 102496. [CrossRef]
24. Zhou, S.; Wang, K.; Yang, S.; Li, W.; Zhang, Y.; Zhang, B.; Fu, Y.; Liu, X.; Run, Y.; Chubwa, O.; et al. Warming Effort and Energy Budget Difference of Various Human Land Use Intensity: Case Study of Beijing, China. *Land* **2020**, *9*, 280. [CrossRef]
25. Wang, Z.; Meng, Q.; Allam, M.; Hu, D.; Zhang, L.; Menenti, M. Environmental and anthropogenic drivers of surface urban heat island intensity: A case-study in the Yangtze River Delta, China. *Ecol. Indic.* **2021**, *128*, 107845. [CrossRef]
26. Xiong, Y.; Zhang, F. Effect of human settlements on urban thermal environment and factor analysis based on multi-source data: A case study of Changsha city. *J. Geogr. Sci.* **2021**, *31*, 819–838. [CrossRef]
27. Liu, Y.; Yuan, Z.; Kong, W.; Sun, B.; An, B. The Changing Trend of Heat Island Intensity and Main Influencing Factors during 1993–2012 in Xi'an City. *J. Nat. Resour.* **2015**, *30*, 974–985.
28. Zhao, C.; Jensen, J.; Weng, Q.; Weaver, R. A Geographically Weighted Regression Analysis of the Underlying Factors Related to the Surface Urban Heat Island Phenomenon. *Remote Sens.* **2018**, *10*, 1428. [CrossRef]
29. Giannaros, T.M.; Melas, D. Study of the urban heat island in a coastal Mediterranean City: The case study of Thessaloniki, Greece. *Atmos. Res.* **2012**, *118*, 103–120. [CrossRef]
30. Tian, L.; Lu, J.; Li, Y.; Bu, D.; Liao, Y.; Wang, J. Temporal characteristics of urban heat island and its response to heat waves and energy consumption in the mountainous Chongqing, China. *Sustain. Cities Soc.* **2021**, *75*, 103260. [CrossRef]
31. Teri, K.; Sian, P.; Diana, B.; Amy, H.; Sian, K.; Ko, K.; Lorena, R. How effective is 'greening' of urban areas in reducing human exposure to ground-level ozone concentrations, UV exposure and the 'urban heat island effect'? An updated systematic review. *Environ. Evid.* **2021**, *10*, 1–38.
32. Ma, Y.; Zhao, M.; Li, J.; Wang, J.; Hu, L. Cooling Effect of Different Land Cover Types: A Case Study in Xi'an and Xianyang, China. *Sustainability* **2021**, *13*, 1099. [CrossRef]
33. Yi, Y.; Shen, G.; Zhang, C.; Sun, H.; Zhang, Z.; Yin, S. Quantitative analysis and prediction of urban heat island intensity on urban-rural gradient: A case study of Shanghai. *Sci. Total Environ.* **2022**, *829*, 154264. [CrossRef]
34. Alibakhshi, Z.; Ahmadi, M.; Asl, M. Modeling Biophysical Variables and Land Surface Temperature Using the GWR Model: Case Study—Tehran and Its Satellite Cities. *J. Indian Soc. Remote Sens.* **2020**, *48*, 59–70. [CrossRef]
35. Mohd, S.; Mohd, R.; Naikoo, W.; Ali, M.; Usmani, T.; Rahman, A. Urban Heat Island Dynamics in Response to Land-Use/Land-Cover Change in the Coastal City of Mumbai. *J. Indian Soc. Remote Sens.* **2021**, *49*, 2227–2247.
36. Liu, X.; Ming, Y.; Liu, Y.; Yue, W.; Han, G. Influences of landform and urban form factors on urban heat island: Comparative case study between Chengdu and Chongqing. *Sci. Total Environ.* **2022**, *820*, 153395. [CrossRef] [PubMed]
37. Wang, J.; Liu, J.; Wang, Y.; Liao, Z.; Sun, P. Spatiotemporal variations and extreme value analysis of significant wave height in the South China Sea based on 71-year long ERA5 wave reanalysis. *Appl. Ocean Res.* **2021**, *113*, 102750. [CrossRef]
38. Rahaman, Z.; Kafy, A.; Saha, M.; Rahim, A.; Almulhim, A.; Rahaman, S.; Fattah, M.; Rahman, M.; Kalaivani, S.; Faisal, A.; et al. Assessing the impacts of vegetation cover loss on surface temperature, urban heat island and carbon emission in Penang city, Malaysia. *Build. Environ.* **2022**, *222*, 109335. [CrossRef]
39. Zhao, Y.; Wu, Q.; Wei, P.; Zhao, H.; Zhang, X.; Pang, C. Explore the Mitigation Mechanism of Urban Thermal Environment by Integrating Geographic Detector and Standard Deviation Ellipse (SDE). *Remote Sens.* **2022**, *14*, 3411. [CrossRef]
40. Zhao, H.; Zhang, H.; Miao, C.; Ye, X.; Min, M. Linking Heat Source–Sink Landscape Patterns with Analysis of Urban Heat Islands: Study on the Fast-Growing Zhengzhou City in Central China. *Remote Sens.* **2018**, *10*, 1268. [CrossRef]
41. Geeta, S.J.; Payal, M. Assessment of seasonal climate transference and regional influential linkages to land cover–Investigation in a river basin. *J. Atmos. Sol.-Terr. Phys.* **2020**, *199*, 105209.
42. Mostafa, A.M.; Gamal, S.E.A.; Mohamed, E.E. Impact of climate change on rainfall variability in the Blue Nile basin. *Alex. Eng. J.* **2021**, *61*, 3265–3275.
43. Zhu, M.; Liu, D.; Tang, W.; Chi, Q.; Zhao, X.; Xu, S.; Ye, S.; Wang, Y.; Cui, Y.; Zhou, S. Exploring the Ecological Climate Effects Based on Five Land Use Types: A Case Study of the Huang-Huai-Hai River Basin in China. *Land* **2022**, *11*, 265. [CrossRef]

44. Li, W.; Yang, J.; Li, X.; Zhang, J.; Li, S. Extraction of urban impervious surface information from TM image. *Remote Sens. Nat. Resour.* **2013**, *25*, 66–70.
45. Hou, Y.; Ding, W.; Liu, C.; Li, K.; Cui, H.; Liu, B.; Chen, W. Influences of impervious surfaces on ecological risks and controlling strategies in rapidly urbanizing regions. *Sci. Total Environ.* **2022**, *825*, 153823. [CrossRef]
46. He, J.; Zhang, H.; Su, H.; Zhou, X.; Chen, Q.; Xie, B.; You, T. Study on long-term change of global spectral surface albedo. *J. Atmos. Environ. Opt.* **2022**, *17*, 279–293.
47. Mao, X.; Zhang, L.; Mao, T.; Wang, Y.; Huang, J. Research on the Influence of Fund Based on Pearson Study of Fund Influence Based on Pearson Correlation Analysis and Regression Analysis Method. *J. Zhejiang Sci-Tech Univ. (Soc. Sci. Ed.)* **2017**, *38*, 306–311.
48. He, P.; Chen, H.; Li, H.; Xi, W. Grey analysis of the urban heat island effect factors of the medium-sized city of Chuxiong on Yunnan Plateau. *Prog. Geogr.* **2009**, *28*, 25–32.
49. Li, X.; Zhang, H.; Qu, Y. Land Surface Albedo Variations in Sanjiang Plain from 1982 to 2015: Assessing with GLASS Data. *Chin. Geogr. Sci.* **2020**, *30*, 876–888. [CrossRef]
50. Gao, T.; Shen, R.; Li, L.; Wang, Y.; Huang, A. Spatial and Temporal Variations of Land Surface Albedo and Its Influencing Factors Based on MODIS Data. *Clim. Environ. Res.* **2021**, *26*, 648–662.
51. Feng, Y.; Hong, Z.; Cheng, J.; Jia, L.; Tan, J. Low Carbon-Oriented Optimal Reliability Design with Interval Product Failure Analysis and Grey Correlation Analysis. *Sustainability* **2017**, *9*, 369. [CrossRef]
52. Guo, J.; Dong, Y.; Zhang, H. Misuse of gray incidence analysis in variable selection. *Syst. Eng. Theory Pract.* **2002**, *22*, 126–128.
53. He, W.; Cao, S.; Du, M.; Hu, D.; Mo, Y.; Liu, M.; Zhao, J.; Cao, Y. How Do Two- and Three-Dimensional Urban Structures Impact Seasonal Land Surface Temperatures at Various Spatial Scales? A Case Study for the Northern Part of Brooklyn, New York, USA. *Remote Sens.* **2021**, *13*, 3283. [CrossRef]
54. Chang, Y.; Xiao, J.; Li, X.; Zhou, D.; Wu, Y. Combining GOES-R and ECOSTRESS land surface temperature data to investigate diurnal variations of surface urban heat island. *Sci. Total Environ.* **2022**, *823*, 153652. [CrossRef] [PubMed]
55. Maryam, M.; Masoud, M.; Zarkesh, K.; Alireza, M.; Ali, J. Achieving sustainable development goals through the study of urban heat island changes and its effective factors using spatio-temporal techniques: The case study (Tehran city). In *Natural Resources Forum*; Blackwell Publishing Ltd.: Oxford, UK, 2022; Volume 46, pp. 88–115.

remote sensing

MDPI

Article

Assessing Local Climate Change by Spatiotemporal Seasonal LST and Six Land Indices, and Their Interrelationships with SUHI and Hot–Spot Dynamics: A Case Study of Prayagraj City, India (1987–2018)

Md. Omar Sarif [1,*], Rajan Dev Gupta [2] and Yuji Murayama [3]

1 Department of Geography, Lovely Professional University, Phagwara 144411, India
2 Civil Engineering Department, Motilal Nehru National Institute of Technology Allahabad, Prayagraj 211004, India
3 Faculty of Life and Environmental Sciences, University of Tsukuba, 1-1-1 Tennodai, Tsukuba City 305-8572, Ibaraki, Japan
* Correspondence: omar.28426@lpu.co.in or mdomarsarif@gmail.com

Abstract: LST has been fluctuating more quickly, resulting in the degradation of the climate and human life on a local–global scale. The main aim of this study is to examine SUHI formation and hotspot identification over Prayagraj city of India using seasonal Landsat imageries of 1987–2018. The interrelationship between six land indices (NDBI, EBBI, NDMI, NDVI, NDWI, and SAVI) and LST (using a mono-window algorithm) was investigated by analyzing correlation coefficients and directional profiling. NDVI dynamics showed that the forested area observed lower LST by 2.25–4.8 °C than the rest of the city landscape. NDBI dynamics showed that the built-up area kept higher LST by 1.8–3.9 °C than the rest of the city landscape (except sand/bare soils). SUHI was intensified in the city center to rural/suburban sites by 0.398–4.016 °C in summer and 0.45–2.24 °C in winter. Getis–Ord G_i* statistics indicated a remarkable loss of areal coverage of very cold, cold, and cool classes in summer and winter. MODIS night-time LST data showed strong SUHI formation at night in summer and winter. This study is expected to assist in unfolding the composition of the landscape for mitigating thermal anomalies and restoring environmental viability.

Keywords: LST; mono-window algorithm; land indices; correlation coefficients; directional profiling; SUHI; hotspots (Getis–Ord G_i* statistics); MODIS night-time LST; Prayagraj city

Citation: Sarif, M.O.; Gupta, R.D.; Murayama, Y. Assessing Local Climate Change by Spatiotemporal Seasonal LST and Six Land Indices, and Their Interrelationships with SUHI and Hot–Spot Dynamics: A Case Study of Prayagraj City, India (1987–2018). *Remote Sens.* **2023**, *15*, 179. https://doi.org/10.3390/rs15010179

Academic Editor: Weiqi Zhou

Received: 5 November 2022
Revised: 24 December 2022
Accepted: 26 December 2022
Published: 28 December 2022

1. Introduction

Globally, 55% of the total populace resided in urban areas in 2018, and prediction statistics show that if this trend continues, then the urban population will account for 68% of the total in 2050 [1]. Recently, the IPCC revealed in its report that the global mean surface temperature (GMST) increased by 1.53 °C and the global mean air temperature (GMAT) [both land and ocean] increased by 0.87 °C during the preindustrial period (1850–1900) and recent postindustrial period (2006–2015) [2]. The rise in land surface temperature (LST) has severe environmental consequences because unplanned urbanization will deteriorate climate equilibrium and hamper human life and health from microscale to macroscale [3,4]. In its AR6 2021 report, IPCC revealed that we are on the way to reaching 1.5 °C more global warming in the next 20 years. The unprecedented changes in the recent past have been highly challenging and alarming, leading to uncertain precipitation, increased glacier melting, mean sea level rising, floods, droughts, damage to agricultural land, and food shortage affecting every region of the globe [5].

The presence of Earth surface objects, such as asphalt, stones, pebbles, and sand, over the city landscape has diverse electromagnetic behavior in terms of evaporation, absorption, and radiation. Longwave radiation, as well as prevailing winds, assimilates into massive heat discharge from the Earth's surface [6]. These surface objects weaken evapotranspiration and accelerate sensitivity [7]. Consequently, a difference in LST has been observed in city landscapes where core city space experiences higher LST than suburban/rural sites. This distinctive LST characteristic is defined by the surface urban heat island (SUHI). The SUHI has emerged because of the conversion of natural land into built-up space at the cost of water bodies, bare land, and forest [8,9].

LST intensification in urban setups is a perilous factor responsible for deteriorating urban climate and degrading human life and living [10,11]. Much attention is now given to mitigating its severity and threats to varied aspects of the environment by policymakers, health authorities, urban planners, climatologists, and environmentalists [12–14]. In the recent past, researchers have been engrossed in the interrelationship between LST and different land indices, such as the normalized difference built-up index (NDBI), enhanced built-up and bareness index (EBBI), normalized difference vegetation index (NDVI), and normalized difference moisture index (NDMI). Worldwide, scientists have been intensively focused on how, where, and what magnitude of land use/land cover (LULC) or land indices dynamics have been influencing the climatic conditions as a result of LST intensification and SUHI, which directly or indirectly make the environment uncomfortable and unhealthy for all animals and plants. Some studies rigorously found similar facts in Taipei city of Taiwan [15], Phoenix city of the United States of America (USA) [16], Singapore [10], Dhaka city of Bangladesh [17], Kathmandu valley of Nepal [18], Nanjing city of China [19], Beijing city of China [20], Tokyo city of Japan [21], Tehran city of Iran [13], 70 selected cities of Europe [22], Hong Kong [23], Baltimore–DC metropolitan area of the USA [24], and Cairo city of Egypt [25]. At the same time, researchers have discussed how their changing aspects affect and transform the environment of the city landscape in various Indian cities, such as Kolkata [26], Delhi and Mumbai [27], Chandigarh [28], Hyderabad [29], Noida [30], Lucknow [31], and Raipur [32].

Mal et al. (2020) conducted a study on the relationship of LST with LULC and elevation in the Ganga River basin, which includes major cities such as Kolkata, Patna, Allahabad (now Prayagraj), Varanasi, Lucknow, Kanpur, New Delhi, and Kathmandu during 2001–2019 using 1 km of MODIS Terra datasets [33]. Other studies over the Ganga River basin were also carried out for different cities such as Delhi [34], Kolkata [35], Kanpur and Patna [36], and Lucknow [30] using the MODIS/Landsat database. However, this study lacked a city-level analysis of LST profiling and SUHI information, especially for Prayagraj city. Furthermore, this study lacked effective land indices such as the NDBI, EBBI, NDVI, NDMI, normalized difference water index (NDWI), and soil-adjusted vegetation index (SAVI) to unfold the land dynamics and their role in the LST increase. Furthermore, the analytical results did not show details for thermal state analysis, including land indices dynamics and SUHI information, on a long spatiotemporal scale for Prayagraj city of India.

UN-Habitat (2018) introduced 17 sustainable development goals (SDGs). SDG-11 has stresses the city's resilience and sustainability and focuses on the significance of greenery and open spaces in bringing environmental viability and prosperity by coping with adverse local climate change and intense landscape transformation [37]. Various studies have shown that two main approaches help study the LST of urban climate, i.e., ground observations (GOBs) and satellite observations (SOBs). The GOB involves conventional data calculation of air temperature using urban and rural meteorological stations. The SOB has spatiotemporal resolutions with a mathematical background for estimating LST, and they are required to study the spatial variations of SUHI [3]. Therefore, in this study, we plan to use spatiotemporal Landsat imageries (1987–2018) to derive land indices, LST, and SUHI information in Prayagraj city using summer season (May–June) and winter season (December–January) datasets. We selected these times for our study on the basis of the finest spectral signature availability and albedo because of haze-free and cloud-free skies [38].

The primary aim of this present work is to examine the interrelations with LST dynamics using directional profiling on the summer/winter seasons during 1987–2018 in

Prayagraj city of India by investigating local level climate change through a long spatiotemporal analysis of six land indices (NDBI, EBBI, NDVI, NDMI, NDWI, and SAVI). In this connection, the tasks are (i) to assess the thermal state over the city landscape along with six land indices, (ii) to explore the interrelationship between six land indices and LST, (iii) to delineate the role of the six different land indices in LST increase using urban–rural directional profiling, (iv) to extract the SUHI state both at daytime and night-time, and (v) to extract the hotspots using Getis–Ord G_i* statistics. The dynamics of six land indices, LST, and their correlation are investigated to achieve these tasks. Then, the scenario of SUHI formation is discussed using directional profiling of LST by assessing how these six land indices impact the dynamics of LST in eight directions, namely, west, east, north, south, southwest, northeast, northwest, and southeast (center of the city to periphery). Next, the interrelationships between land indices and LST dynamics are validated using Google Earth images. Lastly, we delineate the hotspots to find warming or cooling spaces scattered over the city landscape.

2. Materials and Methods

2.1. Study Area

We selected Prayagraj city as a study area because this city was selected as a smart city by the MoHUA, i.e., the Ministry of Housing and Urban Affairs, Government of India (GoI), in 2015 [39]. This city is one of the biggest in terms of size, as well as historically enriched cities, in Uttar Pradesh state in India. Its location ranges from 25°23′7″N to 25°32′14″N latitude and 81°43′57″E to 81°53′59″E longitude, where the mean elevation is 93.77 m (Figure 1). The study area covers 72.98 km^2. This city is located over the holy place called Sangam (confluence of Ganga, Yamuna, and invisible Saraswati rivers) [40,41]. The sides of these rivers (Ganga and Yamuna) have been enriched by eroded materials from the Vidhyan uplands and Himalayas mountains [42]. The study area has a CWG-type climate, i.e., monsoon type with dry winters based on Koppen's climatic regions (KCR) scheme in India [43], and it has 744.1 mm of mean annual rainfall and 20–32.6 °C of mean annual temperature [44]. This city is widely known for religious gatherings at 6 year intervals, Maha Kumbh Mela and Ardha Kumbh Mela, where >100 million pilgrims congregate to make it the largest congregation in Asia [40].

2.2. Data Used

This study uses Landsat 5 (TM) and Landsat 8 (OLI/TIRS) satellite imageries with a spatial resolution of 30 m. These are employed for four distinct time points for distinctive seasons in summer and winter. The summer time points (STP) are (i) 4 June 1988 (S1) of Landsat 5 (TM), (ii) 12 May 1997 (S2) of Landsat 5 (TM), (iii) 10 May 2008 (S3) of Landsat 5 (TM), and (iv) 22 May 2018 (S4) of Landsat 8 (OLI/ TIRS). The winter time points (WTP) are (i) 11 December 1987 (W1) of Landsat 5 (TM), (ii) 3 December 1996 (W2) of Landsat 5 (TM), (iii) 16 January 2007 (W3) of Landsat 5 (TM), and (iv) 16 December 2018 (W4) of Landsat 8 (OLI/TIRS). We selected about 10 years of the gap to depict the dynamics of LST and land indices seasonally (summer and winter).

This study uses night-time LST derived from MODIS (Terra) satellite datasets to investigate summer/winter seasonal LST dynamics and SUHI state at night-times from 2007 to 2018 only as this satellite has been providing imagery since 2000. Before 2000, there were no available data on night-time LST. Table 1 shows the satellite datasets used in this study. The software, ERDAS IMAGINE 2014 was used for preprocessing these satellite images. The same dry summer and dry winter seasons were selected for obtaining the cloud-free data with the finest spectral information.

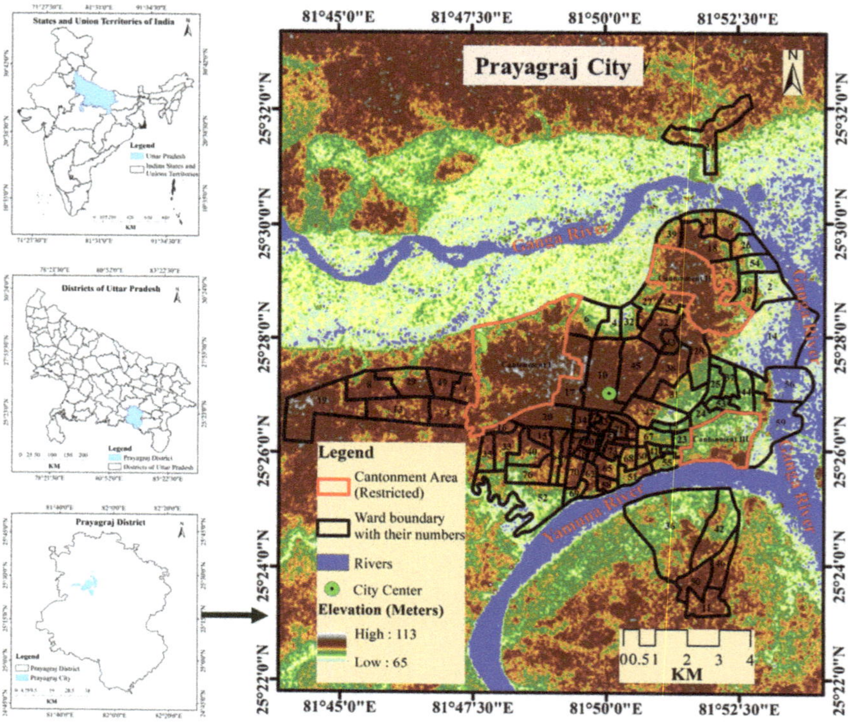

Figure 1. Map of Prayagraj city (India).

Table 1. Summary of the data used.

Satellite (Sensor)/Ancillary Data	Path/Row	Resolution/Scale	Season	Acquisition Date	Time (GMT)	Constants of Thermal Conversion		Source
						K_1	K_2	
Landsat-5 (TM)	143/42	30 m	Summer	04-06-1988	04:31:36	607.76 (Band 6)	1260.56 (Band 6)	United States Geological Survey (USGS) web portal (https://earthexplorer.usgs.gov/, accessed on 15 January 2019)
				12-05-1997	04:29:21	607.76 (Band 6)	1260.56 (Band 6)	
				10-05-2008	04:49:45	607.76 (Band 6)	1260.56 (Band 6)	
Landsat-8 (OLI/TIRS)				22-05-2018	05:00:01	774.8853 (Band 10)	1321.0789 (Band 10)	
Landsat-5 (TM)	143/42	30 m	Winter	11-12-1987	04:29:34	607.76 (Band 6)	1260.56 (Band 6)	
				03-12-1996	04:22:45	607.76 (Band 6)	1260.56 (Band 6)	
				16-01-2007	04:55:59	607.76 (Band 6)	1260.56 (Band 6)	
Landsat-8 (OLI/TIRS)				16-12-2018	05:00:58	480.8883 (Band 11)	1201.1442 (Band 11)	
MODIS (Terra)	-	1 km	Summer	11-05-2008	Night–time	-	-	
	-			22-05-2018		-	-	
	-	1 km	Winter	17-01-2007		-	-	
	-			15-12-2018		-	-	
ASTER	-	30 m	-	13-09-2017	-	-	-	
Ward boundary map	-	1:21,600	-	-	-	-	-	Prayagraj Nagar Nigam
Political map	-	1:4 M	-	2014	-	-	-	Survey of India

2.3. Methods

Figure 2 shows the overall methodological framework for the execution process. All the methods were discussed under subsequent heads.

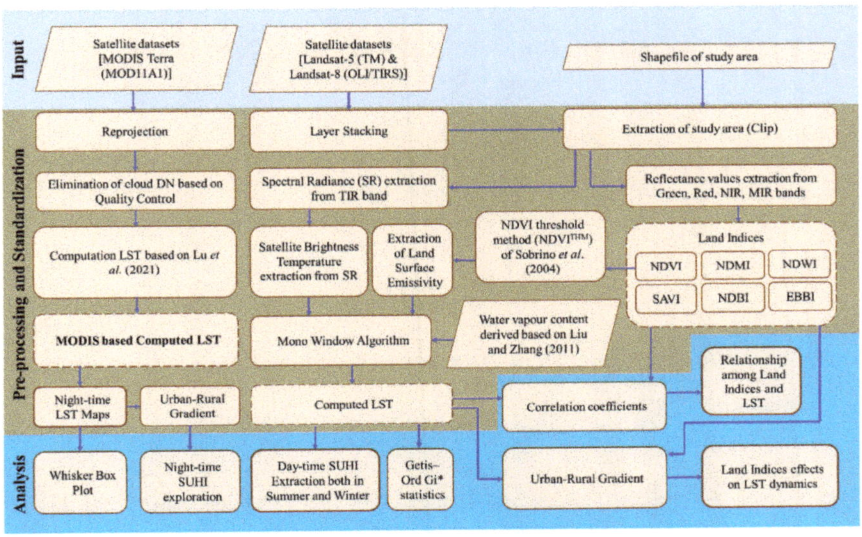

Figure 2. Overall framework of the present research procedure [23,45,46].

2.3.1. Land Indices

NDBI

In the evaluation of urban climate, NDBI is an important indicator. It ranges from −1 to +1. A higher positive value specifies bare soils. Lower positive and negative values adjacent to 0 show the dominance of built-up space. Higher negative values specify vegetation and water bodies. NDBI is calculated using Equation (1).

$$\text{NDBI} = \left[\frac{\text{MIR}_{\text{Band}} - \text{NIR}_{\text{Band}}}{\text{MIR}_{\text{Band}} + \text{NIR}_{\text{Band}}}\right], \tag{1}$$

where, in Landsat 5 TM, NIR_{Band} is band 4, and MIR_{Band} is band 5, whereas, in Landsat 8, NIR_{Band} is band 5 and MIR_{Band} is band 6.

EBBI

EBBI is widely used as a significant indicator in assessing urban climate. It ranges from 0 to +1. A higher positive value (perceived threshold > 0.1) specifies bare soils. A lower positive value (perceived threshold between 0.06 and 0.1) specifies built-up space. m positive value (perceived threshold < 0.06) adjacent to 0 specifies water bodies and vegetation. EBBI is calculated using Equation (2) [17].

$$\text{EBBI} = \left[\frac{\text{MIR}_{\text{Band}} - \text{NIR}_{\text{Band}}}{10\sqrt{\text{MIR}_{\text{Band}} + \text{TIR}_{\text{Band}}}}\right], \tag{2}$$

where, in Landsat 5 TM, NIR_{Band} is band 4, MIR_{Band} is band 5, and TIR_{Band} is band 6, whereas, in Landsat 8, NIR_{Band} is band 5, MIR_{Band} is band 6, and TIR_{Band} is band 10.

NDMI

Another critical indicator in urban climate assessment is NDMI. It ranges from −1 to +1. A positive value specifies vegetation and water bodies. A negative value indicates bare soils and built-up areas. It provides information on the moisture present in the landscape. NDMI is calculated using Equation (3) [47].

$$\text{NDMI} = \left[\frac{\text{NIR}_{\text{Band}} - \text{MIR}_{\text{Band}}}{\text{NIR}_{\text{Band}} + \text{MIR}_{\text{Band}}}\right] \tag{3}$$

where, in Landsat 5 TM, NIR_{Band} is band 4, and MIR_{Band} is band 5, whereas, in Landsat 8, NIR_{Band} is band 5, and MIR_{Band} is band 6.

NDVI

NDVI ranges from −1 to +1. A higher positive value specifies vegetation. A lower positive value indicates bare soils or built-up areas. The negative values specify water bodies. NDVI is calculated using Equation (4) [12].

$$\text{NDVI} = \left[\frac{\text{NIR}_{\text{Band}} - \text{Red}_{\text{Band}}}{\text{NIR}_{\text{Band}} + \text{Red}_{\text{Band}}}\right] \tag{4}$$

where Landsat 5 TM defines NIR_{Band} as band 4 and Red_{Band} as band 3, whereas Landsat 8 OLI/TIRS defines NIR_{Band} as band 5 and Red_{Band} as band 4.

NDWI

NDWI ranges from −1 to +1. A positive value specifies water bodies. A lower positive value adjacent to 0 specifies vegetation space. Negative values specify bare soils and built-up areas. NDWI is calculated using Equation (5) [48].

$$\text{NDWI} = \left[\frac{\text{Green}_{\text{Band}} - \text{NIR}_{\text{Band}}}{\text{Green}_{\text{Band}} + \text{NIR}_{\text{Band}}}\right] \tag{5}$$

where, in Landsat 5 TM, $\text{Green}_{\text{Band}}$ is band 2, and NIR_{Band} is band 4, whereas, in Landsat 8, $\text{Green}_{\text{Band}}$ is band 3, and NIR_{Band} is band 5.

SAVI

SAVI ranges from −1 to +1. A positive value specifies vegetation. A lower positive value adjacent to 0 specifies water bodies. Negative values specify bare soils and built-up space. It is calculated using Equation (6) [49].

$$\text{SAVI} = \left[\frac{\text{NIR}_{\text{Band}} - \text{Red}_{\text{Band}}}{\text{NIR}_{\text{Band}} + \text{Red}_{\text{Band}} + L} \times (L+1)\right] \tag{6}$$

where Landsat 5 TM defines NIR_{Band} as band 4 and Red_{Band} as band 3, whereas Landsat 8 OLI/TIRS defines NIR_{Band} as band 5 and Red_{Band} as band 4. L is the soil brightness correction factor, and it is a constant 0.5.

2.3.2. LST Retrieval

Landsat-Based LST Calculation

Several algorithms are available for retrieving LST for distinct satellite sensors. The most popular algorithms are the mono-window algorithm (MWA) [50], radiative transfer equation (RTE) algorithm [51], split-window algorithm (SWA) [52–54], and single-channel algorithm (SCA) [55,56] for retrieving LST using thermal bands of Landsat. RTE cannot be considered for use if the information of the atmospheric profile is not available on in situ parameters at the satellite pass [57]. The SWA provides accurate results but was not selected for use in the study area as it is specific to band 10 of Landsat 8 (OLI/TIRS) data only for LST computation because of its better calibration. The MWA and SCA also give good results [33,58]. In this study, we selected MWA for LST computation from multitemporal Landsat images because MWA shows significant accuracy for computing LST, with three essential parameters being indispensable to LST retrieval, i.e., the ground emissivity, the atmospheric transmittance (AT), and the effective mean temperature of the atmosphere [59], calculated using Equations (7)–(9), respectively.

$$T_s = \frac{\{a(1-C-D) + [b(1-C-D) + C + D] \times T_b - D \times T_a\}}{C} \quad (7)$$

$$C = \varepsilon \times \tau \quad (8)$$

$$D = (1-\tau) \times [1 + (1-\varepsilon)\tau] \quad (9)$$

where T_s defines the LST (K), T_a defines the mean atmospheric temperature (K), T_b defines the at-sensor pixel brightness temperature (K), C and D algorithm parameters are estimated through land surface emissivity (LSE) and AT, ε defines LSE, τ defines AT, and a and b are constants of the algorithm (−67.355351 and 0.458606, respectively). The following steps are needed to calculate Equations (7)–(9).

Step 1: The TIR band's pixels are converted into radiance. The radiance is computed using Equation (10) for band 6 of Landsat 5 (TM) and Equation (11) for band 10 of Landsat 8 (OLI/TIRS).

$$L_\lambda = \Lambda \times QCAL + \Gamma \quad (10)$$

$$L_\lambda = \frac{L_{max} - L_{min}}{QCAL_{max} - QCAL_{min}} \times (\mathrm{QCAL} - \mathrm{QCAL_{min}}) - \mathrm{L_{min}} \quad (11)$$

where L_λ represents spectral radiance at the top of atmosphere (TOA) ($W/(m^2{\cdot}sr{\cdot}\mu m)$), Λ represents the multiplicative rescaling factor for each specific band in the metadata, Γ represents the additive rescaling factor for each specific band in the metadata, and $QCAL$ represents the quantized and calibrated digital number (DN) values of standard product. $QCAL_{max}$ and $QCAL_{min}$ represent the maximum and minimum DN values of the images, respectively. L_{max} and L_{min} represent the TIR band's spectral radiance at $QCAL_{max}$ and $QCAL_{min}$, respectively. These values of the rescaling factor are available in the metadata of respective Landsat images.

Then, the at-sensor brightness temperature (BT) is computed using Equation (12).

$$T_b = \left[\frac{K_2}{\ln\left(\frac{K_1}{L_\lambda} + 1\right)}\right] \quad (12)$$

where T_b represents the at-sensor BT (K), and K_1 and K_2 (Wm^{-2}) are thermal conversion constants (prelaunch calibration) mentioned in the metadata (Table 1) of the respective sensors of Landsat datasets.

Step 2: To compute LST, LSE is one of the indispensable parameters [60]. The NDVI threshold ($NDVI^{THR}$) method was selected to calculate the LSE because of its significance in segregating pixels of vegetation, water, and soil [45]. LSE can be computed using Equations (13)–(15).

$$\varepsilon = \varepsilon_v P_v + \varepsilon_s(1 - P_V) + C \quad (13)$$

$$P_v = \left[\frac{NDVI - NDVI_S}{NDVI_V - NDVI_S}\right]^2 \quad (14)$$

$$C = (1-\varepsilon_s)\varepsilon_v F(1 - P_V) \quad (15)$$

where ε represents LSE, ε_v represents vegetation emissivity, ε_s represents soil emissivity, P_V represents proportionate of vegetation, C represents constant of surface characteristics, $NDVI_s$ represents NDVI of pure soil, $NDVI_v$ represents NDVI of pure vegetation, and F represents a geometric factor commonly considered as 0.55 [45,61].

The constant values of LSE were calculated using Equation (16) for Landsat 5 (TM) and Equation (17) for Landsat 8 (OLI/TIRS) [61], where ρ_{Red} represents the reflectance value of respective red bands of the imageries.

$$\varepsilon = \begin{cases} 0.979 + 0.035\rho_{\text{Red}} \rightarrow NDVI < 0.2 \\ 0.004P_V + 0.986\rho_{\text{Red}} \rightarrow 0.2 \leq NDVI \geq 0.5 \\ 0.99 \rightarrow NDVI > 0.5 \end{cases} \quad (16)$$

$$\varepsilon = \begin{cases} 0.979 + 0.046\rho_{\text{Red}} \rightarrow NDVI < 0.2 \\ 0.989P_V + 0.977\rho_{\text{Red}} \rightarrow 0.2 \leq NDVI \geq 0.5 \\ 0.987 + C \rightarrow NDVI > 0.5 \end{cases} \quad (17)$$

Step 3: The AT is another indispensable parameter to calculate LST. Before calculating AT, water vapor content should be calculated on the basis of the atmospheric profile (Table 2) using Equation (18) [59,61]. Then, AT is computed using Equation (19).

$$w = 0.0981 \times \left[10 \times 0.6108 \times \exp\left(\frac{17.27 \times (T_0 - 273.15)}{237.3 + (T_0 - 273.15)}\right) \times RH\right] + 0.1697 \quad (18)$$

where w represents the water vapor content (g/cm^2), RH represents the relative humidity, and T_0 represents the near-surface temperature. These atmospheric parameters (RH and T_0) were obtained from the Prediction of Worldwide Energy Resource (POWER) Project of the National Aeronautics and Space Administration (NASA) (https://power.larc.nasa.gov/, accessed on 27 November 2022).

$$\tau = 1.031412 - 0.11536w \quad (19)$$

Table 2. Estimation equations of atmospheric transmittance.

Atmospheric Profile	Water Vapor (w) (g/cm^2)	Equation for Transmittance Estimation	Squared Correlation (R^2)	Standard Error
High air temperature (summer)	0.4–1.6	τ = 0.974290 − 0.08007w	0.99611	0.002368
High air temperature (summer)	1.6–3.0	τ = 1.031412 − 0.11536w	0.99827	0.002539
Low air temperature (winter)	0.4–1.6	τ = 0.982007 − 0.09611w	0.99463	0.003340
Low air temperature (winter)	1.6–3.0	τ = 1.053710 − 0.14142w	0.99899	0.002375

Step 4: The effective mean atmospheric temperature is another indispensable parameter for computing LST (Table 3). It can be computed using Equation (20).

$$T_a = 16.0110 + 0.92621T_0 \quad (20)$$

Table 3. The estimation equation for effective mean atmospheric temperature in four standard atmospheres.

Standard Atmosphere	Estimation Equation (Kelvin)
For USA 1976	$T_a = 25.9396 + 0.88045T_0$
For tropical	$T_a = 17.9769 + 0.91715T_0$
For mid-latitude summer	$T_a = 16.0110 + 0.92621T_0$
For mid-latitude winter	$T_a = 19.2704 + 0.91118T_0$

Lastly, Equation (21) is applied to get LST in degrees Celsius, where $T_s(Kelvin)$ is converted into $T_s(°\text{C})$.

$$T_s(°\text{C}) = T_s(Kelvin) - 273.15 \quad (21)$$

Furthermore, we validated the LST and weather information with NASA's POWER project over selected sample location (Latitude: 25.4495 and Longitude: 81.8417) of selected time points between 1987 and 2018 for this study area, Prayagraj city, which are available in an excel sheet provided as a Supplementary Materials.

MODIS-Based Night-Time LST Calculation

Night-time LST was retrieved using MODIS night-time datasets. First of all, the MODIS night-time dataset reprojection was changed. Then, cloud-affected areas were eliminated using preprocessed quality control. Each pixel DN value was then converted into LST (°C) using Equation (22) [46].

$$T_s(^\circ C) = DN \times 0.02 - 273.15 \tag{22}$$

2.3.3. Influence of Land Indices on LST

Pearson's correlation coefficient (r)-based analysis was incorporated to assess the distinct effect of land indices (NDBI, EBBI, NDVI, NDMI, NDWI, and SAVI) on the intensification of LST. Accordingly, scatter plots were prepared for all four distinct summer time points, i.e., S1, S2, S3, and S4, and all four distinct winter time points, i.e., W1, W2, W3, and W4. In this analysis, 'r' represents the relationships, i.e., LST vs. NDBI, LST vs. EBBI, LST vs. NDMI, LST vs. NDVI, LST vs. NDWI, and LST vs. SAVI, where the dependent variable is the LST, and the independent variables are land indices (NDBI, EBBI, NDVI, NDMI, NDWI, and SAVI). Pearson's 'r' is calculated using Equation (23) [62,63].

$$r = \frac{\sum_i^n (x_i - \bar{x})(y_i - \bar{y})}{\sqrt{\sum_{i=1}^n (x_i - \bar{x})^2}\sqrt{\sum_{i=1}^n (y_i - \bar{y})^2}} \tag{23}$$

where x_i defines the values of land indices (NDBI, EBBI, NDVI, NDMI, NDWI, and SAVI), and y_i defines the LST values.

2.3.4. Intensity of SUHI Calculation

The SUHI is defined by the observed difference of LST between the urban space and the suburban/rural space over a city landscape. It was previously well defined by Oke and East [64] and Oke [65] using Equation (24), which is very popular in the literature [9,13].

$$T_{U-R} = T_U - T_R \tag{24}$$

where T_{U-R} is the intensity of the SUHI, T_U represents the LST of urban space, and T_R represents the LST of suburban/rural space.

2.3.5. Hotspot Analysis (Getis–Ord G_i*)

The spatial LST distribution over the city landscape was examined using hotspot analysis (Getis–Ord G_i*) to characterize both hot and cold spots over the city using each feature (LST value) on the basis of its neighboring features. Hotspots are the clustered areas of high values of the feature, whereas cold spots are the clustered areas of low values of the feature. The Getis–Ord G_i* statistic is derived from Equations (25)–(27) [66].

$$G_i^* = \frac{\sum_{j=1}^{n} w_{i,j}x_j - \overline{X}\sum_{j=1}^{n} w_{i,j}}{S\sqrt{\frac{n\sum_{j=1}^{n} w_{i,j}^2 - \left(\sum_{j=1}^{n} w_{i,j}\right)^2}{n-1}}} \tag{25}$$

$$\overline{X} = \frac{\sum_{j=1}^{n} x_j}{n} \tag{26}$$

$$S = \sqrt{\frac{\sum_{j=1}^{n} x_j^2}{n} - \overline{X}^2} \tag{27}$$

where x_j is defined by the feature attribute value of j, $w_{i,j}$ is defined by the spatial weight between features i and j, and n is defined by the total number of features.

3. Results

3.1. Seasonal Spatiotemporal LST Dynamics

The summer spatiotemporal distribution of LST dynamics for Prayagraj city is shown in Figure 3a for summer time points S1, S2, S3, and S4 with boxplots in Figure 3b and their statistics in Table 4. The mean LST witnessed was 38.20 °C in S1, which increased to 40.44 °C in S2, but declined to 37.49 °C in S3, before again inclining to 38.09 °C in S4. In S1, the foremost area of warm temperature was northeast, and that of cool temperature was southeast. In S2, the foremost area of warm temperature was west (except barren land in the northwest at 8–8.5 km), and that of cool temperature was southeast (except for the area of Ganga River flow in the east at 8–9 km). In S3, the foremost area of warm temperature was southwest, and that of cool temperature was northwest (except for the area of the Ganga River flow in the northeast at 8–9 km). However, in S4, the foremost area of warm temperature was the northeast. The foremost area of cool temperature was the northwest (except for vegetation coverage area in the northeast at 1.5–2.5 km). We recommend further research on the surface types or land use/land cover (LULC) classes of the city landscape in Sarif and Gupta (2022) to additionally determine their distribution over the city landscape [67].

Table 4. Summer/winter seasonal LST dynamics of Prayagraj city (1987–2018).

		Summer LST Dynamics		
Date	**Minimum (°C)**	**Maximum (°C)**	**Mean (°C)**	**Standard Deviation**
04-06-1988	29.72	42.90	38.20	1.83
12-05-1997	26.67	46.66	40.44	2.47
10-05-2008	27.52	44.85	37.49	2.15
22-05-2018	30.84	44.21	38.09	1.66
		Winter LST Dynamics		
Date	**Minimum (°C)**	**Maximum (°C)**	**Mean (°C)**	**Standard Deviation**
11-12-1987	13.26	24.11	19.72	1.14
03-12-1996	13.80	23.68	19.41	1.15
16-01-2007	13.31	24.11	18.06	1.47
16-12-2018	13.77	25.11	19.84	1.24

The winter spatiotemporal LST is mapped in Figure 4a showing the distribution of LST dynamics. Their statistics are presented in Table 4 and Figure 4b. The mean LST witnessed was 19.72 °C in W1, which declined to 19.41 °C in W2. It further declined to 18.06 °C in W3. However, again, it inclined to 19.84 °C in W4. In W1, the foremost area of warm temperature was the northwest, and that of cool temperature was North (except for the area of the Ganga River flow in the northeast at 8–9 km). In W2, the foremost area of warm temperature again was the northwest, and that of cool temperature was the east (except for the area of the Ganga River flow in the northeast at 8–9 km). In W3, the foremost area of warm temperature was the northwest, and that of cool temperature was the northeast. In W4, the foremost area of warm temperature was the southeast, and that of cool temperature was the northwest (except for the area of vegetation coverage in the northeast at 1.5–2.5 km).

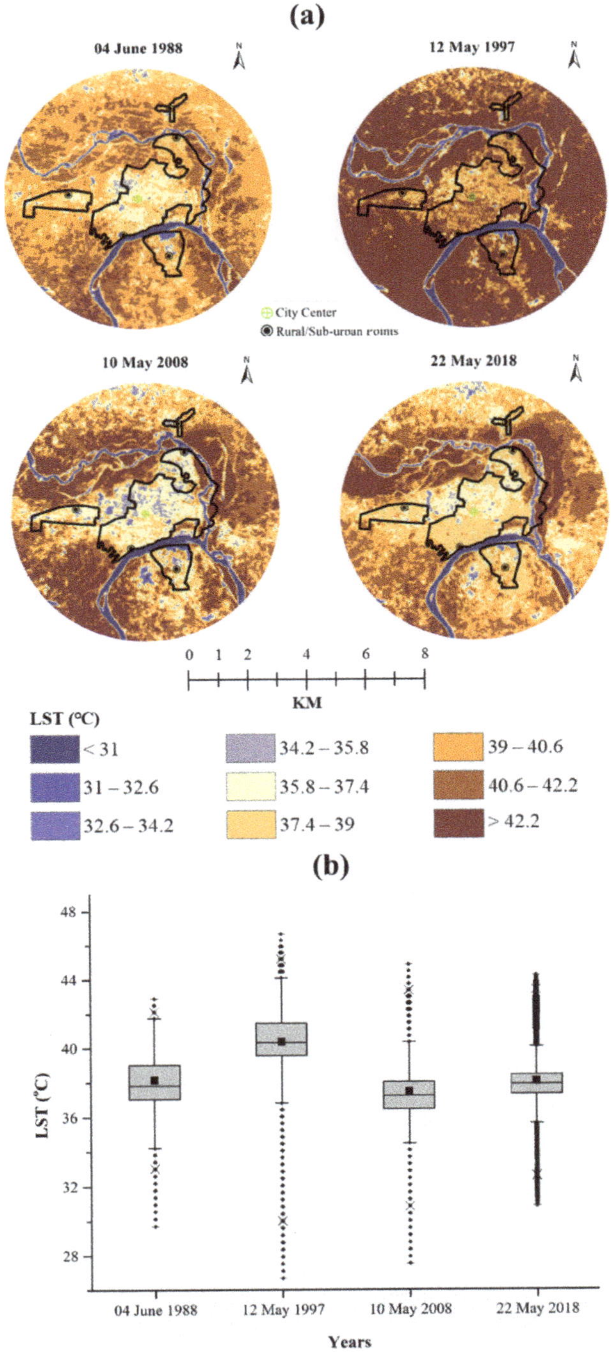

Figure 3. Spatiotemporal dynamics of LST in the summer season over Prayagraj city: (**a**) LST maps during 1988–2018 (five rural/suburban areas shown here used for computing SUHI) and (**b**) boxplots of LST dynamics during 1988–2018.

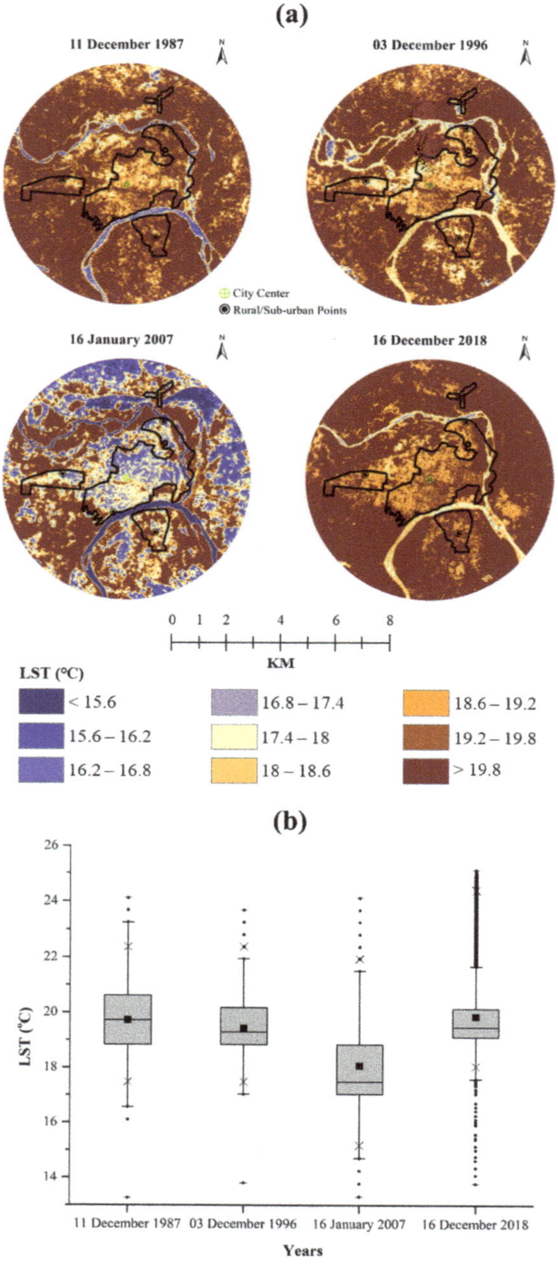

Figure 4. Spatiotemporal dynamics of LST (°C) in winter season over Prayagraj city: (**a**) LST maps during 1987–2018 (five rural/suburban areas shown here used for computing SUHI) and (**b**) boxplots of LST dynamics during 1987–2018.

3.2. Seasonal Magnitude of LST Based on Multiple Ring Profiling

The magnitude of the summer mean LST difference between the periods based on multiple ring profiling was extracted at 0.5 km intervals from the city's center to the city's periphery (Table 5 and Figure 5). It was detected that each zone at each period witnessed a higher temperature in comparison to its preceding time points (except for the period between S2 and S3) by a significant amount. Substantial temperature intensification was experienced in each distinct zone from the city center to the periphery. However, a declining trend was observed in the period of S1–S2.

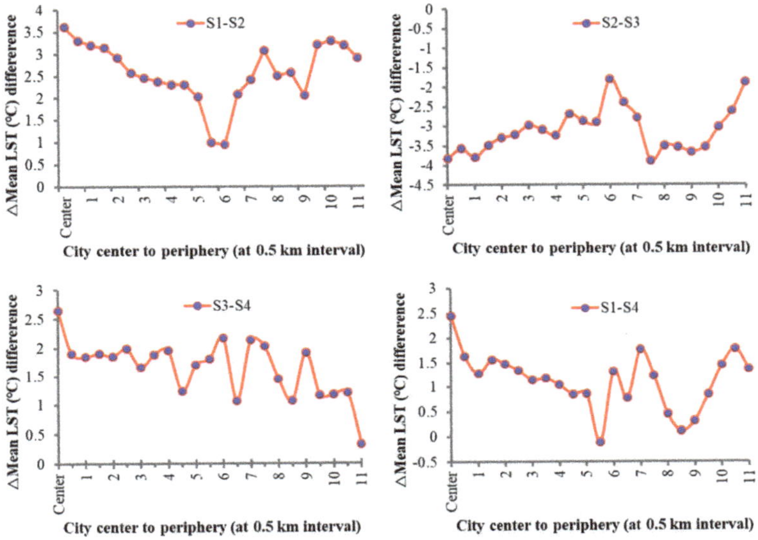

Figure 5. Magnitude of the summer mean LST in different periods (1988–2018).

Between S1 and S2, the mean LST observed amplifying pattern ranged from 0.94 °C to 3.62 °C, wherein the lowest intensified zone was 12 (5.5–6 km) by 0.94 °C because there was existence of grass and forest land. The highest intensified zone was 0 (city center) by 3.62 °C because of the presence of highly dense, impervious land. **Between S2 and S3**, the difference in mean LST observed a declining pattern which ranged from −1.80 °C to −3.88 °C, wherein the lowest decline zone was 12 (5.5–6 km) by −1.80 °C because of the dominance of grass and forest land, but the highest declined zone was 15 (7–7.5 km) by −3.88 °C because of the dominance of sand and barren land. **Between S3 and S4**, the difference in mean LST was found in an amplifying pattern that ranged from 0.33 °C to 2.65 °C, wherein the lowest intensified zone was 22 (10.5–11 km) by 0.33 °C because of the presence of grassland. The highest intensified zone was again 0 (city center) by 2.65 °C because of the presence of highly dense, impervious land. **Between S1 and S4**, the difference in mean LST showed an increasing pattern which ranged from 0.12 °C to 2.45 °C except for zone 11 (5–5.5 km), in which the magnitude of mean LST declined by −0.10 °C because of forest and grassland, wherein the lowest intensified zone was 17 (8–8.5 km) by 0.12 °C because of the presence of the Ganga River flow. The highest intensified zone again was 0 (city center) by 2.45 °C because of the presence of highly dense, impervious land.

The magnitude of the winter mean LST difference distributional pattern between distinctive periods on the basis of multiple ring profiling was also extracted at 0.5 km intervals from the center of the city to the periphery of the city (Table 5 and Figure 6). It was detected that each zone at each period witnessed lower temperatures in comparison to its preceding time points by a significant amount except for the period between W3 and W4. This means that a substantial temperature reduction was experienced in each zone from the city center to the periphery. However, an amplifying trend was observed during the period of W3–W4.

Table 5. Periodical LST magnitudes based on multiple ring buffers from the city center to the periphery at 0.5 km of intervals in Prayagraj city (1987–2018).

Zones	Distance from the City Center at 0.5 km of Interval	Periodical Difference of Mean LST (°C)			
	Summer Magnitude	**S1–S2**	**S2–S3**	**S3–S4**	**S1–S4**
0	Center	3.62	−3.82	2.65	2.45
1	0.5	3.31	−3.57	1.90	1.64
2	1	3.21	−3.78	1.85	1.28
3	1.5	3.14	−3.48	1.91	1.57
4	2	2.91	−3.29	1.86	1.48
5	2.5	2.57	−3.21	1.99	1.34
6	3	2.46	−2.98	1.66	1.14
7	3.5	2.37	−3.09	1.89	1.18
8	4	2.30	−3.22	1.96	1.04
9	4.5	2.30	−2.69	1.24	0.85
10	5	2.02	−2.87	1.71	0.86
11	5.5	0.99	−2.90	1.81	−0.10
12	6	0.94	−1.80	2.18	1.32
13	6.5	2.09	−2.39	1.07	0.77
14	7	2.42	−2.78	2.14	1.77
15	7.5	3.08	−3.88	2.03	1.23
16	8	2.50	−3.50	1.46	0.45
17	8.5	2.58	−3.53	1.07	0.12
18	9	2.05	−3.65	1.92	0.32
19	9.5	3.20	−3.52	1.17	0.84
20	10	3.30	−3.03	1.18	1.45
21	10.5	3.20	−2.61	1.21	1.79
22	11	2.91	−1.88	0.33	1.36
	Winter Magnitude	**W1–W2**	**W2–W3**	**W3–W4**	**W1–W4**
0	Center	0.45	−2.24	2.44	0.64
1	0.5	0.22	−1.49	1.89	0.62
2	1	−0.03	−1.62	1.88	0.22
3	1.5	−0.05	−1.47	1.83	0.31
4	2	−0.09	−1.59	1.96	0.29
5	2.5	−0.15	−1.74	2.09	0.20
6	3	−0.16	−1.61	1.87	0.10
7	3.5	−0.25	−1.72	2.00	0.03
8	4	−0.03	−1.71	1.76	0.02
9	4.5	−0.54	−0.98	1.15	−0.36
10	5	−0.17	−1.20	1.72	0.34
11	5.5	−0.76	−1.34	1.88	−0.21
12	6	−0.49	−1.38	2.45	0.58
13	6.5	−0.20	−0.78	1.41	0.42
14	7	−0.28	−1.04	1.81	0.49
15	7.5	−0.71	−1.04	1.20	−0.55
16	8	−0.68	−0.90	1.17	−0.42
17	8.5	−0.51	−1.02	0.64	−0.89
18	9	−0.70	−1.33	1.35	−0.67
19	9.5	−0.65	−0.86	0.95	−0.56
20	10	−0.36	−0.48	1.11	0.27
21	10.5	−0.53	−1.07	1.99	0.38
22	11	−0.28	−1.14	2.48	1.06

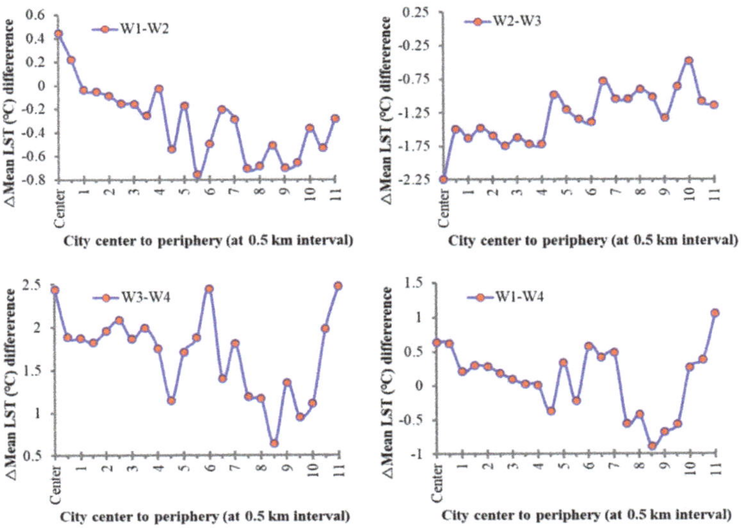

Figure 6. Magnitude of the winter mean LST in different periods (1987–2018).

Between W1 and W2, the difference in mean LST showed a declining pattern (except zone 1 by 0.45 °C and zone 2 by 0.22 °C) which ranged from −0.76 °C to −0.03 °C, wherein the lowest decline zone was 2 (0.5–1 km) by −0.03 °C because there was the existence of high-density impervious land. The highest decline zone was 11 (5–5.5 km) by −0.76 °C because of the presence of forest and grassland. **Between W2 and W3**, the difference in mean LST was found to decline, which ranged from −2.24 °C to −0.48 °C, wherein the lowest decline zone was 20 (city center) by −0.48 °C because of the presence of bare soil with scattered grassland. The highest decline zone was 0 (city center) by −2.24 °C because of the high moisture content over scattered vegetation in the high-density impervious land. **Between W3 and W4**, the difference in mean LST was found amplified in the range 0.64 °C to 2.48 °C, wherein the lowest intensified zone was 17 (8–8.5 km) by 0.64 °C because of the presence of Ganga River flow. The highest intensified zone was 22 (10.5–11 km) by 2.48 °C because of the dominant barren land. **Between W1 and W4**, the difference in mean LST exhibited a very interesting amplifying pattern [except for zone 9 (4–4.5 km), zone 11 (5–5.5 km), and zone 15 to zone 19 (7–9.5 km), which ranged from −0.89 °C to −0.21 °C as these zones had forest and grassland, as well as Ganga River flow] which ranged from 0.02 °C to 1.06 °C, wherein the lowest intensified zone was 8 (3.5–4 km) by 0.02 °C because there was the existence of impervious land. The highest intensified zone again was 22 (10.5–11 km) by 1.06 °C due to the dominant barren land.

3.3. Spatiotemporal Dynamics of Land Indices and LST, and Their Relationships

3.3.1. NDBI Dynamics and Its Connection with LST

The spatial NDBI distributional dynamics maps of Prayagraj are shown in Figure 7a,b for the summer season for S1, S2, S3, and S4 time points and the winter season for W1, W2, W3, and W4, respectively. The statistics of all the six land indices and their relationship with LST for all the summer and winter time points are presented in Table 6. Furthermore, Figure A1 shows whisker boxplots of all the land indices for the summer/winter seasons. The analysis found that summer mean NDBI witnessed a decrease of −0.023 in S1 which amplified to −0.015, −0.02, and 0.02 in S2, S3 and S4, respectively. The winter mean NDBI witnessed −0.036 in W1 which amplified to −0.012 in W2 but declined to −0.018 in W3 and further to −0.039 in W4.

Table 6. Statistics of the six land indices and their relationship with LST for all the summer and winter time points.

Season	Time Points	Land Indices	Minimum	Maximum	Mean	Standard Deviation	Correlation with LST®	Significance (p)
Summer	S1	NDBI	−0.324	0.130	−0.023	0.055	0.668	<0.001
		EBBI	0.051	0.240	0.158	0.027	0.623	<0.001
		NDMI	−0.130	0.324	−0.023	0.055	−0.668	<0.001
		NDVI	−0.098	0.521	0.134	0.068	−0.459	<0.001
		NDWI	−0.463	0.132	−0.168	0.057	0.285	<0.001
		SAVI	−0.048	0.363	0.088	0.043	−0.425	<0.001
	S2	NDBI	−0.407	0.184	−0.015	0.073	0.6758	<0.001
		EBBI	0.010	0.276	0.149	0.038	0.640	<0.001
		NDMI	−0.184	0.407	−0.015	0.073	−0.6758	<0.001
		NDVI	−0.196	0.661	0.180	0.093	−0.266	<0.001
		NDWI	−0.575	0.274	−0.202	0.080	0.070	<0.001
		SAVI	−0.070	0.462	0.111	0.057	−0.259	<0.001
	S3	NDBI	−0.366	0.189	−0.020	0.058	0.6757	<0.001
		EBBI	0.042	0.263	0.153	0.033	0.751	<0.001
		NDMI	−0.189	0.366	−0.020	0.058	−0.6757	<0.001
		NDVI	−0.096	0.562	0.143	0.080	−0.376	<0.001
		NDWI	−0.492	0.136	−0.160	0.070	0.227	<0.001
		SAVI	−0.041	0.391	0.091	0.049	−0.345	<0.001
	S4	NDBI	−0.339	0.188	0.020	0.060	0.636	<0.001
		EBBI	0.043	0.292	0.154	0.034	0.751	<0.001
		NDMI	−0.188	0.339	0.020	0.060	−0.636	<0.001
		NDVI	0.003	0.538	0.205	0.077	−0.277	<0.001
		NDWI	−0.445	0.215	−0.202	0.056	0.272	<0.001
		SAVI	0.002	0.392	0.138	0.051	−0.215	<0.001
Winter	W1	NDBI	−0.783	0.249	−0.036	0.107	0.308	<0.001
		EBBI	0.000	0.184	0.074	0.026	0.520	<0.001
		NDMI	−0.249	0.783	−0.036	0.107	−0.308	<0.001
		NDVI	−0.305	0.683	0.217	0.117	0.113	<0.001
		NDWI	−0.477	0.526	−0.065	0.107	−0.259	<0.001
		SAVI	−0.072	0.358	0.089	0.051	0.191	<0.001
	W2	NDBI	−0.814	0.243	−0.012	0.115	0.467	<0.001
		EBBI	0.000	0.190	0.075	0.027	0.564	<0.001
		NDMI	−0.243	0.814	−0.012	0.115	−0.467	<0.001
		NDVI	−0.271	0.611	0.183	0.103	−0.072	<0.001
		NDWI	−0.512	0.399	−0.159	0.094	−0.074	<0.001
		SAVI	−0.060	0.314	0.072	0.041	−0.003	<0.001
	W3	NDBI	−0.597	0.210	−0.018	0.098	0.536	<0.001
		EBBI	0.000	0.203	0.081	0.029	0.685	<0.001
		NDMI	−0.210	0.597	−0.018	0.098	−0.536	<0.001
		NDVI	−0.165	0.594	0.134	0.083	0.159	<0.001
		NDWI	−0.512	0.249	−0.110	0.080	−0.369	<0.001
		SAVI	−0.046	0.322	0.058	0.037	0.215	<0.001
	W4	NDBI	−0.694	0.352	−0.039	0.108	0.503	<0.001
		EBBI	0.000	0.554	0.079	0.032	0.749	<0.001
		NDMI	−0.352	0.694	−0.039	0.108	−0.503	<0.001
		NDVI	−0.282	0.623	0.163	0.110	0.032	<0.001
		NDWI	−0.524	0.367	−0.148	0.102	−0.186	<0.001
		SAVI	−0.145	0.560	0.125	0.081	0.080	<0.001

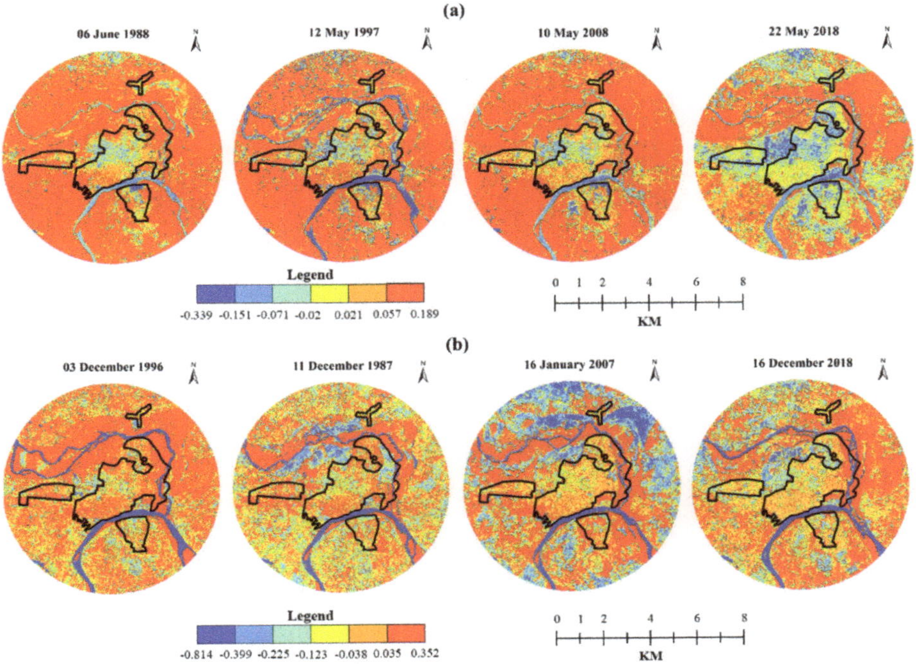

Figure 7. Seasonal NDBI dynamics in Prayagraj city (1987–2018): (**a**) summer and (**b**) winter.

Most of the highly dense impervious/built-up area was intensely concentrated in the city center to the south in S1 and W1, which further expanded in the south in S2, W2, S3, and W3. In S4 and W4, the highly dense built-up area spread in the southwest and northeast in S4 and W4, respectively. Therefore, it is apparent that the built-up area is mostly concentrated in the southwest and northeast up to 8 km.

The correlations between LST and all six land indices for the summer and winter seasons are presented in Figures 13 and 14, respectively. In the summer/winter seasons, the correlations between LST vs. NDBI were found to be positive at all distinctive time points. On the basis of the values of correlation coefficients, it can be found that the NDBI's role in LST intensification in the summer was higher than in the winter. Further, the built-up intensity effect was amplified over the city landscape, resulting in increased temperature growth because of the conversion of forests and water bodies into built-up land.

3.3.2. EBBI Dynamics and Its Connection with LST

The spatiotemporal maps of EBBI dynamics are shown in Figure 8a,b for the summer/winter seasons, respectively, for all the distinctive summer/winter time points (however, summer/winter seasonal statistics are shown in Figure A1). The summer mean EBBI witnessed an amplifying trend in the summer/winter seasons at all time points. The highly dense impervious/built-up area was mostly concentrated in the city center to 4 km of its periphery, and the bare land was mostly concentrated to 6–8 km and 9.5–11 km in both S1 and W1. Then, the highly dense impervious/ built-up area expanded in the south and north up to 6 km in S2 and W2, and bare land mostly remained concentrated in the northeast (6–8 km) and southwest (6–8 km). In S3 and W3, the highly-dense impervious/built-up area expanded in the south and north up to 7 km, and bare land mostly remained concentrated in northeast (6–8 km) and southwest (7–9 km). However, the highly dense impervious/built-up area expanded in the southwest, northeast, and northwest up to 8 km in S4 and W4, and bare land mostly remained concentrated on northeast (6–8 km) and southwest (8–9 km). Therefore, it is apparent that the impervious/built-up area mostly existed in the southwest and northeast up to 7 km.

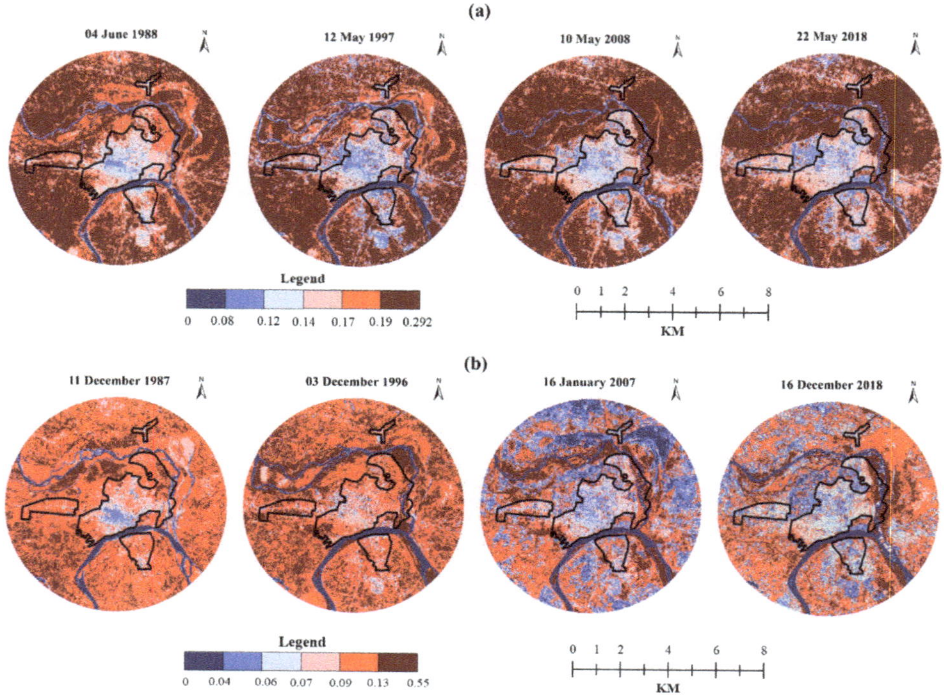

Figure 8. Seasonal EBBI dynamics in Prayagraj city (1987–2018): (**a**) summer and (**b**) winter.

A positive correlation was detected between LST vs. EBBI at all distinctive summer/winter time points (Figures 13 and 14). The role of EBBI in LST intensification remained high in both the summer and the winter (Table 6). This indicates that the impervious/built-up and bare land intensity effect was amplified over the city landscape, resulting in increased temperature growth due to the conversion of forests, water bodies, and agricultural land into built-up land and bare land settings.

3.3.3. NDMI Dynamics and Its Relationship with LST

Seasonal NDMI dynamics are shown in Figure 9a,b for the summer and winter seasons, respectively (however, summer/winter seasonal statistics are shown in Figure A1). The summer mean NDMI witnessed a nonuniform pattern where it first amplified in S1 and S2 but declined in S3, before again amplifying in S4. In contrast, the winter mean NDMI shows an increasing pattern in W1 and W2 but a decreasing one in W3 and W4.

This may be attributed to the fact that the high moisture content area was concentrated in the northwest and northeast in S1 and W1. A decrease was observed in the moisture content area in the northwest in S2 and W2 which further diminished in S3 and W3. However, the high moisture content in the northwest and southeast was again amplified in S4 and W4. In fact, the forest, grassland, and Ganga River with high moisture content were primarily present in the city landscape.

In both seasons, the correlation between LST vs. NDMI was found to be negative at all time points (Figures 13 and 14), while the correlation coefficient values (Table 6) indicate that the role of NDMI in decreasing LST was higher in the summer season than in the winter season.

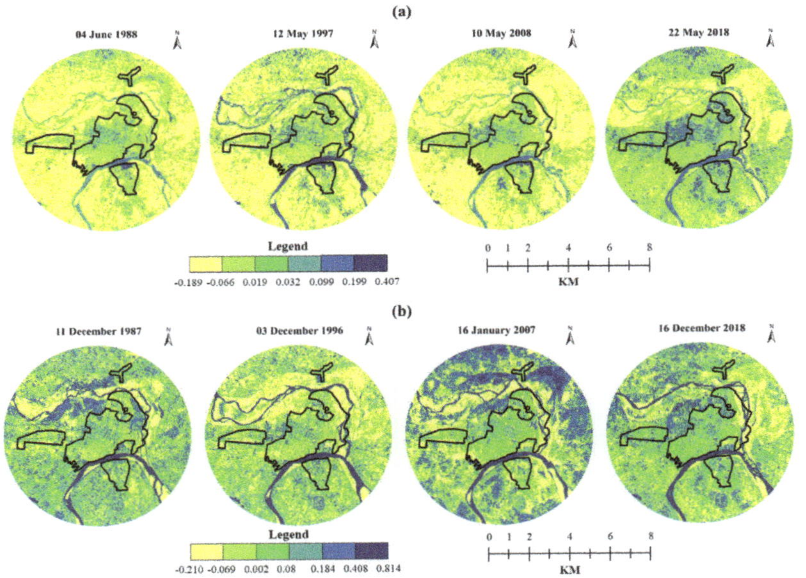

Figure 9. Seasonal NDMI dynamics in Prayagraj city (1987–2018): (**a**) summer and (**b**) winter.

3.3.4. NDVI Dynamics and Its Relationship with LST

The NDVI maps of Prayagraj are shown in Figure 10a,b for the summer and winter seasons, respectively (however, summer/winter seasonal statistics are shown in Figure A1). An increase was observed in the summer mean NDVI at S2, followed by a decrease in S3 but again an increase in S4. However, the winter mean NDVI witnessed a decline in W2 and W3 but an increase in W4.

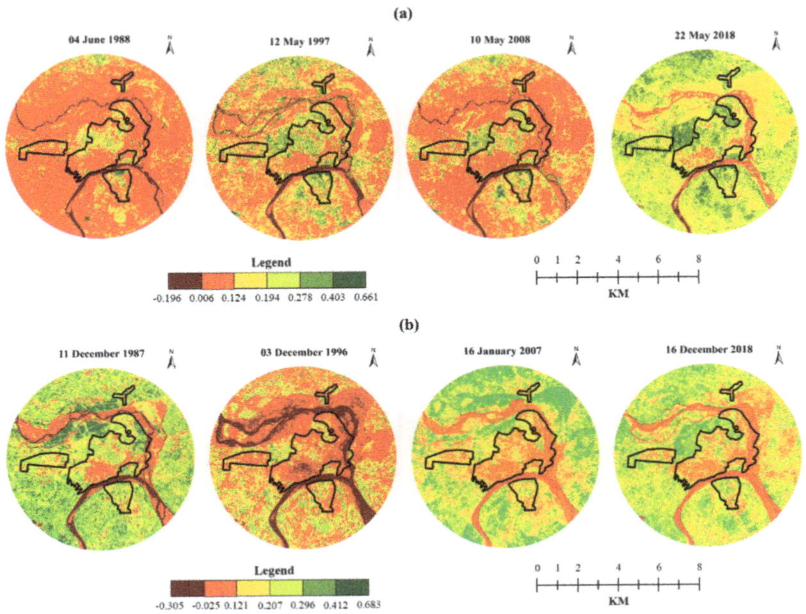

Figure 10. Seasonal NDVI dynamics in Prayagraj city (1987–2018): (**a**) summer and (**b**) winter.

The analysis shows that high-density vegetation was concentrated in the northwest in S1 and W1, showing a diminishing trend in S2, W2, S3, and W3 but an amplification in S4 and W4. It was found that the forest and grassland were mainly dominant in the northwest direction from the city center.

In the summer season, the correlations between LST vs. NDVI were found to be negative at all summer time points (Figure 13), while, in the winter season, the correlation between LST vs. NDVI was also positive at all winter time points except for W2 (Figure 14). This reflects that NDVI played a significant role in decreasing LST in the summer season, but its role was very weak in the winter.

3.3.5. NDWI Dynamics and Its Relationship with LST

The NDWI maps of Prayagraj are shown in Figure 11a,b for the summer and winter seasons for all time points, respectively (however, summer/winter seasonal statistics are shown in Figure A1). The summer and winter mean NDWI followed a similar pattern, first declining in S2 and W2, then amplifying in S3 and W3 before again declining in S4 and W4.

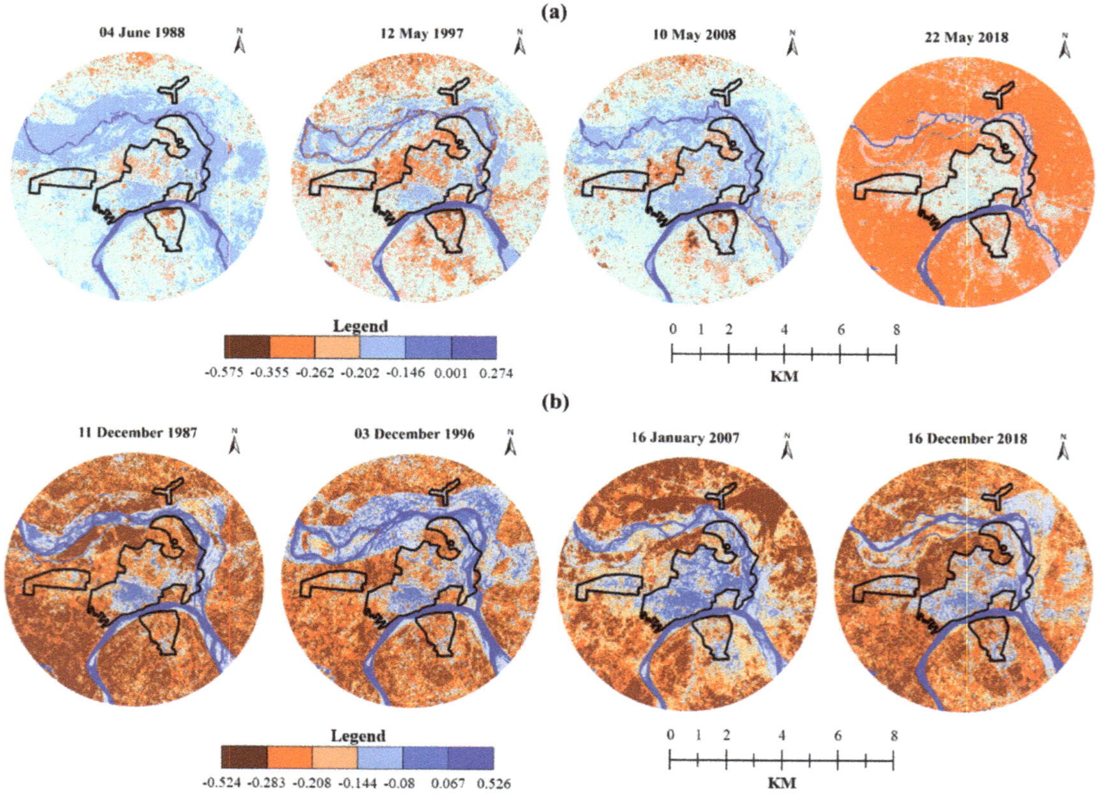

Figure 11. Seasonal NDWI dynamics in Prayagraj city (1987–2018): (**a**) summer and (**b**) winter.

High-density water bodies were concentrated in the northeast (8–9.5 km) due to the Ganga River flow in S1 and W1. The high-density vegetation was concentrated in the northwest and southeast, which decreased in S2, W2, S3, and W3. Then, high-density vegetation increased in the northwest and southeast in S4 and W4, respectively. Therefore, it is apparent that the forest and grassland were mainly dominant in the northwest direction from the city center.

In both summer and winter seasons, the correlations between LST vs. NDWI was positive at all times except in W2 but with relatively low correlation coefficient values (Figures 13 and 14). This means that NDWI played an insignificant role in decreasing LST in both summer and winter because the whole city exhibited a lack of water bodies except for 8–9.5 km in the northeast direction of the Ganga River flow.

3.3.6. SAVI Dynamics and Its Relationship with LST

The distributional dynamics maps of SAVI are shown in Figure 12a,b for the summer and winter seasons, respectively (however, summer/winter seasonal statistics are shown in Figure A1). The summer mean SAVI amplified in S2 but declined in S3 before again amplifying in S4. The winter mean SAVI declined in W2 and W3 but amplified in W4.

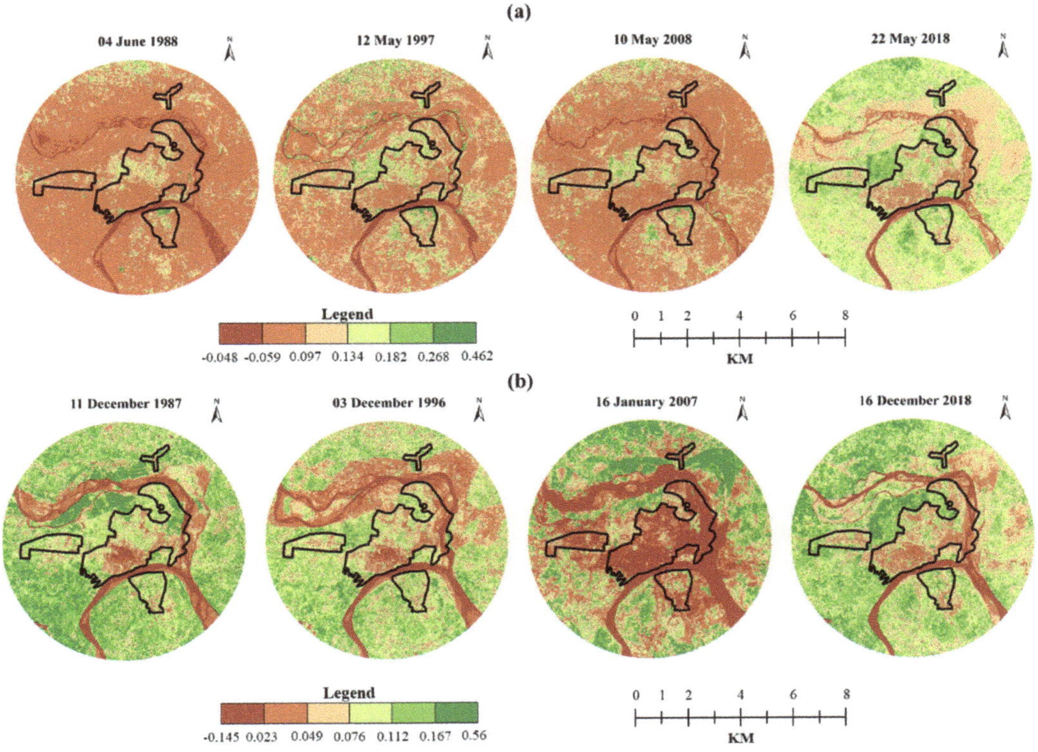

Figure 12. Seasonal SAVI Dynamics in Prayagraj city (1987–2018): (**a**) summer and (**b**) winter.

It can be observed that high-density forest and grassland were concentrated in the northwest and southeast in S1 and W1, which was reduced in S2 and W2. The high-density forest and grassland concentrated in the northwest diminished in S3 and W3. However, the high-density forest and grassland concentrated within the northwest and southeast increased in S4 and W4. Therefore, it was observed that the forest and grassland mostly dominated in the northwest direction from the city center.

In the summer season, the correlation between LST vs. SAVI was found to be negative at each distinctive summer time point (Figure 13), while, in the winter season, the correlation between LST vs. SAVI was found to be positive at all winter time points except in W2 (Figure 14). This means that SAVI played a noteworthy role in decreasing LST in the summer season, but its role was very weak in the winter.

3.4. Effects of Land Indices on LST Distribution

The effect of spatiotemporal seasonal dynamics of all six land indices (i.e., NDBI, EBBI, NDMI, NDVI, NDWI, and SAVI) on LST profiling for summer/winter seasons from the city center to the periphery in eight different directions, i.e., north to south, northeast to southwest, northwest to southeast, and west to east was extracted on the city landscape. The effect of these six land indices on the LST in the summer and winter seasons is presented in Figures 15 and 16, respectively.

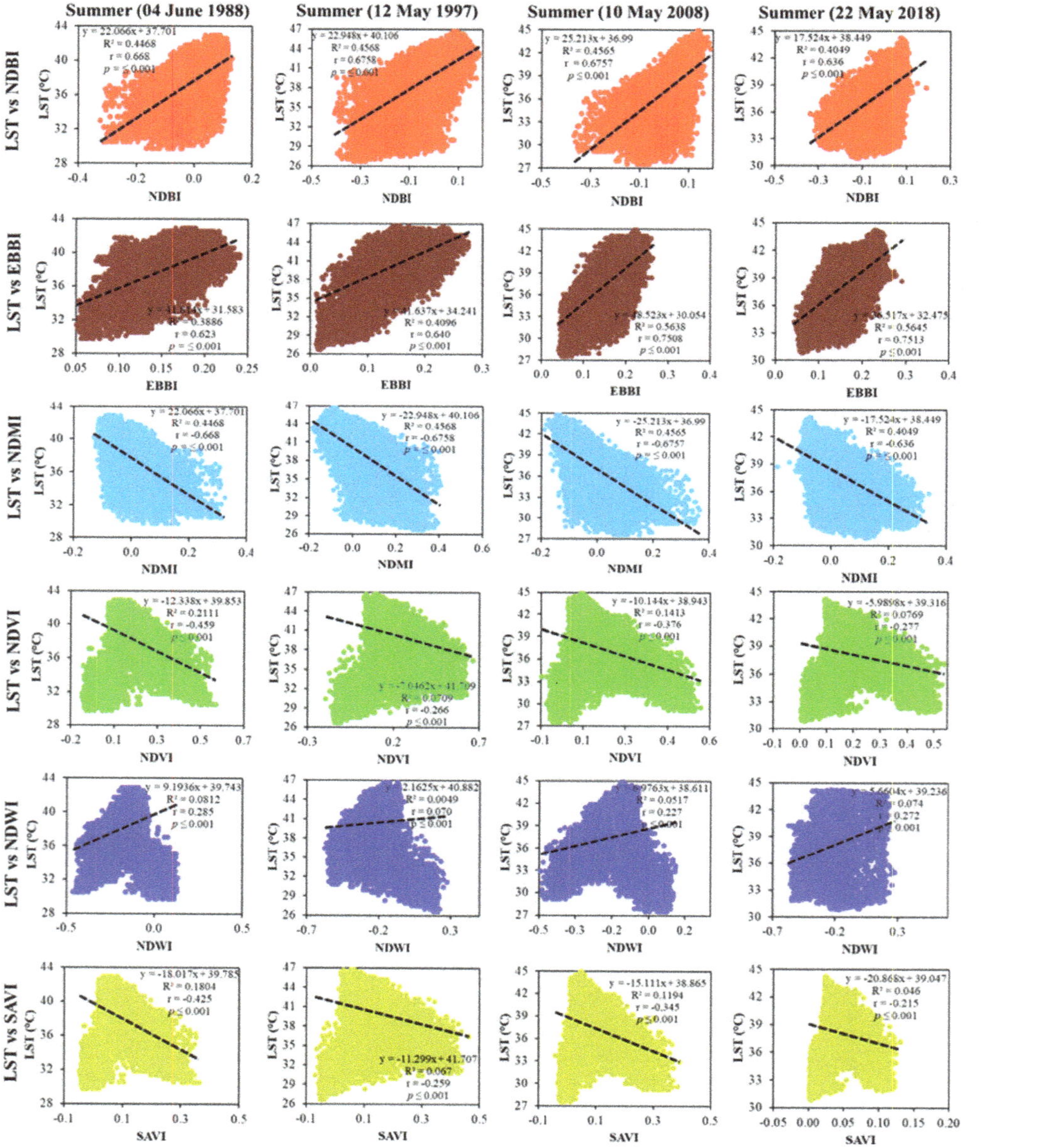

Figure 13. Correlation between the LST and the six land indices of Prayagraj city in the summer season (1988–2018).

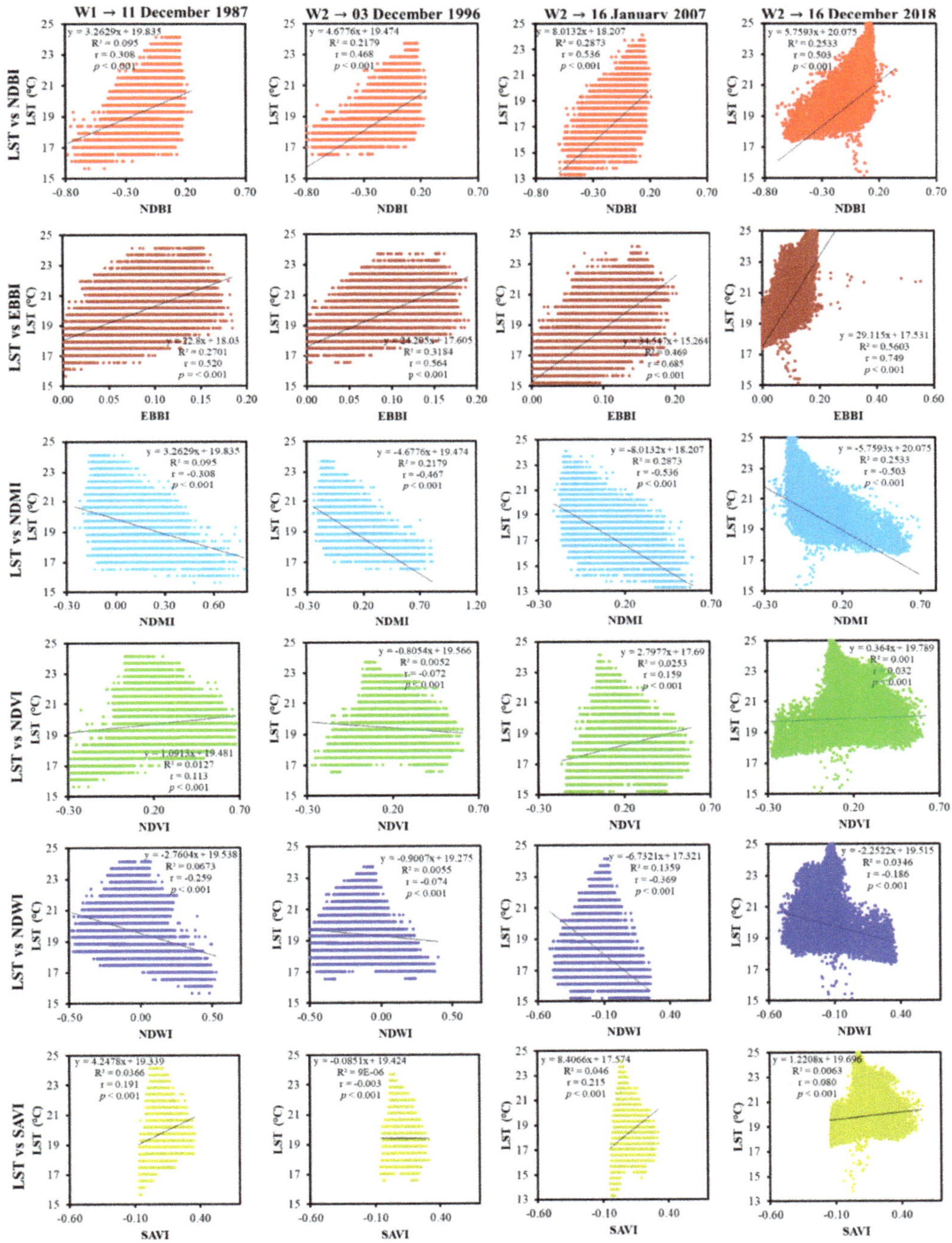

Figure 14. Correlation between the LST and the six land indices of Prayagraj city in the winter season (1987–2018).

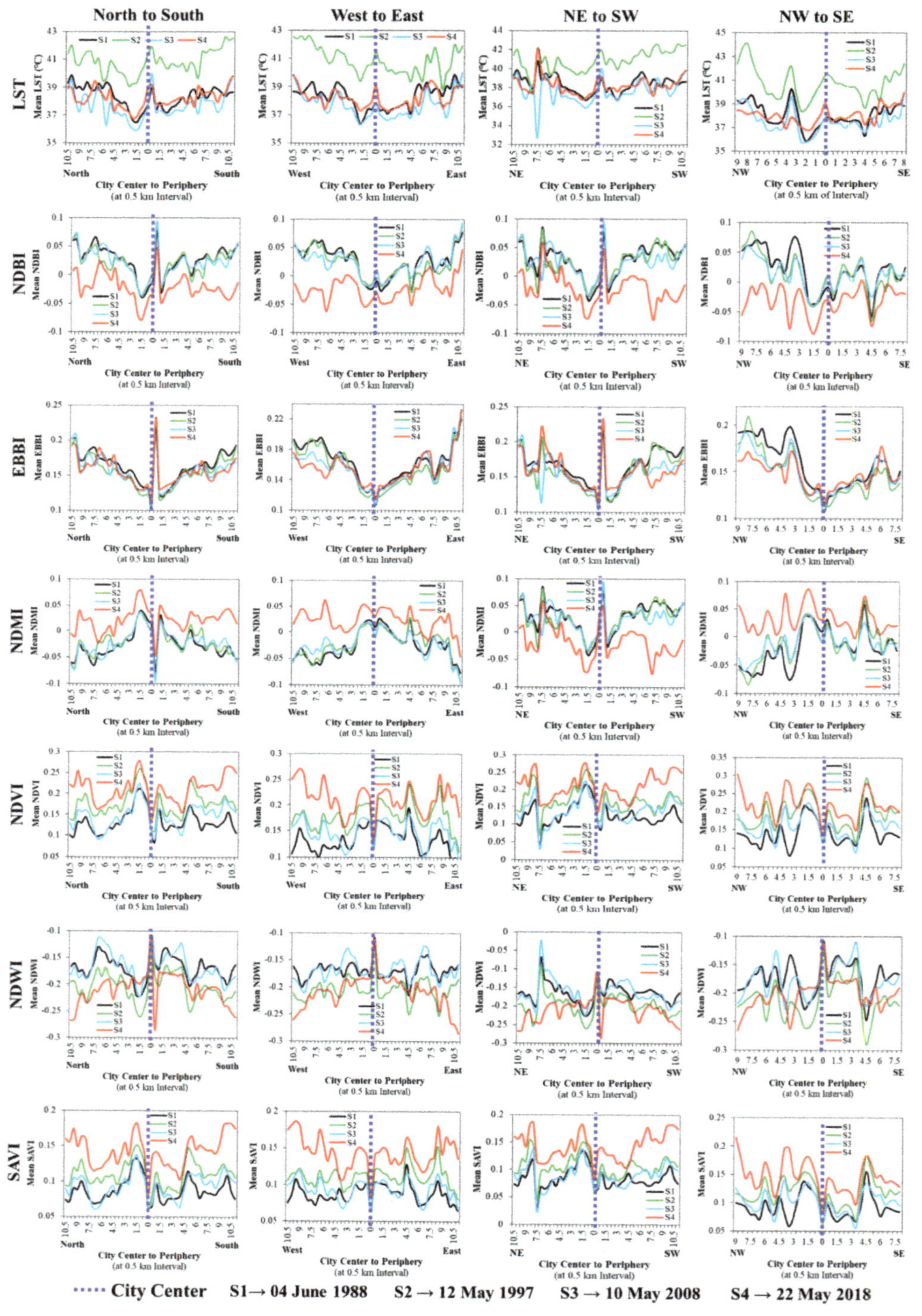

Figure 15. Effects of the land indices on the distribution of LST dynamics in the summer season in Prayagraj city (1988–2018).

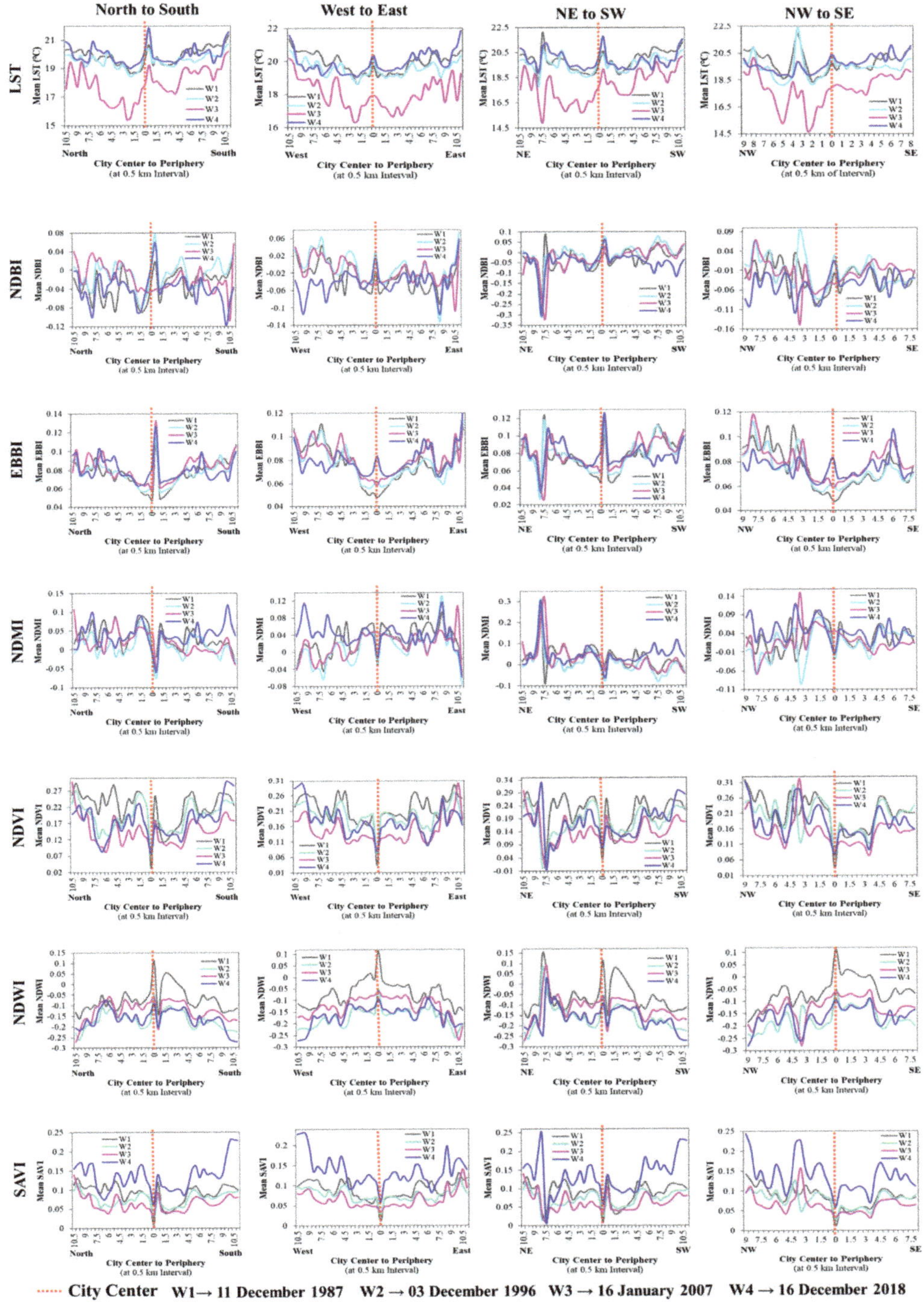

······ City Center W1 → 11 December 1987 W2 → 03 December 1996 W3 → 16 January 2007 W4 → 16 December 2018

Figure 16. Effects of the land indices on the distribution of LST dynamics in the winter season in Prayagraj city (1987–2018).

3.4.1. North to South

In summer, from the city center in both the north and the south directions, the mean LST followed a declining trend (up to 11 km) of 0.2 °C–2.75 °C except for 6.5–11 km in the north direction and 9–11 km in the south direction (Figure 15). The higher mean LST at the city center was because of the impervious/built-up land (high mean NDBI by −0.04 to 0.04). The decline in mean LST witnessed at 1–2 km in the north and 4.5–6.5 km in the south was due to the existence of forest land/grassland (higher mean NDVI by 0.02–0.10), as well as the availability of higher moisture content (higher mean NDMI and mean SAVI). However, the increase in mean LST at 6.5–7.5 km and 10–11 km in the north and 6.5–11 km in the south was because of the existence of barren land/bare soils (higher mean EBBI by 0.01–0.08). The mean NDWI was found to be negative (<−0.1) at all consecutive summer time points, which means that its role was insignificant due to the unavailability of water bodies of large size (except rivers) to impact the reduction in mean LST.

In winter, from the city center in both the north and the south directions, the mean LST witnessed declining inclination up to 11 km by 0.2 °C–3.6 °C except for 6.5–9.5 km in the north, and 5–9 km and 10–11 km in the south (Figure 16). The higher mean LST in the city center was due to the built-up land presence (high mean NDBI by −0.04 to 0.04). In contrast, the decline in mean LST witnessed at 1–2 km and 9.5–11 km in the north and 9–10 km in the south was because of the presence existence of grassland/forest land (higher mean NDVI by 0.02–0.25) coupled with high moisture content availability, as reflected by high mean NDMI and mean SAVI values. The existence of barren land/bare soils (higher mean EBBI by 0.01–0.08) at 6.5–9.5 km in the north and at 5–9 km and 10–11 km in the south resulted in a spike in mean LST. However, as in the summer season, mean NDWI had negative values in the winter season also [<0 in the north and south (except at city center and 1.5–2 km in the south)] at all consecutive winter time points, signifying its insignificant role in the reduction in mean LST.

3.4.2. Northeast (NE) to Southwest (SW)

In summer, in both the NE and the SW directions from the city center, mean LST detected a declining pattern up to 11 km by 0.2 °C–5.8 °C except for 6–7.5 km in the NE and 6.5–11 km in the SW, where a reverse trend was witnessed. The decline in mean LST witnessed at 1–3 km in the NE and 6.5–7.5 km and 9.5–10.5 km in the SW was because of the availability of grassland/forest land (higher mean NDVI by 0.02–0.13) and high moisture, as the mean NDMI and mean SAVI also witnessed a similar pattern to mean NDVI. The availability of barren land/bare soils (higher mean EBBI by 0.01–0.1) resulted in the increase in mean LST at 6.5–7.5 km in the NE and 7.5–9.5 km in the SW. The mean NDWI showed negative values in the summer, indicating that its role was insignificant due to the unavailability of water bodies. However, 7.5–8 km in the NE showed the reverse trend due to the Ganga River flow.

In winter, in both the NE and the SW directions from the city center, the mean LST witnessed a declining pattern up to 11 km by 0.2 °C–4.5 °C except for 6–7.5 km in the NE direction and 10–11 km in the SW direction where the mean LST witnessed an increasing trend (Figure 16). The city center had high-density built-up land (higher mean NDBI by −0.04 to 0.04), leading to higher mean LST which witnessed a decline at 1–3 km in the NE and 4.5–6 km and 10–11 km in the SW because of the existence of grassland/forest land (higher mean NDVI by 0.02–0.2). The mean NDMI and mean SAVI also witnessed a similar pattern to that of mean NDVI, indicating the availability of higher moisture due to the presence of forest/grassland, which helped in the decline in mean LST. However, the mean LST high spikes at 6–7.5 km in the NE direction and at 10–11 km in the SW direction were due to barren land/bare soils (higher mean EBBI by 0.01–0.08) at these locations. However, in summer, the role of NDWI was insignificant except for 7.5–8.5 km in the NE due to Ganga River.

3.4.3. Northwest (NW) to Southeast (SE)

In summer, in both the NW and the SE directions from the city center, it was detected that the mean LST witnessed a declining pattern up to 9 km in the NW and 8 km in the SE by 0.1 °C–2.6 °C because of the existence of grassland/forest land (higher mean NDVI by 0.01–0.15) coupled with high moisture availability due to high mean NDMI and mean SAVI (Figure 15). However, an increasing pattern was witnessed in the zones of 3.5–4.5 km and 6–9 km in the NW and 6.5–8 km in the SE because of the presence of barren land/bare soils (higher mean EBBI by 0.01–0.2). The role of the mean NDWI was insignificant due to negative values (<0.1) at all consecutive summer time points.

In winter, in both the NW and the SE directions from the city center, the mean LST witnessed a declining pattern up to 11 km by 0.1 °C–3.5 °C except for 3–5 km and 6.5–9 in the NW direction and 5–8 km in the SE direction, where the mean LST witnessed an increasing pattern. The decline in mean LST at 1–3 km in the NW and 4.5–6.5 km and 3–5 km in the SE was because of the presence of grassland/forest land (higher mean NDVI by 0.02–0.23) and the high peaks of mean NDMI and mean SAVI (Figure 16). The barren land/bare soils resulted in an increase in mean LST inclination at 3–5 km and 6.5–9 km in the NE and at 5–8 km in the SE (higher mean EBBI by 0.01–0.06). Again, the role of NDWI was insignificant due to negative values of the mean NDWI [except the city center to 1.5 km in the SE at W1, where the mean LST was >0] at all consecutive time points.

3.4.4. West to East

In summer, the mean LST witnessed a higher peak at the city center in comparison to nearby surroundings because of the presence incidence of high-density impervious/built-up land (higher mean NDBI by −0.04 to 0.04). Then, the mean LST followed a declining pattern up to 11 km by 0.1 °C–1.75 °C in both the west (in particular, at 1–3 km and 5.5–6.5 km) and the east (in particular, 3–5 km and 8–9.5 km) due to forest land/grassland (higher mean NDVI by 0.02–0.10) coupled with high moisture availability (Figure 15). The mean LST witnessed an increasing pattern in 4–5.5 km and 6.5–11 km in the west and 6–7.5 and 9.5–11 km in the east because of barren land/bare soils (higher mean EBBI by 0.01–0.2). However, mean NDWI was found to have negative values (<–0.1) at all consecutive summer time points, which means that its role was insignificant due to the unavailability of many water bodies.

In winter, the city center again witnessed higher LST compared to its surroundings. However, the mean LST witnessed declining inclination up to 11 km by 0.1 °C–1.6 °C [except 4–5.5 km and 6.5–11 km in the west and 5.5–7.5 km and 9.5–11 km in the east] in both the west and the east directions. The decline in mean LST witnessed at 1.5–3 km and 5.5–6.5 km in the west and 4–6 km and 8.5–9.5 km in the east was because of the existence of grassland/forest land (higher mean NDVI by 0.02–0.22). The mean NDMI and mean SAVI also witnessed a similar pattern to the mean NDVI, which means that higher moisture was available where forest/grassland was present, consequently helping the decline in the mean LST (Figure 16). However, the mean LST inclination at 4.5–5 km and 6.5–11 km in the west direction and at 5.5–7.5 km and 9.5–11 km in the east direction was because of barren land/bare soils presence (higher mean EBBI by 0.01–0.05). However, the mean NDWI was found to have negative values [<0 in the north and south (except at city center to 0.5 km in the west)] at all consecutive winter time points, which means that its role was insignificant due to the unavailability of water bodies large in size, which may have impacted the reduction in the mean LST.

3.5. SUHI Dynamics

3.5.1. Urban and Rural/Suburban Point Location-Based SUHI

The distribution of SUHI dynamics was evaluated in the summer season at all four time points (S1, S2, S3, and S4), as well as in the winter season at all four time points (W1, W2, W3, and W4) over Prayagraj city. For this purpose, five rural/suburban point locations, well distributed over the study area (shown in Figures 3 and 4) for summer/winter seasons, were selected to study the actual variation in LST between urban and rural/suburban areas. The seasonal SUHI dynamics for all time points of the summer and winter seasons at five rural/suburban locations in Prayagraj city are shown in Figure 17. Furthermore, the statistics of point location-based SUHI are presented in Table 7. It can be observed that these five case points showed strong SUHI evidence for the summer and the winter.

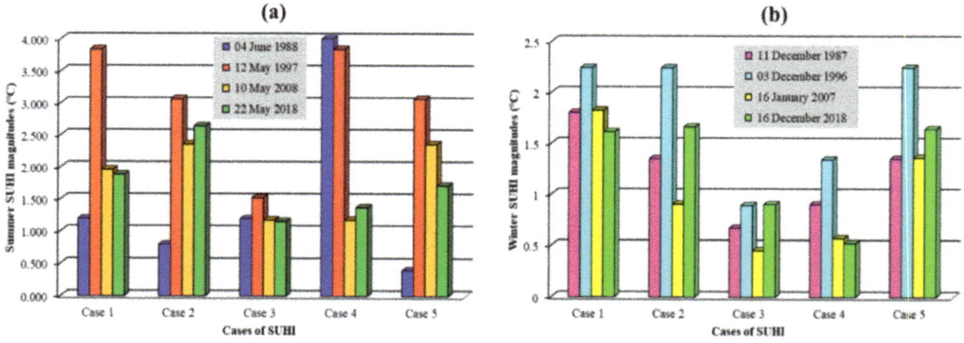

Figure 17. Seasonal SUHI dynamics at five case point locations in Prayagraj city (1987–2018) for all time points: (**a**) summer and (**b**) winter.

Table 7. Seasonal SUHI magnitude based on rural/suburban point locations in Prayagraj city.

Season	**LST (°C) Difference [T_{U-R}]**			
Summer SUHI	**S1**	**S2**	**S3**	**S4**
Case Point 1	1.195	3.840	1.962	1.887
Case Point 2	0.794	3.064	2.358	2.639
Case Point 3	1.194	1.524	1.175	1.151
Case Point 4	4.016	3.840	1.174	1.368
Case Point 5	0.398	3.064	2.358	1.709
Winter SUHI	**W1**	**W2**	**W3**	**W4**
Case Point 1	1.80	2.24	1.82	1.61
Case Point 2	1.35	2.24	0.90	1.66
Case Point 3	0.67	0.89	0.45	0.90
Case Point 4	0.90	1.34	0.57	0.52
Case Point 5	1.35	2.24	1.36	1.64

3.5.2. Directional Ring Profiling of LST for Investigation of SUHI

In the present work, directional ring profiling was used to delineate the actual difference of LST from the city center to its periphery by taking eight-directional ring profiling, i.e., north to south, NE to SE, NW to SE, and west to east. The distribution of LST dynamics was evaluated to investigate the SUHI state in the summer and winter at all four time points over Prayagraj city. Figure A2 shows the seasonal LST profiling for SUHI formation in Prayagraj city (1987–2018) for both the summer and the winter seasons.

North to South

In summer, *at S1*, the mean LST was higher in the city center than in both the north and the south directions by ~0.75–1.5 °C [except for 5–11 km in the north and 6.5–11 km in the south because these areas were constituted mostly of sands and bare soil]. *At S2*, the mean LST was higher in the city center than in both the north and the south directions by ~0.2–2.75 °C [except for 9.5–10.5 km in the north and 7–11 km in the south because these areas were mostly sands and bare soils]. *At S3*, the mean LST was higher at the city center than in both the north and the south directions by ~0.2–2 °C [except for 6.5–7 km, 8–8.5 km, and 9.5–10.5 km in the north, and 10–11 km in the south because these areas were mostly sands and bare soils]. *At S4*, the mean LST was higher at the city center than in both the north and the south directions by ~0.2–2.2 °C except for 6.5–7.5 km and 9.5–10.5 km in the north, and 6.5 km and 10–11 km in the south because these areas were mostly sands and bare soils.

In winter, *at W1*, the mean LST was higher in the city center than in both the north and the south directions by ~0.2–1.7 °C [except for 7.5–11 km in the north, and 6.5–11 km in the south because these areas were mostly sands and bare soil]. *At W2*, the mean LST was higher at the city center than in both the north and the south directions by ~0.2–1.5 °C [except at 6 km in the north and 10–11 km in the south because these areas were mostly sands and bare soils]. *At W3*, the mean LST was higher in the city center than in both the north and the south directions by ~0.2–3.6 °C [except for 8–8.5 km and 9.5–10.5 km in the north, and 10–11 km in the south because these areas were covered mostly with sands and bare soils]. *At W4*, the mean LST was higher in the city center than in both the north and the south directions by ~0.3–2.1 °C except for 10.5–11 km in the south because of the existence of barren land/bare soils.

NE to SE

In summer, the mean LST showed a higher peak at the city center than in both the NE and the SW directions at all four time points, i.e., S1, S2, S3, and S4, by ~0.1–1.2 °C, ~0.1–4.1 °C, ~0.12–5.8 °C, and ~0.2–1.8, respectively. However, because of the existence of sand, barren land, and bare soil, a reverse pattern was observed in zones 5–7.5 km and 9–11 km in the NE, and 5–11 km in the SW at the S1 timepoint, in zones 6–6.5 km and 10–11 km in the NE, and 5.5 km and 6.5–11 km in the SW at the S2 timepoint, in zones 7 km and 9.5–10.5 km in the NE and 10–11 km in the SW at the S3 timepoint, and in zones 6.5–8 km in the NE and 6.5 km and 10–11 km in the SW at the S4 timepoint.

In winter, the mean LST in the city center also followed the same pattern as that of the summer, with higher temperatures observed as compared to suburban areas in both the NE and the SW directions. The mean LST in the city center was higher by ~0.1–3.25 °C, ~0.1–2.4 °C, ~0.2–4.5 °C, and ~0.3–2.1 °C at the W1, W2, W3, and W4 time points. However, in the NE direction, the zones 7–7.5 km at W1, 5.0 and 7.5 km at W2, 6.5–7 km, 8.5 km, and 10 km at W3, and 6–7 km at W4 exhibited a higher LST peak as compared to the city center due to the presence of sand and bare soil. Along the same line, in the SW direction, zones 4.5–11 at W1, 7–8.5 km and 9.5–11 km at W2, 4.5–5.5 km, 8 km, and 9.5–11 km at W3, and 10–11 km at W4 exhibited the same trend.

NW to SE

In summer, *at S1*, the mean LST was higher at the city center than in both the NW and the SE directions by ~0.15–1.8 °C except for zones having predominantly bare soil at 3.5–5 km and 6.5–9 km in the NW and 4.5–8 in the SE. The same pattern was also observed at the S2, S3, and S4 time points. The mean LST at the city center was higher by ~0.1–2.6 °C, ~0.2–1.3 °C, and ~0.3–2 °C in both directions at S2, S3, and S4, respectively. The zones with a reverse trend, due to the presence of predominantly bare soil, were 3.5–4.5 km and 7.5–9 km in the NW and 6–8 km in the SE at S2, 3.5–4.5 km and 7.5–9 km in the NW and 5 km and 6–8 km in the SE at S3, and 3.5–4.5 km and 7.5–9 km in the NW and 5.5–8 km in the SE at S4.

In winter, the city center observed a higher LST peak than the suburban areas in both the NW and the SE directions, as observed in the summer season. The mean LST in the city center was higher by ~0.1–1 °C, ~0.1–1.9 °C, ~0.1–3.5 °C, and ~0.3–1.7 °C at the W1, W2, W3, and W4 time points, respectively. However, exceptions to this trend were also observed in certain zones in both the NW and the SE directions at each distinct timepoint, as can be clearly witnessed in Figure A2, because these areas were predominantly exposed to bare soil.

West to East

In summer, the city center observed higher LST than suburban areas by ~0.1–1.2 °C at S1, ~0.1–1.75 °C at S2, ~0.1–1.7 °C at S3 and ~0.1–1.5 °C at S4 in both the west and the east directions. However, because of the existence of mostly sand, barren land, and bare soil, a few zones observed a reverse trend, namely, 3.5–11 km in the west and 5–11 km in the east at S1, 4.5–5 km and 7–11 km in the west, and 7.5–8 km and 9.5–11 km in the east at S2, 4.5–5.5 km and 7–11 km in the west, and 6.5–7 km and 9.5–11 km in the east at S3, and 10–11 km in the west, and 7–8 km and 10.5–11 km in the east at S4.

In winter, the mean LST in the city center also followed the same pattern as that of the summer, with higher temperatures observed as compared to suburban areas in both the west and the east directions. The mean LST in the city center was higher by ~0.1–0.85 °C, ~0.1–1.6 °C, ~0.1–1.5 °C, and ~0.1–1.3 °C at the W1, W2, W3, and W4 time points. However, exceptions to this trend were also observed in certain zones in both the west and the east directions at each distinctive timepoint, as can be clearly seen in Figure A2, as these areas mostly had sand and bare soil.

3.6. Hotspot Identification

The hotspot analysis was performed using the Getis–Ord G_i* approach to analyze the spatial distribution of LST over Prayagraj city. This approach uses LST values of neighboring features and delineates both hot and cold spots over the city landscape. Hotspots are the clusters of features of high values of LST, while cold spots aggregate the features of low LST values. On the basis of this analysis, the city landscape was categorized into seven classes: very cold, cold, cool, not significant, warm, hot, and very hot.

The summer spatiotemporal hotspot maps of Prayagraj city are shown in Figure 18a for summer time points S1, S2, S3, and S4 to present the clustering distribution of hotspot dynamics, and their statistics are compiled in Table 8. The very cold spot class experienced a high loss of 1.29 km^2 of areal coverage during S1–S4. The cold spot class experienced a loss of 0.07 km^2 of areal coverage during S1–S4. The cool spot class also experienced a loss of 0.44 km^2 of areal coverage during S1–S4. However, the not significant class experienced an enormous gain of 7.56 km^2 of areal coverage during S1–S4. The warm spot class experienced a loss of 1.81 km^2 of areal coverage during S1–S4. The hot spot class experienced a loss of 2.06 km^2 of areal coverage during S1–S4. The very hot spot class experienced a loss of 1.88 km^2 of areal coverage during S1–S4. This summer hotspot pattern indicates that the comfort level of living space intensively decreased in the city landscape as the areas of very cold, cold, and cool spots severely declined.

The winter spatiotemporal hotspot maps of Prayagraj city are shown in Figure 18b for the W1, W2, W3, and W4 winter time points to present the clustering distribution of hotspot dynamics. Their statistics are shown in Table 8. The very cold spot class experienced a severe loss of 6.49 km^2 of areal coverage during S1–S4. The cold spot class experienced a loss of 0.80 km^2 of areal coverage during S1–S4. The cool spot class also experienced a loss of 2.08 km^2 of areal coverage during S1–S4. However, the not significant class has experienced a considerable gain of 15 km^2 of areal coverage during S1–S4. The warm spot class experienced a loss of 1.46 km^2 of areal coverage during S1–S4. The hot spot class experienced a loss of 1.30 km^2 of areal coverage during S1–S4. The very hot spot class experienced a loss of 2.87 km^2 of areal coverage during S1–S4. This winter hotspot pattern also indicates that the comfort level of living space was on the decline in the city landscape as the areas of very cold, cold, and cool spots severely declined.

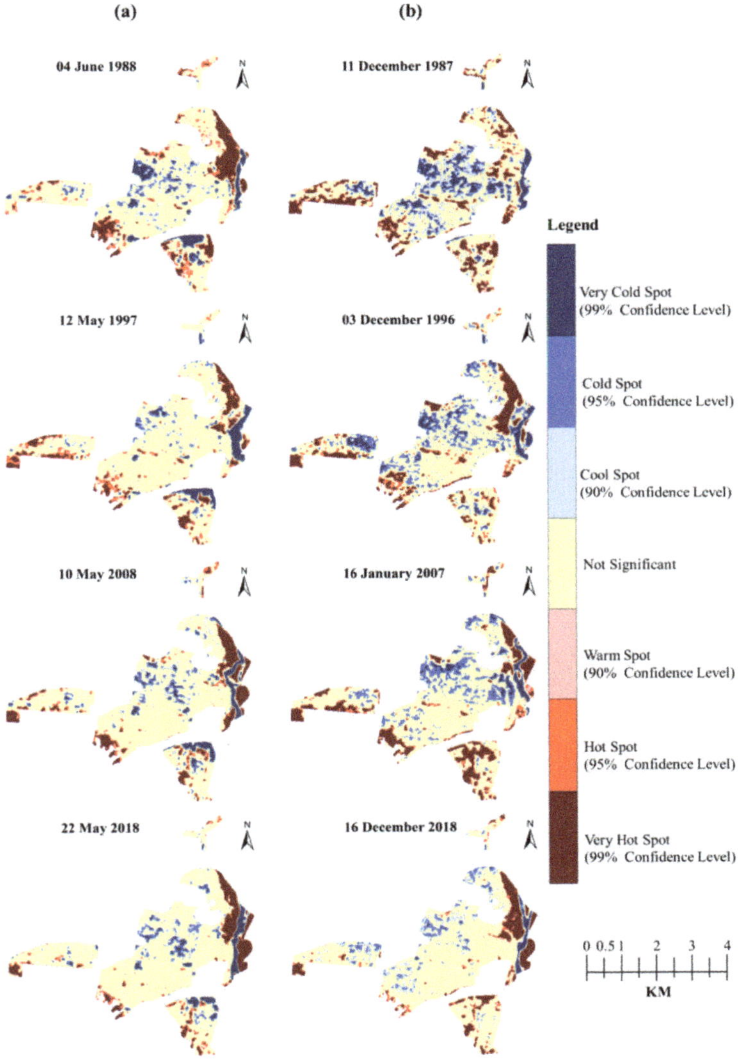

Figure 18. Seasonal hotspot maps based on Getis–Ord G_i* in Prayagraj city (1987–2018): (**a**) summer and (**b**) winter.

Table 8. Seasonal hotspot classes based on Getis–Ord G_i* statistics.

Hot-Spot Classes Based on Getis–Ord Gi* Analysis	Area (km^2) [Area (%)]				
Summer Time Points	**S1**	**S2**	**S3**	**S4**	**Change during S1–S4**
Very cold spot (99% of confidence level)	5.51 (7.55%)	4.76 (6.52%)	4.71 (6.45%)	4.22 (5.78%)	−1.29 (−1.77%)
Cold spot (95% of confidence level)	3.05 (4.18%)	2.03 (2.78%)	3.08 (4.22%)	2.98 (4.08%)	−0.07 (−0.10%)
Cool spot (90% of confidence level)	2.95 (4.04%)	2.21 (3.03%)	2.86 (3.92%)	2.52 (3.45%)	−0.44 (−0.60%)

Table 8. *Cont.*

Hot-Spot Classes Based on Getis–Ord Gi* Analysis	Area (km^2) [Area (%)]				
Not significant	45.35 (62.14%)	50.67 (69.43%)	49.98 (68.48%)	52.90 (72.49%)	7.56 (10.36%)
Warm spot (90% of confidence level)	3.17 (4.34%)	2.62 (3.59%)	1.63 (2.23%)	1.36 (1.86%)	−1.81 (−2.48%)
Hot spot (95% of confidence level)	4.00 (5.48%)	3.96 (5.43%)	2.65 (3.63%)	1.94 (2.66%)	−2.06 (−2.82%)
Very hot spot (99% of confidence level)	8.94 (12.25%)	6.73 (9.22%)	8.07 (11.06%)	7.06 (9.67%)	−1.88 (−2.58%)
Winter Time Points	**W1**	**W2**	**W3**	**W4**	**Change during W1–W4**
Very cold spot (99% of confidence level)	7.97 (10.92%)	5.40 (7.40%)	3.35 (4.59%)	1.49 (2.04%)	−6.49 (−8.89%)
Cold spot (95% of confidence level)	4.16 (5.70%)	7.83 (10.73%)	7.67 (10.51%)	3.36 (4.60%)	−0.80 (−1.10%)
Cool spot (90% of confidence level)	7.44 (10.19%)	4.31 (5.91%)	2.94 (4.03%)	5.36 (7.34%)	−2.08 (−2.85%)
Not significant	36.32 (49.77%)	41.30 (56.59%)	43.62 (59.77%)	51.32 (70.32%)	15.00 (20.55%)
Warm spot (90% of confidence level)	2.83 (3.88%)	0.83 (1.14%)	1.81 (2.48%)	1.37 (1.88%)	−1.46 (−2.00%)
Hot spot (95% of confidence level)	3.42 (4.69%)	3.88 (5.32%)	3.22 (4.41%)	2.12 (2.90%)	−1.30 (−1.78%)
Very hot spot (99% of confidence level)	10.82 (14.83%)	9.42 (12.91%)	10.35 (14.18%)	7.95 (10.89%)	−2.87 (−3.93%)

4. Discussion

4.1. Urbanization: An Assessment for Effective Urban Planning

This work assessed the seasonal (summer and winter) thermal state over the city landscape of Prayagraj city (India). The effects of land indices, namely, NDBI, EBBI, NDMI, NDVI, NDWI, and SAVI, on the thermal state were extensively examined to investigate how water bodies, forest land, wetland, and barren soils control intensification and/or cooling of LST over the landscape of the study area. The study area was delineated using eight-directional ring profiling of land indices, including LST to explore how, where, and what magnitude the LST changed either in increasing or in decreasing patterns due to different land indices. This can help policymakers and planners to conduct sustainable planning and enrich the carrying capacities of the landscape. As per the IPCC AR6 report of 2021, at the local to the global level, an extreme transformation of landscape has occurred, especially in the postindustrial era, which needs to be monitored and mitigated to control the rise in global mean temperature, leading to a decrease in the adverse consequences of climate change [5]. In connection to this issue, multiple cities (such as Taipei city of Taiwan [15], Phoenix city of the United States of America (USA) [16], Singapore [10], Dhaka city of Bangladesh [17], Kathmandu valley of Nepal [18], Nanjing city of China [19], Beijing city of China [20], Tokyo city of Japan [21], Tehran city of Iran [13], 70 selected cities of Europe [22], Hong Kong [23], Baltimore–DC metropolitan area of the USA [24], and Cairo city of Egypt [25]) have witnessed a similar amplifying pattern, which is very concerning and critical for our blue planet. UN-Habitat (2018), through SDG-11, categorically acknowledged the significance of spatial identification of land coverage and their possible effects on the safety, resilience, and sustainability of the city landscape, especially using greenery and open spaces [37].

Worldwide, to minimize the adverse effects of SUHI, including the rise in energy consumption, water scarcity, air pollution, and health problems, e.g., sunstroke, cardiac, and respiratory issues [68,69], populaces are converging on different mitigation strategies, such as the use of light paints and materials, planting of trees on streets, and cool and green roof creations [70,71]. Other strategies, such as designing the size, orientation, and shape of buildings, could improve the wind flow in the city landscape [72]. Moreover, distinctive mitigation approaches, such as preserving wetlands, small to large water bodies, and greenery plantations on barren spaces, can significantly improve the local climate and ecosystem of the city landscape [38,73,74]. For this strategy and subsequent planning and creation, remote sensing and GIS-based information using a multitemporal spatial database can be applied to recognize the ground reality. Along with this, public awareness regarding the adoption and implementation of the above strategies is a prerequisite for their success in sustainable development and ecosystem restoration of the Prayagraj city landscape.

4.2. An Overview of Night-Time LST for SUHI Exploration

The spatiotemporal summer night-time LST maps are shown in Figure 19a, where summer time points, such as May 2008 and May 2018, were incorporated to present the night-time LST dynamics along with their whisker boxplot statistics. The mean night-time LST was severely intensified by 6.94 °C during the summer time periods of 2008–2018. At night-time, the central urban core area intensively experienced higher LST than its periphery.

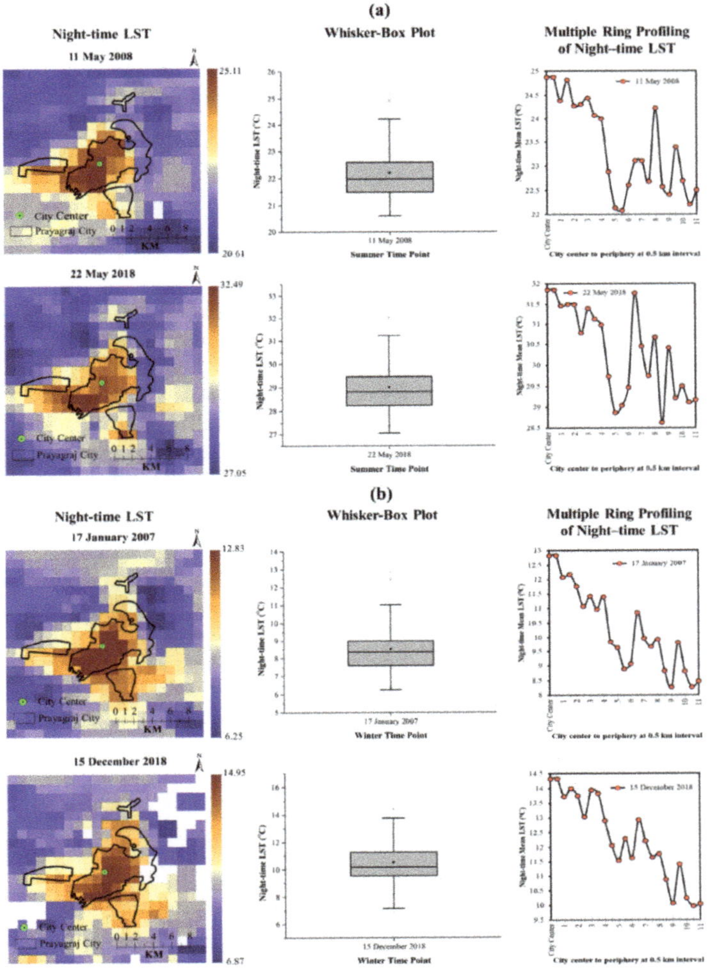

Figure 19. Maps of night-time LST dynamics with whisker boxplots for exploring SUHI at night-time: (**a**) summer and (**b**) winter.

The spatiotemporal winter night-time LST maps are shown in Figure 19b. Winter time points, such as January 2007 and December 2018, were incorporated to present the night-time LST dynamics and their whisker boxplot statistics. The mean night-time LST was also intensified by 1.69 °C during the winter periods from 2007 to 2018. At night-time, the central urban core area intensively experienced higher LST than its periphery. It was further detected that the SUHI phenomenon at night-time was severely intensified by 0–2.98 °C and 0–4.56 °C in summer and winter, respectively, according to multiple ring profiling. Other major cities over the Indo-Gangetic plain, such as Delhi [34], Lucknow [30], Patna [36], and Kolkata [35] have also been experiencing a strong SUHI phenomenon, whereby the city center has exhibited a higher LST than the periphery in the last few decades; a similar pattern was observed in our study area.

The above findings of night-time LST dynamics indicate a severe SUHI occurrence in both the summer and the winter seasons at night-time. These findings further strengthen the results of the daytime LST of Landsat imagery from 2008 to 2018. In the daytime, impervious surfaces absorb albedo extensively, and a long time is taken to radiate it back to the atmosphere due to its physical properties. As a result of increasing urbanization, this phenomenon has been further intensified. Both daytime and night-time LST experience a strong SUHI phenomenon; however, at night-time, SUHI becomes more vulnerable. As per this study, it is suggested to take immediate attention to reduce the SUHI severity using an effective mitigation strategy after urban planners and policymakers consider the aforementioned spatial thermal anomalies in the city.

5. Conclusions

This study explored the dynamics of seasonal (summer and winter) land indices [namely, NDBI, EBBI, NDMI, NDVI, NDWI, and SAVI] and LST dynamics using Landsat 5 (TM) and Landsat 8 (OLI/TIRS) imagery for Prayagraj city of India. The multitemporal spatial pattern of land indices and LST and their correlation dynamics with directional ring profiling over the city landscape were investigated, including the formation of SUHI and its dynamics over the city landscape. It was found that summer periodical LST magnitudes were highly intensified during S1–S4 by 0.32–2.45 °C (except for zone 11 km) at 0.5 km intervals from the city center to the periphery. Winter periodical LST magnitudes were also intensified by 0.02–1.06 °C (except for zones 9, 11, and 15–19 km). It was witnessed that the northeast and southwest directions had a high growth of LST distribution, while the northwest direction had a low growth of LST distribution in both seasons.

The results based on directional ring profiling of the effect of land indices on LST found that most of the vegetation/forest land available, at 1–3 km in the northwest and 5–6 km in the southeast direction, were depleted during the selected period. The impervious/built-up land expanded from the city center to 8 km in all directions during the study time, whereas bare soils and sand were primarily present in the northeast and the northwest (6–11 km). The presence of different land covers significantly controlled the LST distribution as forested area decreased the LST distribution whereas built-up area, bare soils, and sands increased the LST distribution.

Forest cover played a crucial role in declining the LST by 2.25–4.8 °C (except for water bodies), whereas bare soils and sand played a critical role in amplifying the LST by 1.9–5.6 °C. At the same time, built-up land amplified the LST by 1.8–3.9 °C (except for sand and bare soils). These results were further strengthened by the findings that LST vs. NDVI, LST vs. SAVI, LST vs. NDMI, and LST vs. NDWI had a positive correlation in the summer season. However, the LST vs. NDWI relationship had a very weak positive correlation due to the unavailability of water bodies of a significant size, which may have reduced the LST. In contrast, LST vs. NDBI and LST vs. EBBI had a strong positive correlation.

In the winter season, a positive correlation was observed for LST vs. NDVI, LST vs. SAVI, LST vs. NDMI, and LST vs. NDWI. However, the NDVI vs. LST, SAVI vs. LST, and NDWI vs. LST relationships showed a weak positive correlation to reducing mean LST. In contrast, EBBI vs. LST and NDBI vs. LST had a strong positive correlation. These findings indicate that forest and water body coverage played a vital role in reducing the LST, whereas bare soils and sands played a substantial role in amplifying the LST in the Prayagraj city landscape.

The SUHI results confirmed that the urban center observed a higher LST than rural/suburban points in the range of 0.398–4.016 °C and 0.45–2.24 °C in the summer and winter, respectively. Furthermore, according to the directional ring profiling analysis, it was detected that the center of the city had a higher LST than the periphery up to 11 km, mainly in the northwest and the southeast (except for zones 6 and 8 km) directions by 0.1–4.1 °C and 0.1–4.5 °C in the summer and winter seasons, respectively.

Hotspot analysis (using Getis–Ord G_i^* statistics) revealed that very cold spot, cold spot, and cool spot areal coverage declined over the study period in both the summer and the winter seasons. Hotspot analysis revealed the forested and Ganga River areas for very cold, cold, and cool spots, which were also observed in the directional ring profiling of land indices. This further strengthens our findings. It was further detected that the SUHI phenomenon at night-time was severely intensified by 0–2.98 °C and 0–4.56 °C in the summer and winter seasons. Therefore, to minimize the adverse effects of LST intensification and environmental sustainability of the local climate and ecosystem in the Prayagraj city landscape, the aforementioned delineated spatiotemporal seasonal thermal state and the hot and cold spot areas need to be prioritized by urban planners and policymakers for the design of suitable mitigation strategies.

Supplementary Materials: The following supporting information can be downloaded at: https://www.mdpi.com/article/10.3390/rs15010179/s1.

Author Contributions: Conceptualization, M.O.S. and R.D.G.; methodology, M.O.S.; software, M.O.S.; validation, M.O.S.; formal analysis, M.O.S.; investigation, M.O.S.; resources, M.O.S. and R.D.G.; data curation, M.O.S.; writing—original draft preparation, M.O.S.; writing—review and editing, M.O.S., R.D.G. and Y.M.; visualization, M.O.S. and R.D.G.; supervision, R.D.G.; project administration, M.O.S. and R.D.G.; funding acquisition, Y.M. All authors have read and agreed to the published version of the manuscript.

Funding: This research was partly supported by the Japan Society for the Promotion of Science (JSPS) grants of 21K01027 and 18H00763.

Data Availability Statement: Not applicable.

Acknowledgments: The authors are thankful to the USGS for providing the Landsat datasets for daytime seasonal variation and MODIS datasets in the night-time. The authors are also thankful to NASA Langley Research Center (LaRC) POWER Project for weather information. The first author is grateful to the UGC for providing a financial assistantship through a Maulana Azad National Fellowship for Minority Students (MANF) by the Ministry of Minority Affairs during his PhD research (Award Letter No. F1-17.1/2017-18/MANF-2017-18-WES-84175/(SA-III/Website)).

Conflicts of Interest: The authors declare no conflict of interest. The funders had no role in the design of the study, in the collection, analyses, or interpretation of data, in the writing of the manuscript, or in the decision to publish the results.

Appendix A

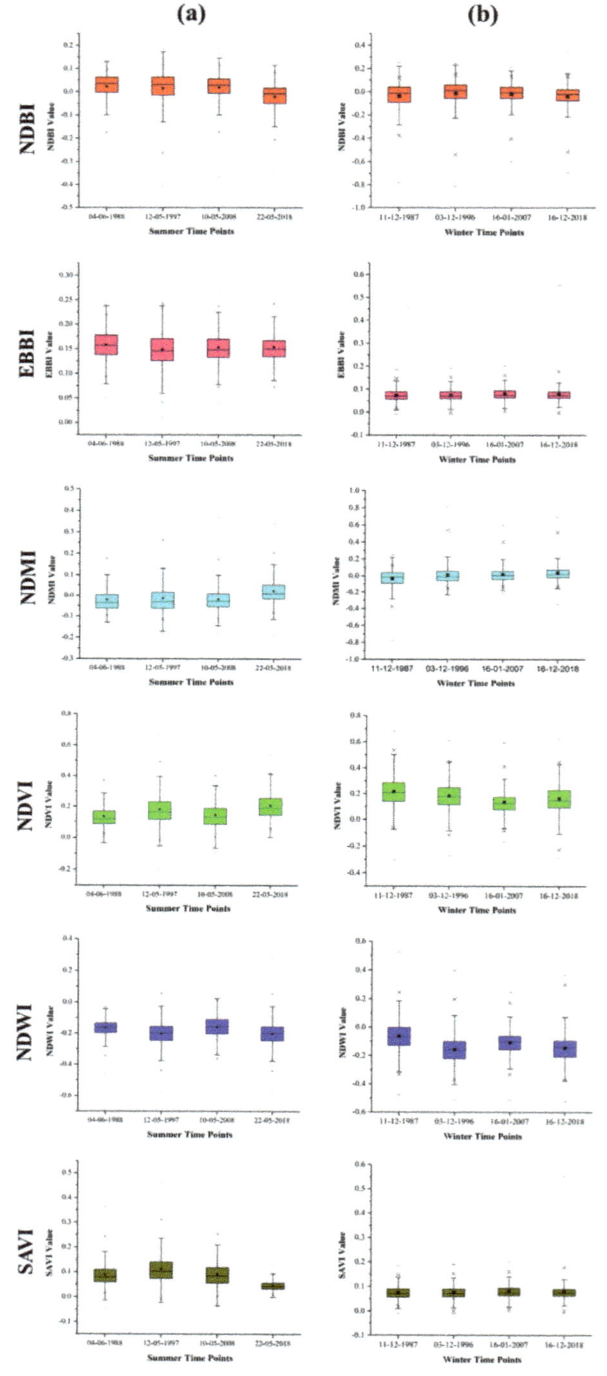

Figure A1. Seasonal whisker boxplots of the six land indices: (**a**) summer and (**b**) winter.

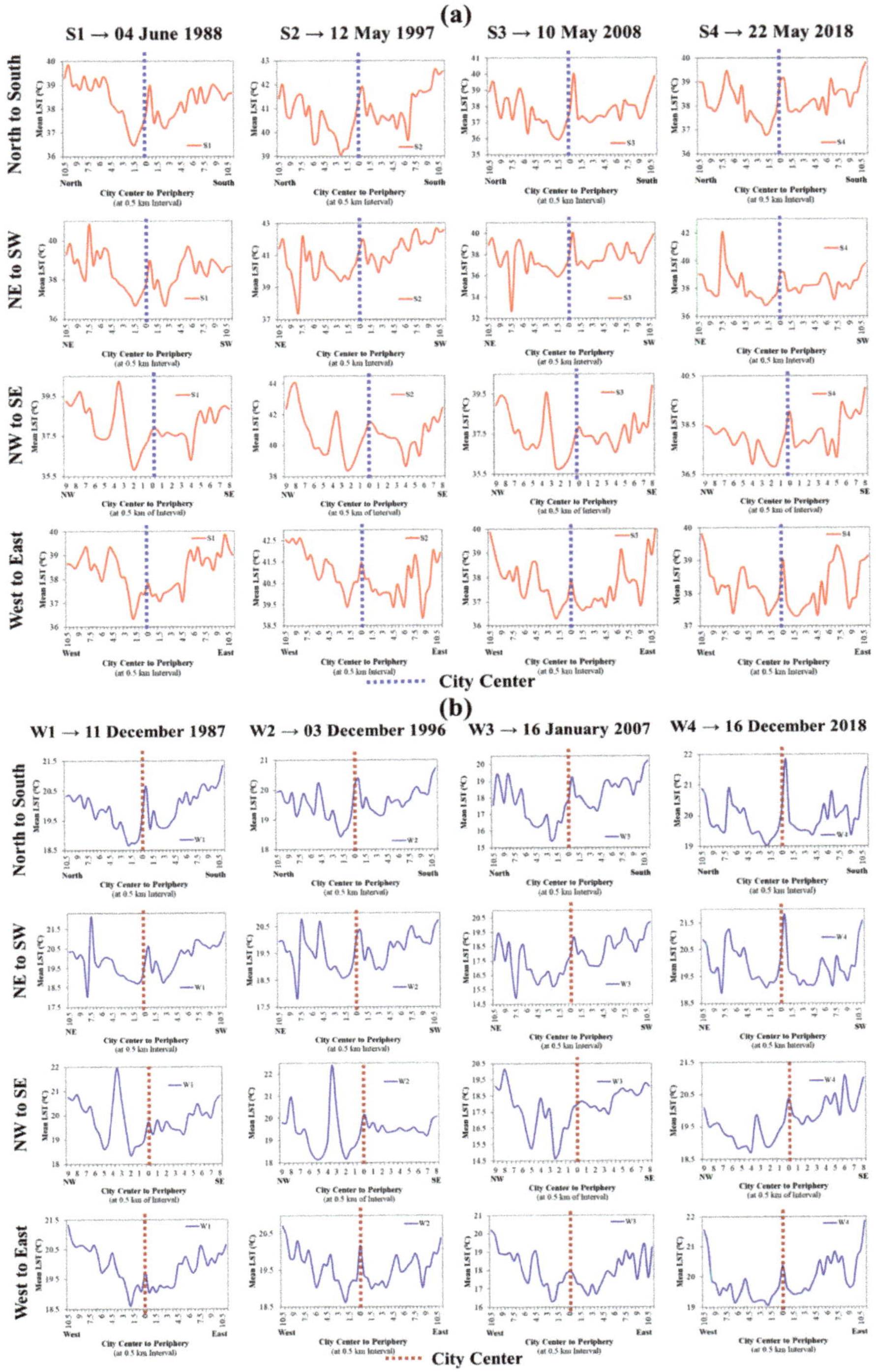

Figure A2. Seasonal LST profiling for SUHI formation in Prayagraj city (1987–2018): (**a**) summer and (**b**) winter.

References

1. United Nations. *The World's Cities in 2018—Data Booklet (ST/ESA/SER.A/417)*; Department of Economic and Social Affairs, Population Division: New York, NY, USA, 2018; pp. 1–34.
2. IPCC. *Climate Change and Land. An IPCC Special Report on Climate Change, Desertification, Land Degradation, Sustainable Land Management, Food Security, and Greenhouse Gas Fluxes in Terrestrial Ecosystems. Summary for Policymakers*; IPCC: Geneva, Switzerland, 2019.
3. Rosa, A.; de Oliveira, F.S.; Gomes, A.; Gleriani, J.M.; Gonçalves, W.; Moreira, G.L.; Silva, F.G.; Ricardo, E.; Branco, F.; Moura, M.M.; et al. Spatial and Temporal Distribution of Urban Heat Islands. *Sci. Total Environ.* **2017**, *605–606*, 946–956. [CrossRef]
4. Thomas, G.; Sherin, A.P.; Ansar, S.; Zachariah, E.J. Analysis of Urban Heat Island in Kochi, India, Using a Modified Local Climate Zone Classification. *Procedia Environ. Sci.* **2014**, *21*, 3–13. [CrossRef]
5. IPCC. *Climate Change 2021: The Physical Science Basis*; IPCC: Geneva, Switzerland, 2021.
6. Ranagalage, M.; Estoque, R.C.; Murayama, Y. An Urban Heat Island Study of the Colombo Metropolitan Area, Sri Lanka, Based on Landsat Data (1997–2017). *ISPRS Int. J. Geo-Inf.* **2017**, *6*, 189. [CrossRef]
7. Son, N.T.; Chen, C.F.; Chen, C.R.; Thanh, B.X.; Vuong, T.H. Assessment of Urbanization and Urban Heat Islands in Ho Chi Minh City, Vietnam Using Landsat Data. *Sustain. Cities Soc.* **2017**, *30*, 150–161. [CrossRef]
8. Liu, H.; Weng, Q. Seasonal Variations in the Relationship between Landscape Pattern and Land Surface Temperature in Indianapolis, USA. *Environ. Monit. Assess.* **2008**, *144*, 199–219. [CrossRef]
9. Sultana, S.; Satyanarayana, A.N. V Urban Heat Island Intensity during Winter over Metropolitan Cities of India Using Remote-Sensing Techniques: Impact of Urbanization. *Int. J. Remote Sens.* **2018**, *39*, 6692–6730. [CrossRef]
10. He, Y.; Lin, E.S.; Zhang, W.; Tan, C.L.; Tan, P.Y.; Wong, N.H. Local Microclimate above Shrub and Grass in Tropical City: A Case Study in Singapore. *Urban Clim.* **2022**, *43*, 101142. [CrossRef]
11. Estoque, R.C.; Murayama, Y. Quantifying Landscape Pattern and Ecosystem Service Value Changes in Four Rapidly Urbanizing Hill Stations of Southeast Asia. *Landsc. Ecol.* **2016**, *31*, 1481–1507. [CrossRef]
12. Dissanayake, D.; Morimoto, T.; Murayama, Y.; Ranagalage, M. Impact of Landscape Structure on the Variation of Land Surface Temperature in Sub-Saharan Region: A Case Study of Addis Ababa Using Landsat Data. *Sustainability* **2019**, *11*, 2257. [CrossRef]
13. Rousta, I.; Sarif, M.O.; Gupta, R.D.; Olafsson, H.; Ranagalage, M.; Murayama, Y.; Zhang, H.; Mushore, T.D. Spatiotemporal Analysis of Land Use/Land Cover and Its Effects on Surface Urban Heat Island Using Landsat Data: A Case Study of Metropolitan City Tehran (1988–2018). *Sustainability* **2018**, *10*, 4433. [CrossRef]
14. Sharma, R.; Chakraborty, A.; Joshi, P.K. Geospatial Quantification and Analysis of Environmental Changes in Urbanizing City of Kolkata (India). *Environ. Monit. Assess.* **2015**, *187*, 4206. [CrossRef] [PubMed]
15. Chen, T.L.; Lin, Z.H. Impact of Land Use Types on the Spatial Heterogeneity of Extreme Heat Environments in a Metropolitan Area. *Sustain. Cities Soc.* **2021**, *72*, 103005. [CrossRef]
16. Myint, S.W.; Brazel, A.; Okin, G.; Buyantuyev, A. Combined Effects of Impervious Surface and Vegetation Cover on Air Temperature Variations in a Rapidly Expanding Desert City. *GIScience Remote Sens.* **2010**, *47*, 301–320. [CrossRef]
17. Faisal, A.A.; Kafy, A.-A.; Al Rakib, A.; Akter, K.S.; Jahir, D.M.A.; Sikdar, M.S.; Ashrafi, T.J.; Mallik, S.; Rahman, M.M. Assessing and Predicting Land Use/Land Cover, Land Surface Temperature and Urban Thermal Field Variance Index Using Landsat Imagery for Dhaka Metropolitan Area. *Environ. Challenges* **2021**, *4*, 100192. [CrossRef]
18. Sarif, M.O.; Rimal, B.; Stork, N.E. Assessment of Changes in Land Use/Land Cover and Land Surface Temperatures and Their Impact on Surface Urban Heat Island Phenomena in the Kathmandu Valley (1988–2018). *ISPRS Int. J. Geo-Inf.* **2020**, *9*, 726. [CrossRef]
19. Kong, F.; Yin, H.; James, P.; Hutyra, L.R.; He, H.S. Effects of Spatial Pattern of Greenspace on Urban Cooling in a Large Metropolitan Area of Eastern China. *Landsc. Urban Plan.* **2014**, *128*, 35–47. [CrossRef]
20. Li, L.; Zha, Y.; Zhang, J. Spatial and Dynamic Perspectives on Surface Urban Heat Island and Their Relationships with Vegetation Activity in Beijing, China, Based on Moderate Resolution Imaging Spectroradiometer Data. *Int. J. Remote Sens.* **2020**, *41*, 882–896. [CrossRef]
21. Wang, R.; Derdouri, A.; Murayama, Y. Spatiotemporal Simulation of Future Land Use/Cover Change Scenarios in the Tokyo Metropolitan Area. *Sustainability* **2018**, *10*, 2056. [CrossRef]
22. Ward, K.; Lauf, S.; Kleinschmit, B.; Endlicher, W. Heat Waves and Urban Heat Islands in Europe: A Review of Relevant Drivers. *Sci. Total Environ.* **2016**, *569–570*, 527–539. [CrossRef]
23. Liu, L.; Zhang, Y. Urban Heat Island Analysis Using the Landsat TM Data and ASTER Data: A Case Study in Hong Kong. *Remote Sens.* **2011**, *3*, 1535–1552. [CrossRef]
24. Tang, J.; Di, L.; Xiao, J.; Lu, D.; Zhou, Y. Impacts of Land Use and Socioeconomic Patterns on Urban Heat Island. *Int. J. Remote Sens.* **2017**, *38*, 3445–3465. [CrossRef]
25. Abutaleb, K.; Ngie, A.; Darwish, A.; Ahmed, M.; Arafat, S.; Ahmed, F. Assessment of Urban Heat Island Using Remotely Sensed Imagery over Greater Cairo, Egypt. *Adv. Remote Sens.* **2015**, *4*, 35–47. [CrossRef]
26. Ghosh, S.; Chatterjee, N.D.; Dinda, S. Relation between Urban Biophysical Composition and Dynamics of Land Surface Temperature in the Kolkata Metropolitan Area: A GIS and Statistical Based Analysis for Sustainable Planning. *Model. Earth Syst. Environ.* **2018**, *5*, 307–329. [CrossRef]
27. Grover, A.; Singh, R.B. Analysis of Urban Heat Island (UHI) in Relation to Normalized Difference Vegetation Index (NDVI): A Comparative Study of Delhi and Mumbai. *Environments* **2015**, *2*, 125–138. [CrossRef]

28. Sultana, S.; Satyanarayana, A.N.V. Assessment of Urbanisation and Urban Heat Island Intensities Using Landsat Imageries during 2000–2018 over a Sub-Tropical Indian City. *Sustain. Cities Soc.* **2020**, *52*, 101846. [CrossRef]
29. Chakraborti, S.; Banerjee, A.; Sannigrahi, S.; Pramanik, S.; Maiti, A.; Jha, S. Assessing the Dynamic Relationship among Land Use Pattern and Land Surface Temperature: A Spatial Regression Approach. *Asian Geogr.* **2019**, *36*, 93–116. [CrossRef]
30. Singh, P.; Kikon, N.; Verma, P. Impact of Land Use Change and Urbanization on Urban Heat Island in Lucknow City, Central India: A Remote Sensing Based Estimate. *Sustain. Cities Soc.* **2017**, *32*, 100–114. [CrossRef]
31. Sarif, M.O.; Gupta, R.D. Land Surface Temperature Profiling and Its Relationships with Land Indices: A Case Study on Lucknow City. In *ISPRS Annals of Photogrammetry, Remote Sensing and Spatial Information Sciences*; Copernicus Publications: Dhulikhel, Nepal, 2019; Volume IV-5/W2, pp. 89–96.
32. Guha, S.; Govil, H.; Mukherjee, S. Dynamic Analysis and Ecological Evaluation of Urban Heat Islands in Raipur City, India. *J. Appl. Remote Sens.* **2017**, *11*, 036020. [CrossRef]
33. Mal, S.; Rani, S.; Maharana, P. Estimation of Spatio-Temporal Variability in Land Surface Temperature over the Ganga River Basin Using MODIS Data. *Geocarto Int.* **2020**, *37*, 3817–3839. [CrossRef]
34. Shahfahad; Naikoo, M.W.; Islam, A.R.M.T.; Mallick, J.; Rahman, A. Land Use/Land Cover Change and Its Impact on Surface Urban Heat Island and Urban Thermal Comfort in a Metropolitan City. *Urban Clim.* **2022**, *41*, 101052. [CrossRef]
35. Mandal, J.; Patel, P.P.; Samanta, S. Examining the Expansion of Urban Heat Island Effect in the Kolkata Metropolitan Area and Its Vicinity Using Multi-Temporal MODIS Satellite Data. *Adv. Sp. Res.* **2022**, *69*, 1960–1977. [CrossRef]
36. Barat, A.; Parth Sarthi, P.; Kumar, S.; Kumar, P.; Sinha, A.K. Surface Urban Heat Island (SUHI) Over Riverside Cities Along the Gangetic Plain of India. *Pure Appl. Geophys.* **2021**, *178*, 1477–1497. [CrossRef]
37. UN-Habitat. *Tracking Progress Towards Inclusive, Safe, Resilient and Sustainable Cities and Human Settlements*; United Nations: New York, NY, USA, 2018.
38. Fonji, S.F.; Taff, G.N. Using Satellite Data to Monitor Land-Use Land-Cover Change in North-Eastern Latvia. *Springerplus* **2014**, *3*, 61. [CrossRef] [PubMed]
39. MoHUA. *Smart Citie: Ministry of Housing and Urban Affairs Reports, Government of India*; MoHUA: New Delhi, India, 2015.
40. PNN. Prayag Kumbh. Prayagraj Nagar Nigam, Government of Uttar Pradesh. Available online: allahabadmc.gov.in/kumbh_mela.html (accessed on 22 October 2019).
41. Singh, S.K.; Mustak, S.; Srivastava, P.K.; Szabó, S.; Islam, T. Predicting Spatial and Decadal LULC Changes Through Cellular Automata Markov Chain Models Using Earth Observation Datasets and Geo-Information. *Environ. Process.* **2015**, *2*, 61–78. [CrossRef]
42. Chaturvedi, R. Application of Remote Sensing and GIS in Land Use/Land Covers Mapping in Allahabad District. *Int. J. Adv. Inf. Eng. Technol.* **2014**, *4*, 1–9.
43. Nanda, M.K. Climatic Classification. In *Environmental Science*; Khan, D.K., Ed.; e-Pathsala: New Delhi, India, 2018; pp. 1–16.
44. IMD. Allahabad Climatological Table (Period: 1981–2010). Indian Meteorological Department, Government of India. Available online: http://www.imd.gov.in/section/climate/extreme/allahabad2.htm (accessed on 22 October 2019).
45. Sobrino, J.A.; Jiménez-Muñoz, J.C.; Paolini, L. Land Surface Temperature Retrieval from LANDSAT TM 5. *Remote Sens. Environ.* **2004**, *90*, 434–440. [CrossRef]
46. Lu, Y.; He, T.; Xu, X.; Qiao, Z. Investigation the Robustness of Standard Classification Methods for Defining Urban Heat Islands. *IEEE J. Sel. Top. Appl. Earth Obs. Remote Sens.* **2021**, *14*, 11386–11394. [CrossRef]
47. Renard, F.; Alonso, L.; Fitts, Y.; Hadjiosif, A.; Comby, J. Evaluation of the Effect of Urban Redevelopment on Surface Urban Heat Islands. *Remote Sens.* **2019**, *11*, 299. [CrossRef]
48. Haque, M.I.; Basak, R. Land Cover Change Detection Using GIS and Remote Sensing Techniques: A Spatio-Temporal Study on Tanguar Haor, Sunamganj, Bangladesh. *Egypt. J. Remote Sens. Sp. Sci.* **2017**, *20*, 251–263. [CrossRef]
49. Hasanlou, M.; Mostofi, N. Investigating Urban Heat Island Estimation and Relation between Various Land Cover Indices in Tehran City Using Landsat 8 Imagery. In Proceedings of the 1st International Electronic Conference on Remote sensing, Basel, Switzerland, 22 June–5 July 2015; MDPI: Basel, Switzerland, 2015; pp. 1–11.
50. Qin, Z.; Karnieli, A.; Berliner, P. A Mono-Window Algorithm for Retrieving Land Surface Temperature from Landsat TM Data and Its Application to the Israel-Egypt Border Region. *Int. J. Remote Sens.* **2001**, *22*, 3719–3746. [CrossRef]
51. Zhu, X.; Duan, S.B.; Li, Z.L.; Zhao, W.; Wu, H.; Leng, P.; Gao, M.; Zhou, X. Retrieval of Land Surface Temperature with Topographic Effect Correction from Landsat 8 Thermal Infrared Data in Mountainous Areas. *IEEE Trans. Geosci. Remote Sens.* **2021**, *59*, 6674–6687. [CrossRef]
52. Yang, L.; Cao, Y.; Zhu, X.; Zeng, S.; Yang, G.; He, J.; Yang, X. Land Surface Temperature Retrieval for Arid Regions Based on Landsat-8 TIRS Data: A Case Study in Shihezi, Northwest China. *J. Arid Land* **2014**, *6*, 704–716. [CrossRef]
53. Shahfahad; Kumari, B.; Tayyab, M.; Ahmed, I.A.; Baig, M.R.I.; Khan, M.F.; Rahman, A. Longitudinal Study of Land Surface Temperature (LST) Using Mono- and Split-Window Algorithms and Its Relationship with NDVI and NDBI over Selected Metro Cities of India. *Arab. J. Geosci.* **2020**, *13*, 1040. [CrossRef]
54. Wang, L.; Lu, Y.; Yao, Y. Comparison of Three Algorithms for the Retrieval of Land Surface Temperature from Landsat 8 Images. *Sensors* **2019**, *19*, 5049. [CrossRef] [PubMed]
55. Jiménez-Munoz, J.C.; Sobrino, J.A. A Generalized Single-Channel Method for Retrieving Land Surface Temperature from Remote Sensing Data. *J. Geophys. Res. Atmos.* **2004**, *108*, 1–9. [CrossRef]

56. Rozenstein, O.; Qin, Z.; Derimian, Y.; Karnieli, A. Derivation of Land Surface Temperature for Landsat-8 TIRS Using a Split Window Algorithm. *Sensors* **2014**, *14*, 5768–5780. [CrossRef]
57. Guha, S.; Govil, H. Annual Assessment on the Relationship between Land Surface Temperature and Six Remote Sensing Indices Using Landsat Data from 1988 to 2019. *Geocarto Int.* **2021**, *37*, 4292–4311. [CrossRef]
58. Majumder, A.; Setia, R.; Kingra, P.K.; Sembhi, H.; Singh, S.P.; Pateriya, B. Estimation of Land Surface Temperature Using Different Retrieval Methods for Studying the Spatiotemporal Variations of Surface Urban Heat and Cold Islands in Indian Punjab. *Environ. Dev. Sustain.* **2021**, *23*, 15921–15942. [CrossRef]
59. Guha, S.; Govil, H. An Assessment on the Relationship between Land Surface Temperature and Normalized Difference Vegetation Index. *Environ. Dev. Sustain.* **2021**, *23*, 1944–1963. [CrossRef]
60. Nimish, G.; Bharath, H.A.; Lalitha, A. Exploring Temperature Indices by Deriving Relationship between Land Surface Temperature and Urban Landscape. *Remote Sens. Appl. Soc. Environ.* **2020**, *18*, 100299. [CrossRef]
61. Sekertekin, A.; Bonafoni, S. Land Surface Temperature Retrieval from Landsat 5, 7, and 8 over Rural Areas: Assessment of Different Retrieval Algorithms and Emissivity Models and Toolbox Implementation. *Remote Sens.* **2020**, *12*, 294. [CrossRef]
62. Kikon, N.; Singh, P.; Singh, S.K.; Vyas, A. Assessment of Urban Heat Islands (UHI) of Noida City, India Using Multi-Temporal Satellite Data. *Sustain. Cities Soc.* **2016**, *22*, 19–28. [CrossRef]
63. Ziaul, S.; Pal, S. Image Based Surface Temperature Extraction and Trend Detection in an Urban Area of West Bengal, India. *J. Environ. Geogr.* **2016**, *9*, 13–25. [CrossRef]
64. Oke, T.R.; East, C. The Urban Boundary Layer in Montreal. *Bound.-Lay. Meteorol.* **1971**, *1*, 411–437. [CrossRef]
65. Oke, T.R. City Size and the Urban Heat Island. *Atmos. Environ.* **1973**, *7*, 769–779. [CrossRef]
66. Tran, D.X.; Pla, F.; Latorre-Carmona, P.; Myint, S.W.; Caetano, M.; Kieu, H.V. Characterizing the Relationship between Land Use Land Cover Change and Land Surface Temperature. *ISPRS J. Photogramm. Remote Sens.* **2017**, *124*, 119–132. [CrossRef]
67. Sarif, M.O.; Gupta, R.D. Spatiotemporal Mapping of Land Use/Land Cover Dynamics Using Remote Sensing and GIS Approach: A Case Study of Prayagraj City, India (1988–2018). *Environ. Dev. Sustain.* **2022**, *24*, 888–920. [CrossRef]
68. Karakuş, C.B. The Impact of Land Use/Land Cover (LULC) Changes on Land Surface Temperature in Sivas City Center and Its Surroundings and Assessment of Urban Heat Island. *Asia-Pac. J. Atmos. Sci.* **2019**, *55*, 669–684. [CrossRef]
69. Barat, A.; Kumar, S.; Kumar, P.; Sarthi, P.P. Characteristics of Surface Urban Heat Island (SUHI) over the Gangetic Plain of Bihar, India. *Asia-Pacific J. Atmos. Sci.* **2018**, *54*, 205–214. [CrossRef]
70. Chen, R.; You, X. Reduction of Urban Heat Island and Associated Greenhouse Gas Emissions. *Mitig. Adapt. Strateg. Glob. Chang.* **2019**, *25*, 689–711. [CrossRef]
71. Hien, W.N.; Tan, P.Y.; Yu, C. Study of Thermal Performance of Extensive Rooftop Greenery Systems in the Tropical Climate. *Build. Environ.* **2007**, *42*, 25–54. [CrossRef]
72. Sen, S.; Roesler, J.; Ruddell, B.; Middel, A. Cool Pavement Strategies for Urban Heat Island Mitigation in Suburban Phoenix, Arizona. *Sustainability* **2019**, *11*, 4452. [CrossRef]
73. Estoque, R.C.; Estoque, R.S.; Murayama, Y. Prioritizing Areas for Rehabilitation by Monitoring Change in Barangay-Based Vegetation Cover. *ISPRS Int. J. Geo-Inf.* **2012**, *1*, 46–68. [CrossRef]
74. Simwanda, M.; Ranagalage, M.; Estoque, R.C.; Murayama, Y. Spatial Analysis of Surface Urban Heat Islands in Four Rapidly Growing African Cities. *Remote Sens.* **2019**, *11*, 1645. [CrossRef]

remote sensing

MDPI

Article

Relationships between Land Surface Temperatures and Neighboring Environment in Highly Urbanized Areas: Seasonal and Scale Effects Analyses of Beijing, China

Qingyan Meng [1,2,3], Wenxiu Liu [1,2,*], Linlin Zhang [1,2,3], Mona Allam [1,4], Yaxin Bi [5], Xinli Hu [1,2,3], Jianfeng Gao [1,2], Die Hu [1,2] and Tamás Jancsó [6]

1 Aerospace Information Research Institute, Chinese Academy of Sciences, Beijing 100094, China
2 University of Chinese Academy of Sciences, Beijing 100049, China
3 Key Laboratory of Earth Observation of Hainan Province, Hainan Aerospace Information Research Institute, Sanya 572029, China
4 Environment & Climate Changes Research Institute, National Water Research Centre, El Qanater El Khairiya 13621/5, Egypt
5 School of Computing, Ulster University, Shore Rd., Newtownabbey BT37 0QB, UK
6 Alba Regia Technical Faculty, Óbuda University, Budai ut 45., H-8000 Szekesfehervar, Hungary
* Correspondence: liuwx@radi.ac.cn

Citation: Meng, Q.; Liu, W.; Zhang, L.; Allam, M.; Bi, Y.; Hu, X.; Gao, J.; Hu, D.; Jancsó, T. Relationships between Land Surface Temperatures and Neighboring Environment in Highly Urbanized Areas: Seasonal and Scale Effects Analyses of Beijing, China. *Remote Sens.* **2022**, *14*, 4340. https://doi.org/10.3390/rs14174340

Academic Editors: Yuji Murayama and Ruci Wang

Received: 19 July 2022
Accepted: 30 August 2022
Published: 1 September 2022

Abstract: Urban environments have a strong influence on the land surface temperature (LST) in urban areas. Understanding the relationship between LST and urban environmental factors can help develop effective strategies to reduce high LSTs in urban areas, which is critical for mitigating the urban heat island effect. Previous studies have focused on the correlation between LST and the environmental factors that drive its formation, without considering the influences of the neighboring environment and the vertical expansion of highly urbanized areas. Notably, the correlation between LST and its neighboring environment in different seasons remains unclear. In this study, we selected central Beijing in China as our study area and employed the moving window method to characterize the environmental factors of the neighboring environment of the central LST cell. We explored eight environmental factors from three layers: normalized difference vegetation index (NDVI), normalized difference built-up index (NDBI), modified normalized difference water index (MNDWI), building density (BD), building height (BH), building volume (BV), sky view factor (SVF), and road density (RD). The Pearson correlation and extreme gradient boosting (XGB) regression methods were applied to measure the correlation between LST and the different factors in moving windows of different sizes. The results indicated that the correlation between NDVI, MNDWI, and LST was considerably different in the winter and other seasons. However, NDBI was positively correlated with LST in all seasons, although the correlation was strongest/weakest in summer/winter. Among building-related factors, BD and BH were more strongly correlated with LST, and the positive/negative correlation between BD/BH and LST was stronger in summer/winter. The correlation between LST and its neighboring environment varied with increasing window size, and this variation differs significantly between winter and other seasons. In spring, summer, and autumn, the strength of the correlation between LST and its neighboring environment showed an "inverted V" pattern with increasing window size. The optimal spatial scales to explore the influence of neighboring environments on the LST of 30-m cells were 210 m and 270 m. This study revealed the seasonal correlation between LST and its neighboring environment while explaining the variation at a spatial scale. Notably, this study can provide a new perspective for understanding the driving mechanism of the urban thermal environment, while contributing to its scientific optimization and management.

Keywords: land surface temperature; neighboring environment; seasonal effect; scale effect; optimal spatial scale; urban heat island; extreme gradient boosting regression

1. Introduction

With the acceleration of urbanization, the development of urban heat islands (UHIs) in cities being warmer than surrounding rural areas has been observed frequently on a global scale [1–3]. High urban temperatures not only facilitate energy consumption for cooling [4], contributing to air-pollutant emissions and global warming [5], but also increase the risk of heat-related morbidity and mortality, especially during extreme heat events [6–8]. At present, over 50% of the global population lives in urban areas, and this proportion is expected to reach 70% by 2050 [9]. The negative impacts of UHIs pose a great threat to sustainable urban development [2,3]. Land surface temperature (LST), an effective indicator of surface UHI, is generally affected by various factors related to urbanization [10,11]. Disclosing the correlation between LST and urban environmental factors is critical in developing reasonable strategies to optimize the urban thermal environment, prevent the further intensification of the UHI effect, and achieve sustainable development.

Urbanization is a complex process characterized by the transformation of natural land cover types (e.g., vegetation, water bodies, forests, and croplands) to impervious surfaces, such as roads, buildings, pavements, and parking lots [12,13]. Previous studies have confirmed that impervious surfaces can absorb and retain heat, resulting in higher surface temperatures in urban areas [2]. In addition, roads in urban areas are associated with traffic activities that lead to waste heat emissions [10]. Hardened pavement areas and the number of private vehicles are related to road density (RD) [14]. Therefore, urban road networks cannot be excluded when analyzing the potential driving factors of urban surface temperature. Vegetation and water bodies have an important cooling effect that can reduce the LST in urban areas [15]. Widely studied surface biophysical indexes, including the normalized difference built-up index (NDBI) [16], normalized difference vegetation index (NDVI) [17], and modified normalized difference water index (MNDWI) [18], exhibited different correlations with LST. Reducing impervious surface areas and increasing urban green vegetation/water areas are ideal strategies to mitigate high temperatures in urban areas. However, in highly urbanized areas, the amount of space available for greening is limited and it is impractical to considerably reduce impervious surfaces, such as buildings and roads. Meanwhile, studies in 3D urban morphology have indicated that optimizing urban building forms can help mitigate the UHI effect [19–23]. As a result, there is increasing interest regarding the effects of multidimensional factors of urban environments on LST [9,24–26].

In highly urbanized areas, buildings reflect urban expansion in the vertical dimension. This has also been the most important urban landscape in 3D urban morphology studies. The combination of high buildings and narrow streets can trap long-wave radiation and increase the degree of thermal stress in summer and solar radiation absorption [27,28]. However, the shadows of high-rise buildings can reduce the absorption of solar radiation by the ground [29]. Thus, a comprehensive investigation of the relationship between LST and building-related factors is crucial for urban planning and intra-urban thermal environment management. Although recent studies examine the relationship between LST and building factors, including building density (BD), building height (BH), building volume (BV), and sky view factor (SVF) factors [19,20,25,30,31], most of them focus on specific seasons, with the summer receiving the most attention. More studies are required to understand the correlations between building factors and LST during different seasons. Moreover, previous studies on the relationship between LST and building features have yielded inconsistent results. A study conducted at the residential scale suggested that building height had a greater impact on LST than building density [19]. However, a seasonal analysis at a geographical scale of 500 m $\times$ 500 m indicated that building density had a stronger influence on LST than building height [21]. A recent study based on a road-block scale also indicated that the influence of building cover ratio on LST was stronger than building height [20]. These inconsistent conclusions suggest that the correlations between building factors and LST may be affected by the spatial scale. However, at present, only a few studies have comprehensively discussed the seasonal and scale characteristics of the correlation between building factors and LST [9].

The neighboring environment plays an important role in the driving mechanism of the LST in highly urbanized areas. Previous studies have confirmed that the cooling effects of urban vegetation and water bodies are largely regulated by neighboring vegetation and impervious surface landscapes [32,33]. The heat sources in urban areas, such as industrial parks, contribute to the increase in the surface temperatures within a certain spatial extent around them [34]. Thus, these findings suggest that the local LST is related to the neighboring environment. However, the strength of the correlation between LST and land cover varies with spatial resolution, indicating the spatial range of the effect of the neighboring environment on LST [25,35]. Determining the effect of neighboring environments on LST provides a new perspective for understanding the potential driving factors of the urban thermal environment in a comprehensive manner. Thus, there is an increasing interest in the thermal effects of the neighboring environment. However, in investigating the thermal effects of neighboring environments, previous studies mainly focused on special urban scenarios, such as public and industrial parks and meteorological station areas, without considering the general situation. Therefore, more case studies are required to measure the influence of neighboring environments on fine-scale LSTs in highly urbanized areas.

In this study, we analyzed the central urban area of Beijing; the moving window method was adopted to measure the neighboring environmental factors of LST at different spatial scales. Eight environmental factors were selected from the urban greenness, urban wetness, and urban grayness layers: NDVI, NDBI, MNDWI, BD, BH, BV, SV, and RD. The Pearson correlation and extreme gradient boosting (XGB) regression methods were used to measure the influence of these factors on the LST from the perspectives of individual and combined effects. The aim of this study is to address the following questions: (1) What are the seasonal differences in the correlations between LST and the neighboring environment, in terms of multidimensional factors? (2) How does a change in spatial scale affect the correlation between LST and the neighboring environment? What are the optimal spatial scales? (3) Finally, is there any inter-seasonal difference in the effects of spatial scale on the correlation between LST and its neighboring environment?

This study aims to reveal the seasonal and scale effects on the correlations between LST and multidimensional environmental factors from the perspective of neighboring environments to improve the understanding of the driving mechanism of the urban thermal environment.

2. Study Area and Dataset

2.1. Study Area

Beijing is located on the North China Plain and experiences a continental monsoon climate (hot and rainy in the summer and cold and dry in the winter). As the capital of China, Beijing has undergone rapid urbanization in the last four decades. Its permanent population increased gradually from 10.47 million in 1987 to 21.94 million in 2017 (URL: http://nj.tjj.beijing.gov.cn/nj/main/2021-tjnj/zk/indexch.htm, accessed on 1 June 2022). The percentage of the urban population was 86.45% in 2017. Rapid urbanization has dramatically changed urban landscape patterns and caused a variety of environmental problems, including water and air pollution and the UHI effect. The deterioration of the thermal environment has become a major challenge for sustainable urban development in Beijing. Notably, urbanization in Beijing occurs in a ring-shaped pattern, that is, a concentric expansion from the urban center to the periphery. The region within the Fourth Ring Road covers most of the central urban areas that have the highest urbanization rates and population densities and contains various land cover types and complicated landscape patterns. Therefore, in this study, we investigated the thermal environment of the region within the Fourth Ring Road of Beijing, which covers an area of 340.26 km^2, and covers all of the Core Functional Zone and a small part of the Urban Function Extended Zone (Figure 1).

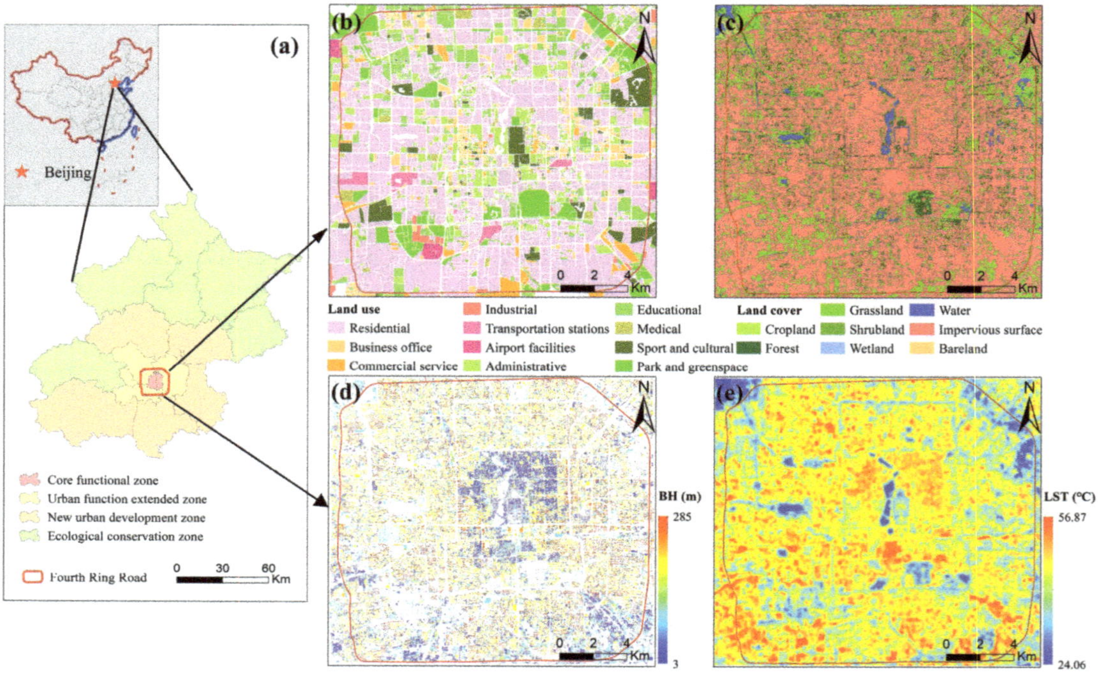

Figure 1. Geographic location of the study area: (**a**) the location of the Fourth Ring Road of Beijing; (**b**) the land use map of study area; (**c**) the land cover map of study area; (**d**) building height map of study area; (**e**) the land surface temperature (summer) distribution map of study area (the land use data in 2018 and land cover data in 2017 were derived from the website: http://data.ess.tsinghua.edu.cn/, accessed on 21 August 2022).

2.2. Dataset

The dataset used in this study included Landsat-8 Operational Land Imager/Thermal Infrared Sensor (OLI/TIRS) images and building and road network data. The Landsat-8 OLI/TIRS images obtained from the Geospatial Data Cloud official website (http://www.gscloud.cn/, accessed on 28 November 2020) were used to estimate the LST and surface biophysical factors. After screening all of the available images for 2017, four Landsat-8 OLI/TIRS images collected on May 23 (spring), July 10 (summer), September 28 (autumn), and December 17 (winter) were employed to characterize the thermal environment for different seasons [36]. The building data used to acquire building-related factors were obtained from the Baidu, Inc. (https://map.baidu.com, accessed on 1 October 2017) with a spatial resolution of 10 m. The building data can describe the geographical location and height attributes of buildings in 2017 and cover the area within the Fourth Ring Road of Beijing. The road data in 2017 were obtained from the OpenStreetMap (http://www.openstreetmap.org, accessed on 1 June 2020), a substantial global spatial database, and were used to explore the thermal effects of road density on the LST in this study.

3. Methods

3.1. Land Surface Temperature (LST) Estimation

In this study, the radiative transfer equation (RTE) method was employed to estimate the LST from the Landsat-8 OLI/TIRS images. The RTE method is more accurate than the split window method and single channel method and could guarantee an accuracy of 0.6 K for the retrieved LST [37,38]. First, the RTE method was used to estimate the black-body radiation brightness by removing the influence of the atmosphere. This can be expressed in an equation, as follows:

$$B(T_S) = [L_\lambda - L\uparrow - \tau(1-\varepsilon)L\downarrow]/L_\lambda\varepsilon \tag{1}$$

where $B(T_S)$ is the black-body radiation; L_λ (W/m^2·sr·µm) is the spectral radiation brightness for band 10; τ is the transmittance of thermal infrared bands in the atmosphere; $L\uparrow$ (W/m^2·sr·µm) and $L\downarrow$ (W/m^2·sr·µm) indicate the upwelling and downwelling atmospheric radiance (http://atmcorr.gsfc.nasa.gov, accessed on 28 November 2020), respectively; and ε denotes the land surface emissivity, which is calculated based on vegetation proportion [38]. Finally, according to the Plank function, the LST was derived from $B(T_S)$, using the following equation:

$$T_s = K_2/\ln(K_1/B(T_S)+1) - 273 \tag{2}$$

where T_s is the LST (°C), K_2 is 1321.08 K, and K_1 is 774.89 (W/m^2·sr·µm).

The spatial distribution of the LST in spring, summer, autumn, and winter is shown in Figure 2. The high-temperature area is significant for understanding the spatiotemporal pattern of LST and optimizing the thermal environment. We observed three seasonal stable high-temperature areas in the study area, mainly distributed within the Second Ring Road. Compared with the high-temperature areas inside the Second Ring Road, the distribution of the high-temperature areas located outside the road was relatively fragmented.

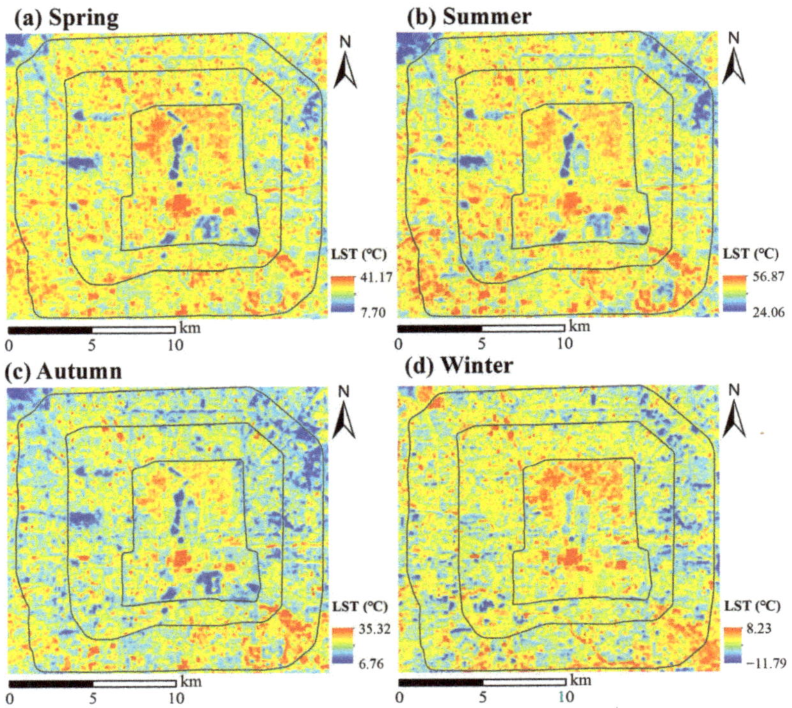

Figure 2. Spatial distribution of land surface temperature (LST) in: (**a**) spring, (**b**) summer, (**c**) autumn, and (**d**) winter.

3.2. Selection and Calculation of Multidimensional Environmental Factors

The LST in urban areas is mainly determined by complex urban landscapes. According to previous studies, the driving factors of LST can be categorized into three layers: urban greenness, wetness, and grayness (Table 1). In this study, three surface biophysical factors (NDVI, NDBI, and MNDWI) were selected to represent the overall spatial distributions of urban greenness, grayness, and wetness, respectively, from a two-dimensional (2D) perspective. Buildings and roads are two typical artificial construction types in urban areas, as well as the main components of urban grayness. In this study, we employed RD, BD, BH, BV, and SVF to describe the detailed spatial characteristics of the urban grayness components from 2D and 3D perspectives. The BD was defined as the proportion of the ground building area per unit of analysis, and RD was defined as the total length of the roads in an analysis unit divided by the area of the unit. The BH was defined as the average height of the buildings per unit of analysis, and BV was defined as the total building volume per unit of analysis. Furthermore, SVF_j was defined as the proportion of visible sky at a certain observation point j and was calculated based on the buildings, according to the method proposed by [39]. Then, all the SVF_j values per analysis unit were averaged. Since seasonal effects were considered in this study, the three surface biophysical factors chosen in this study were calculated separately from four Landsat-8 images captured during the different seasons, whereas the building and road-related factors were considered to be the same for all seasons. The spatial distributions of the selected environmental factors are shown in Figure 3.

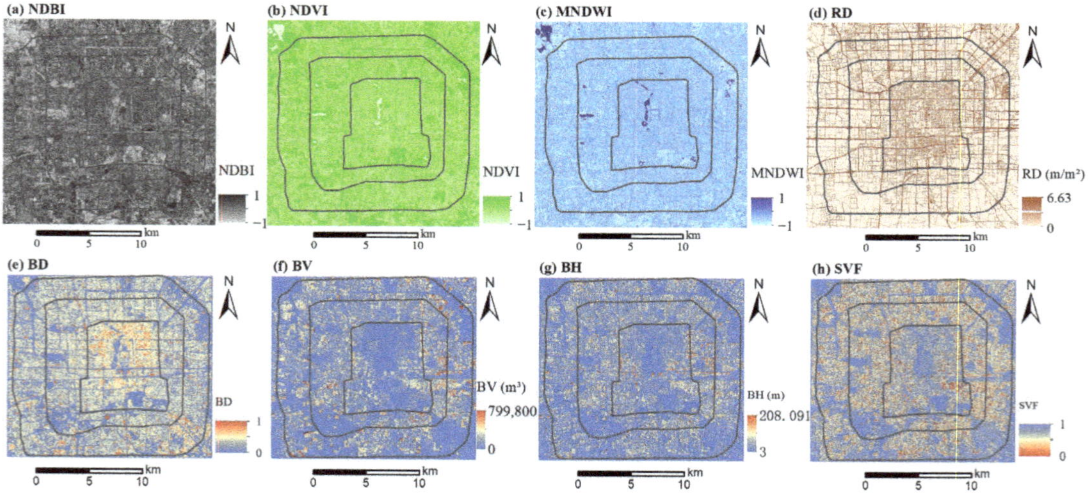

Figure 3. Spatial distribution of environmental factors considered in this study using different indexes: normalized difference built-up index (NDBI), normalized difference vegetation index (NDVI), modified normalized difference water index (MNDWI), building density (BD), building height (BH), building volume (BV), sky view factor (SVF), and road density (RD). Images for NDBI, NDVI, and MNDWI factors were captured in summer.

Table 1. Description of environmental factors considered in this study.

Type	Layers	Factors (Abbreviation)	Equation	Reference	Unit
Surface biophysical factors	Urban greenness	Normalized difference vegetation index (NDVI)	$NDVI = \frac{\rho_{NIR} - \rho_{Red}}{\rho_{NIR} + \rho_{Red}}$	[40]	-
	Urban grayness	Normalized difference built-up index (NDBI)	$NDBI = \frac{\rho_{SWIR1} - \rho_{NIR}}{\rho_{SWIR1} + \rho_{NIR}}$	[16]	-
	Urban wetness	Modified normalized difference water index (MNDWI)	$MNDWI = \frac{\rho_{Green} - \rho_{SWIR1}}{\rho_{Green} + \rho_{SWIR1}}$	[18]	-
Multidimensional factors of two typical components of urban grayness	Urban grayness (Road network)	Road density (RD)	$RD = \frac{Len_{roads}}{\text{Unitarea}}$	[41]	m/m^2
	Urban grayness (Buildings)	Building density (BD)	$BD = \frac{\sum_{i=1}^{n} Bi_{area}}{\text{Unitarea}}$	[30]	-
		Building height (BH)	$BH = \frac{\sum_{i=1}^{n} Bi_{height}}{n}$	[42]	m
		Building volume (BV)	$BV = \sum_{i=1}^{n} Bi_{volume}$	[42]	m^3
		Sky view factor (SVF)	$SVF = \sum_{i=1}^{k} SVF_j / k$	[9]	-

Note: Bi_{area} is the land coverage area of the i-th building in the analysis unit, Bi_{height} is the height of building i, Bi_{volume} is the volume of building i, n is the total number of buildings in the analysis unit, Len_{roads} is the total length of all roads in an analysis unit, $\text{Unit}_{\text{area}}$ is the area of the analysis unit, SVF_j is the SVF value of cell j in the analysis unit, and k is the total number of cells in the analysis unit. ρ_{Green}, ρ_{Red}, ρ_{NIR}, and ρ_{SWIR1} represents the band 3, band 4, band 5, and band 6 of Landsat-8, respectively.

3.3. *Moving Window Samples for Analysis*

The moving window method is a common multiscale analysis approach [43]. The method can be used to capture the continuous spatial variation in specific characteristics of the urban environment at custom scales [44]. In this study, for each analysis sample, the dependent variable was the LST value of the central cell of the moving window, and each independent variable was the average value of the corresponding environmental factor within the moving window. To investigate the influence of the spatial scale, 20 window sizes were used to characterize the urban environmental factors and measure their relationship with the LST of 30-m cells. Since the spatial resolution of the LST data in this study is 30 m, the minimum window size is 90 m $\times$ 90 m, which can ensure a clear central pixel in the window. Moreover, the growth interval of the window size is twice the spatial resolution of the LST data, which helps to capture the spatially continuous neighboring environmental characteristics. In addition, a sufficient range of moving window size is necessary to adequately explore the correlation between LST and multiple environmental factors, especially to explore how the correlation changes with increasing window size. Therefore, in this study, the window size was increased from 90 m $\times$ 90 m to 1230 m $\times$ 1230 m, with an interval of 60 m. The samples obtained from the moving windows of different sizes were used in the subsequent analysis to reveal the response of the LST to its neighboring environment.

3.4. *Correlations between Land Surface Temperature (LST) and Environmental Factors*

In this study, the Pearson correlation method [45,46], which has been widely used in previous studies, was used to measure the correlation direction and intensity between LST and the selected environmental factors. Twenty correlation coefficients were estimated for each factor in each season, corresponding to twenty sizes of the moving windows. One of the aims of this study was to investigate the seasonal variations in the effects of the urban environment on LST. Therefore, the mean values of the correlation coefficients were calculated for the individual factors with respect to different seasons. In each season, three environmental factors were identified as the dominant driving factors of LST, according to the correlation intensity. Then, the spatial characteristics of the effects of the dominant driving factors on LST were explored by analyzing the trend of correlation intensity with increasing window size. Notably, we focused on the effects of dominant driving factors, as they exhibited relatively strong correlations with LST.

3.5. Application of Regression Model to Analyze Correlation between Land Surface Temperature (LST) and Environmental Factors

Regression models are an efficient method for measuring the combined effects of multiple driving factors on LST. The XGB regression model was used in this study according to its advantages and our research needs. In this study, we aimed to explore the variation in combined effects of multiple environmental factors on LST with increasing window size in different seasons. Compared with statistical linear regression models, the XGB regression model, a tree machine learning regression method, is better adapted to the complex nonlinear relationships between LST and its driving factors. In addition, the XGB regression model is not affected by multicollinearity among driving factors, which means that all environmental factors can be entered into the model [47]. At present, XGB regression is the most effective stochastic gradient boosting algorithm. The idea of the XGB regression model is to build a series of shallow regression trees using a gradient boosting technique, where each tree attempts to correct the residuals in the predictions made by previous trees [48]. The XGB regression model can calculate the optimal solution for the whole model and reduce the overfitting phenomenon [49]. Notably, previous study has confirmed that the XGB regression model outperforms random forest, support vector, and decision tree regressions with higher accuracy in LST prediction [50]. Moreover, the XGB regression model has shown good application effects in many fields, including crime prediction [47], vegetation mapping [51], algal biochar yield prediction [52], flood susceptibility modeling [53], and urban thermal environment [54]. Therefore, in this study, we developed XGB regression models to measure the combined effects of multiple environmental factors on LST. The samples were randomly partitioned into 80% and 20% for training and validation, respectively. The performance of the regression model was measured using the coefficient of determination (R^2) metric, which indicated the goodness of fit. It also represented the proportion of variance of the LST, explained by the independent variables in the model. In addition, the root mean square error (RMSE) and the mean absolute error (MAE) were used to evaluate the accuracy of the regression model. The measured R^2, RMSE, and MAE were calculated using the following equations:

$$R^2(y,\widehat{y}) = 1 - \frac{\sum_{i=1}^{n}(y_i - \widehat{y}_i)^2}{\sum_{i=1}^{n}(y_i - \overline{y})^2} \quad (3)$$

$$\overline{y} = \frac{1}{n}\sum_{i=1}^{n} y_i \quad (4)$$

$$RMSE(y,\widehat{y}) = \sqrt{\frac{1}{n}\sum_{i=1}^{n}(y_i - \widehat{y}_i)^2} \quad (5)$$

$$MAE(y,\widehat{y}) = \frac{1}{n}\sum_{i=1}^{n}|y_i - \widehat{y}_i| \quad (6)$$

where $\widehat{y}_i$ is the predictive value of the i-th sample, and y_i is the corresponding true value for n samples.

4. Results

4.1. Seasonal Correlations between Land Surface Temperature (LST) and Neighboring Environmental Factors

The average correlation between the LST and environmental factors with respect to different seasons is shown in Figure 4. The Pearson correlation coefficient indicated the direction and intensity of the correlation between the LST and environmental factors. The average correlation results were at the 95% significance level. In terms of the direction of correlation, all factors, except BV, SVF, and NDVI, exhibited consistent correlations with the LST in Beijing across all seasons. In addition, the correlation strengths between the urban environmental factors and LST were strong in summer and weak in winter, in terms of the strongest correlation factors.

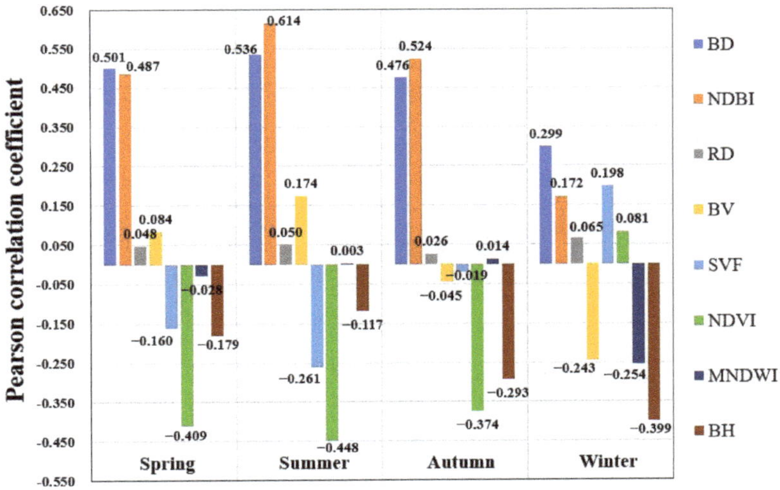

Figure 4. Pearson correlations between land surface temperature (LST) and different environmental factors, normalized difference built-up index (NDBI), normalized difference vegetation index (NDVI), modified normalized difference water index (MNDWI), building density (BD), building height (BH), building volume (BV), sky view factor (SVF), and road density (RD).

We observed positive correlations between the BD, NDBI, and RD and the LST in all of the seasons. The correlation between BD and LST was strong (0.501, 0.536, 0.476, and 0.299 in four seasons, respectively), whereas that between RD and LST was weak (0.048, 0.050, 0.026, and 0.065 in four seasons, respectively). These results indicated that the positive association between LST and building density was stronger, whereas the influence of urban road networks on LST was limited. The NDBI factor represented the overall grayness features of the urban area, including both building and road features; therefore, its correlation with LST was similar to that of the BD factor. Notably, the positive correlations between the NDBI and BD factors and the LST of the city were consistent in the spring (0.487), summer (0.614), and autumn (0.524) seasons; their correlation strengths with LST were stronger in spring, summer, and autumn than in winter. This indicated that the warming effect of urban grayness factors was more prominent during warm seasons. Furthermore, BH was negatively correlated with the LST in all seasons, which was completely different from the BD factor, although the latter also represented building characteristics. This result indicated that BD increased the LST of the area, and conversely, BH decreased the LST. In addition, the results also showed that the strength of the correlation between the BH and LST of the city was highest in winter (−0.399). The correlation of the MNDWI factor with the LST was weak and almost negligible in spring, summer, and autumn, with its negative correlation with LST being evident in winter (−0.254).

The direction of the correlation between the urban environmental factors and LST was also reversed during seasonal changes. The BV factor was positively correlated with the LST in the spring (0.084) and summer (0.174) seasons, less correlated with the LST in autumn (−0.045), and negatively correlated with it in winter (−0.243). The changes in the correlations of the SVF and NDVI factors with the LST were similar. Specifically, they both showed negative correlations with the LST in spring (−0.160 and −0.490) and summer (−0.261 and −0.448); the negative correlations weakened in autumn. The factors exhibited positive correlations with the LST in winter (0.198 and 0.081).

In general, there were no significant differences in the correlation results for spring, summer, and autumn. In contrast, the direction or strength of the correlation between each environmental factor and LST changed markedly in winter, compared to that in the other three seasons. These findings imply that the main environmental factors affecting LST in spring, summer, and autumn were consistent to some extent, whereas the factors affecting LST in winter differed from those in warmer seasons.

4.2. Spatial Characteristics of the Correlations between Land Surface Temperature (LST) and Dominant Driving Factors

We selected three factors for spring, summer, and autumn (BD, NDBI, and NDVI) and winter (BD, BH, and MNDWI) for further analysis, according to their strong correlations with the LST (Figure 4). This exploration was conducted with various window sizes, i.e., from 90 m to 1230 m, with an interval of 60 m, to reveal the effect of scale through the change in the correlation strength. The influence of the BD, NDBI, and NDVI factors on the LST showed a similar trend, i.e., an inverted V shape with increasing window sizes in relatively warmer seasons (spring, summer, and autumn) (Figure 5a–c). First, the correlation intensity increased with the moving window size and then decreased gradually, which indicated the optimal spatial scales to study the influence of the neighboring environments on the LST. For the BD factor, the intensity of its positive correlation with LST showed an initial sharp rise, followed by a gradual decline after reaching the peak; the peak correlation intensity was obtained when the window size was set at 270 m (summer) and 330 m (spring and autumn). The intensity curve of the NDVI factor associated with LST showed a trend similar to the intensity curve of the BD–LST correlation. The spring curve was located between the summer and autumn curves, and the maximum correlation intensity corresponded to a window size of 210 m (spring and summer) and 270 m (autumn), although the NDVI was negatively correlated with the LST. Similar to the results of BD and NDVI, the summer curve of the NDBI was also at the top, indicating that these environmental factors were most associated with the LST in summer. The difference between the first two was that the gap between the spring and autumn curves gradually increased with the size of the moving window. However, the strongest correlations between the NDBI and LST still occurred at window sizes of 210 m (spring and summer) and 270 m (autumn). The general and intuitive correlations between the environmental factors and LST are shown in Figure 4. The comparative analysis between the scales further revealed that the scales of 210–270 m were more effective in capturing the spatial characteristics of these correlations, regardless of whether the factors were positively (NDBI and BD) or negatively (NDVI) correlated with the LST.

In winter, the BH and MNDWI were negatively correlated with the LST. The curves of these two factors also portrayed an inverted V shape. For example, the correlation intensity between BH and LST first increased sharply, followed by a gradual decrease. The size of the optimal scale at which the strongest correlation strength appeared was 450 m, but the correlation strength at the optimal scale did not differ markedly from that observed for the window size of 1230 m. The correlation intensity curve of the MNDWI also had an inverted V shape, but it differed from that of BH, portraying a sharp decrease with increasing window size. The apex of the curve (corresponding to the strongest correlation) was located at a window size of 150 m. The curve of BD, which was positively correlated to the LST, behaved differently; it continued to rise and then remained flat, without a clear inflection point. The influence of the building characteristics (BH and BD) on the LST was more prominent in the winter. In general, the trend of the winter curve was markedly different from that of the warmer seasons, implying that there was a significant difference in the formation mechanism of the urban thermal environment between the warm (spring, summer, and autumn) and winter seasons.

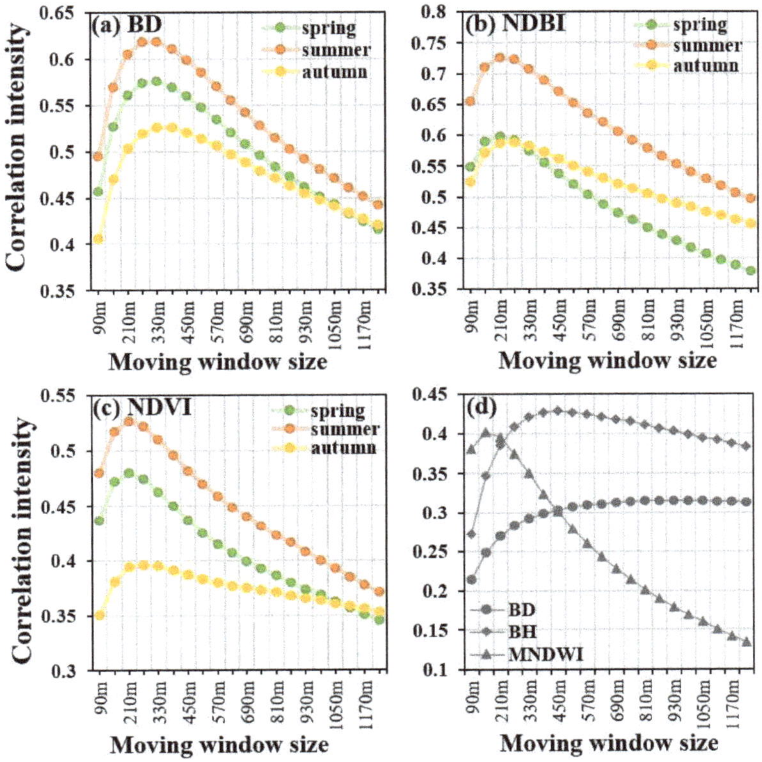

Figure 5. Correlation intensities of dominant factors: (**a**) correlation intensities of building density (BD) in spring, summer, and autumn; (**b**) correlation intensities of normalized difference built-up index (NDBI) in spring, summer, and autumn; (**c**) correlation intensities of normalized difference vegetation index (NDVI) in spring, summer, and autumn; (**d**) correlation intensities of BD, building height (BH); and modified normalized difference water index (MNDWI) in winter.

4.3. Combined Effect of Environmental Factors on Land Surface Temperature (LST)

The LST is the result of the combined effects of multiple factors. Therefore, in this study, the XGB regression method was adopted to construct the LST regression models for different spatial scales in order to investigate the combined effects of these factors on the LST of the city. The explanatory rate indicator, i.e., the percentage of LST variance explained by the regression model, was used to measure the combined effects of these factors on the LST [48]. Figure 6 shows that the RMSE and MAE metrics are lower than 1.6 and 1.2, respectively. Table 2 summarizes the explanatory rates at different scales. For each season, the same color scheme was used to portray the change in the explanatory rate with scale. Green indicates a low explanatory rate and yellow indicates a high explanatory rate. For the spring and summer LST, the XGB regression models established at 210 m had the highest explanatory rate (74.94% and 77.55%, respectively), followed by the model established at 270 m (74.62% and 77.34%, respectively). For the autumn LST, the model established at 270 m had the highest explanatory rate (68.04%), followed by that established at 210 m (67.53%). These results indicated that, when studying the LST during the warm season, the multiple regression model had the strongest interpretation of the LST at a spatial scale of 210–270 m.

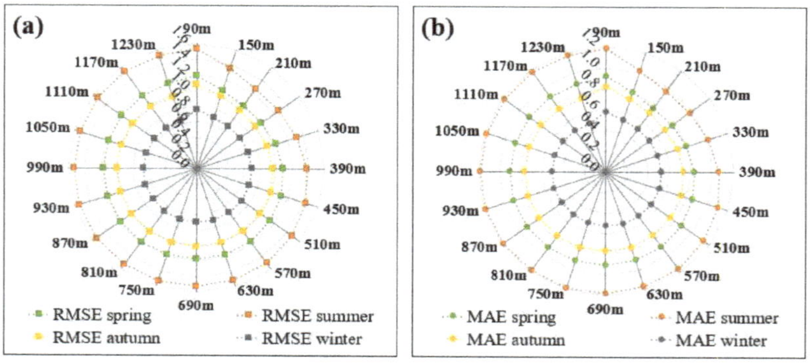

Figure 6. The root mean square error (RMSE) (**a**) and mean absolute error (MAE) results (**b**).

Table 2. Percentage of explained variance of the land surface temperature (LST) for different seasons.

Scales (m)	Percentage of Explained Variance of LST (%)			
	Spring	Summer	Autumn	Winter
90	64.29	66.94	56.52	46.09
150	72.11	74.64	64.20	54.49
210	74.94	77.55	67.53	57.12
270	74.62	77.34	68.04	57.13
330	72.98	75.59	66.92	55.84
390	70.74	73.34	65.56	54.48
450	68.86	71.23	64.07	53.92
510	67.17	69.47	63.05	53.43
570	65.98	68.18	62.09	53.10
630	65.27	67.59	61.52	53.30
690	65.06	67.22	61.53	54.31
750	65.05	67.09	61.79	55.20
810	64.88	66.93	61.59	55.92
870	64.88	66.62	61.94	56.54
930	65.29	66.36	62.21	57.14
990	65.69	66.75	62.42	57.52
1050	66.05	66.60	62.87	57.92
1110	66.13	66.66	63.33	58.52
1170	66.15	66.94	63.53	59.69
1230	66.44	67.26	63.91	60.23
Average	67.63	69.51	63.23	55.59

Similar to the findings in Section 4.2, the explanatory rate of the XGB regression model and its variation differed markedly between the winter and the warm seasons (spring, summer, and autumn). For the winter LST, the explanatory rate of the model first increased until 57.13% (270 m) and then decreased gradually until 53.10% (570 m); the rate increased again until 60.23% (1230 m). The XGB regression model, including the eight environmental factors, explained the LSTs in spring, summer, and autumn more effectively than that in winter, both in terms of the mean and peak values. These findings suggested that the influencing factors, formation mechanism, and scale effect of the thermal environment in winter were more complex than those observed in the spring, summer, and autumn seasons.

5. Discussion

5.1. Seasonal Characteristics of the Thermal Effects of Urban Environmental Factors

5.1.1. Cooling Effects of Urban Green Vegetation and Water Bodies

Urban vegetation and water bodies are the two main landscape types that provide cooling effects in urban areas and have attracted much attention for improving the urban thermal environment. However, the difference in the cooling effect between the seasons cannot be ignored. In general, urban vegetation decreases the LST through evapotranspiration and shading. The cooling effect of evaporation and shading greatly varies among species and depends on the canopy density in the region [55,56]. It has been confirmed that the cooling effect of tree-cover vegetation is better than that of grass-cover vegetation [57,58]. The results of this study indicated significant inter-seasonal differences in the thermal effects of vegetation and water bodies. The NDVI exhibited a significant negative correlation with the LST in the spring, summer, and autumn seasons, which is consistent with the findings of [59,60]. However, the NDVI exhibited a weak positive correlation with the LST in winter (Figure 4). This finding was consistent with that of a study conducted in North America [61]. Notably, deciduous broad-leaved trees occupy the largest area in Beijing [30]. Additionally, there is an inevitable reduction in green vegetation in winter, resulting in the weakening of the evapotranspiration and shading processes [62]. In addition, the distribution pattern and area of the vegetation may also affect its cooling effect. Therefore, in winter, it is difficult for highly fragmented vegetation to generate a significant cooling effect [19].

Previous studies have indicated that urban water bodies are negatively correlated with LST [63,64]. However, in this study, we observed that the MNDWI factor, an effective indicator of water bodies in urban areas, was weakly correlated to the LST in the spring, summer, and autumn seasons (Figure 4). This discrepancy may be because the total water area in the city is small; therefore, the water body may not be able to provide an effective cooling effect [10]. The MNDWI exhibited a relatively stronger negative correlation with the LST in winter, compared to the NDVI (Figure 4). This finding was consistent with a previous study that reported that the cooling effect of water was stronger than that of vegetation in November [65].

5.1.2. Thermal Effects of Urban Grayness Factors

Buildings, roads, and other impervious surfaces absorb solar radiation, resulting in higher temperatures in urban areas [21,66–68]. In this study, the NDBI, BD, and RD were positively correlated to the LST in all the seasons. The strength of the correlation between RD and LST indicated the limited impact of RD on LST, which was consistent with the findings of [41]. A recent study showed that the correlation between the NDBI and LST was weaker in winter than in other seasons [69]. Similarly, in this study, the NDBI and BD exhibited stronger correlations with the LST in summer than that in winter (Figures 4 and 5). Buildings in urban areas are typically closely related to a variety of human activities. For example, in summer, buildings with air conditioning release large amounts of heat, contributing to an increase in the LST [70]. Similarly, the cold air emitted from the air conditioner during winter can reduce the LST. Therefore, the seasonal characteristics of the effects of the NDBI and BD on the LST may be influenced by the amount of solar radiation absorbed by the impervious surface areas and the heat/cooling emissions from buildings. Although the positive correlations of BD and NDBI with the LST have been widely reported, our study revealed seasonal variations in the strength of their correlations with the LST in urban areas at fine scales.

In this study, the BV, an indicator of the total volume of buildings, was positively/negatively correlated to LST in summer/winter. The larger values of BV indicate the higher intensity of the anthropogenic activity and possible larger amount of anthropogenic heat emissions, which is expected to increase LST [71]. In addition, in winter, there may be a larger amount of cooling emissions from buildings in the region with larger BV. A previous study conducted in Berlin and Cologne also observed the positive correlation between BV and

LST in summer [42], however, the correlation between BV and LST in winter was positive. This difference could be explained by the different climate backgrounds of urban areas. The influence of SVF on LST is complicated and conflicting conclusions on the correlation between LST and SVF have been reported in previous studies [72–74]. Larger SVF indicated more effective air circulation and heat dissipation, and it also means that there may be more incoming solar radiation. Smaller SVF hinders airflow movement and is not conducive to heat dissipation, and it also means the possibility of a larger area of shadow. In our study, the correlation between SVF and LST was season-varying. Similarly, a study conducted in Wuhan city also found a season-varying correlation between SVF and LST [9]. Notably, in this study, the SVF was negatively correlated to LST in summer, which is consistent with a recent study conducted in Beijing [72]. In addition, the negative/positive correlation between SVF and LST in summer/winter may be associated to the anthropic activities, as the smaller SVF is not conducive to the dissipation of the heat/cooling emissions from buildings in summer/winter.

In particular, BH was negatively correlated with the LST in all seasons, which was in line with the results of previous studies [9,19,20]. This may be because the shadows provided by high-rise buildings prevented the nearby ground from absorbing solar radiation, thus, reducing the LST [20,29]. Therefore, higher buildings are expected to formulate lower temperatures owing to their possible larger shadows [20]. Additionally, the negative effect of BH on LST was stronger in winter and autumn than in other seasons, which can also be explained by the larger shadow area in the winter and autumn seasons due to the lower solar elevation [9]. The results related to building features can provide reliable references for the planning and construction of urban buildings. In addition to urban vegetation and water body planning, the season-stable negative effect of BH on LST could provide new insights for the mitigation of the UHI effect for effective urban building planning.

5.2. Spatial Characteristic of the Thermal Effects of Neighboring Environment

Previous studies have confirmed that the cooling effects of urban vegetation and water bodies are largely regulated by the vegetation and impervious surface landscapes of the neighboring regions [32,33]. The heat sources in urban areas, such as industrial parks, will continue to contribute to the increase in the LST within a certain spatial extent around them [34]. These findings indicate that the neighboring environment plays an important role in the driving mechanism of LSTs in urban areas. In addition, the correlations between LST and surface coverage indicators generally vary with the geographical scale [66,75]. This may be related to the diversity of the surface cover features at different geographic scales [62]. Therefore, an appropriate spatial scale is vital for the quantitative analysis of the influence of the neighboring environment on the LST, balancing useful surrounding environment information with the irrelevant information of the distant surrounding environment. Notably, LST is the result of a combination of driving factors, which implies that it is necessary to consider the individual and combined effects of various factors. Therefore, we used both the correlation analysis and the XGB regression method to measure the correlation between the urban environmental factors and LST at different spatial scales.

In this study, the correlation between the urban environmental factors and LST portrayed an initial increasing trend with increasing spatial scales, followed by a decreasing trend (Figure 5 and Table 2), which implied the spatial range of the influence of the neighboring environment on the LST of the 30-m cell. The strongest correlation between the NDVI and LST was obtained within a 210-m moving window size in spring and summer and a 270-m size in autumn (Figure 5). This finding is in line with that of [25], in which the vegetation cover exhibited the highest correlation with the LST within a 210-m moving window in summer. Another study conducted in India also indicated that the median cooling range of urban green spaces was approximately 270 m outside the boundary [76]. The correlations between the MNDWI and LST in winter varied with the sizes of the moving windows and peaked at 150 m (Figure 5). This finding was consistent with the findings of [32], which concluded that the cooling range of water bodies was approximately 74 m. A previous study conducted in the Phoenix metropolitan area indicated that the correlation between impervious surfaces and maximum air temperature in the warm season decreased after a 210-m spatial scale, and the strongest negative correlation between vegetation cover and maximum air temperature occurred at 210 m and 270 m spatial scales [35]. In this study, the optimal spatial scale for exploring the correlation relationship between the 30-m LST and its neighboring environment was 210–270 m in warm seasons (spring, summer, and autumn), considering the individual and combined correlations between the LST and multidimensional factors of the urban environment (Figure 5 and Table 2). These findings suggest that 210–270 m could be an appropriate spatial scale for characterizing the relationship between the LST and its neighboring environment in Beijing. These findings could provide a reference for spatial-scale selection in future LST studies in Beijing and other large cities in northern China.

5.3. Limitations and Scope for Future Work

This study established a viable framework to explore the influence of neighboring environments on LST, considering the multidimensional features of dominant urban landscape types. However, this study has a few limitations. First, although we considered the seasonal variation in the LST, using only one daytime scene collected in each season to obtain LST and surface biophysical indices could be insufficient. In addition, the acquisition time of the images should be considered to ensure that the images can accurately reflect the characteristics of LST in different seasons. Adopting multiple daytime and nighttime images from different seasons to address this aspect may prove to be more efficient in future studies. Second, this study was conducted only in the urban areas of Beijing, and more cities in different climate zones should be considered in future studies [30]. Third, this study explored the thermal effects of buildings and roads. However, the spatial resolution of LST was coarser than that of the building data in the study area. LST data with higher spatial resolution are expected to reveal more accurate relationships between LST and urban 3D landscapes. Moreover, the positional accuracy of roads data from the OpenStreetMap is mainly determined by the positioning technologies and references used by volunteers while digitizing these data [77]. The roads data with high positioning accuracy and accurate information on the width attributes should be considered in future works, although the accuracy of OSM road network data (+/− 20 m) is better than that of other publicly available global datasets such as Global Roads Open Access Data Set (+/− 500 m) [78]. In addition, more factors such as building types/patterns, tree species/patterns, and anthropogenic heat [79], are worthy of future study. Fourth, this study focused on the combined effect of multiple factors on LST in the regression analysis, and the contribution of each environmental factor is worth exploring in future studies. Finally, the thermal effects of urban environmental factors and their spatial scale characteristics differed markedly between seasons, mainly between winter and the warmer seasons. Further studies are required to explain the different driving mechanisms of the thermal environment in winter and other seasons.

6. Conclusions

In recent years, rapid urbanization has caused a series of significant changes in green vegetation, impervious surfaces, human activities, energy consumption, and thermal emissions, and the UHI effect has become increasingly severe. This study was conducted in the central urban areas of Beijing and investigated how the LST of 30-m cells correlated with their neighboring environment in different seasons. Moving windows were applied to characterize the environmental factors of neighboring environments, with a total of 20 spatial scales ranging from 90–1230 m. Eight environmental factors were explored from three layers of urban greenness, wetness, and grayness: NDVI, MNDWI, NDBI, BD, BH, BV, SVF, and RD. The Pearson correlation and XGB regression methods were used to measure the correlation between these factors and the LST, while considering the individual and combined effect.

This study confirmed the significant seasonal differences in the correlation between the LST of 30-m cells and their neighboring environments. The correlation between the NDVI and MNDWI and the LST was considerably different between the winter and the other seasons, indicating that the cooling effects of urban vegetation and water bodies were associated with the season. The NDBI, as a comprehensive indicator of urban grayness, was positively correlated to the LST in all four seasons, indicating that the increasing impervious surfaces can promote an increase in the LST. Reasonable control of the growth of impervious surfaces can help prevent further deterioration of the UHI effect. Road density had a limited impact on LST, whereas buildings had a significant impact on LST. Notably, BD and BH were more strongly correlated to LST than other building factors. BD was positively correlated with LST in all four seasons, indicating that the increasing building density promoted higher LST. BD and NDBI were more strongly correlated with LST in summer than in winter, implying that anthropogenic heat production due to seasonal changes may also affect the correlation between urban grayness and LST. Owing to the shadow effect, high BH values contributed to low LSTs in urban areas. Notably, there was a negative correlation between LST and BH in all four seasons, and this negative correlation was stronger in winter than in summer. In addition to increasing urban green space, the building density and height characteristics can be optimized to help mitigate the UHI effect.

The correlation between LST and its neighboring environment varied with spatial scale, with inter-seasonal differences in the scale effect, mainly between winter and other seasons. In spring, summer, and autumn, the strength of correlation between LST and its neighboring environment portrayed an "inverted V" pattern with increasing spatial scale in terms of the Pearson correlation and XGB regression results, indicating the spatial range of the strongest influence of the neighboring environment on LST of 30-m cells, i.e., the optimal spatial scale to explore the relationship between LST and its neighboring environment. Considering the correlation of individual factors with LST and the explanatory rate of the XGB regression model for LST, 210 m and 270 m were considered as the optimal scales to explore the relationship between LST and its neighboring environment in this study, which can provide a reference for the selection of spatial scales in quantitative LST driving force studies.

These findings contribute to the understanding of the correlation between LST and multidimensional environmental factors in urban areas from the perspective of the neighboring environment. Additionally, the framework developed in this study can be applied to other urban areas. The results of these studies can help urban planners develop rational strategies for optimizing urban thermal environments.

Author Contributions: Conceptualization, W.L. and Q.M.; methodology, W.L. and J.G.; software and writing—original draft, W.L.; data curation, W.L. and Q.M.; writing—review, Q.M., L.Z., M.A., Y.B., X.H., T.J. and D.H.; writing—review and editing, W.L., M.A. and Q.M.; visualization, W.L.; supervision, Q.M.; funding acquisition, Q.M. All authors have read and agreed to the published version of the manuscript.

Funding: This research was funded by the National Natural Science Foundation of China Major Program, grant number 42192580, 42192584; the National Natural Science Foundation of China, grant number 42171357; the National Key Research and Development Program, grant number 2020YFC0833100; and the Bilateral Chinese-Hungarian Project, grant numbers 2019-2.1.11-TÉT-2020-00171.

Data Availability Statement: The data presented in this study are available on request from the corresponding website.

Conflicts of Interest: The authors declare no conflict of interest. The funders had no role in the design of the study; in the collection, analyses, or interpretation of data; in the writing of the manuscript, or in the decision to publish the results.

References

1. Oke, T.R. The energetic basis of the urban heat island. *Q. J. R. Meteorol. Soc.* **1982**, *108*, 1–24. [CrossRef]
2. Ziter, C.D.; Pedersen, E.J.; Kucharik, C.J.; Turner, M.G. Scale-dependent interactions between tree canopy cover and impervious surfaces reduce daytime urban heat during summer. *Proc. Natl. Acad. Sci. USA* **2019**, *116*, 7575–7580. [CrossRef]
3. Mora, C.; Dousset, B.; Caldwell, I.R.; Powell, F.E.; Geronimo, R.C.; Bielecki, C.R.; Counsell, C.W.; Dietrich, B.S.; Johnston, E.T.; Louis, L.V. Global risk of deadly heat. *Nat. Clim. Chang.* **2017**, *7*, 501–506. [CrossRef]
4. Frayssinet, L.; Merlier, L.; Kuznik, F.; Hubert, J.-L.; Milliez, M.; Roux, J.-J. Modeling the heating and cooling energy demand of urban buildings at city scale. *Renew. Sustain. Energy Rev.* **2018**, *81*, 2318–2327. [CrossRef]
5. Akbari, H.; Kurn, D.M.; Bretz, S.E.; Hanford, J.W. Peak power and cooling energy savings of shade trees. *Energy Build.* **1997**, *25*, 139–148. [CrossRef]
6. Taylor, J.; Wilkinson, P.; Davies, M.; Armstrong, B.; Chalabi, Z.; Mavrogianni, A.; Symonds, P.; Oikonomou, E.; Bohnenstengel, S.I. Mapping the effects of urban heat island, housing, and age on excess heat-related mortality in London. *Urban Clim.* **2015**, *14*, 517–528. [CrossRef]
7. Anderson, G.B.; Bell, M.L. Heat waves in the United States: Mortality risk during heat waves and effect modification by heat wave characteristics in 43 US communities. *Environ. Health Perspect.* **2011**, *119*, 210–218. [CrossRef]
8. Jenerette, G.D.; Harlan, S.L.; Buyantuev, A.; Stefanov, W.L.; Declet-Barreto, J.; Ruddell, B.L.; Myint, S.W.; Kaplan, S.; Li, X. Micro-scale urban surface temperatures are related to land-cover features and residential heat related health impacts in Phoenix, AZ USA. *Landsc. Ecol.* **2016**, *31*, 745–760. [CrossRef]
9. Li, H.; Li, Y.; Wang, T.; Wang, Z.; Gao, M.; Shen, H. Quantifying 3D building form effects on urban land surface temperature and modeling seasonal correlation patterns. *Build. Environ.* **2021**, *204*, 108132. [CrossRef]
10. Peng, J.; Jia, J.; Liu, Y.; Li, H.; Wu, J. Seasonal contrast of the dominant factors for spatial distribution of land surface temperature in urban areas. *Remote Sens. Environ.* **2018**, *215*, 255–267. [CrossRef]
11. Do Nascimento, A.C.L.; Galvani, E.; Gobo, J.P.A.; Wollmann, C.A. Comparison between air temperature and land surface temperature for the city of São Paulo, Brazil. *Atmosphere* **2022**, *13*, 491. [CrossRef]
12. Chen, J.; Zhan, W.; Jin, S.; Han, W.; Du, P.; Xia, J.; Lai, J.; Li, J.; Liu, Z.; Li, L. Separate and combined impacts of building and tree on urban thermal environment from two-and three-dimensional perspectives. *Build. Environ.* **2021**, *194*, 107650. [CrossRef]
13. Wollmann, C.A.; Hoppe, I.L.; Gobo, J.P.A.; Simioni, J.P.D.; Costa, I.T.; Baratto, J.; Shooshtarian, S. Thermo-hygrometric variability on waterfronts in negative radiation balance: A case study of balneário Camboriú/SC, Brazil. *Atmosphere* **2021**, *12*, 1453. [CrossRef]
14. Hu, X.; Wu, Z.; Wu, C.; Ye, L.; Lan, C.; Tang, K.; Xu, L.; Qiu, R. Effects of road network on diversiform forest cover changes in the highest coverage region in China: An analysis of sampling strategies. *Sci. Total Environ.* **2016**, *565*, 28–39. [CrossRef]
15. Gunawardena, K.R.; Wells, M.J.; Kershaw, T. Utilising green and bluespace to mitigate urban heat island intensity. *Sci. Total Environ.* **2017**, *584*, 1040–1055. [CrossRef]
16. Guha, S.; Govil, H.; Dey, A.; Gill, N. Analytical study of land surface temperature with NDVI and NDBI using Landsat 8 OLI and TIRS data in Florence and Naples city, Italy. *Eur. J. Remote Sens.* **2018**, *51*, 667–678. [CrossRef]
17. Alexander, C. Normalised difference spectral indices and urban land cover as indicators of land surface temperature (LST). *Int. J. Appl. Earth Obs. Geoinf.* **2020**, *86*, 102013. [CrossRef]
18. Xu, H. Modification of normalised difference water index (NDWI) to enhance open water features in remotely sensed imagery. *Int. J. Remote Sens.* **2006**, *27*, 3025–3033. [CrossRef]
19. Zheng, Z.; Zhou, W.; Yan, J.; Qian, Y.; Wang, J.; Li, W. The higher, the cooler? Effects of building height on land surface temperatures in residential areas of Beijing. *Phys. Chem. Earth Parts A/B/C* **2019**, *110*, 149–156. [CrossRef]
20. Sun, F.; Liu, M.; Wang, Y.; Wang, H.; Che, Y. The effects of 3D architectural patterns on the urban surface temperature at a neighborhood scale: Relative contributions and marginal effects. *J. Clean. Prod.* **2020**, *258*, 120706. [CrossRef]
21. Guo, G.; Zhou, X.; Wu, Z.; Xiao, R.; Chen, Y. Characterizing the impact of urban morphology heterogeneity on land surface temperature in Guangzhou, China. *Environ. Model. Softw.* **2016**, *84*, 427–439. [CrossRef]
22. Das, M.; Das, A. Assessing the relationship between local climatic zones (LCZs) and land surface temperature (LST)—A case study of Sriniketan-Santiniketan Planning Area (SSPA), West Bengal, India. *Urban Clim.* **2020**, *32*, 100591. [CrossRef]
23. Du, P.; Chen, J.; Bai, X.; Han, W. Understanding the seasonal variations of land surface temperature in Nanjing urban area based on local climate zone. *Urban Clim.* **2020**, *33*, 100657. [CrossRef]
24. Alavipanah, S.; Schreyer, J.; Haase, D.; Lakes, T.; Qureshi, S. The effect of multi-dimensional indicators on urban thermal conditions. *J. Clean. Prod.* **2018**, *177*, 115–123. [CrossRef]
25. Alexander, C. Influence of the proportion, height and proximity of vegetation and buildings on urban land surface temperature. *Int. J. Appl. Earth Obs. Geoinf.* **2021**, *95*, 102265. [CrossRef]
26. Hu, Y.; Dai, Z.; Guldmann, J.-M. Modeling the impact of 2D/3D urban indicators on the urban heat island over different seasons: A boosted regression tree approach. *J. Environ. Manag.* **2020**, *266*, 110424. [CrossRef]
27. Oke, T.R. Canyon geometry and the nocturnal urban heat island: Comparison of scale model and field observations. *J. Climatol.* **1981**, *1*, 237–254. [CrossRef]
28. Oke, T.R. Street design and urban canopy layer climate. *Energy Build.* **1988**, *11*, 103–113. [CrossRef]
29. Nichol, J.E. High-resolution surface temperature patterns related to urban morphology in a tropical city: A satellite-based study. *J. Appl. Meteorol. Climatol.* **1996**, *35*, 135–146. [CrossRef]

30. Song, J.; Chen, W.; Zhang, J.; Huang, K.; Hou, B.; Prishchepov, A.V. Effects of building density on land surface temperature in China: Spatial patterns and determinants. *Landsc. Urban Plan.* **2020**, *198*, 103794. [CrossRef]
31. Wu, Z.; Yao, L.; Zhuang, M.; Ren, Y. Detecting factors controlling spatial patterns in urban land surface temperatures: A case study of Beijing. *Sustain. Cities Soc.* **2020**, *63*, 102454. [CrossRef]
32. Du, H.; Song, X.; Jiang, H.; Kan, Z.; Wang, Z.; Cai, Y. Research on the cooling island effects of water body: A case study of Shanghai, China. *Ecol. Indic.* **2016**, *67*, 31–38. [CrossRef]
33. Qiu, K.; Jia, B. The roles of landscape both inside the park and the surroundings in park cooling effect. *Sustain. Cities Soc.* **2020**, *52*, 101864. [CrossRef]
34. Meng, Q.; Hu, D.; Zhang, Y.; Chen, X.; Zhang, L.; Wang, Z. Do industrial parks generate intra-heat island effects in cities? New evidence, quantitative methods, and contributing factors from a spatiotemporal analysis of top steel plants in China. *Environ. Pollut.* **2022**, *292*, 118383. [CrossRef]
35. Myint, S.W.; Brazel, A.; Okin, G.; Buyantuyev, A. Combined effects of impervious surface and vegetation cover on air temperature variations in a rapidly expanding desert city. *GIScience Remote Sens.* **2010**, *47*, 301–320. [CrossRef]
36. Fu, X.; Yao, L.; Xu, W.; Wang, Y.; Sun, S. Exploring the multitemporal surface urban heat island effect and its driving relation in the Beijing-Tianjin-Hebei urban agglomeration. *Appl. Geogr.* **2022**, *144*, 102714. [CrossRef]
37. Yu, X.; Guo, X.; Wu, Z. Land surface temperature retrieval from Landsat 8 TIRS—Comparison between radiative transfer equation-based method, split window algorithm and single channel method. *Remote Sens.* **2014**, *6*, 9829–9852. [CrossRef]
38. Sobrino, J.A.; Jiménez-Muñoz, J.C.; Paolini, L.J.R.S.O.E. Land surface temperature retrieval from LANDSAT TM 5. *Remote Sens. Environ.* **2004**, *90*, 434–440. [CrossRef]
39. Zakšek, K.; Oštir, K.; Kokalj, Ž.J.R.S. Sky-view factor as a relief visualization technique. *Remote Sens.* **2011**, *3*, 398–415. [CrossRef]
40. Weng, Q.; Lu, D.; Schubring, J. Estimation of land surface temperature–vegetation abundance relationship for urban heat island studies. *Remote Sens. Environ.* **2004**, *89*, 467–483. [CrossRef]
41. Taripanah, F.; Ranjbar, A. Quantitative analysis of spatial distribution of land surface temperature (LST) in relation Ecohydrological, terrain and socio-economic factors based on Landsat data in mountainous area. *Adv. Space Res.* **2021**, *68*, 3622–3640. [CrossRef]
42. Berger, C.; Rosentreter, J.; Voltersen, M.; Baumgart, C.; Schmullius, C.; Hese, S. Spatio-temporal analysis of the relationship between 2D/3D urban site characteristics and land surface temperature. *Remote Sens. Environ.* **2017**, *193*, 225–243. [CrossRef]
43. Hagen-Zanker, A. A computational framework for generalized moving windows and its application to landscape pattern analysis. *Int. J. Appl. Earth Obs. Geoinf.* **2016**, *44*, 205–216. [CrossRef]
44. Wang, X.; Meng, Q.; Zhang, L.; Hu, D. Evaluation of urban green space in terms of thermal environmental benefits using geographical detector analysis. *Int. J. Appl. Earth Obs. Geoinf.* **2021**, *105*, 102610. [CrossRef]
45. Guo, F.; Wu, Q.; Schlink, U. 3D building configuration as the driver of diurnal and nocturnal land surface temperatures: Application in Beijing's old city. *Build. Environ.* **2021**, *206*, 108354. [CrossRef]
46. Peng, J.; Xie, P.; Liu, Y.; Ma, J. Urban thermal environment dynamics and associated landscape pattern factors: A case study in the Beijing metropolitan region. *Remote Sens. Environ.* **2016**, *173*, 145–155. [CrossRef]
47. Zhang, X.; Liu, L.; Lan, M.; Song, G.; Xiao, L.; Chen, J. Interpretable machine learning models for crime prediction. *Comput. Environ. Urban Syst.* **2022**, *94*, 101789. [CrossRef]
48. Yu, S.; Chen, Z.; Yu, B.; Wang, L.; Wu, B.; Wu, J.; Zhao, F. Exploring the relationship between 2D/3D landscape pattern and land surface temperature based on explainable eXtreme Gradient Boosting tree: A case study of Shanghai, China. *Sci. Total Environ.* **2020**, *725*, 138229. [CrossRef]
49. Mousa, S.R.; Bakhit, P.R.; Osman, O.A.; Ishak, S. A comparative analysis of tree-based ensemble methods for detecting imminent lane change maneuvers in connected vehicle environments. *Transp. Res. Rec.* **2018**, *2672*, 268–279. [CrossRef]
50. Mohammad, P.; Goswami, A.; Chauhan, S.; Nayak, S. Machine learning algorithm based prediction of land use land cover and land surface temperature changes to characterize the surface urban heat island phenomena over Ahmedabad city, India. *Urban Clim.* **2022**, *42*, 101116. [CrossRef]
51. Zhang, H.; Eziz, A.; Xiao, J.; Tao, S.; Wang, S.; Tang, Z.; Zhu, J.; Fang, J. High-resolution vegetation mapping using eXtreme gradient boosting based on extensive features. *Remote Sens.* **2019**, *11*, 1505. [CrossRef]
52. Pathy, A.; Meher, S.; Balasubramanian, P. Predicting algal biochar yield using eXtreme Gradient Boosting (XGB) algorithm of machine learning methods. *Algal Res.* **2020**, *50*, 102006. [CrossRef]
53. Linh, N.T.T.; Pandey, M.; Janizadeh, S.; Bhunia, G.S.; Norouzi, A.; Ali, S.; Pham, Q.B.; Anh, D.T.; Ahmadi, K. Flood susceptibility modeling based on new hybrid intelligence model: Optimization of XGboost model using GA metaheuristic algorithm. *Adv. Space Res.* **2022**, *69*, 3301–3318. [CrossRef]
54. Sun, Y.; Gao, C.; Li, J.; Wang, R.; Liu, J. Evaluating urban heat island intensity and its associated determinants of towns and cities continuum in the Yangtze River Delta Urban Agglomerations. *Sustain. Cities Soc.* **2019**, *50*, 101659. [CrossRef]
55. Shahidan, M.F.; Jones, P.J.; Gwilliam, J.; Salleh, E. An evaluation of outdoor and building environment cooling achieved through combination modification of trees with ground materials. *Build. Environ.* **2012**, *58*, 245–257. [CrossRef]
56. Rahman, M.A.; Armson, D.; Ennos, A. A comparison of the growth and cooling effectiveness of five commonly planted urban tree species. *Urban Ecosyst.* **2015**, *18*, 371–389. [CrossRef]

57. Yang, G.; Yu, Z.; Jørgensen, G.; Vejre, H. How can urban blue-green space be planned for climate adaption in high-latitude cities? A seasonal perspective. *Sustain. Cities Soc.* **2020**, *53*, 101932. [CrossRef]
58. Yu, Z.; Yao, Y.; Yang, G.; Wang, X.; Vejre, H. Strong contribution of rapid urbanization and urban agglomeration development to regional thermal environment dynamics and evolution. *For. Ecol. Manag.* **2019**, *446*, 214–225. [CrossRef]
59. Chen, L.; Li, M.; Huang, F.; Xu, S. Relationships of LST to NDBI and NDVI in Wuhan City based on Landsat ETM+ image. In Proceedings of the 2013 6th International Congress on Image and Signal Processing (CISP), Hangzhou, China, 16–18 December 2013; pp. 840–845.
60. Mukherjee, S.; Joshi, P.; Garg, R.D. Evaluation of LST downscaling algorithms on seasonal thermal data in humid subtropical regions of India. *Int. J. Remote Sens.* **2015**, *36*, 2503–2523. [CrossRef]
61. Sun, D.; Kafatos, M. Note on the NDVI-LST relationship and the use of temperature-related drought indices over North America. *Geophys. Res. Lett.* **2007**, *34*. [CrossRef]
62. Zhou, W.; Cao, F. Effects of changing spatial extent on the relationship between urban forest patterns and land surface temperature. *Ecol. Indic.* **2020**, *109*, 105778. [CrossRef]
63. Deilami, K.; Kamruzzaman, M.; Liu, Y. Urban heat island effect: A systematic review of spatio-temporal factors, data, methods, and mitigation measures. *Int. J. Appl. Earth Obs. Geoinf.* **2018**, *67*, 30–42. [CrossRef]
64. Dai, Z.; Guldmann, J.-M.; Hu, Y. Spatial regression models of park and land-use impacts on the urban heat island in central Beijing. *Sci. Total Environ.* **2018**, *626*, 1136–1147. [CrossRef] [PubMed]
65. Yu, Z.W.; Guo, Q.; Sun, R. Impacts of urban cooling effect based on landscape scale: A review. *J. Appl. Ecol.* **2015**, *26*, 636–642.
66. Guo, J.; Han, G.; Xie, Y.; Cai, Z.; Zhao, Y. Exploring the relationships between urban spatial form factors and land surface temperature in mountainous area: A case study in Chongqing city, China. *Sustain. Cities Soc.* **2020**, *61*, 102286. [CrossRef]
67. Weng, Q.; Rajasekar, U.; Hu, X. Modeling urban heat islands and their relationship with impervious surface and vegetation abundance by using ASTER images. *IEEE Trans. Geosci. Remote Sens.* **2011**, *49*, 4080–4089. [CrossRef]
68. Morabito, M.; Crisci, A.; Guerri, G.; Messeri, A.; Congedo, L.; Munafò, M. Surface urban heat islands in Italian metropolitan cities: Tree cover and impervious surface influences. *Sci. Total Environ.* **2021**, *751*, 142334. [CrossRef]
69. Guha, S.; Govil, H.; Taloor, A.K.; Gill, N.; Dey, A. Land surface temperature and spectral indices: A seasonal study of Raipur City. *Geod. Geodyn.* **2022**, *13*, 72–82. [CrossRef]
70. Hwang, R.-L.; Lin, T.-P.; Lin, F.-Y. Evaluation and mapping of building overheating risk and air conditioning use due to the urban heat island effect. *J. Build. Eng.* **2020**, *32*, 101726. [CrossRef]
71. Qian, J.; Meng, Q.; Zhang, L.; Hu, D.; Hu, X.; Liu, W. Improved anthropogenic heat flux model for fine spatiotemporal information in Southeast China. *Environ. Pollut.* **2022**, *299*, 118917. [CrossRef]
72. Gu, Y.; You, X.-Y. A spatial quantile regression model for driving mechanism of urban heat island by considering the spatial dependence and heterogeneity: An example of Beijing, China. *Sustain. Cities Soc.* **2022**, *79*, 103692. [CrossRef]
73. Yang, X.; Li, Y. The impact of building density and building height heterogeneity on average urban albedo and street surface temperature. *Build. Environ.* **2015**, *90*, 146–156. [CrossRef]
74. Scarano, M.; Mancini, F. Assessing the relationship between sky view factor and land surface temperature to the spatial resolution. *Int. J. Remote Sens.* **2017**, *38*, 6910–6929. [CrossRef]
75. Guo, L.; Liu, R.; Men, C.; Wang, Q.; Miao, Y.; Zhang, Y. Quantifying and simulating landscape composition and pattern impacts on land surface temperature: A decadal study of the rapidly urbanizing city of Beijing, China. *Sci. Total Environ.* **2019**, *654*, 430–440. [CrossRef]
76. Shah, A.; Garg, A.; Mishra, V. Quantifying the local cooling effects of urban green spaces: Evidence from Bengaluru, India. *Landsc. Urban Plan.* **2021**, *209*, 104043. [CrossRef]
77. Haklay, M. How good is volunteered geographical information? A comparative study of OpenStreetMap and Ordnance Survey datasets. *Environ. Plan. B Plan. Des.* **2010**, *37*, 682–703. [CrossRef]
78. Gong, P.; Chen, B.; Li, X.; Liu, H.; Wang, J.; Bai, Y.; Chen, J.; Chen, X.; Fang, L.; Feng, S.; et al. *Mapping Essential Urban Land Use Categories in China (EULUC-China): Preliminary Results for 2018*; Lanzhou University: Lanzhou, China, 2020; Volume 65, pp. 182–187.
79. Liu, W.; Meng, Q.; Allam, M.; Zhang, L.; Hu, D.; Menenti, M. Driving factors of land surface temperature in urban agglomerations: A case study in the Pearl River Delta, China. *Remote Sens.* **2021**, *13*, 2858. [CrossRef]

remote sensing

MDPI

Article

Assessing Surface Urban Heat Island Related to Land Use/Land Cover Composition and Pattern in the Temperate Mountain Valley City of Kathmandu, Nepal

Siri Karunaratne [1,2,*], Darshana Athukorala [1], Yuji Murayama [3] and Takehiro Morimoto [3]

1 Graduate School of Life and Environmental Sciences, University of Tsukuba, Tsukuba 305-8572, Japan
2 Land Use Policy Planning Department, Ministry of Lands, Colombo 00500, Sri Lanka
3 Faculty of Life and Environmental Sciences, University of Tsukuba, Tsukuba 305-8572, Japan
* Correspondence: s1830203@u.tsukuba.ac.jp

Abstract: Rapid urban growth has coincided with a substantial change in the environment, including vegetation, soil, and urban climate. The surface urban heat island (UHI) is the temperature in the lowest layers of the urban atmosphere; it is critical to the surface's energy balance and makes it possible to determine internal climates that affect the livability of urban residents. Therefore, the surface UHI is recognized as one of the crucial global issues in the 21st century. This phenomenon affects sustainable urban planning, the health of urban residents, and the possibility of living in cities. In the context of sustainable landscapes and urban planning, more weight is given to exploring solutions for mitigating and adapting to the surface UHI effect, currently a hot topic in urban thermal environments. This study evaluated the relationship between land use/land cover (LULC) and land surface temperature (LST) formation in the temperate mountain valley city of Kathmandu, Nepal, because it is one of the megacities of South Asia, and the recent population increase has led to the rapid urbanization in the valley. Using Landsat images for 2000, 2013, and 2020, this study employed several approaches, including machine learning techniques, remote sensing (RS)-based parameter analysis, urban-rural gradient analysis, and spatial composition and pattern analysis to explore the surface UHI effect from the urban expansion and green space in the study area. The results revealed that Kathmandu's surface UHI effect was remarkable. In 2000, the higher mean *LST* tended to be in the city's core area, whereas the mean *LST* tended to move in the east, south, north, and west directions by 2020, which is compatible with urban expansion. Urban periphery expansion showed a continuous enlargement, and the urban core area showed a predominance of impervious surface (IS) on the basis of urban-rural gradient analysis. The city core had a lower density of green space (GS), while away from the city center, a higher density of GS predominated at the three time points, showing a lower surface UHI effect in the periphery compared to the city core area. This study reveals that landscape composition and pattern are significantly correlated with the mean *LST* in Kathmandu. Therefore, in discussing these findings in order to mitigate and adapt to prominent surface UHI effects, this study provides valuable information for sustainable urban planning and landscape design in mountain valley cities like Kathmandu.

Keywords: urbanization; surface urban heat island; land surface temperature; sustainable cities; green space; impervious surface; Kathmandu

Citation: Karunaratne, S.; Athukorala, D.; Murayama, Y.; Morimoto, T. Assessing Surface Urban Heat Island Related to Land Use/Land Cover Composition and Pattern in the Temperate Mountain Valley City of Kathmandu, Nepal. *Remote Sens.* **2022**, *14*, 4047. https://doi.org/10.3390/rs14164047

Academic Editor: Weiqi Zhou

Received: 12 June 2022
Accepted: 15 August 2022
Published: 19 August 2022

1. Introduction

Urbanization and associated LULC changes [1,2] significantly impact the urban thermal environment of cities and their neighboring areas. This can result in numerous environmental problems, such as deforestation, ecological degradation [3], air pollution [4], energy imbalance [5], and hydrological stress [6], while the most apparent environmental problem is the increase in the urban heat island (UHI) [7–11]. The urban heat island (UHI) refers to urban core areas with temperatures that are higher than those in surrounding rural areas. Rapid urban expansion exerts substantial pressure on the natural environment [2,3], and as a result, the area has become built-up land [12]. Built-up land is mainly covered by impervious surfaces and can modify the surface energy and hydrological balance in urban areas [13]. Due to the UHI effect, increasing temperatures in urban areas can lead to higher energy and water consumption, air pollution, and a greater health risk for urban dwellers [14,15].

UHI can be categorized as either surface UHI or atmospheric UHI [16]. Surface UHI is estimated using land surface temperature, derived from remotely sensed thermal infrared (TIR) data. Atmospheric UHI is calculated using in situ data and is often categorized into the canopy and boundary-layer UHI [16,17]. This study focuses on surface UHI to more deeply understand the processes underlying changes in land surface temperature in accordance with LULC composition and pattern. Surface UHI includes both the daytime and nighttime UHI; the daytime surface UHI is stronger than the nighttime UHI due to solar radiation [17,18].

Many studies have shown that local climate zones for urban heat island studies standardize the global exchange of urban thermal observations [19–22]. Moreover, a growing body of literature shows the advantages of satellite remote sensing (RS) for monitoring urban LULC patterns and *LST* [11,23–25]. For example, Estoque et al., 2017 [17] studied the influence of landscape composition and *LST* patterns in three megacities in Southeast Asia. Athukorala and Murayama 2020 [18] examined the spatial variation in land use/land cover and composition and its impact on surface urban heat islands in the tropical sub-Saharan city of Accra, Ghana. Jiang et al., 2015 [26] assessed the effects of urbanization-associated land use cover changes on land surface temperature and surface moisture in the midwestern United States. Yan et al., 2022 [27] evaluated the warming effect of urbanization and agriculture in highly developed urban agglomerations in China, considering both daytime and nighttime. These geospatial analyses provided critical insights for increasing and understanding surface UHI research in order to implement proper urban planning to reduce the surface UHI effect in many cities worldwide [11,16,17,24,28], thus improving the living conditions for urban residents [16,17].

Generally, there is a contrast between heat absorption on impervious surfaces (concrete, asphalt, and other heat-absorbing substances) and heat absorption in the natural environment in urban areas [28–31]. Many studies have shown that *LST* in urban areas can be reduced by increasing urban green space, because such areas produce a cooling effect and enhance humidity and emissivity [32–34]. Moreover, they can create a shadow effect that covers land, limiting the direct heat applied to urban land surfaces from solar radiation [13,16]. Therefore, many researchers have investigated the relationship between spatial variations in impervious surfaces and green space in urban areas and cities on a local [13,18], regional [35,36], and global scale [37], as well as the use of modeling [38–40] to understand this phenomenon.

Many studies have investigated the relationship between spatial variations of impervious surface and green space in many cities and in various climatic regions, including tropical [41,42], tropical mountain [13], subtropical [18,43], temperate [9,44], and hot desert [16] regions. However, a study of temperate mountain valley cities that provide a favorable climate for their residents is still lacking [45–47].

Kathmandu is one of the most developed cities in Nepal with respect to population and economic development, with built-up areas covering the majority of its land. Recent research has revealed that thermal comfort is a significant problem in Kathmandu [48]. For example, Maharjan et al., 2021 [49] studied urban heat islands in densely populated cities of South Asia, including Kathmandu Valley, using the normalized difference vegetation index (NDVI) and normalized difference built-up index (NDBI). Their study found an increase in surface temperature of up to 30 °C between 2015 and 2018. Therefore, our study explores the relationship between *LST* and the spatial variation of impervious surface and green space in the temperate mountain valley city of Kathmandu, Nepal, to provide valuable insights for urban planners and policymakers in order to achieve the proper management of Kathmandu. We used Landsat data and various geospatial approaches such as machine learning techniques, urban-rural analysis, index-based analysis, and landscape configuration analysis to deeply understand the relationship between the *LST* and LULC patterns of Kathmandu, Nepal.

2. Materials and Methods

2.1. Study Area

Kathmandu, the capital of Nepal, is located in the central area of Nepal (Figure 1). The study area consists of three districts: Bhaktapur, Kathmandu, and Lalitpur, which contain five municipalities: Kathmandu metropolitan, Bhaktapur, Madyapur Thimi, Lalitpur sub-metropolitan, and Kirtipur [50]. The rate of urbanization during the 1990s was 6.6% per annum, which was among the highest in the Asia Pacific region [50]. As a result, the total population of Kathmandu reached 2.5 million by 2016, of which about 1,465,254 were considered urban residents [50]. The altitude range of the area is 1026 m through 2547 m (Figure 1). Kathmandu City stretches along the Bugmathi river basin [50]. According to the Koppen classification, the study area belongs to a temperate, dry winter, hot summer climate (Cwa) [51]. The area experiences four seasons: winter (December–February), pre-monsoon (March–May), monsoon (June–September), and post-monsoon (October–November) [50,52]. The average summer and winter temperatures in Kathmandu are 29 °C and 10.1 °C, respectively [50,52]. Kathmandu is regarded as a high urban accumulated area, and is the industrial and administrative hub in Nepal. Therefore, Kathmandu Valley has accelerated in terms of both population and urban development, expanding to the outer areas. After considering the urban development process and the potential restricted effects of administrative boundaries, we defined our study area as 20 × 20 km, with a 10 km radial from the city center in Kathmandu's central hub (Figure 1).

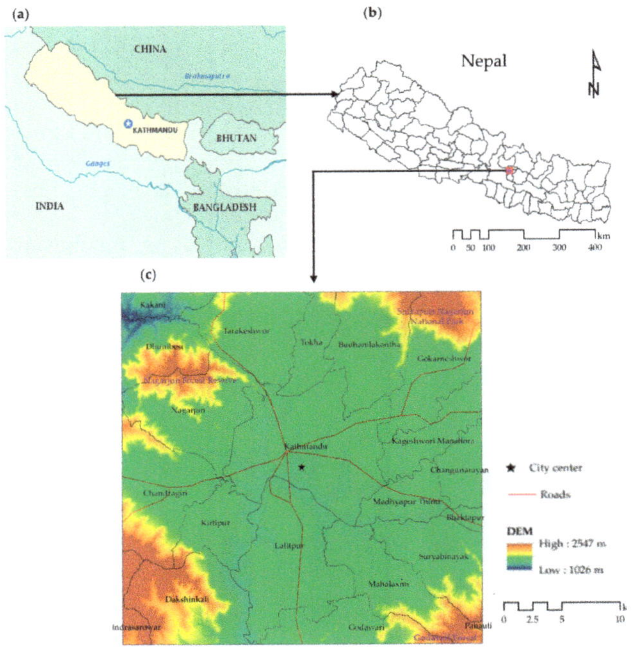

Figure 1. Location of the study area. (**a**) Nepal (28°23′42″N/84°7′40″E) and some other countries in South Asia [53]; (**b**) Kathmandu City and its immediate surrounding areas [54]; and (**c**) study area of 20 × 20 km with a 10 km radius.

2.2. LULC Classifications

In assessing the relationship between LULC and *LST*, spatially and temporally consistent LULC maps are required. However, we could not find spatially and temporally consistent LULC maps for Kathmandu. Therefore, we used earth-observing satellite remote sensing data to classify LULC maps in 2000, 2013, and 2020, considering urbanization and data availability. We used Landsat images in this study [55]: Bands 4, 3, and 2 for Landsat 5 and bands 5, 4, and 3 for Landsat 8 were used for LULC classification. The target years were selected on the basis of the urban development of the Kathmandu Valley and previous studies [52,56]. We employed machine learning techniques—the random forest classification using R software [57]—to classify the LULC maps for the study area. Many studies have shown that the random forest classification method has higher LULC classification accuracy in many regions of the world [58–60]. Six land use and land cover categories were identified: impervious surface (IS), green space 1 (GS1), green space 2 (GS2), bare land (BL), water (W), and other land (OL). The IS category consists of buildings, roads, airports, schools, industrial areas, and asphalt areas. The GS1 category consists of all types of forests, and GS2 consists of croplands, grasslands, and small types of bushes. BL comprises exposure areas both natural and man-made. The water (W) category includes rivers, lakes, and sub-channels, and OL mainly contains clouds, snow, and shadows in the study area.

We used 600 sample points to evaluate the accuracy of each map in 2000, 2013, and 2020. Google Earth historical images were used for the accuracy assessment. The accuracy of the maps was determined by automatic sampling in the algorithm for each year using the software. The spatial resolution of the classified LULC maps was 30 × 30 m.

2.3. Estimation of LST

This study used Landsat collection 2 level-1 product data (one TM 5 image for 2000 and two OLI/TIRS images for 2013 and 2020) to estimate *LST* in Kathmandu. Captured Landsat TM 5 images in 2000 (4 April; 10:07 local solar time) and 2013, and 2020 Landsat 8 OLI/TIRS (26 March; 10:33 local solar time, and 11 April; 10:22 local solar time) images were utilized for the analysis. Before assessing the *LST* in the area, we pre-processed the data on the basis of methods presented in our previous studies (surface reflectance values for the multispectral bands and at-satellite brightness temperature (T_b) values for the thermal bands) [18,61]. Generally, the frequently applied process of obtaining unprocessed Landsat data requires the DN value of thermal bands (Landsat thematic mapper TM = band 6, and Landsat 8 OLI/TIRS = bands 10 and 11) [18]. First, we obtained absolute radiance values and performed the derivation of satellite brightness temperature [17,18]. We used the pre-processed bands (band 6 for Landsat 5 TM and bands 10 and 11 of Landsat 8 OLI/TIRS) and the normalized difference vegetation index (NDVI) method to estimate land surface emissivity values [17,18]. Consequently, Kelvin values of the top-of-atmosphere brightness temperature were shifted to Celsius (°C). Equation (1) [13,18] was used to estimate the *LST* for Kathmandu:

$$LST = \frac{T_b}{1 + (\lambda \times T_b/\rho)\, IN_\varepsilon} \tag{1}$$

where T_b refers to Landsat 5 TM band 6, Landsat 8 OLI/TIRS band 10 brightness temperature, λ refers to the wavelength of emitted radiance (11.5 μm for band 6 and 10.8 μm for band 10), $\rho = h \times c/\sigma$ (1.438×10^{-2} m K), σ = Boltzmann constant (1.38×10^{-23} J/k), h = Planck's constant (6.626×10^{-34} Js), c = velocity of light (2.998×108 m/s), *IN* refers to the pre-launch logarithm, and ε refers to the emissivity of the land surface.

2.4. Spatial Profile of Surface UHI in Kathmandu City

A typical UHI formation represents the temperature distribution from the urban core. The urban core area shows a heating condition and comes to the middle. It shows a decreasing trend in basins, plateaus, and valleys. Estoque and Murayama 2017 [13] introduced surface UHI profiling for the South Asian mountain city of Baguio, Philippines, considering urban-rural gradient surface UHI formation using Landsat data. Athukorala and Murayama 2020 [18] investigated the surface UHI formation in Accra, Ghana, observing cross-sectional surface UHI profiles based on Landsat data. This study also focused on surface UHI definition based on 210 × 210 m grid size because this grid size is robust for predicting methodological factors and environmental elements utilizing statistical connections in surface UHIor atmospheric UHI, and LULC categories, as well as spatial configurations and patterns (210 × 210 m grid size used based on previous studies [13,16,18]); then, a surface UHI profile of the Kathmandu was created.

2.5. Remote Sensing-Related Parameter Analysis

Geologically, a valley is considered as an extended depression on the earth's surface that is usually surrounded by mountain ridges. The depth and natural landscape features of the valley strongly impact its local climate, and valley cities develop under these conditions. Kathmandu is a bowl-shaped temperate mountain valley city (Figure 1). Therefore, it is crucial to understand the relationship between natural parameters such as NDVI, modified normalized difference water index (MNDWI), normalized difference bareness index (NDBal), and elevation and *LST*. Understanding the relationship between natural parameters and *LST* and their interconnectivity with each variable provides valuable insights for urban planners and policymakers for the purposes of sustainable city planning. To do this, we used four remote sensing parameters: NDVI, MNDWI, NDBal, and elevation for Kathmandu City, and each variable was selected on the basis of the knowledge of the study area and previous studies [13,62–65]. Elevation (digital elevation model (DEM)) data were obtained from the Advanced Spaceborne Thermal Emission and Reflection Radiometer (ASTER) with 30 × 30 m resolution [55]. We resampled NDVI, MNDWI, and NDBal into 0 to 100 using the resample tool in the ArcGIS 10.5 software. For the final analysis, we used a 210 × 210 m grid size, as explained in Section 2.4. The multiple linear regression (MLR) model [66,67] was applied to demonstrate the relationship between remote sensing-related parameters and the mean *LST* using 9025 analytical grids. Some previous studies have also employed the same statistical method [13]. The objective of MLR is to model the linear relationship between response (dependent) variables and explanatory (independent) variables. Several important assumptions had to be tested during the regression analysis, including the presence of linearity between the dependent and independent variables. The R-squared was used to calculate how much of the variation in the independent variables can be attributed to the variation in the outcome.

2.6. Spatial Analysis

2.6.1. Characteristics in Surface UHI of Kathmandu

We assessed the characteristics of SUHII along the urban-rural gradient in Kathmandu to understand its formation in the temperate mountain valley condition. As described in Section 2.4, all 210 × 210 m grids were aimed in the same direction for this analysis (see Estoque and Murayama 2017 [13]). Accordingly, 48 urban-rural buffers were demarcated as urban-rural zones in this study (URZs), i.e., URZ_1, URZ_2, URZ_3, ... , URZ_{48} (Appendix A). The zones show the mean *LST* and LULC densities of IS, GS1, and GS2 (we calculated the IS, GS1, and GS2 densities along each URZs), and the formation of the remote sensing-related parameters, i.e., NDVI, MNDWI, and NDBal, ranging from 0 to 100 at 210 × 210 m intervals (see Section 2.5). The surface UHI intensity changes were estimated between URZ_1 (URZ_1 was recognized as a high urban intensity zone) and other URZs. We applied the same procedure for 2000, 2013, and 2020. On the basis of previous studies [16,18], we considered high-IS-density zones (URZ_1) as urban zones and URZs with <15% IS density as rural zones. We excluded bare land (BL), water (W), and other land (OL) categories in this analysis, because they possess relatively low areas compared to the other LULC categories.

2.6.2. Landscape Composition and Pattern Analysis

This investigation aims to determine which spatial traits of the IS, GS1, and GS2 patches are likely to have impacted the spatial formation of *LST* in Kathmandu. The 210 × 210 m grid size used in Section 2.4 was inadequate for this composition and pattern analysis due to the grid size. Therefore, we used a relatively large grid size for this analysis. We applied a 4 × 4 km fishnet to divide the entire study area (20 × 20 km) into 25 sub-parts. All sub-parts were considered in this analysis. For further investigation, the LULC and *LST* maps were clipped with the corresponding polygon grid years 2000, 2013, and 2020.

We used five class-level spatial matrices: mean patch area (AREA_MN), number of patches (NP), largest patch index (LPI), percentage of landscape (PLANND), and cohesion (COHESION) (Table 1) (more information—Fragstats [68]). These class-level spatial metrics have been widely applied in previous UHI studies. The 8-cell neighbor rule was employed to estimate the five metrics. BL, W, and OL were excluded in this analysis due to the same conditions explained in Section 2.5. Finally, the resulting metric values were correlated with the mean *LST* of the LULC category of each sub-part to determine the influence of landscape composition and pattern on the mean *LST* in Kathmandu.

Table 1. Class-level spatial metrics used in this study [68].

Index	Description	Unit	Measure
Mean Patch Area (AREA_MN)	The average patch size of LULC classes. The spatial pattern and heterogeneity of the area.	Hectare	Composition of each LULC class in the study area (LULC classes).
Number of Patches (NP)	Derived using the total landscape area.	Number of patches per hectare	Estimation of the fragmentation of each LULC class.
Largest Patch Index (LPI)	Quantifies the percentage of the total landscape area taken up by the largest patch at the class level. It is a simple gauge of dominance.	0–100	LPI has the ability to detect the advantages of the LULC.
Percentage of Landscape (PLANND)	Sum of the LULC classes divided by the total landscape area × 100.	Percentage	Measurement of the abundance of the corresponding LULC class.
Cohesion (COHESION)	The physical connectivity of the corresponding patch type of the LU class increases with more clustering of the patch type in its configuration, resulting in more physical amalgamation.	0–100	The physical connectivity of the equivalent patches of LULC class.

3. Results

3.1. LULC and LST Changes in Kathmandu from 2000 to 2020

The overall accuracy of the classified LULC maps was greater than 85% (Appendix B). The classified LULC maps show that Kathmandu has undergone rapid urbanization over the last two decades (2000–2020) (Figure 2). The area of IS increased by 762 ha between 2000 and 2013, and it expanded by 4240 ha between 2013 and 2020, corresponding to a total of 5002 ha in the last 20 years (Table 2). In the visual interpretation, most of the IS area was accumulated in the central part of the study area in 2000, and it gradually increased in the outer area by 2020. The results show a total net loss of GS1 and GS2 recorded at 2807 ha and 2202 ha in the 20 years, respectively (Figure 2). We observed that most GS1 areas were located in the north-eastern, south-western, north-western, and south-eastern parts of the study area, and GS2 was mainly concentrated in the middle part of the study area.

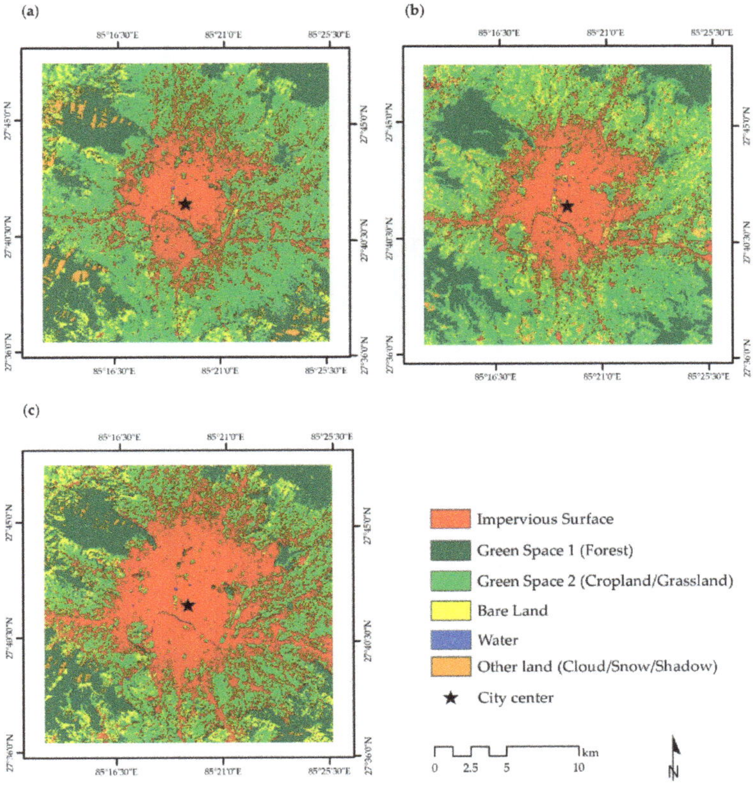

Figure 2. LULC maps of Kathmandu City, Nepal, and its surrounding areas classified using machine learning techniques (see Section 2.2): (**a**) LULC map in 2000; (**b**) LULC map in 2013; and (**c**) LULC map in 2020.

Table 2. LULC area matrix of Kathmandu City from 2000 to 2020.

LULC Type	2000 km^2	%	2013 km^2	%	2020 km^2	%
Impervious Surface	86.96	21.74	94.58	23.65	136.98	34.25
Green Space 1 (Forest)	116.69	29.17	99.7	24.93	88.62	22.16
Green Space 2 (Cropland/Grassland)	149.66	37.42	153.84	38.46	127.64	31.91
Bare Land	31.52	7.88	38.64	9.66	39.31	9.83
Water	1.93	0.48	1.79	0.45	1.7	0.43
Other Land (Cloud/Snow/Shadow)	13.24	3.31	11.44	2.86	5.75	1.44

Figure 3 shows the *LST* distribution of Kathmandu City from 2000 to 2020. The mean *LST* in 2000 was 18.94 °C, and the *LST* was mainly concentrated in the central part, east, northeast, and the periphery of southern and north-western parts. In 2013, the mean *LST* was 25.19 °C and mainly accumulated in the central, south, and east parts. In 2020, the mean *LST* was 26.11 °C, and the overall higher *LST* was observed in the central, east, south, and north-western parts of the study area. In particular, we observed significantly higher *LST* values in the Tribhuvan International Airport in Kathmandu at each time point (mean *LST* was 25.11 °C in 2000, 27.43 °C in 2013, and 30.48 °C in 2020) (Figure 3).

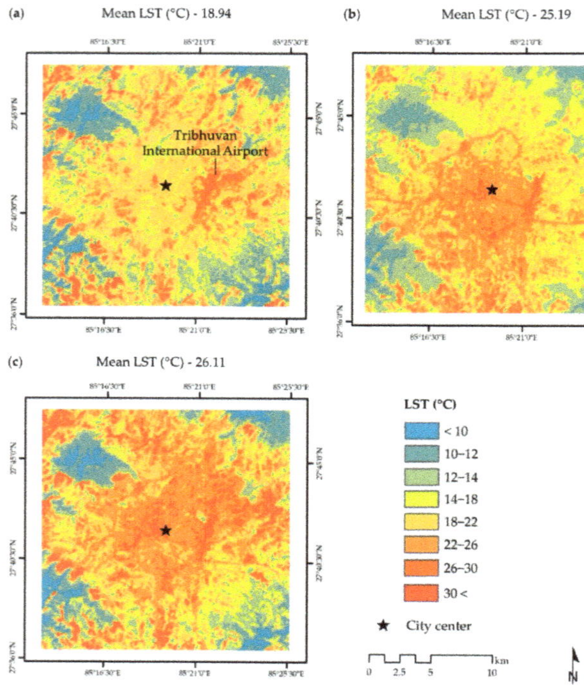

Figure 3. *LST* maps of Kathmandu City, Nepal, and its surrounding areas derived from Landsat imagery: (**a**) *LST* map in 2000; (**b**) *LST* map in 2013; and (**c**) *LST* map in 2020.

Figure 4 shows the mean *LST* of each LULC category in Kathmandu. The results show that the mean *LST* of IS was 19.80°C in 2000, 26.52 °C in 2013, and 27.40 °C in 2020. The mean *LST* of GS 1 was recorded at 16.88 °C in 2000 and 22.39 °C in 2020, while GS2's was 18.66 °C in 2000 and 24.68 °C in 2020, indicating a lower mean *LST* than the IS category at each time point. As mentioned before, the BL of the study area shows a relatively low area, and most areas were located in the periphery of the study area (Figure 2). The mean *LST* of BL was 21.49 °C in 2000, 25.82 °C in 2013, and 27.60 °C in 2020. The mean *LST* of the water category was recorded at 18.89 °C in 2020 and 20.03°C by 2020 over the study period.

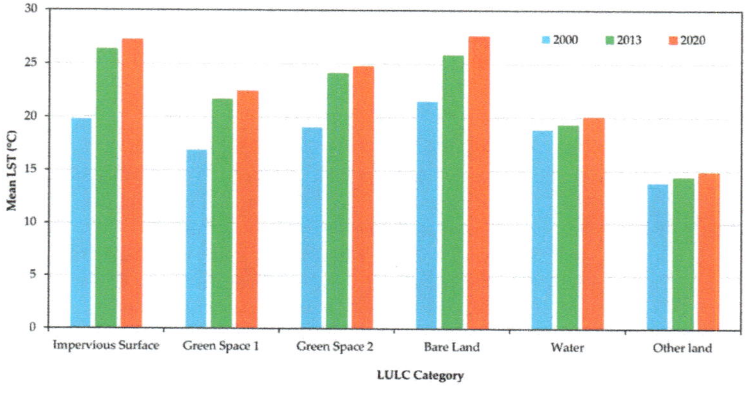

Figure 4. Mean *LST* of each LULC category in Kathmandu City, Nepal in 2000, 2013, and 2020.

3.2. Characteristics of RS-Based Spatial Parameters

3.2.1. Changes in NDVI, MNDWI, and NDBal

Figure 5 indicates the normalized difference vegetation index (NDVI) values in Kathmandu City in 2000, 2013, and 2020. High NDVI values were concentrated in the north-east, south-east, south-west, and north-west regions, while lower NDVI values were concentrated on a substantial portion of the study areas in 2000 and 2013. However, we observed lower NDVI values by 2020 compared to in the years 2000 and 2013, and NDVI in 2020 was mainly concentrated in the north-east, south-east, south-west, and north-west areas. The central part of Kathmandu showed lower NDVI values ranging from 1 to 20 during the study period.

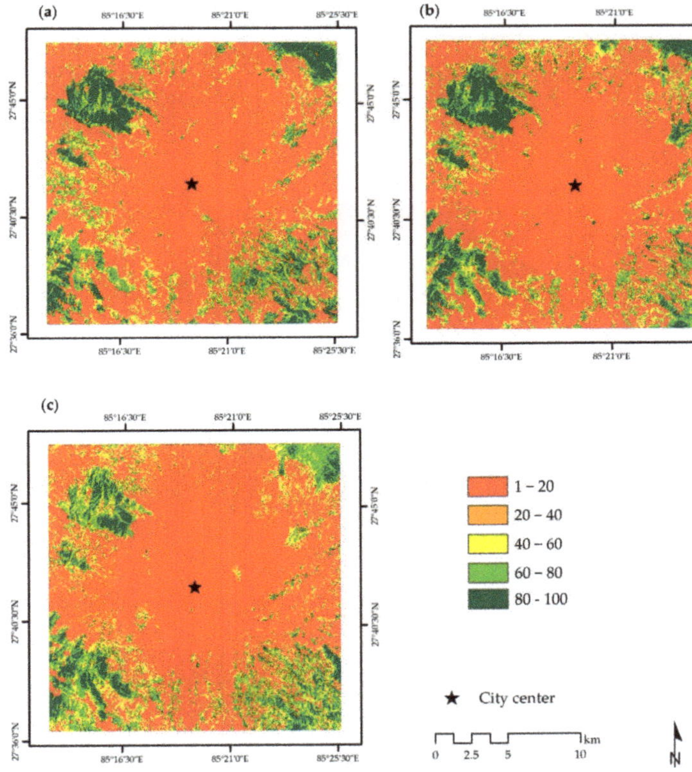

Figure 5. NDVI maps of Kathmandu City, Nepal: (**a**) NDVI in 2000; (**b**) NDVI in 2013; and (**c**) NDVI in 2020.

Figure 6 reveals the modified normalized difference water index (MNDWI) of Kathmandu City in 2000, 2013, and 2020. The MNDWI is used extensively as the RS parameter in surface UHI studies. In 2000, higher values of MNDWI were recorded in the central, north-western, south, and south-eastern areas (Figure 6). The MNDWI was more apparent in the central, north-east, south-western, and north-east areas by 2013. Our results show the MNDWI concentration in the central part of the study area over three time points, with the effect of the Bhagmathi river basin and several tributaries (Bishnumati and Manamati) flowing through the central part of the study area. Most areas at the three time points show an MNDVI ranging between 1 and 20, indicating some water stress in the study area.

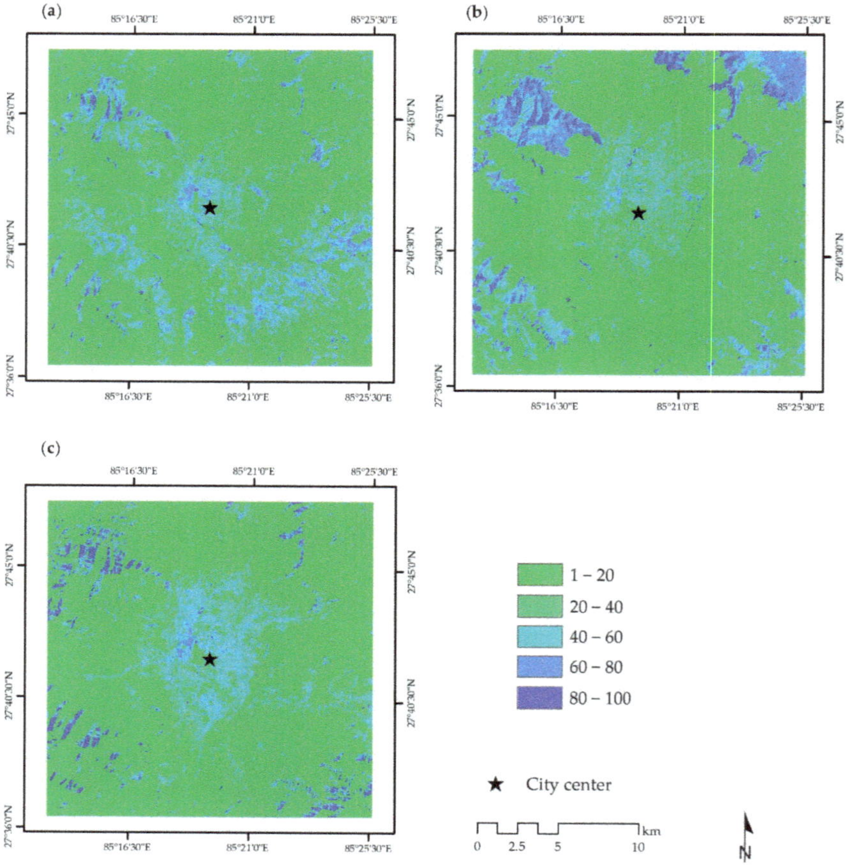

Figure 6. MNDWI maps of Kathmandu, Nepal: (**a**) MNDWI in 2000; (**b**) MNDWI in 2013; and (**c**) MNDWI in 2020.

Figure 7 shows the normalized difference bareness index (NDBal) of Kathmandu in 2000, 2013, and 2020. According to the three maps (Figure 7), higher NDBal values were identified in the middle, north, east, south, and west parts of the study area. We observed that most of the NDBal values of Kathmandu were spread out and located as a ring away from and around the city center at the given time points, indicating rapid LULC change due to the urban process in the area. The three maps indicate that substantial-high NDBal values were located around the top of mountain areas compared to the city area, and the LULC maps (Figure 2) also corroborated this condition. Moreover, the three *LST* maps of Kathmandu show relatively high *LST* values related to the high NDBal around the top of mountain areas (Figure 3).

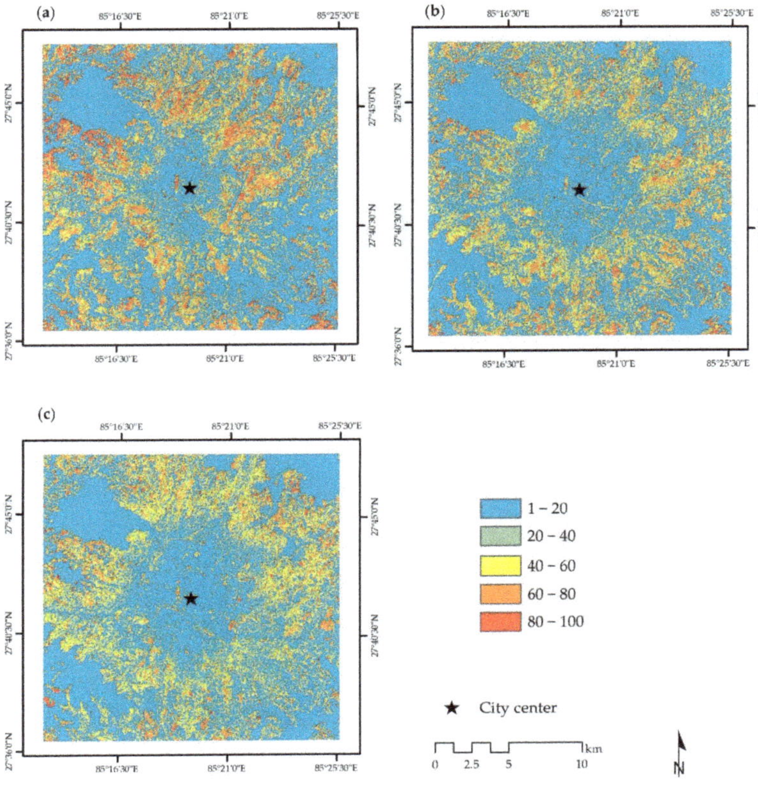

Figure 7. NDBal maps of Kathmandu, Nepal: (**a**) NDBal in 2000; (**b**) NDBal in 2013; and (**c**) NDBal in 2020.

3.2.2. Relationship between Mean *LST* and RS-Based Parameters

The results show a relationship between the mean *LST* and RS-based parameters at a grid size of 210 × 210 m (Table 3). Overall, the MLR analysis results revealed that combining the RS-based parameters used in this analysis makes it possible to explain the significance of the parameters used in the mean *LST* during the study period. Moreover, each regression coefficient β of the RS-based parameters showed statistical significance ($p < 0.001$) at the three time points. For example, in 2000, the standardized regression coefficients showed that the mean NDBal had a significant positive correlation with the mean *LST*, indicating the heating power of the area. The mean NDVI, mean DEM, and mean MNDWI had a significant negative relationship with mean *LST*, indicating the cooling power of the area. In 2013 and 2020, the mean NDVI and mean elevation had the highest negative relationship with the mean *LST*. In contrast, the mean NDBal had the highest negative relationship with the mean *LST* in Kathmandu in 2013 and 2020. Our results indicate that the standardized regression coefficients between the mean NDVI, mean elevation, and mean MNDWI and the mean *LST* increased (negatively) during the study period. In contrast, the relationship between mean NDBal and mean *LST* increased in 2020; however, the mean NDBal in 2013 showed a lower value than that in the years 2000 and 2020.

Table 3. Results of MLR analysis in Kathmandu (dependent variable: mean *LST*; 210 × 210 m grid size; and N = 9025).

RS-Based Parameters	Coefficients			
	Unstandardized β	**Std. Error**	**Standardized β**	**Sig.**
2000				
(Constant)	22.590	0.144		
Mean NDVI	−0.031	0.001	−0.319	0.000
Mean MNDWI	−0.017	0.001	−0.116	0.000
Mean NDBal	0.057	0.001	0.434	0.000
Mean elevation	−0.003	0.000	−0.218	0.000
		R^2 = 0.696; Adjusted R^2 = 0.695		
2013				
(Constant)	35.163	0.124		
Mean NDVI	−0.043	0.001	−0.506	0.000
Mean MNDWI	−0.024	0.001	−0.170	0.000
Mean NDBal	0.042	0.001	0.410	0.000
Mean elevation	−0.006	0.000	−0.453	0.000
		R^2 = 0.781; Adjusted R^2 = 0.780		
2020				
(Constant)	31.154	0.112		
Mean NDVI	−0.046	0.001	−0.516	0.000
Mean MNDWI	−0.045	0.001	−0.410	0.000
Mean NDBal	0.059	0.001	0.457	0.000
Mean elevation	−0.006	0.000	−0.506	0.000
		R^2 = 0.729; Adjusted R^2 = 0.729		

3.3. Characteristics of Surface UHI in Kathamndu

Our results reveal that the density of IS and the mean *LST* have regular characteristics (Figure 8a). The URZs near the central business district (CBD) exhibited the highest mean *LST*, which declined along the urban−rural gradient in 2000, 2013, and 2020 (Figure 8a). Conversely, the density of GS1 and GS2 indicates the lowest mean *LST* as being near the CBD. The density of GS1 and GS2 gradually increased from the CBD, corresponding to a large part of the rural area in the study period (Figure 8a). The density of IS increased from the center to URZ_4, and a quick drop was identified from URZ_4 to URZ_5. This drop exhibited an ascending trend from URZ_5 to URZ_9, which continued until URZ_{12} with small fluctuations (Figure 8a). From URZ_{12} to URZ_{48}, the density of IS showed a decreasing trend that is compatible with the mean *LST* at the three time points.

In contrast, the density of GS2 had a higher value than that of GS1 around the city center at the three time points, and it gradually declined until URZ_4. From URZ_4 to URZ_6, the density of GS2 increased until URZ_6, before again declining between URZ_6 and URZ_{12}, apart from in 2013. By 2020, the density of GS2 showed an increasing trend until URZ_{33}, and then decreased until URZ_{48}, indicating changes in the urban structure in the suburbs (Figure 2). The density of GS1 decreased from the center grid to URZ_{20}, indicating that urban pressure in Kathmandu was reflected from the CBD to the suburbs. We observed that from URZ_{20} to URZ_{48}, the density of GS1 increased across the three time points. Moreover, we found that the density of GS2 was higher from URZ_{12} to URZ_{42} than that of GS1, showing a significant impact on mean *LST* in Kathmandu (see mean *LST* values in each LULC at three time points) (Figure 8a).

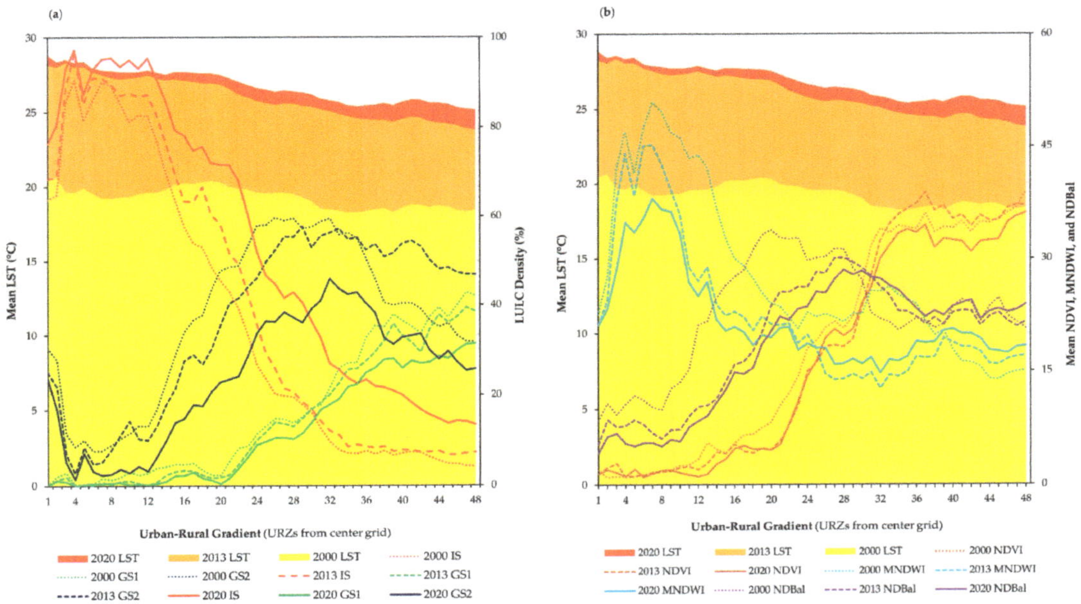

Figure 8. Urban–rural gradient in Kathmandu City: (**a**) mean *LST* and the density of IS, GS1, and GS2 along the urban–rural gradient; and (**b**) mean *LST* and NDVI, MNDWI, and NDBal along the urban–rural gradient.

The mean NDVI did not reveal the highest mean *LST* to be in the CBD. Our results revealed that the mean MNDWI had higher values near the urban core area than the mean NDVI and NDBal values because the area flourished in rivers and sub-waterways. The mean MNDWI value showed an increasing trend from the center grid to URZ8, with some fluctuations, and it gradually declined along the urban-rural gradient during the study period. We observed that the mean MNDWI values increased from URZ_{25} to URZ_{48} compared to 2013, indicating enhanced green cover (forest, cropland, and grassland) in the urban-rural transition area and the rural area in comparison to the LULC maps of the study area (Figure 2). The decrease in the mean NDVI reported around the city core area indicated rapid urban growth in the central part of Kathmandu and an abundance in the urban fringe (Figure 8b). The results revealed that mean NDVI values increased gradually from URZ_{29} to URZ_{48}, indicating a greater cooling effect in these zones.

The mean NDBal of the study area showed an increasing trend until URZ_{26} and declined from that point until the final zone, with few fluctuations. However, we observed a slight increase in the mean NDBal between URZ_{25} and URZ_{29} (Figures 7 and 8b), indicating that the BL was mostly located between the hinterland and the peripheral in the study area in 2020 (Figure 2). We observed an increase in NDBal from URZ_{32} to URZ_{48} in 2020 compared to in 2013. This enhancement may have been due to the urbanization process near hill areas and increased deforestation activities in this area, because the density of the GS1 also decreased in this area (Figures 2 and 8a).

On the basis of the differences between Figure 8a,b, it can be concluded that the density of GS1 and GS2 decreased over the three time points. This pattern corresponds to the mean NDVI, MNDWI, and NDBal from URZ_1 to URZ_{48}, URZ_{25}, and URZ_{31}, respectively, indicating the urbanization characteristics (i.e., the density of IS) in the area influenced the formation of UHI in the area.

3.4. LULC Composition and Pattern vs. Mean LST in Kathmandu

The results indicate that IS, GS1, and GS2 had a significant relationship with mean *LST* in Kathmandu in 2000, 2013, and 2020 (Table 4). The correlation coefficients of IS showed a positive relationship with mean *LST* during the study period. Conversely, GS1 and GS2 had a negative relationship with mean *LST* during the study period. The values of the five IS metrics revealed the rapid expansion of IS in Kathmandu. These results correspond to our classified LULC maps for the area (Figure 2). The Area_MN, LPI, and COHESION of GS2 had a negative relationship at the three time points. The PD and PLAND of GS1 possessed lower values in 2013, but higher values in 2020. The AREA_MN, PD, LPI, and PLAND of GS2 had a negative relationship with mean *LST* in Kathmandu. All values were statistically significant ($p < 0.001$).

Table 4. Results of LULC composition vs. mean *LST*.

	2000			2013			2020		
	IS	**GS1**	**GS2**	**IS**	**GS1**	**GS2**	**IS**	**GS1**	**GS2**
AREA_MN	0.425	−0.235	−0.357	0.454	−0.629	−0.204	0.532	−0.670	−0.103
PD	0.108	−0.730	−0.087	0.257	−0.123	−0.138	0.477	−0.560	−0.194
LPI	0.127	−0.464	−0.039	0.481	−0.614	−0.34	0.505	−0.686	−0.024
PLAND	0.173	−0.549	−0.151	0.351	−0.800	−0.124	0.421	−0.660	−0.112
COHESION	0.209	−0.512	−0.196	0.539	−0.789	−0.173	0.611	−0.897	−0.262

Note 1: Area_MN—mean patch size; PD—patch density; LPI—largest patch index; PLAND—percentage of landscape; and COHESION—cohesion. Note 2: IS—impervious surface; GS1—green space 1; and GS2—green space 2.

Overall, five matrices—AREA_MN, PD, LPI, PLAND, and COHESION, and IS, GS1, and GS2—have a confirmed potential influence on the mean *LST* in the temperate mountain valley city of Kathmandu. A growing body of literature indicates that widespread patches of green space can promote a cooling effect compared to the small and scattered patches of green space. We observed widespread IS patches in Kathmandu to promote the surface UHI effect. However, small and scattered patches of IS promoted a relatively lower surface UHI effect in Kathmandu.

4. Discussion

4.1. Change in Urban Structure in Kathmandu

To investigate the effect on surface UHI of urban expansion and green space distribution in Kathmandu, this study tried various approaches, including machine learning techniques, remote sensing (RS)-based parameter analysis, urban-rural gradient analysis, and spatial composition and pattern analysis. Here, we offer sustainable landscape and urban planning strategies for use by city planners and policymakers in Kathmandu to mitigate the UHI effect and improve the quality of life of city dwellers.

Our results provide strong evidence of the rapid urban growth in Kathmandu, Nepal, showing an exponential expansion of IS over a time span of 20 years (from 2000 to 2020) (Figure 2). Some studies have shown that the Kathmandu Valley is the most densely populated area, the major economic core, and one of the fastest-blooming urban agglomerations in South Asia [69,70]. This city is critical, because Kathmandu is a mountain valley city with unique geophysical characteristics in a land-locked country that has no land connected to an ocean or coastlines [71]. Kathmandu's urbanization has primarily been driven by religion, tourism, and the city's pleasant cool climate [72,73]. The population of Kathmandu City was 2.5 million in 2016, and is expected to increase rapidly with these impacts and drivers [52,71–79]. These results reveal that IS has rapidly encroached in the east, north, south, and west with increasing population and infrastructure from the urban core to the suburbs (Figure 2). This condition indicates that the central part of the Kathmandu Valley and its neighborhoods are critical determinants of its urban development.

4.2. Linking Surface SUHI Formation with LULC

In this study, we derived three remote sensing-based *LST* maps for 2000, 2013, and 2020, and the mean *LST* of Kathmandu was 18.94 °C in 2000, 25.19 °C in 2013, and 26.11 °C in 2020 (Figure 3). However, we discovered that the *LST* values at the three time points not only closely corresponded to LULC, but those other environmental variables, such as surface moisture, humidity, solar radiation, wind speed, precipitation, and anthropogenic heat release, may not have been temporally stable across the three time points when the thermal images were acquired.

Hence, our research focused on the temporal fluctuations of surface UHI in Kathmandu. Here, two factors are considered, i.e., surface UHI and the density of LULC difference between climate zones (between LULC categories and URZs) during the study period. This approach enables the comparison of surface urban heat intensities in Kathmandu from 2000 to 2020. The results revealed the increasing trend of surface UHI intensity between 2000 and 2020. We found that the surface UHI intensity between GS2 and GS1 increased by 1.37 °C during the study period. Considering the urban-rural gradient, an average increase in surface UHI intensity between 2000 and 2020 based on URZ1 (urban zone with the highest IS density at the three time points) and the rural zone (the first URZ with <15% IS density) generated a higher value, at 3.89 °C. Therefore, the underlying mechanisms of the increased surface UHI intensity in Kathmandu need to be understood in order to achieve sustainable city planning.

In 2000 and 2013, the proportions of the study area accounted for by IS in Kathmandu were 21.74% and 23.65%, respectively. However, during the period between 2013 and 2020, the proportion of IS was 34.25%, an increase of 10.6% compared to the 2000–2013 period. Several studies have discovered that the surface UHI intensity is positively correlated with city size [9,11,80–82]. Therefore, we selected a 10 km buffer to restrict the study area. We discovered that the increasing trend of surface UHI intensity in Kathmandu had been impacted by changes in the natural landscape caused by rapid urban processes arising from the considerable expansion of IS and the visible degradation of green space in the area. The natural landscape change from GS1 (forest) to GS2 (cropland/grassland) cannot be neglected, because the mean *LST* difference between GS2 and GS1 shows an increasing trend, as mentioned before, and this can enhance the surface UHI effect in the area. Urban landscape transformation also influences changes in the values of the RS-based parameters (NDVI, MNDWI, and NDBal) considered in this study. Therefore, future urban planning should pay more attention to this condition.

Our study found a high surface UHI value near Tribhuvan International Airport at three time points, with increasing intensity between 2000 and 2020 (Figure 3). We gave more attention to this area because the airport is the most critical place for the country from a socio-economic perspective. The results revealed that GS2 and GS1 near the airport area had declined with the rapid expansion of IS (sub-urbanization), promoting more surface UHI in this vicinity. Therefore, urban planners should pay more attention to reducing *LST* by means of possible and practical treatments in this area.

Along the urban-rural gradient, we observed a slightly decreased IS density between URZ_4 and URZ_7 and an improved GS2 density in the same area. However, the decline in IS density and increase in GS2 density did not fully correspond to the mean *LST* in this region (see Figure 8a). As mentioned above, the MNDWI in the URZs exhibited lower values, and NDBal showed relatively high values in these URZs. This effect might have enhanced the mean *LST* in these URZs. Previous studies have reported similar findings [16,18]. For instance, XIAO et al. (2007) revealed that surface UHI was not most pronounced in the CBD. Rather, it was located in the south of the central city near the 4th ring road and the dry Yongding river in the south-western part of the city, which is the biggest area of bare land in Beijing, China [83]. Estoque and Murayama studied the impact of landscape composition and pattern on land surface temperature in the three megacities of Southeast Asia [17]. They identified a significant correlation between land use categories and mean *LST* changes along the urban-rural gradient in Bangkok, Jakarta, and Manila. Athukorala and Murayama discovered a strong relationship between LULC density and tasseled cap transformation (TCT) and the mean *LST* change in the sub-Saharan city of Accra, Ghana [18]. They found that GS2 (cropland and grassland) positively contributes to enhancing surface UHI by combining the effects of the bareness index in the sub-Saharan climate.

4.3. Effect of Landscape Composition and Pattern on Surface UHI Formation

Our study shows that the five spatial metrics were significantly correlated with the mean *LST* in Kathmandu (the density of IS (positive) in 2000, 2013, and 2020, and the density of GS1 and GS2 (negative)) (Table 4). These findings are similar to other studies by Estoque et al. (2017) [17], Myint et al. (2013) [84], Hou and Estoque (2020) [32], Zhou et al. (2011) [66], Athukorala and Murayama (2021) [16], and Zhou et al. (2017) [85]. They revealed that the AREA_MN, PD, LPI, PLAND, and COHESION of IS and GS significantly correlate with the mean *LST*. However, data information, including the magnitude, significance, and angle of the influence, differed between our findings and those of prior studies. It is important to note that rapid urban growth changes natural environments into the IS, receiving more solar energy and little reflected solar radiation [16,86]. In that context, *LST* affects urban thermal environmental change and modifies environmental factors (humidity, evapotranspiration, and energy balance) in the urban area, influencing human health and thermal comfort.

Generally, vegetation and shadow help to reduce the surface temperature [16,87–89]. In Kathmandu, we observed that forest cover, high-rise building shadow, and mountain shadow mitigate surface UHI in certain areas. However, the position of this shadow effect varies with earth rotation and time. For example, many studies have investigated the surface UHI effect using Landsat data [13,17,89]. The local time of data capture was during the morning. The solar incidence angle (the angle between solar rays and the vertical direction) produces the evapotranspiration and shadow effects. However, the magnitude, location, and surface covering of areas with these effects should be fully considered for urban planning, especially the relationship between the effects of shadow and earth rotation.

Geophysically, Kathmandu City is located in a mountain valley and one of the country's major river basins (the Bagmati river basin). Urban planners should pay more attention to these factors with respect to urban planning. Landscape composition and pattern analysis can fill this gap and provide more insights for urban planners [8,66,85,90,91]. Generally, large patches of IS produce more surface UHI, and relatively small patches have a lower surface UHI effect. On the other hand, large GS patches have a more cooling effect, and small patches have less of a cooling effect. Our study reveals that enlarged and continuous patches of GS1 (forest) and GS2 (cropland/grassland) generate a more significant cooling effect in rural areas (Shanti Danda, Chonga Ganesh temple area, Hanumante river area, Gokarna, Chanautipato, Coronation garden, Chhanui military barracks, Swayambhunath, and Shoyembhu areas) than in urban core areas. Similar results have been reported in Bangkok, Jakarta, Manila, Accra, greater Cairo, and Baguio [13,16–18]. However, the magnitude of the cooling effect of GS1 is higher than GS2 in Kathmandu. We discovered that the complex shape of the forest and cropland/grassland was more active in cooling *LST* in Kathmandu. Zhou et al. (2017) [85] revealed the same conditions in Sacramento, but in Baltimore, a simple tree crown with a small margin in the same area performed better in terms of cooling effect. Such differences have emerged in the literature between study areas with different climatic conditions. Therefore, the relationship between landscape metrics and *LST* should be thoroughly considered in order to achieve sustainable city planning.

Overall, for Kathmandu, it is suggested to plant suitable trees or set waterways and protect the existing tributaries (we observed that some sub-water channels were dispersed around the city core area during the study period) near roads and buildings to decrease the surrounding surface UHI. Athukorala and Murayama (2021) [16] revealed that rooftop greening provided a practical cooling effect during both daytime and nighttime in greater Cairo, Egypt. In this respect, Kathmandu is a bowl-shaped city, and reducing surface UHI remains a challenge. Moreover, the lessons learned from rooftop agriculture and vegetation in greater Cairo, Egypt [16] provide a more efficient approach to mitigating the surface UHI effect. Therefore, urban planners could rearrange the green space (trees) in Kathmandu to create adjoining patches, maximizing their cooling ability.

4.4. Implication for Surface UHI Mitigation and Urban Climate Adaptation

Kathmandu has been Nepal's largest urban agglomeration, industrial center, and socio-economic and tourist hub in recent decades [52,71,72], during which period the extensive natural and semi-natural landscape was rapidly transformed into IS. Many researchers across the world have recently focused their attention on the relationship between the significant loss of urban green space and the increasing surface UHI effect [13,17,66]. Our findings revealed that the tremendous growth of IS and the decline of green space made the surface UHI effect more pronounced in downtown Kathmandu. Many urban agglomerations are rapidly approaching mitigation and climatic adaption to the surface UHI effect by means of sustainable city planning [8,16,17]. The growing literature shows that efficiently distributed vertical greenery [92–94], such as rooftop vegetation and green walls [16,95,96], makes a reliable contribution to reducing surface UHI and heat fluxes by means of moist and shading facades, thus facilitating human thermal comfort.

According to the visual interpretation, the IS in the 2000 map shows relatively high fragmentation compared to the IS in the 2020 map. However, the actual situation is that the IS in 2020 is more fragmented than the IS presented in the 2000 map because of urban development with high buildings and rapid urban structure change, resulting in greater fragmentation of IS in Kathmandu by 2020. This condition also substantially affected the increase in surface UHI in Kathmandu. In this context, the implementation of rooftop solar photovoltaic systems (SPVSs) has provided roof cover and improved indoor thermal conditions by minimizing extreme heatwaves in the urban core area [97–99]. Our results reveal that the urban areas in Kathmandu have expanded from the CBD to the urban periphery and suburban areas.

Urban agglomerations can effectively produce a positive cooling effect by optimizing the green space [17]. The cooling effects of PD, PLAND, COHESION, and LPI values on surface UHI were exceptional and robust in Kathmandu. Therefore, it is critical to ensure sufficient size and consistency of green space when planning green landscapes to achieve maximum surface UHI cooling by improving the interconnection between patches [16], such as establishing urban parks and green corridors and decreasing the extent of patches [100].

In response to the formation of surface UHI in Kathmandu from 2000 to 2020, we propose that Kathmandu's urban core should be filled in with green patches with high population density and aggregated urban land, and that more green infrastructure networks, in connection with topographic features and roads, should be planned, thus further promoting cooling conditions in the urban area. Moreover, as explained previously, urban planners in Kathmandu should pay a great deal of attention to the Tribhuvan International Airport area, enhancing greenery around the airport vicinity. The link between the GS and RS parameters indicates that vegetation and blue infrastructure should be connected at micro and macro levels. Similar results and implementations have been reported in previous studies [18,101]. To reduce the heatwave effect on urban city dwellers, urban designers and planners should understand the characteristics of rapid urban growth structure change at various development levels and logically composite patterns of urban green spaces and anthropogenic activities.

5. Conclusions

This study assessed the surface UHI and related it to the LULC composition and pattern in the temperate mountain valley city of Kathmandu, Nepal, from 2000 to 2020, using Landsat images. The study area experienced rapid urban development during the study period. This indicates a substantial expansion of impervious surfaces and loss of green space in Kathmandu, and the surface UHI phenomenon is remarkable in the city. Our results revealed a significant relationship between the mean *LST* and urban expansion (impervious surface and green space change). The mean *LST*, mean NDVI, mean MNDWI, mean NDBal, and elevation were critical spatial parameters for deriving the surface UHI in the study area (the statistical relationships were R^2 = 0.695 in 2000, R^2 = 0.781 in 2013, and R^2 = 0.729 in 2020). On the basis of various geospatial methods, this study provides valuable practical approaches for sustainable urban planning and design in Kathmandu. The results show that the surface UHI in Kathmandu has increasing characteristics. The LULC has exhibited a drastic change in Kathmandu during the past 20 years, and the city core indicated the promotion of surface UHI due to the high density of IS.

In contrast, other land covers (GS1, GS2, W, and OL) had lower surface UHI, with the exception of the bare land category. The urban periphery showed the expansion of the city into more rural areas; these land changes enhanced the surface UHI in the urban fringe. GS1 and GS2 gradually decreased in the urban core areas during the study period, showing more fragmentation and reducing the patch size gap, resulting in a reduced cooling effect in the city core. Based on the above-mentioned conclusions, in order to control the further strengthening of surface UHI, protecting vegetation cover and the urban river system, it may be helpful to increase river basin sustainability (Bagmati river basin) and enhance the ecological characteristics of urban greenery, thus mitigating and adapting to the surface UHI effect in Kathmandu in the future.

Author Contributions: S.K. and D.A. conducted the research, performed the analysis, and wrote the paper. Y.M. and T.M. provided research supervision. All authors have read and agreed to the published version of the manuscript.

Funding: This research was partly supported by the Japan Society for the Promotion of Science (JSPS) grants of 21K01027 and 18H00763.

Data Availability Statement: All data generated and analyzed in this study are included in the published article.

Conflicts of Interest: The authors declare no conflict of interest.

Appendix A

Figure A1. Urban–rural gradient. Grid order from city center to outer (i.e., URZ_1, URZ_2, URZ_3, . . . , URZ_{48}). Background image (Landsat 8, 11 April 2020 image, false color band composite bands 6, 4, and 2).

Appendix B

Table A1. Confusion matrices of the classified LULC maps of this study.

LULC Category	Reference Data						Total	User's Accuracy (%)
	IS	GS 1	GS 2	BL	W	OL		
				2000				
IS	98	1	4	5	1	2	111	88.29
GS 1	3	96	3	1	4	6	113	84.96
GS 2	1	6	89	2	1	2	101	88.12
BL	3	0	3	79	0	1	86	91.86
W	1	3	2	2	81	1	90	90.00
OL	1	4	1	4	5	84	99	84.85
Total	107	110	102	93	92	96	600	
Producer's accuracy (%)	91.59	87.27	87.25	84.95	88.04	87.50		
Overall accuracy (%) = 87.83								
				2013				
IS	103	2	3	4	3	1	116	88.79
GS 1	2	79	2	3	3	6	95	83.16
GS 2	3	3	98	5	5	2	116	84.48
BL	3	2	1	84	2	5	97	86.60
W	1	1	4	3	72	4	85	84.71
OL	4	3	3	2	3	76	91	83.52
Total	116	90	111	101	88	94	600	
Producer's accuracy (%)	88.79	87.78	88.29	83.17	81.82	80.85		
Overall accuracy (%) = 85.33								
				2020				
IS	93	4	2	5	1	3	108	86.11
GS 1	3	84	1	1	3	2	94	89.36
GS 2	1	2	96	3	2	4	108	88.89
BL	1	3	4	87	3	3	101	86.14
W	2	5	2	1	79	5	94	84.04
OL	3	1	1	4	2	84	95	88.42
Total	103	99	106	101	90	101	600	
Producer's accuracy (%)	90.29	84.85	90.57	86.14	87.78	83.17		
Overall accuracy (%) = 87.16								

Note: impervious surface (IS), green space 1 (GS1), green space 2 (GS2), bare land (BL), water (W), and other land (OL).

References

1. Dewan, A.M.; Yamaguchi, Y. Land use and land cover change in Greater Dhaka, Bangladesh: Using remote sensing to promote sustainable urbanization. *Appl. Geogr.* **2009**, *29*, 390–401. [CrossRef]
2. Athukorala, D.; Estoque, R.C.; Murayama, Y.; Matsushita, B. Ecosystem services monitoring in the Muthurajawela Marsh and Negombo lagoon, Sri Lanka, for sustainable landscape planning. *Sustainability* **2021**, *13*, 11463. [CrossRef]
3. Athukorala, D.; Estoque, R.C.; Murayama, Y.; Matsushita, B. Impacts of urbanization on the Muthurajawela Marsh and Negombo Lagoon, Sri Lanka: Implications for landscape planning towards a sustainable urban wetland ecosystem. *Remote Sens.* **2021**, *13*, 316. [CrossRef]
4. Huang, R.J.; Zhang, Y.; Bozzetti, C.; Ho, K.F.; Cao, J.J.; Han, Y.; Daellenbach, K.R.; Slowik, J.G.; Platt, S.M.; Canonaco, F.; et al. High secondary aerosol contribution to particulate pollution during haze events in China. *Nature* **2015**, *514*, 218–222. [CrossRef]
5. Rizwan, A.M.; Dennis, L.Y.C.; Liu, C. A review on the generation, determination and mitigation of Urban Heat Island. *J. Environ. Sci.* **2008**, *20*, 120–128. [CrossRef]
6. Kifle, A.B.; Mengistu, T.G.; Stoffberg, G.H.; Tadesse, T. Climate change and population growth impacts on surface water supply and demand of Addis Ababa, Ethiopia. *Clim. Risk Manag.* **2017**, *18*, 21–33. [CrossRef]
7. Kalnay, E.; Cai, M. Impact of urbanization and land-use change on climate. *Nature* **2003**, *425*, 102. [CrossRef]

8. Li, J.; Song, C.; Cao, L.; Zhu, F.; Meng, X.; Wu, J. Impacts of landscape structure on surface urban heat islands: A case study of Shanghai, China. *Remote Sens. Environ.* **2011**, *115*, 3249–3263. [CrossRef]
9. Imhoff, M.L.; Zhang, P.; Wolfe, R.E.; Bounoua, L. Remote sensing of the urban heat island effect across biomes in the continental USA. *Remote Sens. Environ.* **2010**, *114*, 504–513. [CrossRef]
10. Yuan, F.; Bauer, M.E. Comparison of impervious surface area and normalized difference vegetation index as indicators of surface urban heat island effects in Landsat imagery. *Remote Sens. Environ.* **2007**, *106*, 375–386. [CrossRef]
11. Chen, X.L.; Zhao, H.M.; Li, P.X.; Yin, Z.Y. Remote sensing image-based analysis of the relationship between urban heat island and land use/cover changes. *Remote Sens. Environ.* **2006**, *104*, 133–146. [CrossRef]
12. Estoque, R.C.; Murayama, Y. Landscape pattern and ecosystem service value changes: Implications for environmental sustainability planning for the rapidly urbanizing summer capital of the Philippines. *Landsc. Urban Plan.* **2013**, *116*, 60–72. [CrossRef]
13. Estoque, R.C.; Murayama, Y. Monitoring surface urban heat island formation in a tropical mountain city using Landsat data (1987–2015). *ISPRS J. Photogramm. Remote Sens.* **2017**, *133*, 18–29. [CrossRef]
14. Santamouris, M. Cooling the cities—A review of reflective and green roof mitigation technologies to fight heat island and improve comfort in urban environments. *Sol. Energy* **2014**, *103*, 682–703. [CrossRef]
15. Luber, G.; McGeehin, M. Climate Change and Extreme Heat Events. *Am. J. Prev. Med.* **2008**, *35*, 429–435. [CrossRef]
16. Athukorala, D.; Murayama, Y. Urban heat island formation in Greater Cairo: Spatio-temporal analysis of daytime and nighttime land surface temperatures along the urban-rural gradient. *Remote Sens.* **2021**, *13*, 1396. [CrossRef]
17. Estoque, R.C.; Murayama, Y.; Myint, S.W. Effects of landscape composition and pattern on land surface temperature: An urban heat island study in the megacities of Southeast Asia. *Sci. Total Environ.* **2017**, *577*, 349–359. [CrossRef]
18. Athukorala, D.; Murayama, Y. Spatial Variation of Land Use / Cover Composition and Impact on Surface Urban Heat Island in a Tropical Sub-Saharan City of Accra, Ghana. *Sustainability* **2020**, *12*, 7953. [CrossRef]
19. Stewart, I.D.; Oke, T.R. Local climate zones for urban temperature studies. *Bull. Am. Meteorol. Soc.* **2012**, *93*, 1879–1900. [CrossRef]
20. Stewart, I.D.; Oke, T.R.; Krayenhoff, E.S. Evaluation of the "local climate zone" scheme using temperature observations and model simulations. *Int. J. Climatol.* **2014**, *34*, 1062–1080. [CrossRef]
21. Emery, J.; Pohl, B.; Crétat, J.; Richard, Y.; Pergaud, J.; Rega, M.; Zito, S.; Dudek, J.; Vairet, T.; Joly, D.; et al. How local climate zones influence urban air temperature: Measurements by bicycle in Dijon, France. *Urban Clim.* **2021**, *40*, 101017. [CrossRef]
22. Leconte, F.; Bouyer, J.; Claverie, R.; Pétrissans, M. Using Local Climate Zone scheme for UHI assessment: Evaluation of the method using mobile measurements. *Build. Environ.* **2015**, *83*, 39–49. [CrossRef]
23. Ibrahim, G.R.F. Urban land use land cover changes and their effect on land surface temperature: Case study using Dohuk City in the Kurdistan Region of Iraq. *Climate* **2017**, *5*, 13. [CrossRef]
24. Bokaie, M.; Zarkesh, M.K.; Arasteh, P.D.; Hosseini, A. Assessment of Urban Heat Island based on the relationship between land surface temperature and Land Use/ Land Cover in Tehran. *Sustain. Cities Soc.* **2016**, *23*, 94–104. [CrossRef]
25. Derdouri, A.; Wang, R.; Murayama, Y.; Osaragi, T. Understanding the links between lulc changes and suhi in cities: Insights from two-decadal studies (2001–2020). *Remote Sens.* **2021**, *13*, 3654. [CrossRef]
26. Jiang, Y.; Fu, P.; Weng, Q. Assessing the impacts of urbanization-associated land use/cover change on land surface temperature and surface moisture: A case study in the midwestern united states. *Remote Sens.* **2015**, *7*, 4880–4898. [CrossRef]
27. Yan, Z.; Zhou, D.; Li, Y.; Zhang, L. An integrated assessment on the warming effects of urbanization and agriculture in highly developed urban agglomerations of China. *Sci. Total Environ.* **2022**, *804*, 150119. [CrossRef]
28. Park, J.; Kim, J.H.; Lee, D.K.; Park, C.Y.; Jeong, S.G. The influence of small green space type and structure at the street level on urban heat island mitigation. *Urban For. Urban Green.* **2017**, *21*, 203–212. [CrossRef]
29. Qin, Y. Urban canyon albedo and its implication on the use of reflective cool pavements. *Energy Build.* **2015**, *96*, 86–94. [CrossRef]
30. Qin, Y.; Hiller, J.E. Understanding pavement-surface energy balance and its implications on cool pavement development. *Energy Build.* **2014**, *85*, 389–399. [CrossRef]
31. Gunawardena, K.R.; Wells, M.J.; Kershaw, T. Utilising green and bluespace to mitigate urban heat island intensity. *Sci. Total Environ.* **2017**, *584*, 1040–1055. [CrossRef] [PubMed]
32. Hou, H.; Estoque, R.C. Detecting Cooling Effect of Landscape from Composition and Configuration: An Urban Heat Island Study on Hangzhou. *Urban For. Urban Green.* **2020**, *53*, 126719. [CrossRef]
33. Zhang, X.; Estoque, R.C.; Murayama, Y. An urban heat island study in Nanchang City, China based on land surface temperature and social-ecological variables. *Sustain. Cities Soc.* **2017**, *32*, 557–568. [CrossRef]
34. Zheng, Y.; Li, Y.; Hou, H.; Murayama, Y.; Wang, R.; Hu, T. Quantifying the cooling effect and scale of large inner-city lakes based on landscape patterns: A case study of Hangzhou and Nanjing. *Remote Sens.* **2021**, *13*, 1526. [CrossRef]
35. Du, T.M.J.; Cilliers, S.S.; Dallimer, M.; Goddard, M.; Guenat, S.; Cornelius, S.F. Urban green infrastructure and ecosystem services in sub-Saharan Africa. *Landsc. Urban Plan.* **2018**, *180*, 249–261. [CrossRef]
36. Mohan, M.; Kandya, A. Impact of urbanization and land-use/land-cover change on diurnal temperature range: A case study of tropical urban airshed of India using remote sensing data. *Sci. Total Environ.* **2015**, *506*, 453–465. [CrossRef]
37. Fischer, E.M.; Oleson, K.W.; Lawrence, D.M. Contrasting urban and rural heat stress responses to climate change. *Geophys. Res. Lett.* **2012**, *39*, 1–8. [CrossRef]
38. Mushore, T.D.; Odindi, J.; Dube, T.; Mutanga, O. Prediction of future urban surface temperatures using medium resolution satellite data in Harare metropolitan city, Zimbabwe. *Build. Environ.* **2017**, *122*, 397–410. [CrossRef]

39. Kolokotroni, M.; Ren, X.; Davies, M.; Mavrogianni, A. London's urban heat island: Impact on current and future energy consumption in office buildings. *Energy Build.* **2012**, *47*, 302–311. [CrossRef]
40. Santamouris, M. Cooling the buildings—Past, present and future. *Energy Build.* **2016**, *128*, 617–638. [CrossRef]
41. Singh, P.; Kikon, N.; Verma, P. Impact of land use change and urbanization on urban heat island in Lucknow city, Central India. A remote sensing based estimate. *Sustain. Cities Soc.* **2017**, *32*, 100–114. [CrossRef]
42. Wong, N.H.; Yu, C. Study of green areas and urban heat island in a tropical city. *Habitat Int.* **2005**, *29*, 547–558. [CrossRef]
43. Manatsa, D.; Chingombe, W.; Matarira, C.H. The impact of the positive Indian Ocean dipole on Zimbabwe droughts Tropical climate is understood to be dominated by. *Int. J. Climatol.* **2008**, *2029*, 2011–2029. [CrossRef]
44. Tran, H.; Uchihama, D.; Ochi, S.; Yasuoka, Y. Assessment with satellite data of the urban heat island effects in Asian mega cities. *Int. J. Appl. Earth Obs. Geoinf.* **2006**, *8*, 34–48. [CrossRef]
45. Liu, C.; Li, Y. Spatio-temporal features of urban heat island and its relationship with land use/cover in mountainous city: A case study in Chongqing. *Sustainability* **2018**, *10*, 1943. [CrossRef]
46. Mishra, B.; Sandifer, J.; Gyawali, B.R. Urban Heat Island in Kathmandu, Nepal: Evaluating Relationship between NDVI and *LST* from 2000 to 2018. *Int. J. Environ.* **2013**, *1*, 9–19. [CrossRef]
47. Sarif, M.O.; Rimal, B.; Stork, N.E. Assessment of changes in land use/land cover and land surface temperatures and their impact on surface Urban heat Island phenomena in the Kathmandu Valley (1988–2018). *ISPRS Int. J. Geo-Inform.* **2020**, *9*, 726. [CrossRef]
48. Aryal, A.; Shakya, B.M.; Maharjan, M.; Talchabhadel, R.; Thapa, B.R. Evaluation of the Land Surface Temperature using Satellite Images in Kathmandu Valley. *Nepal J. Civ. Eng.* **2021**, *1*, 1–10. [CrossRef]
49. Maharjan, M.; Aryal, A.; Man Shakya, B.; Talchabhadel, R.; Thapa, B.R.; Kumar, S. Evaluation of Urban Heat Island (UHI) Using Satellite Images in Densely Populated Cities of South Asia. *Earth* **2021**, *2*, 6. [CrossRef]
50. UN HABITAT. *For a Better Urban Cities and Climate Change Initiative: Kathmandu Valley, Nepal*; United Nations Human Settlements Programme: Nairobi, Kenya, 2015.
51. Karki, R.; Talchabhadel, R.; Aalto, J.; Baidya, S.K. New climatic classification of Nepal. *Theor. Appl. Climatol.* **2016**, *125*, 799–808. [CrossRef]
52. Thapa, R.B.; Murayama, Y. Urban growth modeling of Kathmandu metropolitan region, Nepal. *Comput. Environ. Urban Syst.* **2011**, *35*, 25–34. [CrossRef]
53. Educational Software—Maps, Learn to Read and More. Available online: http://www.yourchildlearns.com/ (accessed on 11 December 2021).
54. DIVA-GIS. Available online: https://www.diva-gis.org/ (accessed on 11 December 2021).
55. EarthExplorer. Available online: https://earthexplorer.usgs.gov/ (accessed on 11 December 2021).
56. Thapa, R.B.; Murayama, Y. Examining Spatiotemporal Urbanization Patterns in Kathmandu Valley, Nepal: Remote Sensing and Spatial Metrics Approaches. *Remote Sens.* **2009**, *1*, 534–556. [CrossRef]
57. The R Project for Statistical Computing. Available online: https://www.r-project.org/ (accessed on 11 December 2021).
58. Rodriguez-Galiano, V.F.; Ghimire, B.; Rogan, J.; Chica-Olmo, M.; Rigol-Sanchez, J.P. An assessment of the effectiveness of a random forest classifier for land-cover classification. *ISPRS J. Photogramm. Remote Sens.* **2012**, *67*, 93–104. [CrossRef]
59. Belgiu, M.; Drăgu, L. Random forest in remote sensing: A review of applications and future directions. *ISPRS J. Photogramm. Remote Sens.* **2016**, *114*, 24–31. [CrossRef]
60. Pal, M. Random forest classifier for remote sensing classification. *Int. J. Remote Sens.* **2005**, *26*, 217–222. [CrossRef]
61. Yu, K.; Chen, Y.; Wang, D.; Chen, Z.; Gong, A.; Li, J. Study of the seasonal effect of building shadows on urban land surface temperatures based on remote sensing data. *Remote Sens.* **2019**, *11*, 497. [CrossRef]
62. Weng, Q.; Lu, D.; Schubring, J. Estimation of land surface temperature-vegetation abundance relationship for urban heat island studies. *Remote Sens. Environ.* **2004**, *89*, 467–483. [CrossRef]
63. Shahfahad; Talukdar, S.; Rihan, M.; Hang, H.T.; Bhaskaran, S.; Rahman, A. Modelling urban heat island (UHI) and thermal field variation and their relationship with land use indices over Delhi and Mumbai metro cities. *Environ. Dev. Sustain.* **2021**, *24*, 3762–3790. [CrossRef]
64. Sun, Q.; Wu, Z.; Tan, J. The relationship between land surface temperature and land use/land cover in Guangzhou, China. *Environ. Earth Sci.* **2012**, *65*, 1687–1694. [CrossRef]
65. Du, J.; Xiang, X.; Zhao, B.; Zhou, H. Impact of urban expansion on land surface temperature in Fuzhou, China using Landsat imagery. *Sustain. Cities Soc.* **2020**, *61*, 102346. [CrossRef]
66. Zhou, W.; Huang, G.; Cadenasso, M.L. Does spatial configuration matter? Understanding the effects of land cover pattern on land surface temperature in urban landscapes. *Landsc. Urban Plan.* **2011**, *102*, 54–63. [CrossRef]
67. Asgarian, A.; Amiri, B.J.; Sakieh, Y. Assessing the effect of green cover spatial patterns on urban land surface temperature using landscape metrics approach. *Urban Ecosyst.* **2015**, *18*, 209–222. [CrossRef]
68. McGarigal, K.; Cushman, S.A.; Ene, E. FRAGSTATS v4: Spatial Pattern Analysis Program for Categorical and Continuous Maps. Computer Software Program Produced by the Authors at the University of Massachusetts Amherst. 2012. Available online: http://www.umass.edu/landeco/research/fragstats/fragstats.html (accessed on 11 June 2022).
69. Shrestha, K.C.S.; Ninsawat, S.; Chonwattana, S. Predicting flood events in Kathmandu Metropolitan City under climate change and urbanisation. *J. Environ. Manag.* **2021**, *281*, 111894. [CrossRef]

70. Lamichhane, S.; Shakya, N.M. Shallow aquifer groundwater dynamics due to land use/cover change in highly urbanized basin: The case of Kathmandu Valley. *J. Hydrol. Reg. Stud.* **2020**, *30*, 100707. [CrossRef]
71. Thapa, R.B.; Murayama, Y.; Ale, S. Kathmandu. *Cities* **2008**, *25*, 45–57. [CrossRef]
72. Thapa, R.B.; Murayama, Y. Drivers of urban growth in the Kathmandu valley, Nepal: Examining the efficacy of the analytic hierarchy process. *Appl. Geogr.* **2010**, *30*, 70–83. [CrossRef]
73. Luan, W.; Li, X. Rapid urbanization and its driving mechanism in the Pan-Third Pole region. *Sci. Total Environ.* **2021**, *750*, 141270. [CrossRef]
74. Lamichhane, S.; Shakya, N.M. Alteration of groundwater recharge areas due to land use/cover change in Kathmandu Valley, Nepal. *J. Hydrol. Reg. Stud.* **2019**, *26*, 100635. [CrossRef]
75. Mitchell, M.; Roca Iglesias, A. Urban agriculture in Kathmandu as a catalyst for the civic inclusion of migrants and the making of a greener city. *Front. Arch. Res.* **2019**, *9*, 169–190. [CrossRef]
76. Haack, B.N.; Rafter, A. Urban growth analysis and modeling in the Kathmandu Valley, Nepal. *Habitat Int.* **2006**, *30*, 1056–1065. [CrossRef]
77. Thapa, R.B.; Murayama, Y. Scenario based urban growth allocation in Kathmandu Valley, Nepal. *Landsc. Urban Plan.* **2012**, *105*, 140–148. [CrossRef]
78. Prajapati, R.; Upadhyay, S.; Talchabhadel, R.; Thapa, B.R.; Ertis, B.; Silwal, P.; Davids, J.C. Investigating the nexus of groundwater levels, rainfall and land-use in the Kathmandu Valley, Nepal. *Groundw. Sustain. Dev.* **2021**, *14*, 100584. [CrossRef]
79. Dahal, A.; Khanal, R.; Mishra, B.K. Identification of critical location for enhancing groundwater recharge in Kathmandu Valley, Nepal. *Groundw. Sustain. Dev.* **2019**, *9*, 100253. [CrossRef]
80. Zhang, H.; Qi, Z.-F.; Ye, X.-Y.; Cai, Y.-B.; Ma, W.-C.; Chen, M.-N. Analysis of land use/land cover change, population shift, and their effects on spatiotemporal patterns of urban heat islands in metropolitan Shanghai, China. *Appl. Geogr.* **2013**, *44*, 121–133. [CrossRef]
81. Oke, T.R. City size and the urban heat island. *Atmos. Env.* **1973**, *7*, 769–779. [CrossRef]
82. Yeon-Hee, K.; Jong-Jin, B. Spatial and Temporal Structure of the Urban Heat Island in Seoul. *J. Appl. Meteorol.* **2005**, *44*, 591–605.
83. Xiao, R.-B.; Ouyang, Z.-Y.; Zheng, H.; Li, W.-F.; Schienke, E.W.; Wang, X.-K. Spatial pattern of impervious surfaces and their impacts on land surface temperature in Beijing, China. *J. Environ. Sci.* **2007**, *19*, 250–256. [CrossRef]
84. Myint, S.W.; Wentz, E.A.; Brazel, A.J.; Quattrochi, D.A. The impact of distinct anthropogenic and vegetation features on urban warming. *Landsc. Ecol.* **2013**, *28*, 959–978. [CrossRef]
85. Zhou, W.; Wang, J.; Cadenasso, M.L. Effects of the spatial configuration of trees on urban heat mitigation: A comparative study. *Remote Sens. Environ.* **2017**, *195*, 1–12. [CrossRef]
86. EPA (US Environmental Protection Agency). *Reducing Urban Heat Islands: Com- Pendium of Strategies*; US Environmental Protection Agency: Washington, DC, USA, 2008.
87. Pérez, G.; Coma, J.; Sol, S.; Cabeza, L.F. Green facade for energy savings in buildings: The influence of leaf area index and facade orientation on the shadow effect. *Appl. Energy* **2017**, *187*, 424–437. [CrossRef]
88. Morakinyo, T.E.; Kong, L.; Lau, K.K.L.; Yuan, C.; Ng, E. A study on the impact of shadow-cast and tree species on in-canyon and neighborhood's thermal comfort. *Build. Environ.* **2017**, *115*, 1–17. [CrossRef]
89. Fu, P.; Weng, Q. A time series analysis of urbanization induced land use and land cover change and its impact on land surface temperature with Landsat imagery. *Remote Sens. Environ.* **2016**, *175*, 205–214. [CrossRef]
90. Li, X.; Zhou, W.; Ouyang, Z.; Xu, W.; Zheng, H. Spatial pattern of greenspace affects land surface temperature: Evidence from the heavily urbanized Beijing metropolitan area, China. *Landsc. Ecol.* **2012**, *27*, 887–898. [CrossRef]
91. Maimaitiyiming, M.; Ghulam, A.; Tiyip, T.; Pla, F.; Latorre-Carmona, P.; Halik, Ü.; Sawut, M.; Caetano, M. Effects of green space spatial pattern on land surface temperature: Implications for sustainable urban planning and climate change adaptation. *ISPRS J. Photogramm. Remote Sens.* **2014**, *89*, 59–66. [CrossRef]
92. Besir, A.B.; Cuce, E. Green roofs and facades: A comprehensive review. *Renew. Sustain. Energy Rev.* **2018**, *82*, 915–939. [CrossRef]
93. Aflaki, A.; Mirnezhad, M.; Ghaffarianhoseini, A.; Ghaffarianhoseini, A.; Omrany, H.; Wang, Z.H.; Akbari, H. Urban heat island mitigation strategies: A state-of-the-art review on Kuala Lumpur, Singapore and Hong Kong. *Cities* **2017**, *62*, 131–145. [CrossRef]
94. Wong, N.H.; Tan, A.Y.K.; Tan, P.Y.; Wong, N.C. Energy simulation of vertical greenery systems. *Energy Build.* **2009**, *41*, 1401–1408. [CrossRef]
95. Bustami, R.A.; Belusko, M.; Ward, J.; Beecham, S. Vertical greenery systems: A systematic review of research trends. *Build. Environ.* **2018**, *146*, 226–237. [CrossRef]
96. Wong, N.H.; Jusuf, S.K.; Syafii, N.I.; Chen, Y.; Hajadi, N.; Sathyanarayanan, H.; Manickavasagam, Y.V. Evaluation of the impact of the surrounding urban morphology on building energy consumption. *Sol. Energy* **2011**, *85*, 57–71. [CrossRef]
97. Millstein, D.; Menon, S. Regional climate consequences of large-scale cool roof and photovoltaic array deployment. *Environ. Res. Lett.* **2011**, *6*, 034001. [CrossRef]
98. Dimond, K.; Webb, A. Sustainable roof selection: Environmental and contextual factors to be considered in choosing a vegetated roof or rooftop solar photovoltaic system. *Sustain. Cities Soc.* **2017**, *35*, 241–249. [CrossRef]
99. Kılkış, Ş.; Krajačić, G.; Duić, N.; Montorsi, L.; Wang, Q.; Rosen, M.A.; Ahmad Al-Nimr, M. Research frontiers in sustainable development of energy, water and environment systems in a time of climate crisis. *Energy Convers. Manag.* **2019**, *199*, 1–21. [CrossRef]

100. Bowler, D.E.; Buyung-Ali, L.; Knight, T.M.; Pullin, A.S. Urban greening to cool towns and cities: A systematic review of the empirical evidence. *Landsc. Urban Plan.* **2010**, *97*, 147–155. [CrossRef]
101. Weng, Q. Thermal infrared remote sensing for urban climate and environmental studies: Methods, applications, and trends. *ISPRS J. Photogramm. Remote Sens.* **2009**, *64*, 335–344. [CrossRef]

remote sensing MDPI

Article

Spatial Effects of Landscape Patterns of Urban Patches with Different Vegetation Fractions on Urban Thermal Environment

Yu Zhang [1], Yuchen Wang [2,*] and Nan Ding [1]

1 School of Geography, Geomatics and Planning, Jiangsu Normal University, Xuzhou 221116, China
2 School of Management Science and Engineering, Xuzhou University of Technology, Xuzhou 221018, China
* Correspondence: wangyuchen@cumt.edu.cn

Abstract: The landscape patterns of urban green spaces have been proven to be important factors that affect urban thermal environments. However, the spatial effect of the landscape patterns of urban patches with different vegetation densities on land surface temperature (LST) has not been investigated in detail. In this study, the built-up area of Xuzhou City was taken as the study region, and the four phases of Landsat 8 images and their corresponding ground observations from 2014 to 2020 were selected as the basic data. Normalized spectral mixture analysis and an improved mono-window algorithm were used to invert the vegetation component fraction (VF) and LST maps of the study area, respectively, and the surface patches were classified into five levels according to the VF values, from low to high. Four landscape-level and five class-level metrics were then selected to represent the landscape characteristics of each VF-level patch. The tested values of 60 and 780 m were regarded as the best grain size and spatial extent, respectively, in the calculation of all landscape metrics of ALL VF-level patches (VFLM) using the moving-window method. The results of bivariate Moran's I for VFLM and LST showed the following: (1) for landscape-level metrics, only the Shannon diversity index and patch diversity have substantial negative spatial correlations with LST (with average |Moran's I| < 0.2), indicating that the types of VF levels and the number of patches exert weak negative effects on the thermal environment for a certain area; (2) for class-level metrics such as percentage of landscape, patch cohesion index, largest patch index, landscape shape index, and aggregation index, only the class-level metrics of sub-high VF (LV4) and extreme-high (LV5) VF levels patches have significant negative spatial correlations with LST (with high Moran's I value, and high–high and low–high distributions in local indications of spatial association cluster maps), indicating that only the patches of high VF levels can effectively alleviate LST and that patch proportion, natural connectivity degree, predominance degree, shape complexity, and aggregation degree are important landscape factors for regulating the thermal environment. Principal component analysis and multiple linear regression were applied to determine the impact weights of the class-level VFLMs of LV4 and LV5 patches on LST, which revealed the contributions of these landscape metrics to mitigating the urban heat island effect (UHI). These results signify the importance of and differences in the spatial patterns of various VF-level patches for UHI regulation; these patterns can provide new perspectives and references for urban green space planning and climate management.

Keywords: urban vegetation; urban heat island effect; landscape patterns; spatial correlation; Landsat 8

Citation: Zhang, Y.; Wang, Y.; Ding, N. Spatial Effects of Landscape Patterns of Urban Patches with Different Vegetation Fractions on Urban Thermal Environment. *Remote Sens.* **2022**, *14*, 5684. https://doi.org/10.3390/rs14225684

Academic Editors: Yuji Murayama and Ruci Wang

Received: 30 September 2022
Accepted: 5 November 2022
Published: 10 November 2022

1. Introduction

In the process of urbanization, many natural surfaces are replaced by dense sidewalks, buildings, and other surfaces that absorb and retain heat; urbanization not only changes the ground thermal radiation characteristics but also generates a large amount of anthropogenic heat, resulting in a condition wherein the atmospheric and surface temperatures in urban areas are higher than those in the surrounding suburban natural and agricultural areas. This phenomenon is known as the urban heat island effect (UHI), which is also a form of air pollution and a major cause of global warming. UHI expansion not only deteriorates regional thermal environments [1] but also leads to increased energy consumption, thermal risk, air pollution, and pollution-related mortality [2], which seriously reduce the quality of living environments for urban residents [3]. The urban thermal environment has become a research hotspot in the field of urban ecology, environment, and climate, and methods that can mitigate the risks and negative consequences of UHI are urgently needed.

The development of open green spaces, also known as "green infrastructure" and "green space cold islands" [4], has become a widely accepted strategy for mitigating UHI. Vegetation can reduce temperature and increase humidity by transpiration and shielding against long- and short-wave radiation [5,6]. As important parts of urban ecosystems, urban forests and greening can effectively improve urban substrates and play a major role in regulating the temperature and mitigating UHI [7]. Green spaces containing a large number of vegetation canopies can cool the atmosphere and air by creating a shading effect [8]. Interconnected shadow areas called "Shadeways" can pass through the city along infrastructure corridors and roads, provide relatively high natural and artificial shadows, and align cool areas with active travel spaces [9,10]. According to the vegetation species and environmental conditions in urban areas, shadow areas have a cooling effect of between 2.3 °C and 2.5 °C [11].

The cooling effect of green spaces on UHI can be analyzed by evaluating the statistical relationship between land surface temperature (LST) and regional landscape patterns or vegetation indices [12]. Some vegetation indices, such as the normalized difference vegetation index (NDVI) and fractional vegetation coverage (FVC), have been proven to exhibit negative correlations with LST [13,14]. Owing to urban land restrictions, it is impossible to mitigate UHI only by increasing urban vegetation. In addition to expanding the number and scale of urban forests (in a limited manner), maximizing the eco-efficiency of urban vegetation is a fundamental strategy to mitigate the thermal environment. Therefore, optimizing the landscape patterns of urban vegetation has become an effective option for alleviating UHI. Landscape patterns describe the landscape features of an ecosystem, such as its spatial arrangement and configuration. Spatial composition represents the proportions of all types of land cover within a certain spatial unit, whereas spatial configuration refers to their physical arrangement and distribution [15]. Spatial composition and configuration can be quantified using various landscape metrics that describe different spatial characteristics [3]. Some reports have described the coverage, patch characteristics, and spatial distribution of urban vegetation using landscape metrics and then discussed the relationship between urban vegetation and UHI. The results show that the size, shape, and spatial distribution of vegetation patches have a significant impact on urban thermal environments and that increasing the area and density of vegetation patches can effectively reduce LST [16,17]. However, most studies on the relationship between urban vegetation landscape patterns and UHI consider urban forests and open green spaces as a single type of vegetation, and few studies have considered the differences in the regulation effects of patches with different vegetation densities on UHI.

In this study, Landsat 8 and ground observation data were used to extract information on the urban vegetation component fraction (VF) and LST, respectively, and the urban surface considered was divided into patches with different vegetation densities according to VF values. The LST and landscape metrics of these patches extracted using the moving-window method were then analyzed via geospatial and statistical approaches. The results revealed the spatial effect of the landscape pattern of urban patches with different VFs on the urban thermal environment. The detailed technical process was shown in Figure 1.

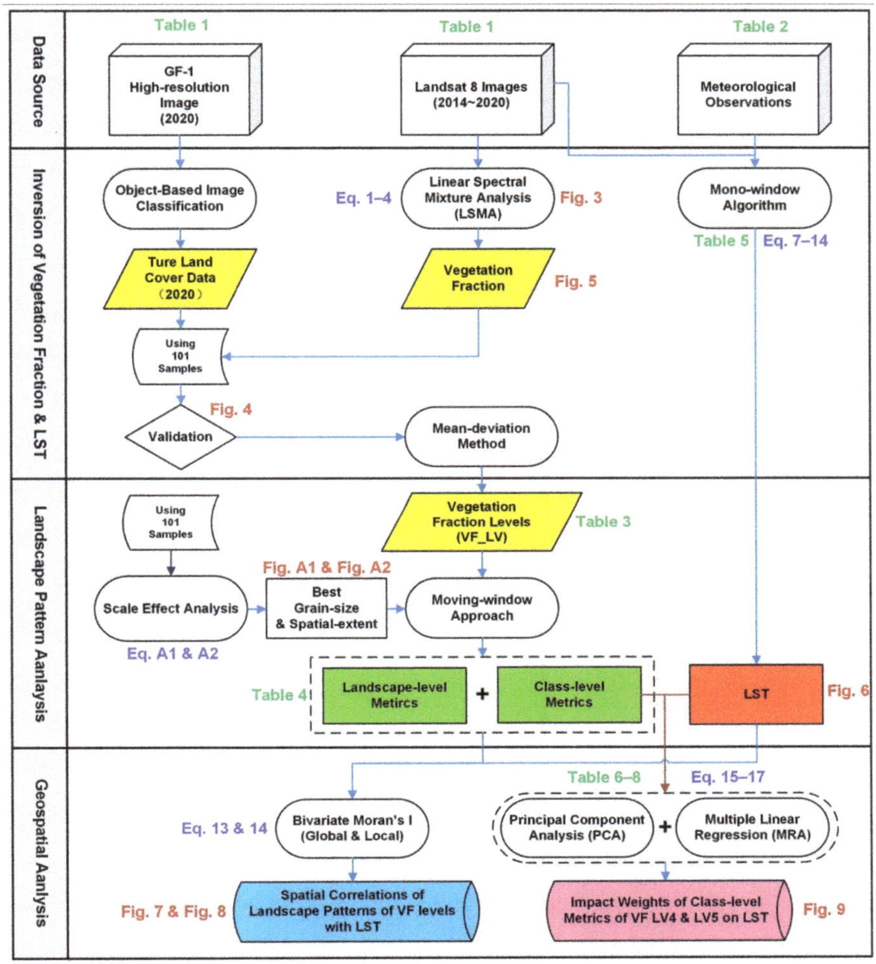

Figure 1. Technical roadmap of this study.

2. Materials and Methods

2.1. Study Area

The built-up area of Xuzhou, Jiangsu Province, was selected as the study location (Figure 2). Xuzhou is located southeast of the North China Plain (between 116°22′–118°40′E and 33°43′–34°58′N) and has a temperate monsoon climate. Xuzhou is an important transportation hub, an important coal-producing area in China, and an electric power base in eastern China. Xuzhou also has a large-scale manufacturing industry. However, the vegetation coverage rate of Xuzhou is as high as 32%, of which the urban vegetation coverage rate is 27.8%, making this city rank first in Jiangsu Province. The developed transportation and energy sectors, industries, and a large area of impervious surfaces have caused the built-up areas of Xuzhou to experience significant UHI. Simultaneously, the large area of urban vegetation has also played an important role in regulating the urban thermal environment. These characteristics of the study area make it suitable for exploring the spatial relationship between landscape patterns of patches with different vegetation densities and the urban thermal environment.

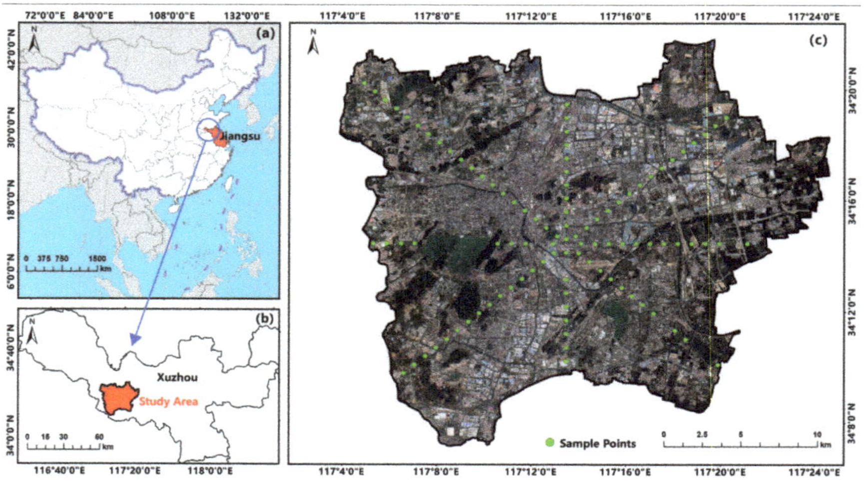

Figure 2. Location and satellite image of the study area. (**a**) Map of Jiangsu Province in China; (**b**) map of the study area in Xuzhou; and (**c**) GF-1 satellite image of the study area on 28 April 2020.

As shown in Figure 2c, a total of 101 sample points were selected in the study for the scale effect analysis and pixel unmixing validation in subsequent sections. At an interval of 900 m, the sample points were distributed on four sample lines from the city center to the suburb along four directions (east–west, south–north, southeast–northwest, and southwest–northeast). This distribution mode provides stable variation in the vegetation component ratio of the pixels along each sample line, owing to the sample pixels along each direction gradually changing from the suburb mainly covered by vegetation to the urban built-up area mainly covered by impervious surface, which ensures the diversity of the component ratios of the sample pixels.

2.2. Data Descriptions

2.2.1. Satellite Data

The remote sensing data used in this study included four Landsat 8 images from May 2014 to 2020 and one GF-1 image from April 2020. Detailed information is shown in Table 1. Visible and near-infrared bands 1–7 of the Landsat 8 OLI sensor were primarily used to extract mixed pixel endmember fractions and calculate other land surface parameters. Thermal infrared band 10 of the TIRS sensor was primarily used to invert LST. The spatial resolution of the fusion of the panchromatic and multispectral bands of the GF-1 image was 2 m; this resolution could be used to extract high-precision surface coverage data to verify the accuracy of the mixed pixel decomposition. As there are no high-resolution satellite data that completely coincide with the transit dates of the four phases of Landsat 8 images, the acquisition date of the GF-1 image used in this study was 28 April 2020, which is only 19 days from the acquisition date of the corresponding Landsat 8 image. Therefore, it is acceptable to use this GF-1 scene as the validation data.

2.2.2. Ground Observations

The necessary meteorological observations for the remote sensing inversion of LST included near-surface temperature and air relative humidity, as shown in Table 2. All the ground observations from 2014 to 2018 were obtained from the Collaborative Observation Test Site of the China University of Mining and Technology in the study area (data acquisition frequency: 30 min), and the meteorological data for 2020 were obtained from the NOAA National Centers for Environmental Information of USA [18] (data acquisition frequency: 1 h).

Table 1. Technical parameters of Landsat 8 and GF-1 data.

Satellite	Sensor Resolution	Scene ID	Acquisition Date	Acquisition Time (GMT + 8)
Landsat 8	OLI: 30 m TIRS: 100 m	LC81210362014121LGN00	1 May 2014	10:42:29
		LC81220362017136LGN00	16 May 2017	10:48:22
		LC81220362018123LGN00	3 May 2018	10:48:04
		LC81210362020138LGN00	17 May 2020	10:42:10
GF-1	PAN: 2 m MSS: 8 m	GF1_PMS1_E117.2_N34.1_20200428_L1A0004767917	28 April 2020	11:14:21
		GF1_PMS1_E117.3_N34.4_20200428_L1A0004767915	28 April 2020	11:14:17

Table 2. Ground meteorological observation data at satellite transit time.

Date	Acquisition Time (GMT + 8)	Air Temperature (K)	Air Relative Humidity (%)
1 May 2014	11:00:00	297.42	55.12
16 May 2017	11:00:00	296.33	39.76
3 May 2018	11:00:00	294.96	48.00
17 May 2020	11:00:00	299.48	53.19

2.3. *Extraction of Vegetation Fraction from Mixed Pixels*

2.3.1. Selection of Spectral Mixture Analysis Model

The pixels of low- and medium-resolution remote sensing images of urban areas are often mixed with endmembers, such as impervious surfaces, vegetation, and bare soil (water bodies generally exist independently). The commonly used methods for extracting the proportion of each endmember in a mixed pixel are divided into two categories: nonlinear and linear models. The theoretical premise of the linear spectral mixture analysis model (LSMA) is that the spectral signal of a pixel is a linear combination of the spectral signals of each endmember; the endmember fractions can be inverted using the least squares method based on the spectral characteristics of each endmember, simplifying the relationship between the components in the mixed pixel [19]. The fully constrained linear spectral mixture analysis model (FCLS), which is based on LSMA, adds two constraints: (1) the sum of all endmember fractions in each pixel is 100%; (2) all endmember fractions are non-negative [20]. The greater the number of endmembers, the more detailed the spectral analysis, which is beneficial to the fitting accuracy of FCLS. However, too many endmembers will increase the sensitivity of the decomposition results to endmember errors [21–23]. Based on a field investigation in the study area, the mixed pixel endmembers were divided into five types: forest, grassland (including grass, shrubs, and farmland), bare soil, and high- and low-albedo impervious surfaces (IPS). In addition, to solve the problems associated with brightness variation and shade, the reflectance values of all bands were normalized before applying FCLS [24]; the algorithms used were as follows.

$$\overline{R_b} = \frac{R_b}{\frac{1}{N}\sum_{b=1}^{N} \times R_b} \times 100 \tag{1}$$

$$\overline{R_b} = \sum_{i=1}^{N} f_i \overline{R_{i,b}} + e_b \tag{2}$$

$$\sum_{i=1}^{N} f_i = 1,\ \ f_i \geq 0 \tag{3}$$

where $\overline{R_b}$ is the normalized reflectance value of the mixed pixel for band b, R_b is the original reflectance of the mixed pixel for band b, N is the number of endmembers, f_i is the fraction of endmember i in the mixed pixel, $\overline{R_{i,b}}$ is the reflectance of endmember i for band b in the mixed pixel, and e_b is the model calculation residual for band b. The final fitting accuracy of FCLS is represented by the RMS calculated using e_b, and the calculation formula is as follows [22].

$$RMS = (\sum_{b=1}^{N} e_b/N)^{0.5} \quad (4)$$

2.3.2. Extraction of Endmember Fraction

In ENVI 5.3, the minimum noise fraction rotation (MNF) was used to calculate the band eigenvalues to select the main bands by reducing the data dimensions (Figure 3a). Next, the pure pixel index was applied to identify and extract pure pixels in the multispectral data. The filtered pure pixels were used as the region of interest of the first four bands of the MNF image to generate a 3D scatter plot in the ENVI n-D visualizer. The scatter plot was rotated, and the clustered scatter clouds (endmembers) were marked (Figure 3b). Finally, the classified endmembers (Figure 3c) were substituted into FCLS to extract the component fractions.

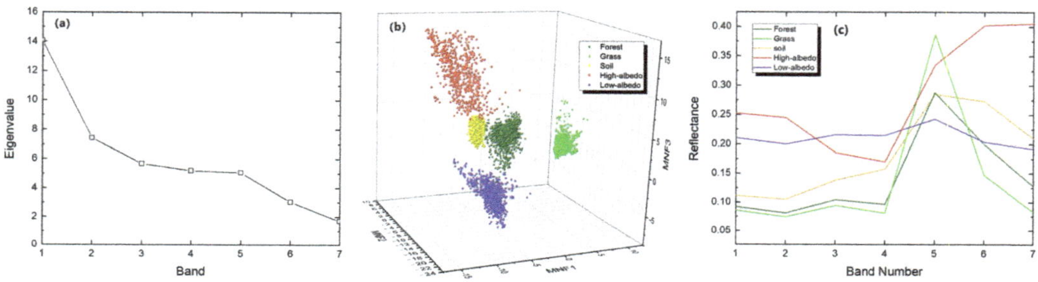

Figure 3. Endmember selection (data of 28 April 2020). (**a**) Eigenvalues of all minimum noise fraction rotation (MNF) bands; (**b**) 3D scatter plot of each endmember in MNF bands 1–3; and (**c**) response curve of all endmembers at each wavelength.

2.3.3. Validation of Endmember Fraction

A total of 101 sample points were generated in the study area (Figure 2c), and 101 validation sample areas of 90 × 90 m^2 (3 × 3 pixels) were established with each sample point as the center. All sample areas were superimposed onto the mixed pixel decomposition image and GF1 image of 2020 to extract the endmember fraction inversion value and high-resolution interpretation value (true value) of each sample area. The endmembers of vegetation (forest and grassland) and IPS (high- and low-albedo) exhibited high correlation coefficients (r > 0.89) and linear goodness of fit (R^2 > 0.80) between the endmember fraction inversion value and the true value (Figure 4), indicating that the FCLS result had a high accuracy. As the four Landsat 8 images are all from May and the principles of endmember selection are completely consistent, it can be considered that the FCLS results of the four-phase data can meet the accuracy requirements.

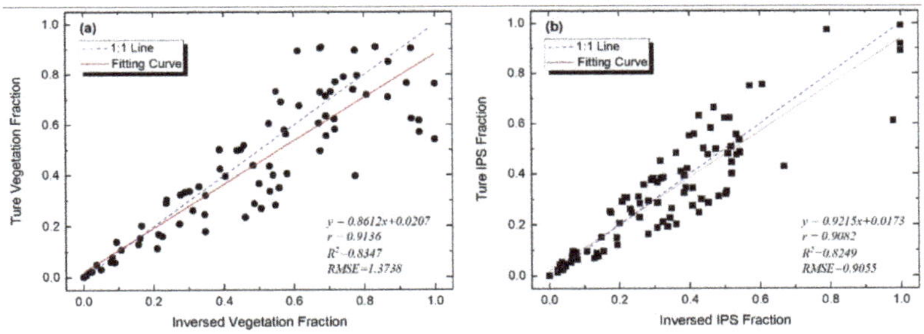

Figure 4. Validation of endmember fraction inversion results. (**a**) Accuracy of inversed vegetation fraction; (**b**) accuracy of inversed impervious surface fraction.

2.4. Analysis of Vegetation Fraction Levels Using Landscape Metrics

2.4.1. Definitions of Vegetation Fraction Levels

The mean-deviation method (Table 3) was used to classify the VFs into five levels: extreme-high VF (Level 5), sub-high VF (Level 4), medium VF (Level 3), sub-low VF (Level 2), and extreme-low VF (Level 1), which represent patches with various vegetation densities.

Table 3. Vegetation fraction levels.

VF Levels	Interval
Extreme-high VF (Level 5, LV5)	$VF > VF_{\_mean} + std$
Sub-high VF (Level 4, LV4)	$VF_{\_mean} + 0.5std < VF \leq VF_{\_mean} + std$
Medium VF (Level 3, LV3)	$VF_{\_mean} - 0.5std < VF \leq VF_{\_mean} + 0.5std$
Sub-low VF (Level 2, LV2)	$VF_{\_mean} - std < VF \leq VF_{\_mean} - 0.5std$
Extreme-low VF (Level 1, LV1)	$VF < VF_{\_mean} - std$

2.4.2. Selection of Landscape Metrics

Landscape metrics are effective tools for quantifying the spatial characteristics of various land-cover types [3]. Class-level and landscape-level metrics were used to analyze the number and spatial distribution characteristics of VF-level patches (Table 4). Five class-level metrics—percentage of landscape (PLAND), largest patch index (LPI), landscape shape index (LSI), aggregation index (AI), and patch cohesion index (COHESION)—were selected to represent the proportion, predominance, shape complexity, aggregation, and natural connectivity of patches at each VF level. Four landscape-level metrics—the Shannon diversity index (SHDI), Shannon evenness index (SHEI), patch diversity (PD), and contagion index (CONTAG)—were selected to represent the diversity, distribution, number, and aggregation degrees of all patches within an analysis unit.

2.4.3. Scale Effect Analysis

Scale effects, including the effects of grain size (spatial resolution) and spatial extent [25], are important factors that affect the results of landscape pattern analyses [26]. Landscape structure information changes with variations in grain size, and some landscape features can only be found at a specific grain size [27]. For the data of 17 May 2020, the VF-level images were resampled from 30 to 200 m at 10 m intervals to obtain the corresponding raster images with 18 spatial resolutions, and the class-level metrics for each grain size were calculated using FRAGSTATS 4.2. As shown in Figure A1a–e in Appendix B, most inflection points in the response curves appear at a grain size of 60 m. Figure A1f in Appendix B describes the land-area information-loss index calculated using Equations (A1) and (A2) in Appendix A for each grain size, indicating that the land area accuracy loss is small when the grain size is 60 m. Therefore, 60 m was selected as the best grain size for landscape pattern analysis.

Table 4. Formulas and descriptions of landscape metrics selected in this study.

Metrics		Equation	Description	Parameter Explanation
Class level	PLAND	$\left(\sum_{j=1}^{n} a_{ij}/A\right) \times 100$	Porportion of the patch type	a_{ij}—the area of patch ij; A—total landscape area; e_{ik}—total length (m) of edge in landscape between patch types (classes) i and k; g_{ii}—number of adjacent patches of the same landscape type i; p_{ij}—perimeter of patch ij in terms of number of cell surfaces; Z—total number of cells in the landscape.
	LPI	$\left[\max\left(a_{ij}\right)/A\right] \times 100$	Porportion of the largest patch type	
	LSI	$0.25 \sum_{j=1}^{n} e_{ik}/\sqrt{A}$	Shape complexity degree of the patch type	
	AI	$\left[\frac{g_{ii}}{max \to g_{ii}}\right] \times 100$	Aggregation degree of the patch type	
	COHESION	$\left\{1 - \left[\sum_{i=1}^{m}\sum_{k=1}^{m} p_{ij} / \sum_{i=1}^{m}\sum_{k=1}^{m}\left(p_{ij} \times \sqrt{a_{ij}}\right)\right]\right\} \times \left(1 - 1/\sqrt{Z}\right)^{-1} \times 100$	Natural connectivity degree between patches	
Landscape level	SHDI	$-\sum_{i=1}^{n}[p_i \times ln(p_i)]$	Diversity indicator of all patch types	p_i—proportion of the landscape occupied by patch type (class) i; m—number of patch types (classes) present in the landscape, excluding the landscape border if present; N—number of patches in the landscape of patch type (class) i; A—total landscape area.
	SHEI	$-\sum_{i=1}^{n}[p_i \times \ln(p_i)]/ln(m)$	Even distribution indicator of all patch types	
	PD	N/A	Number of patches per unit area	
	CONTAG	$\left\{1 + \left[\sum_{i=1}^{m}\sum_{k=1}^{m}\left(p_i \times \frac{g_{ik}}{\sum_{j=1}^{n} g_{ij}}\right) \times ln\left(p_i \times g_{ik}/\sum_{j=1}^{n} g_{ij}\right)\right]/[2ln(m)]\right\} \times 100$	Aggregation degree of all patch types	

The moving-window method [28,29] was used to quantify the landscape patterns of VF-level patches. Using this method, landscape metrics within a spatial extent were assigned to the center pixel for each window movement, and finally, the mapped distributions of the landscape metrics were generated. Owing to the scale dependence of the landscape metrics [30], the size of the moving window used in the calculation may affect the results [28]. When the spatial extent is smaller, the landscape index changes significantly and fluctuates greatly and hence cannot effectively reflect the gradient characteristics of the landscape patterns. When the spatial extent is larger, the landscape metrics change gently, eliminating the interference caused by high-resolution and regional differences but also leading to the loss of some gradient features. Six window sizes ranging from 300 × 300 to 1500 × 1500 m were tested at 240 m intervals. To avoid pixel splitting, the interval for the moving-window sizes was set to multiples of 60 m (the best grain size). Four landscape-level metrics were calculated for each tested window size, and then all metrics were extracted for the 101 sample points (Figure 2c) to record the metric changes under different moving-window sizes. Figure A2 shows that the curve of variation in landscape metrics tends to be stable when the spatial extent reaches 780 m. Therefore, 780 m was selected as the best spatial extent.

2.5. Landscape Surface Temperature Inversion

Landsat 8 has two thermal infrared bands: 10 and 11. However, owing to the uncertainty of band 11 information, the USGS recommends using single-channel algorithms based on band 10 for LST inversion [31,32]. The improved mono-window model [33] was used to determine the inversion parameters for the Landsat 8 sensor, as shown in Equations (5)–(7).

$$LST = \{a(1 - C - D) + [b(1 - C - D) + C + D]T_{10} - DT_{air\_e}\}/C \quad (5)$$

$$C = \varepsilon\tau \tag{6}$$

$$D = (1-\tau)[1+(1-\varepsilon)\tau] \tag{7}$$

where $a = -62.7182$ and $b = 0.4339$ are the linear regression coefficients of the Planck blackbody radiation function for band 10 (0–70 °C), T_{10} is the brightness temperature calculated using band 10, T_{air_e} is the effective atmospheric average temperature, ε is the surface emissivity, and τ is the atmospheric transmissivity.

$$T_{10} = K_2/\ln(1+K_1/L_{10}) \tag{8}$$

$$\varepsilon_{nature} = P_v R_v \varepsilon_v + (1-P_v) R_s \varepsilon_s + d_\varepsilon \tag{9}$$

$$\varepsilon_{construction} = P_v R_v \varepsilon_v + (1-P_v) R_m \varepsilon_m + d_\varepsilon \tag{10}$$

where $K_1 = 774.89$ and $K_2 = 1321.08$, are the thermal conversion constants for Landsat 8 band 10; L_{10} is the thermal spectral radiance calculated from the pixel DN value of band 10 [33]; R_v, R_s, and R_m are the radiation ratios of vegetation, bare soil, and construction land, respectively [34]; $\varepsilon_v = 0.986$, $\varepsilon_s = 0.970$, and $\varepsilon_m = 0.970$, are the emissivity of pure vegetation, bare soil, and construction land, respectively; P_v is the fractional vegetation coverage; and d_ε is the effect of the interaction between vegetation and bare soil on the surface emissivity [34].

Table 5 shows the linear relationship between the mid-latitude summer effective mean atmospheric temperature (T_{air_e}) and near-surface air temperature (T_{air}) [34]. In addition, the regression functions between atmospheric transmissivity (τ) and the atmospheric water content (w) for Landsat 8 TIR band 10 were also simulated using the Moderate Resolution Atmospheric Transmission (MODTRAN 4) program [33]. Since the atmospheric water content data of the study area was not directly obtained, according to the algorithm obtained by fitting the historical observation data of the China Meteorological Stations, water content can be estimated by the average altitude of the study area (H_E = 40 m), the latitude of the study area (φ = 34.24°), and the relative humidity (RH), which have been shown as Equations (11) and (12) [35].

$$w = 0.03exp\left(-1.39{H_E}^2 + 2.74H_E + 0.15\right)e + \frac{0.066}{(\varphi-33)^2+4.41} + 0.17 \tag{11}$$

$$e = 0.6112exp\left(\frac{17.67T_{air}}{T_{air}+243.5}\right) \times RH \tag{12}$$

Table 5. Regression functions for T_{air_e} and τ estimation.

w (g·cm^{-2})	τ Functions	T_{air_e} Function
0.2–1.6	0.9184–0.0725w	
1.6–4.4	1.0163–0.1330w	16.0110 + 0.9262T_{air}
4.4–5.4	0.7029–0.0620w	

2.6. Spatial Correlation Analysis

Spatial autocorrelation analysis was used to determine whether the distribution of spatial variables was related to the adjacent variables. It can be divided into global and local spatial autocorrelations. Bivariate Moran's I [36], proposed by Anselin, can effectively reflect the correlation and dependence characteristics of the spatial distribution of two variables. OpenGeoDa software was used for bivariate spatial autocorrelation analysis. Bivariate global Moran's I was used to analyze the spatial response of LST to landscape metrics of VF-level patches; it was calculated using Equation (13).

$$I = \frac{\sum_{i=1}^{n}\sum_{j=1}^{n} W_{ij}(x_i-\overline{x})(y_j-\overline{y})}{S^2\sum_{i}^{n}\sum_{j=1}^{n} W_{ij}} \tag{13}$$

where I is the bivariate global spatial autocorrelation index, that is, the overall correlation between the spatial distribution of variables x and y; N is the total number of spatial units; W_{ij} is the spatial weight matrix established by the K adjacency method; x_i and y_j are the observed values of the independent and dependent variables in spatial units i and j, respectively; and S^2 is the variance of all samples.

Bivariate local Moran's I was calculated as follows [37]:

$$I_i = Z_i \sum_{j=1}^{n} W_{ij} Z_j \tag{14}$$

where I_i is the local spatial relationship between the independent and dependent variables of spatial unit i, and z_i and z_j are the standardized variance values of the observed values of spatial units i and j. Based on I_i, there are four clustering modes in the Local Indications of Spatial Association (LISA) distribution map, which can intuitively present the agglomeration and differentiation characteristics of independent and dependent variables in local areas. The clustering mode can be classified as high–high (HH) clustering, that is, the independent variable of the spatial unit i and the dependent variable of the adjacent unit j are both high; low–low (LL) clustering, that is, the independent variable of the spatial unit i and the dependent variable values of the adjacent unit j are both low; low–high (LH) aggregation, that is, the independent variable value of spatial unit i is low and the dependent variable value of the adjacent unit j is high; and high–low (HL) aggregation, that is, the independent variable value of the spatial unit i is high and the dependent variable value of the adjacent unit j is low.

2.7. Impact Weight Calculation Based on Principal Component Analysis

To further explore the contribution of the class-level metrics negatively correlated with LST in urban thermal environment mitigation, the impact weight of each metric was calculated by combining the principal component analysis (PCA) and multiple linear regression. The standardized coefficients in multiple linear regression are often used to represent the influence weight of independent variables on dependent variables. However, to prevent multicollinearity between class-level VFLMs, the multiple linear regression method cannot be used directly. In this study, PCA was used to extract the principal component variable F_i [38] for the class-level VFLMs, and standardized regression coefficients (β_i) were extracted by multiple linear regression between F_i and LST. This procedure removes the multicollinearity among the metrics, and the weighted impact (W_j) of each VFLM on LST can be obtained using the eigenvalues (λ_i), PCA component matrix (θ_{ij}), and β_i. The equations for impact weight calculation are shown as follows.

$$\alpha_{ij} = \frac{\theta_{ij}}{\sqrt{\lambda_i}} \tag{15}$$

$$F_i = \sum_{j=1}^{n} \alpha_{ij} Xj \tag{16}$$

$$W_j = \left| \alpha_{1j} \times \beta_1 + \alpha_{2j} \times \beta_2 \right| \tag{17}$$

where Xj is the original variable; n is the total number of the original variables; a_{ij} is the contribution coefficient of original e variable in corresponding principal component variable F_i.

3. Results

3.1. Inversion Results for VF and LST

Figure 5 shows the VF maps of the study area from 2014 to 2020, which indicate the spatial distribution and area ratio of the vegetation patches. LSMA could not only extract large-area and high-proportion vegetation patches in the study area but also fully extract the fragmented and low-proportion vegetation distributions in mixed pixels, such as small green belts near buildings and roads. Complete vegetation quantity information is beneficial for accurately exploring the effects of patches with different VFs on the thermal environment. Figure 6 shows the LST maps of the study area from 2014 to 2020, revealing that the LST values of the areas primarily covered by urban IPS were significantly higher than those of the area covered by vegetation. The superposition of the LST and VF maps indicates that the LST of patches with different VFs varies greatly. The VF and LST values of 101 sample points (Figure 2c) were extracted for correlation and linear regression analyses, as shown in Figure A3 of Appendix B. The Pearson correlation coefficients between VF and LST for the four phases were −0.6923, −0.6226, −0.6820, and −0.6573, indicating a strong negative correlation between them. The linear regression results also showed that there was a significant negative linear correlation between VF and LST ($p < 0.001$) and that for every 0.1 increase in VF, LST decreased by approximately 0.48–0.78 °C. Therefore, a high VF exerts a significant alleviating effect on LST, and it is necessary to further explore the spatial impact of patches with various vegetation densities on the urban thermal environment.

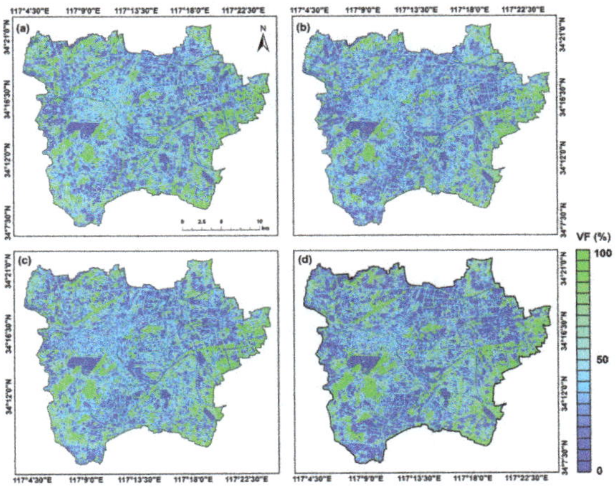

Figure 5. Vegetation fraction inversion maps of (**a**) 1 May 2014, (**b**) 16 May 2017, (**c**) 3 May 2018, and (**d**) 17 May 2020.

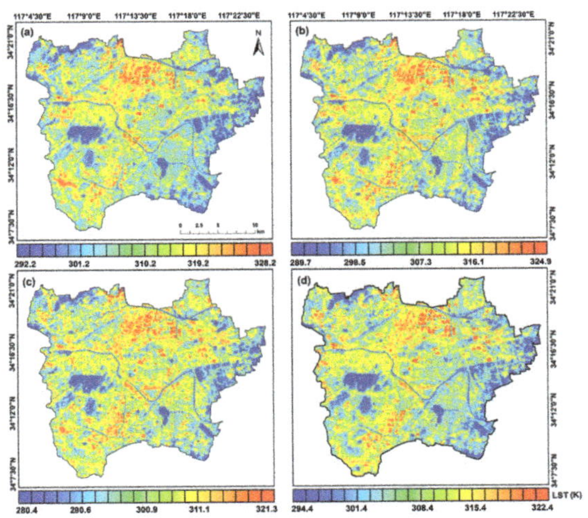

Figure 6. Land surface temperature inversion maps of (**a**) 1 May 2014, (**b**) 16 May 2017, (**c**) 3 May 2018, and (**d**) 17 May 2020.

3.2. Bivariate Moran's I between Landscape Metrics of VF-Level Patches and LST

Figure 7 shows the bivariate global Moran's I between the landscape metrics of VF-level patches (hereinafter referred to as VFLM) and LST. The spatial correlations of VFLM and LST varied widely for different landscape metrics. For the same landscape metric, the spatial correlation between VFLM and LST also varied greatly from VF LV1 to LV5. The landscape-level metrics of SHDI and PD exhibited significant negative spatial correlations with LST (average Moran's I was −0.17 and −0.13, respectively), indicating that the region contained more types of VF levels and that patch numbers have weak effects on regional cooling. Both positive and negative spatial correlations were observed between LST and SHEI, and CONTAG, indicating that the even distribution degree and aggregation degree of the five types of VF levels have no stable influence on regional temperature. Therefore, the impact of landscape-level VFLM on LST is limited.

Class-level VFLMs such as PLAND, LPI, LSI, AI, and COHESION showed a significant spatial correlation with LST ($p < 0.001$), implying that the spatial distribution characteristics of VF-level patches, such as patch proportion, predominance degree, shape complexity degree, aggregation degree, and natural connectivity degree, are important factors for regulating environmental temperature. In addition, the bivariate global Moran's I values between VFLM and LST were positive and gradually increased from VF LV3 to LV1. In contrast, bivariate global Moran's I values between VFLM and LST were negative and gradually increased from VF LV4 to LV5. Therefore, the patch VF can effectively alleviate ambient temperature only when it reaches sub- and extreme-high levels, and the higher the VF level, the stronger the mitigation effect.

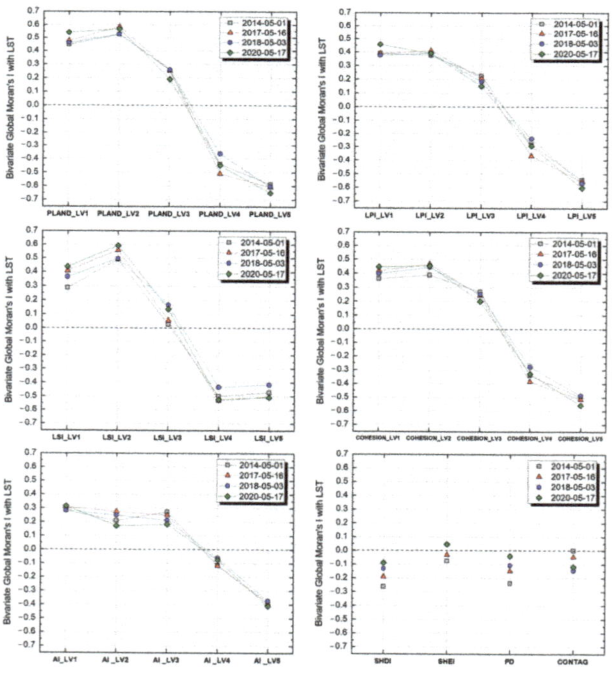

Figure 7. Bivariate global Moran's I between landscape metrics of IPSD levels and LST (the significance levels of all Moran's I were at $p < 0.001$ except for CONTAG of 1 May 2014 with $p > 0.05$).

The average values of bivariate global Moran's I between the VFLM of LV4 and LV5 patches and LST in the four phases were, from high to low, in the order of PLAND (−0.53), LPI (−0.43), LSI (−0.49), COHESION (−0.42), and AI (−0.24), indicating that patch proportion and aggregation degree have the strongest and weakest spatial correlations with LST, respectively. However, Moran's I value cannot directly represent the contributions of these five class-level VFLMs to LST mitigation, which is necessary for further discussion of their impact.

Bivariate local Moran's I was used to further reveal the spatial characteristics of the landscape patterns of VF LV4 and LV5 patches regulating the thermal environment. The LISA maps (Figure 8) show the spatial distribution characteristics of the spatial correlation between VFLM (class level) and LST. The clustering patterns between VFLM (PLAND, LPI, LSI, AI, and COHESION) and LST were primarily HL and LH. The LH areas were primarily distributed in the urban center and mostly covered by IPS. The HL areas were primarily distributed in urban forests, parks, and suburban farmlands and were primarily covered by natural vegetation and farmland.

Figure 8. Local Indications of Spatial Association (LISA) cluster maps between landscape metrics of the vegetation fraction (VFLM; LV4 and LV5) and land surface temperature (LST): (**a**) 1 May 2014; (**b**) 16 May 2017; (**c**) 3 May 2018; (**d**) 17 May 2020 (HH: high VFLM and high LST; LL: low VFLM and low LST; LH: low VFLM and high LST; HL: high VFLM and low LST).

3.3. Impact Weight of Class-Level VFLMs of LV4 and LV5 on LST

The class-level VFLMs of LV4 and LV5 patches have been shown to have a significant negative spatial correlation with LST; however, the contribution rate of each VFLM in mitigating the thermal environment remains unclear. Therefore, it is necessary to discuss the impact weights of PLAND, LPI, LSI, COHESION, and AI of LV4 and LV5 patches in LST regulation.

The five class-level VFLMs of LV4 and LV5 patches were normalized before PCA. As listed in Table 6, the values of the Kaiser–Meyer–Olkin (KMO) measure of sampling adequacy for the four phases were all greater than 0.6, indicating that the PCA method is feasible for analyzing these variables. Two principal components (F_1 and F_2) were extracted from each phase, with a high proportion of cumulative variance (>77%).

Table 7 lists the loadings of the original variables in the principal components (F_1 and F_2). A common feature is that the metrics of LV5 (X1–X5) and LV4 (X6–X10) are the core parameters affecting the principal components F_1 and F_2, respectively (with high θ_i values; values with $\theta_i < 0.3$ have been rejected). Therefore, F_1 and F_2 can be used as new variables to represent the class-level VFLMs of LV5 and LV4 patches, respectively. Equation (15) was used to calculate the contribution coefficient (α_{ij}) of each original variable based on eigenvalues (λ_i) and θ_i, and F_1 and F_2 were calculated using Equation (16).

Table 6. Total variance explained from principal component analysis (PCA).

Date	KMO	Sums of Squared Loadings	F_1 ①	F_2 ①
1 May 2014	0.7190	Eigenvalue (λ_i)	4.02	3.74
		Cumulative percent (%)	77.56	
16 May 2017	0.7109	Eigenvalue (λ_i)	4.00	3.71
		Cumulative percent (%)	77.14	
3 May 2018	0.7096	Eigenvalue (λ_i)	3.89	3.84
		Cumulative percent (%)	77.32	
17 May 2020	0.6855	Eigenvalue (λ_i)	3.89	3.88
		Cumulative percent (%)	77.74	

① F_1 and F_2 are components 1 and 2 extracted using PCA, respectively.

Table 7. Component matrix of principle component analysis ①.

Normalized Original Variables (Class-Level VFLMs)	1 May 2014		16 May 2017		3 May 2018		17 May 2020	
	F_1 (θ_1)	F_2 (θ_2)	F_1 (θ_1)	F_2 (θ_1)	F_1 (θ_1)	F_2 (θ_2)	F_1 (θ_1)	F_2 (θ_2)
(X1) PLAND_LV5	0.914	/	0.910	/	0.914	/	0.921	/
(X2) COHESION_LV5	0.914	/	0.911	/	0.909	/	0.914	/
(X3) LPI_LV5	0.883	/	0.880	/	0.885	/	0.891	/
(X4) AI_LV5	0.827	/	0.838	/	0.826	/	0.823	/
(X5) LSI_LV5	0.627	0.572	0.611	0.586	0.596		0.579	/
(X6) PLAND_LV4	0.341	0.875	0.372	0.868	0.300	0.896	/	0.908
(X7) COHESION_LV4	/	0.923	/	0.924	/	0.924	/	0.935
(X8) LPI_LV4	/	0.873	/	0.883	/	0.880	/	0.883
(X9) AI_LV4	/	0.733	/	0.705	/	0.733	/	0.724
(X10) LSI_LV4	0.555	0.632	0.531	0.624	0.513	0.631	0.519	0.642

① Rotation method: Varimax with Kaiser normalization.

Multiple linear regressions between normalized LST (LST_N) and F_1 and F_2 (Table 8) of the four phases showed significant negative linear correlations between them ($p < 0.001$). The absolute values of the standardized coefficients (β_1) of F1 were all larger than those of F2 (β_2), indicating that the class-level VFLM of LV5 had a stronger regulatory effect on LST than that of LV4. Finally, the impact weights (W_j) of all class-level VFLMs were calculated using Equation (17), and the results are shown in Figure 9.

The heat map in Figure 9 again shows that the class-level VFLM of LV5 has a higher impact weight on LST than that of LV4 and that PLAND_LV5 (X1) and COHESION_LV5 (X2) had the highest impact weights among all metrics, indicating that patch proportion and natural connectivity degree between LV5 patches are the two most important factors affecting LST, followed by LPI_LV5 (X3), AI_LV5 (X4), and LSI_LV5 (X5). In contrast to LV5, the impact weights of the class-level VFLM of LV4 were, from high to low, in the order of LSI_LV4 (X10), PLAND_LV4 (X6), COHESION_LV4 (X7), LPI_LV4 (X8), and AI_LV4 (X9). In urban areas, the land-cover type of the LV5 patch is almost entirely vegetation; however, the LV4 patch is still mixed with more land-cover types, which is the main reason for the difference in the orders of impact weights between the VFLMs of LV4 and LV5.

Table 8. Results of multiple linear regression between dependent variable LST_N and independent variables F_1 and F_2.

Regression Coefficients		1 May 2014	16 May 2017	3 May 2018	17 May 2020
r		0.569 ***	0.612 ***	0.553 ***	0.644 ***
R^2		0.323 ***	0.375 ***	0.306 ***	0.415 ***
Regression Constant		0.949 ***	0.578 ***	0.693 ***	0.635 ***
Unstandardized Coefficients	F_1	−0.010 ***	−0.110 ***	−0.083 ***	−0.109 ***
	F_2	−0.004 ***	−0.066 ***	−0.014 ***	−0.054 ***
Standardized Coefficients (β_i)	β_1	−0.474 ***	−0.463 ***	−0.516 ***	−0.538 ***
	β_2	−0.139 ***	−0.207 ***	−0.060 ***	−0.179 ***

*** Significant: $p < 0.001$.

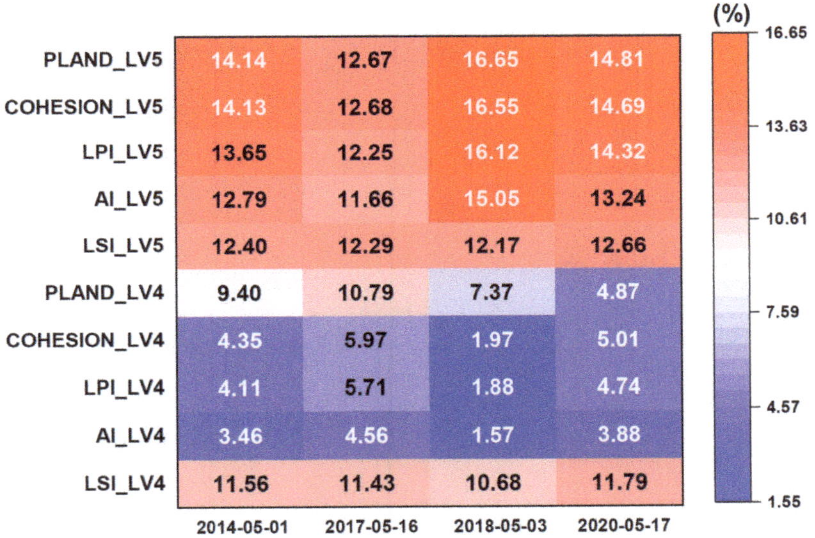

Figure 9. Heatmap of the impact weights of vegetation-fraction landscape metrics on land surface temperature.

4. Discussion

In previous studies on the relationship between green space and LST based on mid- or low-resolution satellite images, vegetation was generally regarded as one or several kinds of homogeneous patches (such as grassland, farmland, and forest) in landscape pattern analysis [12,15,39], without considering the spatial impact of vegetation density on the thermal environment. Since vegetation density has been proven to have significant negative correlations with LST [40,41], an innovative method was explored to incorporate vegetation density into landscape pattern analysis in this study. According to the fraction value of vegetation component from low to high, the land surface of the study area was classified into five levels (or five patch types), which is similar to the land-cover types. A same index is used to classify the urban surface with high heterogeneity, which is conducive to revealing the difference in the impacts of the spatial patterns of patches with various vegetation densities on UHI regulation.

The results showed that there were differences in the impacts of landscape-level and class-level metrics on LST. Particularly, the effects of the landscape-level SHEI and CONTAG on LST were unstable (both positive and negative effects exist). This is mainly because landscape-level metrics represent the regional spatial characteristics of five vegetation density levels, which includes both the mitigation effects of high-vegetation-density patches on the thermal environment and the enhancement impacts of low-vegetation-density patches (with a higher density of impervious surfaces) on the surface temperature. Therefore, the impacts of the even distribution degree and aggregation degree of the five vegetation levels on LST depend on which level is dominant in the region. This phenomenon is consistent with the previous research. On the premise that the land surfaces are divided into grassland, forest, farmland, impervious water, and other types, the correlations of LST with landscape-level metrics such as AI, LPI, CONTAG, SHDI, and PD are unstable (with variable correlation directions and significances) [3,42].

The absolute value of the spatial correlation between class-level metrics and LST (|Moran's I|) gradually decreases from LV1 to LV3 and conversely increases gradually from LV4 to LV5. This phenomenon is also owing to the variation in vegetation density among different patches. From LV1 to LV3, the patches' spatial patterns were positively correlated with LST, because these patches were primarily covered by IPS and because the vegetation cooling effect could not offset the heating effect caused by the IPS thermal radiation and anthropogenic heat. However, as the vegetation density gradually increased and the impervious surface density gradually decreased, resulting in a decreasing trend of |Moran's I|. From LV4 to LV5, the spatial pattern of patches was negatively correlated with LST, because these patches were dominated by vegetation cover and because their cooling effect was sufficient to control the ambient temperature. Furthermore, owing to the gradual increase in vegetation density, |Moran's I| showed an increasing trend. In addition, the land-cover types of the LV5 patches were almost entirely vegetation. However, the LV4 patches were still mixed with more land-cover types. This is an important reason for the difference in the orders of impact weights between the VFLMs of LV4 and LV5. In previous studies, for the patches of forest, grassland, and farmland, their class-level metrics of PLAND, LSI, and AI were found to be negatively correlated with LST, and the correlation between PLAND and LST was far stronger than AI [42], which is also consistent with our results. According to our research, priority should be given to increasing the area ratio, natural connectivity, and shape complexity degrees of high-density vegetation, which can effectively enhance the ability of green space in regional UHI mitigation.

5. Conclusions

In this study, four periods of Landsat 8 OLI/TIRS images and various geospatial methods (including landscape measurement, spatial correlation, and statistics) were used to examine the spatial relationship between LST and the landscape patterns of patches with various vegetation densities. There was a certain spatial correlation between the landscape metrics of the five VF-level patches and LST. Landscape-level metrics such as SHDI and PD only have significant but weak negative spatial correlations with LST, indicating that the types of VF levels and number of patches have limited impacts on regional temperature. However, all class-level metrics, such as PLAND, COHESION, LPI, LSI, and AI, of extreme- and sub-high-level patches had significant negative spatial correlations with LST, and the correlation degree was high. This indicates that spatial distribution characteristics such as patch proportion, natural connectivity degree, predominance degree, shape complexity, and aggregation degree of areas with high vegetation density are important factors in regulating the thermal environment. In addition, patches with high VF levels have a strong ability to alleviate surface temperature; in particular, PLAND and COHESION of VF LV5 patches have the highest impact weights on LST. These findings reveal the importance of and differences in the spatial patterns of areas with various VFs in urban thermal environment regulation, thus providing new perspectives and references for urban green-space planning and climate management.

Special attention should be paid to the fact that different vegetation types have significant differences in evapotranspiration, shading, canopy heat radiation characteristics, etc., so their spatial patterns will have different impact mechanisms on LST. Further research is required on the distinction of the regulatory effects of the spatial patterns of the vegetation types with different canopy characteristics on LST, with the help of Uninhabited Air Vehicles (UAV) or high-resolution satellites equipped with multi-spectral and thermal infrared sensors.

Author Contributions: Conceptualization, Y.Z. and Y.W.; methodology, Y.Z.; software, N.D.; validation, Y.W.; formal analysis, Y.Z.; data curation, Y.Z.; writing—original draft preparation, Y.Z.; writing—review and editing, Y.W.; visualization, N.D.; supervision, N.D.; funding acquisition, Y.Z. and N.D. All authors have read and agreed to the published version of the manuscript.

Funding: This research was funded by the National Natural Science Foundation of China, Grant Nos. 42101256 and 41904013, and is a project funded by the Priority Academic Program Development of Jiangsu Higher Education Institutions (PAPD) (The Fourth Phase).

Data Availability Statement: Not applicable.

Acknowledgments: The authors would like to thank the United States Geological Survey for supporting Landsat 8 data. Further, we thank China University of Mining and Technology and the NOAA National Centers for Environmental Information of USA for providing meteorological data. The comments and suggestions of the editor and the anonymous reviewers are gratefully acknowledged.

Conflicts of Interest: The authors declare no conflict of interest.

Appendix A

$$L_j = \left(A_j - A_{bj}\right) / A_{bj} \times 100 \tag{A1}$$

$$S_j = \sqrt{\frac{\sum_{j=1}^{n} L_j^2}{n}} \tag{A2}$$

where L_j refers to the relative value of area loss; A_j is the area of VF level j at a converted grain size; A_{bj} is the area of VF level j at the original grain size; n is the number of VF levels, and S_j is the land area loss index.

Appendix B

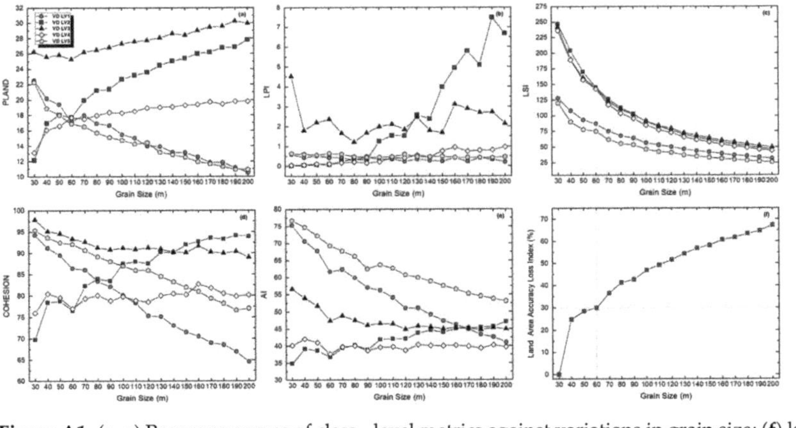

Figure A1. (**a–e**) Response curves of class−level metrics against variations in grain size; (**f**) land area loss index curve corresponding to grain size variation.

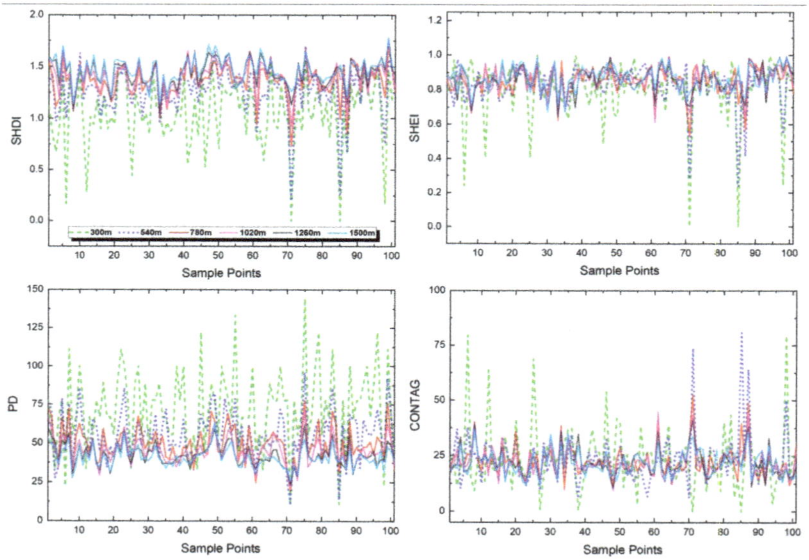

Figure A2. Trends in the variation of landscape−level metrics of 101 sample points for different spatial extents.

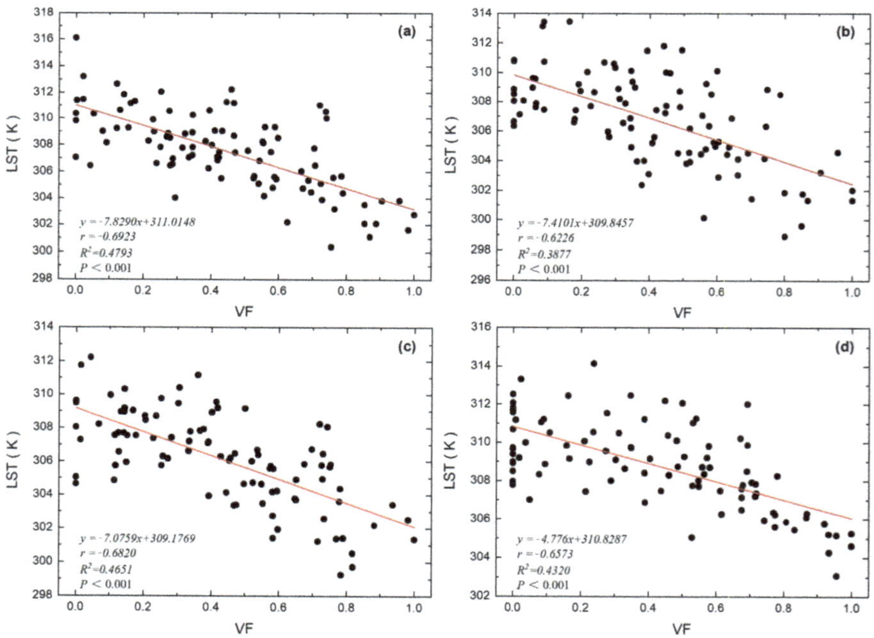

Figure A3. Linear regression between vegetation fraction and land surface temperature of sample points: (**a**) 1 May 2014; (**b**) 16 May 2017; (**c**) 3 May 2018; and (**d**) 17 May 2020.

References

1. He, B.-J.; Wang, J.; Zhu, J.; Qi, J. Beating the Urban Heat: Situation, Background, Impacts and the Way Forward in China. *Renew. Sustain. Energy Rev.* **2022**, *161*, 112350. [CrossRef]
2. Santamouris, M. Recent Progress on Urban Overheating and Heat Island Research. Integrated Assessment of the Energy, Environmental, Vulnerability and Health Impact. Synergies with the Global Climate Change. *Energy Build.* **2020**, *207*, 109482. [CrossRef]
3. Rakoto, P.Y.; Deilami, K.; Hurley, J.; Amati, M.; Sun, Q. (Chayn) Revisiting the Cooling Effects of Urban Greening: Planning Implications of Vegetation Types and Spatial Configuration. *Urban For. Urban Green.* **2021**, *64*, 127266. [CrossRef]
4. Amati, M.; Taylor, L. From Green Belts to Green Infrastructure. *Plan. Pract. Res.* **2010**, *25*, 143–155. [CrossRef]
5. Chen, A.; Yao, X.A.; Sun, R.; Chen, L. Effect of Urban Green Patterns on Surface Urban Cool Islands and Its Seasonal Variations. *Urban For. Urban Green.* **2014**, *13*, 646–654. [CrossRef]
6. Shih, W. Greenspace Patterns and the Mitigation of Land Surface Temperature in Taipei Metropolis. *Habitat Int.* **2017**, *60*, 69–80. [CrossRef]
7. Bowler, D.E.; Buyung-Ali, L.; Knight, T.M.; Pullin, A.S. Urban Greening to Cool Towns and Cities: A Systematic Review of the Empirical Evidence. *Landsc. Urban Plan.* **2010**, *97*, 147–155. [CrossRef]
8. Horváthová, E.; Badura, T.; Duchková, H. The Value of the Shading Function of Urban Trees: A Replacement Cost Approach. *Urban For. Urban Green.* **2021**, *62*, 127166. [CrossRef]
9. Colter, K.R.; Middel, A.C.; Martin, C.A. Effects of Natural and Artificial Shade on Human Thermal Comfort in Residential Neighborhood Parks of Phoenix, Arizona, USA. *Urban For. Urban Green.* **2019**, *44*, 126429. [CrossRef]
10. Deilami, K.; Rudner, J.; Butt, A.; MacLeod, T.; Williams, G.; Romeijn, H.; Amati, M. Allowing Users to Benefit from Tree Shading: Using a Smartphone App to Allow Adaptive Route Planning during Extreme Heat. *Forests* **2020**, *11*, 998. [CrossRef]
11. Hsieh, C.M.; Li, J.J.; Zhang, L.; Schwegler, B. Effects of Tree Shading and Transpiration on Building Cooling Energy Use. *Energy Build.* **2018**, *159*, 382–397. [CrossRef]
12. Wang, X.; Dallimer, M.; Scott, C.E.; Shi, W.; Gao, J. Tree Species Richness and Diversity Predicts the Magnitude of Urban Heat Island Mitigation Effects of Greenspaces. *Sci. Total Environ.* **2021**, *770*, 145211. [CrossRef] [PubMed]
13. Lee, P.S.-H.; Park, J. An Effect of Urban Forest on Urban Thermal Environment in Seoul, South Korea, Based on Landsat Imagery Analysis. *Forests* **2020**, *11*, 630. [CrossRef]
14. Zhang, Y.; Chen, L.L.; Wang, Y.; Chen, L.L.; Yao, F.; Wu, P.; Wang, B.; Li, Y.; Zhou, T.; Zhang, T. Research on the Contribution of Urban Land Surface Moisture to the Alleviation Effect of Urban Land Surface Heat Based on Landsat 8 Data. *Remote Sens.* **2015**, *7*, 10737–10762. [CrossRef]
15. Guo, G.; Wu, Z.; Chen, Y. Complex Mechanisms Linking Land Surface Temperature to Greenspace Spatial Patterns: Evidence from Four Southeastern Chinese Cities. *Sci. Total Environ.* **2019**, *674*, 77–87. [CrossRef]
16. Estoque, R.C.; Murayama, Y.; Myint, S.W. Effects of Landscape Composition and Pattern on Land Surface Temperature: An Urban Heat Island Study in the Megacities of Southeast Asia. *Sci. Total Environ.* **2017**, *577*, 349–359. [CrossRef]
17. Zhou, W.; Cao, F. Effects of Changing Spatial Extent on the Relationship between Urban Forest Patterns and Land Surface Temperature. *Ecol. Indic.* **2020**, *109*, 105778. [CrossRef]
18. NOAA's Satellite and Information Service (NESDIS) NOAA National Centers for Environmental Information. Available online: https://www.ngdc.noaa.gov/ (accessed on 7 September 2022).
19. Li, L.; Canters, F.; Solana, C.; Ma, W.; Chen, L.; Kervyn, M. Discriminating Lava Flows of Different Age within Nyamuragira's Volcanic Field Using Spectral Mixture Analysis. *Int. J. Appl. Earth Obs. Geoinf.* **2015**, *40*, 1–10. [CrossRef]
20. Heinz, D.C.D.C.; Chang, C.I. Fully Constrained Least Squares Linear Spectral Mixture Analysis Method for Material Quantification in Hyperspectral Imagery. *IEEE Trans. Geosci. Remote Sens.* **2001**, *39*, 529–545. [CrossRef]
21. Small, C. Estimation of Urban Vegetation Abundance by Spectral Mixture Analysis. *Int. J. Remote Sens.* **2001**, *22*, 1305–1334. [CrossRef]
22. Wu, C.; Murray, A.T. Estimating Impervious Surface Distribution by Spectral Mixture Analysis. *Remote Sens. Environ.* **2003**, *84*, 493–505. [CrossRef]
23. Ridd, M.K. Exploring a V-I-S (Vegetation-Impervious Surface-Soil) Model for Urban Ecosystem Analysis through Remote Sensing: Comparative Anatomy for Cities. *Int. J. Remote Sens.* **1995**, *16*, 2165–2185. [CrossRef]
24. Wu, C. Normalized Spectral Mixture Analysis for Monitoring Urban Composition Using ETM+ Imagery. *Remote Sens. Environ.* **2004**, *93*, 480–492. [CrossRef]
25. Wu, J. Effects of Changing Scale on Landscape Pattern Analysis: Scaling Relations. *Landsc. Ecol.* **2004**, *19*, 125–138. [CrossRef]
26. Teng, M.; Zeng, L.; Zhou, Z.; Wang, P.; Xiao, W.; Dian, Y. Responses of Landscape Metrics to Altering Grain Size in the Three Gorges Reservoir Landscape in China. *Environ. Earth Sci.* **2016**, *75*, 1055. [CrossRef]
27. Oyana, T.J.; Johnson, S.J.; Wang, G. Landscape Metrics and Change Analysis of a National Wildlife Refuge at Different Spatial Resolutions. *Int. J. Remote Sens.* **2014**, *35*, 3109–3134. [CrossRef]
28. Chefaoui, R.M. Landscape Metrics as Indicators of Coastal Morphology: A Multi-Scale Approach. *Ecol. Indic.* **2014**, *45*, 139–147. [CrossRef]
29. Kong, F.; Nakagoshi, N. Spatial-Temporal Gradient Analysis of Urban Green Spaces in Jinan, China. *Landsc. Urban Plan.* **2006**, *78*, 147–164. [CrossRef]

30. Wu, J.; Shen, W.; Sun, W.; Tueller, P.T. Empirical Patterns of the Effects of Changing Scale on Landscape Metrics. *Landsc. Ecol.* **2002**, *17*, 761–782. [CrossRef]
31. Barsi, J.; Schott, J.; Hook, S.; Raqueno, N.; Markham, B.; Radocinski, R. Landsat-8 Thermal Infrared Sensor (TIRS) Vicarious Radiometric Calibration. *Remote Sens.* **2014**, *6*, 11607–11626. [CrossRef]
32. Montanaro, M.; Lunsford, A.; Tesfaye, Z.; Wenny, B.; Reuter, D. Radiometric Calibration Methodology of the Landsat 8 Thermal Infrared Sensor. *Remote Sens.* **2014**, *6*, 8803–8821. [CrossRef]
33. Wang, F.; Qin, Z.; Song, C.; Tu, L.; Karnieli, A.; Zhao, S. An Improved Mono-Window Algorithm for Land Surface Temperature Retrieval from Landsat 8 Thermal Infrared Sensor Data. *Remote Sens.* **2015**, *7*, 4268–4289. [CrossRef]
34. Qin, Z.; Karnieli, A.; Berliner, P. A Mono-Window Algorithm for Retrieving Land Surface Temperature from Landsat TM Data and Its Application to the Israel-Egypt Border Region. *Int. J. Remote Sens.* **2001**, *22*, 3719–3746. [CrossRef]
35. Zhang, Y.; Li, L.; Chen, L.; Liao, Z.; Wang, Y.; Wang, B.; Yang, X. A Modified Multi-Source Parallel Model for Estimating Urban Surface Evapotranspiration Based on ASTER Thermal Infrared Data. *Remote Sens.* **2017**, *9*, 1029. [CrossRef]
36. Anselin, L. Local Indicators of Spatial Association—LISA. *Geogr. Anal.* **1995**, *27*, 93–115. [CrossRef]
37. Zhang, Y.; Liu, Y.; Zhang, Y.; Liu, Y.; Zhang, G.; Chen, Y. On the Spatial Relationship between Ecosystem Services and Urbanization: A Case Study in Wuhan, China. *Sci. Total Environ.* **2018**, *637–638*, 780–790. [CrossRef]
38. Liu, R.X.; Kuang, J.; Gong, Q.; Hou, X.L. Principal Component Regression Analysis with Spss. *Comput. Methods Programs Biomed.* **2003**, *71*, 141–147. [CrossRef]
39. Dutta, K.; Basu, D.; Agrawal, S. Synergetic Interaction between Spatial Land Cover Dynamics and Expanding Urban Heat Islands. *Environ. Monit. Assess.* **2021**, *193*, 184. [CrossRef]
40. Wetherley, E.B.; McFadden, J.P.; Roberts, D.A. Megacity-Scale Analysis of Urban Vegetation Temperatures. *Remote Sens. Environ.* **2018**, *213*, 18–33. [CrossRef]
41. Zaitunah, A.; Samsuri, S.; Silitonga, A.F.; Syaufina, L. Urban Greening Effect on Land Surface Temperature. *Sensors* **2022**, *22*, 4168. [CrossRef]
42. Zhao, H.; Tan, J.; Ren, Z.; Wang, Z. Spatiotemporal Characteristics of Urban Surface Temperature and Its Relationship with Landscape Metrics and Vegetation Cover in Rapid Urbanization Region. *Complexity* **2020**, *2020*, 1–12. [CrossRef]

remote sensing

MDPI

Article

The Seasonality of Surface Urban Heat Islands across Climates

Panagiotis Sismanidis [1,*], Benjamin Bechtel [1], Mike Perry [2] and Darren Ghent [2]

[1] Institute of Geography, Ruhr-University Bochum, 44801 Bochum, Germany; benjamin.bechtel@rub.de
[2] National Centre for Earth Observation, Department of Physics and Astronomy, University of Leicester, Leicester LE1 7RH, UK; mike.perry@leicester.ac.uk (M.P.); djg20@leicester.ac.uk (D.G.)
* Correspondence: panagiotis.sismanidis@rub.de

Abstract: In this work, we investigate how the seasonal hysteresis of the Surface Urban Heat Island Intensity (SUHII) differs across climates and provide a detailed typology of the daytime and nighttime SUHII hysteresis loops. Instead of the typical tropical/dry/temperate/continental grouping, we describe Earth's climate using the Köppen–Geiger system that empirically maps Earth's biome distribution into 30 climate classes. Our thesis is that aggregating multi-city data without considering the biome of each city results in temporal means that fail to reflect the actual SUHII characteristics. This is because the SUHII is a function of both urban and rural features and the phenology of the rural surroundings can differ considerably between cities, even in the same climate zone. Our investigation covers all the densely populated areas of Earth and uses 18 years (2000–2018) of land surface temperature and land cover data from the European Space Agency's Climate Change Initiative. Our findings show that, in addition to concave-up and -down shapes, the seasonal hysteresis of the SUHII also exhibits twisted, flat, and triangle-like patterns. They also suggest that, in wet climates, the daytime SUHII hysteresis is almost universally concave-up, but they paint a more complex picture for cities in dry climates.

Keywords: surface urban heat island; SUHI; land surface temperature; LST; seasonal hysteresis; MODIS; ESA-CCI; Köppen–Geiger climate zones

Citation: Sismanidis, P.; Bechtel, B.; Perry, M.; Ghent, D. The Seasonality of Surface Urban Heat Islands across Climates. *Remote Sens.* **2022**, *14*, 2318. https://doi.org/10.3390/rs14102318

Academic Editors: Yuji Murayama and Ruci Wang

Received: 3 April 2022
Accepted: 6 May 2022
Published: 11 May 2022

1. Introduction

Cities are generally warmer than their surroundings. This phenomenon is known as the Urban Heat Island (UHI) and is one of the clearest examples of human-induced climate modification [1]. UHIs increase the cooling energy demand, aggravate the feeling of thermal discomfort, and influence air quality [2–6]. As such, they impact the health and welfare of the urban population and increase the carbon footprint of cities [7–9]. The root cause of an UHI is the transformation of the natural landscape to a corrugated, mostly manufactured, and less vegetated surface [10]. The radiative, aerodynamic, thermal, and moisture properties of man-made surfaces are fundamentally different to natural ones, leading to reduced evapotranspiration and the uptake, storage, and release of more heat [1,11]. The relative warmth of the urban atmosphere, surface, and substrate leads to four distinct UHI types that are governed by a different mix of physical processes. These four types are the canopy layer, boundary layer, surface, and subsurface UHI.

Surface UHIs (SUHI) result from modifications of the surface energy balance at urban facets, canyons, and neighborhoods [1,12–14]. They exhibit complex spatial and temporal patterns that are strongly related to land cover and are usually estimated from remotely-sensed Land Surface Temperature (LST) data with a kilometer or sub-kilometer spatial resolution [10,15–17]. These data are retrieved from satellite observations of the surface-emitted thermal infrared radiation and provide a spatially continuous representation of the urban surface at the satellite overpass time. The difference between the urban and rural LST is known as the SUHI Intensity (SUHII) and is controlled primarily by differences in the evapotranspiration and aerodynamic roughness [11,18,19]. The SUHII varies rapidly in space and time as the surface conditions, the weather, and the incoming radiation change, and it is generally strongest during the daytime and summertime [13,20–23]. These variations are driven by changes in both the urban and the rural LST [24,25] and not only by the urban as it is usually assumed. The dependency of the SUHII on the characteristics of the rural reference area makes it inappropriate for urban heat mitigation [24] and complicates the analysis of multi-city SUHII data [25].

The seasonal variation of the SUHII has been extensively studied on local [15,26–32], regional [33–40], and global [13,22,37,41] levels, primarily by analyzing monthly and seasonal means. A global study of 419 cities with more than one million inhabitants estimated that the 2003–2008 mean daytime SUHII is 1.9 K for the summertime and 1.1 K for the wintertime (the corresponding nighttime means are 1.0 K) [22]. A more recent work that examines 9500 cities and a longer time period (2000–2017) reports lower daytime means—1.3 K for the summertime and 0.4 K for the wintertime—and that the majority (87%) of urban areas exhibit positive daytime SUHIIs [23]. It also reports that in warm temperate and snow cities, the SUHII peaks in June–July and is least in November–December, while in arid and equatorial cities, it hardly shows any seasonality. Independent works investigating the seasonal variation of the SUHII in India [35], Europe [33], China [36], and North America [37,38] corroborate most of these findings and show that the SUHII is stronger in temperate and continental cities than in arid and semi-arid. They also show that negative SUHIIs occur primarily in dry areas in summer [23,38].

One of the most interesting findings, however, is that the SUHII, when plotted against the rural LST, exhibits a rate-dependent seasonal hysteresis that strongly depends on local climate conditions [26,29,33,34,42]. This means that the seasonal variation of the SUHII exhibits a looping pattern whose shape is controlled primarily by the local climate and that its magnitude (at any point in time) depends on both present (relative to that timepoint) and recent past effects. This hysteresis is first reported in Zhou et al. [33], where seven distinct and geographically separated types of SUHII hysteresis are identified in Europe. A plausible explanation for the seasonal hysteresis of the SUHII is given in Manoli et al. [42], where it is hypothesized that this behavior is the result of time lags between the surface energy budget of urban and rural areas. Testing this hypothesis using a coarse-grained SUHI model [43], the SUHII hysteresis of wet (London, Milan, and Paris) and dry (Madrid and Nicosia) cities in Europe was replicated and led the authors to conclude that in wet climates, the shape of the daytime hysteresis is controlled by the time lag between solar radiation and the air temperature, whereas in seasonally dry climates, it is controlled by the time lag between solar radiation and the rainfall. The former generates a concave-up loop that is positive throughout the year and peaks in summer, while the latter is a concave-down loop that peaks in spring and is negative during summer and autumn [42].

These findings improve our understanding about the influence of the local climate conditions on the SUHII and it is worth further investigation and testing. The first step toward this direction is to extend the work of Zhou et al. [33] in cities outside of Europe and consistently describe the SUHII seasonal hysteresis in every climate inhabited by humans. This paper aims to address this gap and derive a complete typology of the SUHII hysteresis loops for both the daytime and nighttime. In this regard, we ask: (i) what is the shape of the SUHII hysteresis loops in every densely populated climate; (ii) when is the SUHII strongest and weakest within the year; (iii) what these values are; and (iv) how they relate to the seasonal variation of the precipitation and solar radiation? To answer these questions, we use 18 years (2000–2018) of global LST and land cover data from the European Space Agency's Climate Change Initiative (ESA-CCI). To characterize the climate of each city, we use the Köppen–Geiger classification system [44] that empirically maps Earth's biome distribution (i.e., vegetation zones) in 30 climate classes. The SUHII is a function of both urban and rural features [24,25], and the phenology of the rural surroundings can differ considerably between cities, even within the same climate zone (i.e., tropical, dry, temperate, continental, and polar). Controlling the city biomes is especially important over dry and tropical regions where cities exhibiting positive and negative SUHIIs may coexist (e.g., in

India [35]). In such cases, aggregating multi-city SUHII data can result in temporal means that fail to reflect the actual SUHII characteristics and lead to erroneous conclusions.

Following this introduction, in Section 2, we describe our method for calculating the SUHII and deriving the hysteresis loops. In Sections 3 and 4, we present our results and discuss how they differ between climates, and in Section 5, we provide our conclusions.

2. Materials and Methods

2.1. ESA-CCI MODIS LST

The ESA-CCI project on Land Surface Temperature (LST_cci) provides validated LST products across all land surfaces of the Earth over the past 20 to 25 years. In this work, we use the 0.01° daily 2000–2018 LST_cci Terra MODIS (Moderate Resolution Imaging Spectroradiometer) product (v.1.0) for the focus areas presented in Figure 1. Terra MODIS is a multispectral sun-synchronous satellite instrument that crosses the equator at 10:30 (local solar time) in the descending orbit and 22:30 in the ascending orbit and views almost the entire surface of the Earth every day. The LST_cci MODIS LST are retrieved as a linear combination of the 11.0 μm and 12.0 μm clear-sky brightness temperatures (BT) using a Generalized Split-Window (GSW) algorithm [45]. The GSW coefficients depend on the satellite view angle and the water vapor and are derived by linearly regressing simulated BT and LST for various conditions. The land surface emissivity is used explicitly in the GSW formulation and is obtained from the Baseline Fit Emissivity Database of the Cooperative Institute for Meteorological Satellite Studies (CIMSS) [46], while the cloud masks are from the MOD35_L2 operational product. A strong point of the LST_cci MODIS data is that they provide a detailed quantification of the uncertainty of each pixel using three uncertainty components: the random (u_{ran}), the locally correlated (u_{loc}), and the systematic (u_{sys}). The employed uncertainty model is described in detail in Ghent et al. [47] and is equally applicable across different surface temperature domains. In this work, we use the total LST uncertainty (u_{total}) of each pixel, by combining u_{ran}, u_{loc}, and u_{sys} using Equation (1).

$$u_{\text{total}} = \sqrt{u_{\text{ran}}^2 + u_{\text{loc}}^2 + u_{\text{sys}}^2} \quad (1)$$

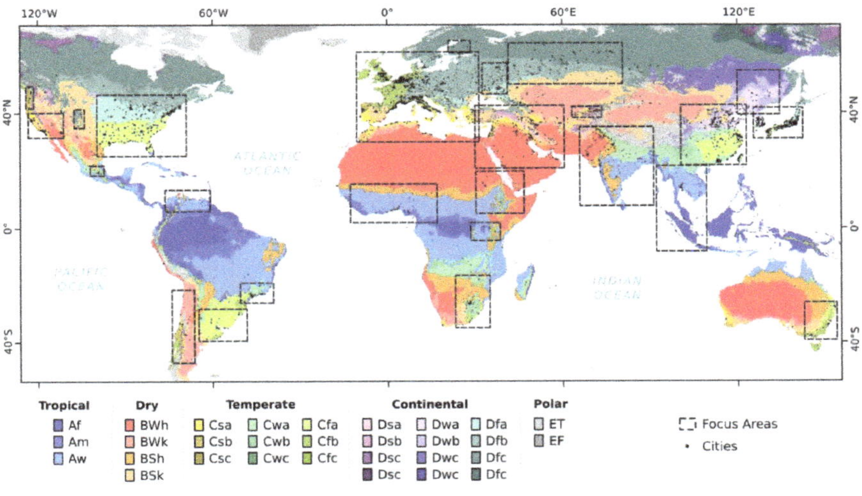

Figure 1. Areas of dense human habitation and the focus areas of this work. The present Köppen–Geiger climate zones are also shown.

2.2. Present Köppen–Geiger Climate Classification Map

The present (1980–2016) Köppen–Geiger climate map from Beck et al. [44] classifies Earth's climate into five main classes and 30 sub-classes (Figure 1). The five main classes are the tropical, dry, temperate, continental, and polar, while the sub-classes reflect differences in seasonal precipitation and level of heat. It is derived from an ensemble of four independent high-resolution (~1 km) climatic maps of air temperature and precipitation that have been topographically corrected and adjusted to reflect the period 1980–2016. The classification is based on threshold values and the seasonality of monthly air temperature and precipitation, and it is applied to each climatic dataset combination separately (12 in total; for full details, see [44]). The final climate map is derived from the 12 individual maps by selecting, for each grid-cell, the most common Köppen–Geiger climate class. A corresponding confidence map is also derived by dividing the frequency of occurrence of the most common class by the ensemble size and converting these fractions to percentages [44]. In general, class confidence levels are generally lower in the vicinity of borders between climate zones, especially in high-latitude regions where the climatic data have higher uncertainty.

2.3. City Delineation and SUHII Calculation

To delineate the cities in the focus areas of Figure 1, we use land cover (LC) data from the ESA-CCI Land Cover project [48]. This data product provides annual high-resolution (300 m) LC maps that classify the global surface in 37 classes according to the United Nations Land Cover Classification System (UNLCCS) with an overall accuracy of 75.4% [48]. To process the LC data, we first resample them to the $0.01^\circ \times 0.01^\circ$ LST grid by calculating the LC fractions of each grid cell. Then, for each year from 2000 to 2018, we create a binary urban mask of all the pixels that have an urban fraction of at least 95%, a water fraction equal to 0%, and are more than ~2 km away from the coastline. To eliminate single pixels and small objects from the resulting masks, we apply a morphological operator that removes any objects with eight or fewer connected pixels using scikit-image v.0.18.1 [49]. Finally, we segment the filtered masks into clusters that correspond to cities and label all the instances of each city with the same unique ID. Our method follows the principle of the City Clustering Algorithm [50], which has been used in several SUHII studies due to its ability to describe the extent of a city more accurately than administrative or population data [18,22,33].

To select appropriate rural pixels for each city, we use the Boundary Generation Algorithm (BGA) [33] that iteratively expands a rural buffer around each city until its size is approximately that of the urban area. To ensure consistency in the SUHII estimates over time, we create a single rural buffer per city that is representative for all the years between 2000 and 2018. In our implementation of BGA, we do not use all the pixels in each new ring but filter them according to the following rules: the rural LC fraction of each candidate pixel is at least 95% for every year between 2000 and 2018; the corresponding urban and water LC fractions are equal to 0%; and the elevation of each candidate pixel does not differ by more than ± 200 m from the median elevation of the corresponding city. Here, we define as rural LC the aggregate class that is derived by summing the LC fractions presented in Table 1 (the resulting class closely resembles LCZ B, C, D, and F [51]). To ensure that only rural pixels adjacent to each city are selected, the search zone of BGA is limited to within 30 pixels from the city boundary.

Following standard practice [1,13,22,33], we calculate SUHII as the difference between the mean urban (LST_{urban}) and rural LST (LST_{rural}) for each date d (Equation (2)).

$$SUHII(d) = LST_{urban}(d) - LST_{rural}(d), \quad (2)$$

Table 1. The rural LC component classes.

ID	ESA-CCI Land Cover Class
10	Rainfed croplands
11	Rainfed croplands with herbaceous cover
12	Rainfed croplands with tree or shrub cover
20	Irrigated croplands
30	Mosaic croplands (>50%) with natural vegetation
40	Mosaic natural vegetation (>50%) with croplands
110	Mosaic herbaceous cover (>50%) with trees and shrubs
120	Shrublands
121	Evergreen shrublands
122	Deciduous shrublands
130	Grasslands
140	Lichens and mosses
150	Sparse vegetation
151	Sparse trees
152	Sparse shrubs
153	Sparse herbaceous cover
200	Bare areas
201	Consolidated bare areas

We apply Equation (2) to each city (separately for daytime and nighttime), using the corresponding urban and rural masks to select the respective MODIS pixels. Before we calculate the urban and rural means, we quality-filter the LST data of each group, keeping only the values that range from 240 K to 360 K and have a total uncertainty equal or better than 2 K. In addition to these two checks, we also apply a median absolute deviation test to remove any remaining outliers (the test statistics are calculated separately for the urban and rural pixels of each city, using the corresponding LST for each year). To evaluate our SUHII estimates, we compare them with reference SUHIIs derived from MOD11A1 v.6.0 LST using the method described above (the results are presented in Appendix A).

2.4. Data Analysis

Our goal is to analyze the seasonal hysteresis of SUHII vs. LST_{rural} as a function of climate. To do this, we use the ~1 km Köppen–Geiger present climate map from Beck et al. [44] and assign each city to a climate class. To reduce misclassifications, we keep only the cities where the climate confidence flag is at least 90%. We also discard any climate classes with insufficient SUHII data (<10% of clear-sky days) and/or fewer than 10 cities. To derive each class's hysteresis loop, we first calculate the 2000–2018 SUHII and LST_{rural} monthly means for each city. We do this using only the dates where at least 70% of the urban and rural pixels are available. Before we calculate the monthly means, we subtract from each date 80 days (i.e., the approximate day-of-year of spring equinox) if it is a north hemisphere city and 265 days if it is south. This, in essence, results in a custom calendar where the corresponding south and north hemisphere equinoxes and solstices are in sync. To derive the SUHII hysteresis loop for each climate class, we average the hysteresis loops from the individual cities located in that class. For this calculation, we use only the cities with complete loops, i.e., loops without missing months. To describe and compare the resulting hysteresis patterns, we use the loop direction, the minimum and maximum SUHII (calculated using the absolute values), and the month when these values occur. In contrast to Zhou et al. [33], we do not approximate the LST_{rural} and SUHII time series of each city with a Fourier series but follow a data-driven approach that relies on the good quality and the large volume of the collected data.

To investigate the influence of precipitation and solar radiation on the shape of the resulting loops, we use precipitation data (v2020) from the Global Precipitation Climatology Centre (GPCC) [52] and at-surface clear-sky downwelling shortwave fluxes (SW) [53] from the CERES Level-3b Energy Balanced and Filled Climate Data Record (v4.1). Both datasets provide global coverage, are available at monthly resolution, and cover the study period (2000–2018). To extract the corresponding precipitation and SW time series for each city, we use the coordinates of the urban polygon centroid. This is because the grid-cell size of the precipitation and SW data ($0.5° \times 0.5°$ for the former and $1° \times 1°$ for the latter) is much coarser than that of the LST data. To derive the climate-class monthly means, we first calculate the 2000–2018 mean precipitation and SW value per month and city and then aggregate the data from the relevant cities, as performed for SUHII and LST_{rural}. To analyze the seasonality of SUHII and LST_{rural} versus that of precipitation, and SW per climate class, we study their co-seasonality and determine the months of minimum and maximum.

3. Results

3.1. Delineated Cities and SUHII Climatology

Using the City Clustering Algorithm [50], we delineate 1511 global cities in one tropical, two dry, five temperate, and three continental Köppen–Geiger classes (Table 2). The location and the number of cities per class is presented in Figure 2, while their characteristics (area, elevation, and percentage of coastal/inland cities) are discussed in Appendix B. From these 11 classes, 10 include more than 50 cities, 6 more than 100 cities, and 1 more than 300 cities. The majority (90.5%) of them are located in Asia (40.0%), Europe (27.5%), and North America (23.0%), and only 9.5% are in Africa (5.9%), South America (2.7%), and Oceania (0.9%). In Figure 2, we also present the daytime (~10:30 local time) and nighttime (~22:30 local time) SUHII climatology of each class as the bivariate distribution of the daily SUHII and LST_{rural} that we have randomly sampled using the months as strata.

The tropical class is the Aw (tropical savanna), which comprises African, Asian, and South American cities (Figure 2A). The Aw climate has two distinct seasons—a wet and a dry—and is warm throughout the year. This weak seasonality is evident in the SUHII climatologies in Figure 2A, which are shaped like convex blobs. The interquartile range (shown as $[Q_{25}, Q_{75}]$, where Q_{25} and Q_{75} are the first and third quartiles) of the Aw SUHII is [−0.4 K, 3.1 K] for the daytime and [0.5 K, 2.2 K] for the nighttime. The corresponding LST_{rural} values are [306 K, 315 K] and [293.7 K, 298.9 K], respectively, which make the Aw the climate with the least intra-annual variation in our analysis (the corresponding means are shown in Table 3). The dry classes are the BSh (hot semi-arid) and the BSk (cold semi-arid). They are intermediate climates between desert and humid climates and are usually dominated by grasslands and shrubs. In our analysis, the BSh is represented mainly by cities in India, Africa, Mexico, and the Middle East, while the BSk is represented by cities in Europe and Asia. The shape of the BSh and BSk SUHII climatologies is more complex than that of the Aw and clearly influenced by seasons (Figure 2B,C). The interquartile range of the daytime SUHII (and LST_{rural}) is [−2.0 K, 0.8 K] ([305.7 K, 319.8 K]) for the BSh and [−1.4 K, 1.2 K] ([283.7 K, 310.3 K]) for the BSk. The corresponding nighttime values are [1.1 K, 2.8 K] ([288.0, K 298.9 K]) and [1.0 K, 2.8 K] ([270.1 K, 290.6 K]), respectively. The key characteristic of these two dry climates is that daytime SUHIIs are mostly negative, especially when the LST_{rural} is maximum.

The five temperate classes are the Csa (hot-summer Mediterranean), Cfa (humid subtropical), Cfb (oceanic), Cwa (dry-winter humid subtropical), and Cwb (dry-winter humid highland). Temperate climates are generally defined as environments with moderate rainfall, sporadic droughts, mild-to-warm summers, and cool-to-cold winter. They occur in mid-latitude regions and have four seasons (winter, spring, summer, and autumn). The Csa exhibits wet winters and hot, dry summers and is water deficient during part of the growing season. This makes the daytime climatology of the Csa (Figure 2D) to be considerably different than that of the other temperate climates. It has a concave-down shape with negative summertime SUHIIs and an interquartile range of [−1.1 K, 1.3 K] for the SUHII and [294.2 K, 313.1 K] for the LST_{rural}. In contrast, the shape of the Csa nighttime climatology is convex and always positive with an interquartile range of [0.9 K, 2.5 K] and [282.4 K, 293.8 K], respectively. The Cfa is represented by cities in North and South America, Europe, Australia, and Asia (Figure 2E), while the Cfb is represented by cities in western Europe (Figure 2F). Their daytime (and nighttime) climatologies exhibit almost identical concave-up shapes, with the SUHII and LST_{rural} peaking almost simultaneously. The daytime interquartile range of the SUHII (and LST_{rural}) is [0.3 K, 3.2 K] ([292.9 K, 309.1 K]) for the Cfa and [0.4 K, 3.0 K] ([282.0 K, 301.5 K]) for the Cfb. The corresponding nighttime values are [0.7 K, 2.2 K] ([279.6 K, 294.8 K]) and [0.4 K, 1.9 K] ([275.1 K, 285.9 K]). The last two temperate classes, namely the Cwa and Cwb, are represented mainly by cities in Asia and Central America, respectively. The Cwa is a monsoon-influenced climate with dry winters and hot summers, while the Cwb is a climate mainly found in tropical and subtropical highlands with cold, dry winters and rainy summers. They both exhibit a completely different daytime SUHII climatology than the other temperate climates. The shape of the Cwa climatology is presented in Figure 2G, while that of the Cwb is in Figure 2H. The interquartile range of the daytime SUHII (and LST_{rural}) is [−0.1 K, 2.5 K] ([298.0, 311.0]) for the Cwa and [−1.0 K, 3.4 K] ([301.1 K, 310.0 K]) for the Cwb. The corresponding nighttime values are [1.0 K, 2.5 K] ([286.2 K, 298.1 K]) and [1.1 K, 3.4 K] ([282.5 K, 290.1 K]), respectively.

The three continental Köppen–Geiger classes are the Dfa (hot-summer humid continental), Dfb (warm-summer humid continental), and Dwa (monsoon-influenced hot-summer humid continental). The Dfa is represented mainly by cities in North America (Figure 2I), while the Dfb (Figure 2J) and the Dwa (Figure 2K) are represented by cities in Europe and Asia, respectively. Continental climates occur within large landmasses away from the moderating effect of oceans and are characterized by an extreme range of annual near-surface air temperatures. They exhibit four distinct seasons (winter, spring, summer, and autumn) with warm-to-hot summers and cold, snowy winters. The shapes of the Dfa, Dfb, and Dwa climatologies (Figure 2I–K) are almost identical and quite similar to that of the Cfa and Cfb (Figure 2E,F). When the LST_{rural} is below 300 K, the daytime SUHII of continental climates is rather constant and close to 1 K. Above 300 K, the SUHII increases considerably and peaks when the LST_{rural} is maximum. The shape of the corresponding nighttime climatologies are rather flat and featureless like all the other climates presented in Figure 2. The interquartile range of the daytime SUHII is [0.3 K, 3.3 K] for the Dfa, [0.2 K, 2.7 K] for the Dfb, and [−0.5 K, 2.5 K] for the Dwa. The corresponding nighttime values are [0.5 K, 2.1 K], [0.4 K, 2.4 K], and [1.1 K, 2.8 K]. The extreme range of continental annual temperatures is evident in the interquartile range of the daytime LST_{rural}, which is [282.9 K, 304.8 K] for the Dfa, [275.2 K, 302.4 K] for the Dfb, and [282.5 K, 306.7 K] for the Dwa. The corresponding nighttime values are [271.5 K, 290.2 K], [268.9 K, 285.7 K], and [270.1 K, 292.3 K].

Table 2. The Köppen–Geiger climate classes analyzed in this work.

ID	Parent Class	Description
Aw	Tropical	Tropical savanna with dry-winter characteristics
BSh	Dry	Hot semi-arid
BSk		Cold semi-arid
Csa	Temperate	Hot-summer mediterranean
Cfa		Humid subtropical
Cfb		Oceanic
Cwa		Dry-winter humid subtropical
Cwb		Dry-winter subtropical highland
Dfa	Continental	Hot-summer humid continental
Dfb		Warm-summer humid continental
Dwa		Monsoon-influenced hot-summer humid continental

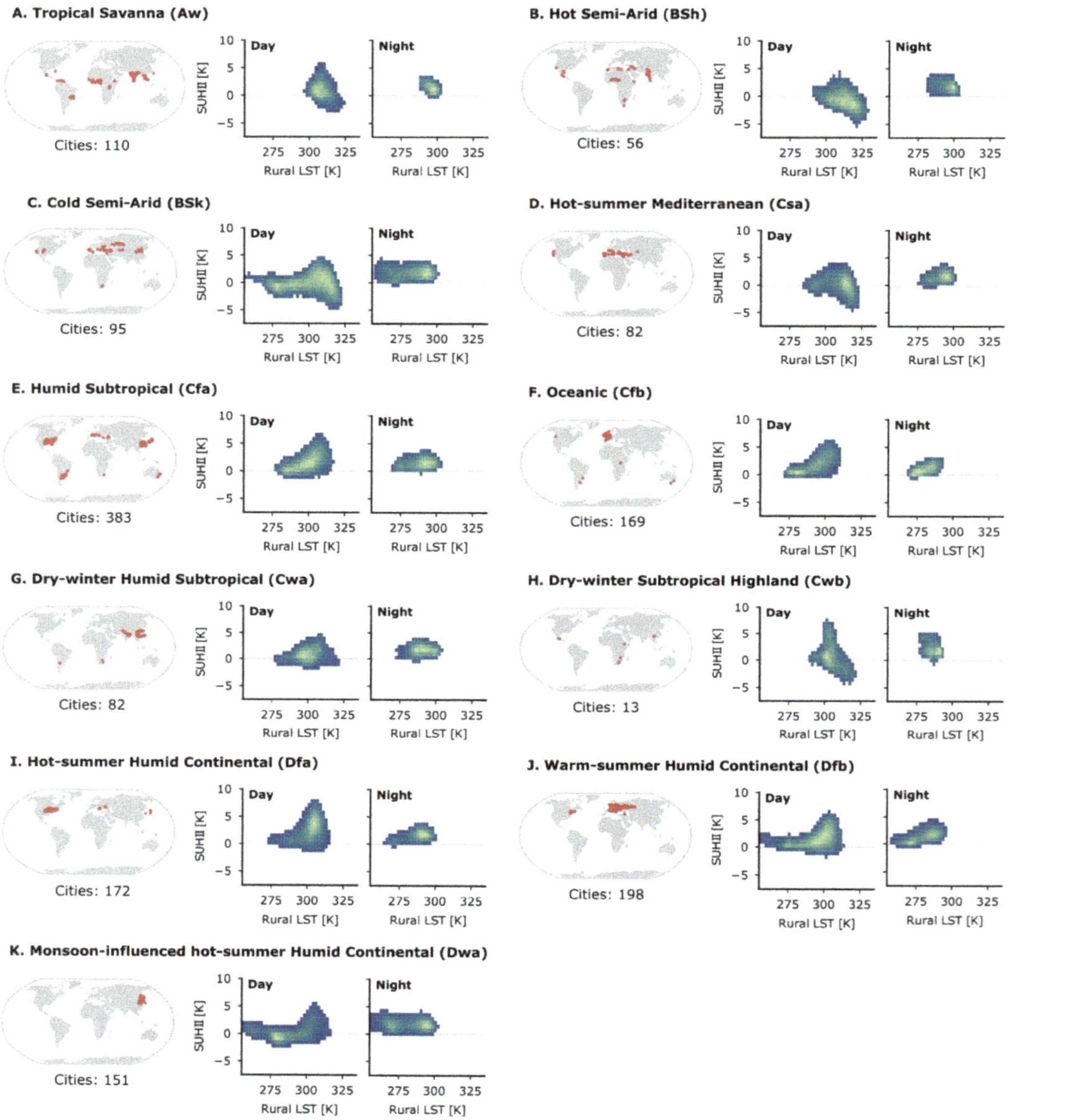

Figure 2. The location and number of delineated cities per Köppen–Geiger climate class and the corresponding daytime and nighttime SUHII climatologies visualized as the bivariate distribution of daily SUHII and rural LST. Each distribution presents all the possible combinations of SUHII and rural LST and is unique and characteristic of the respective climate class. The colors indicate how likely each combination is: yellow indicates the most likely and blue the least.

Table 3. The mean and the standard deviation (SD) of the 2000–2018 daily SUHII and rural LST shown in Figure 2.

Climate	SUHII (K)				Rural LST (K)			
	Daytime		Nighttime		Daytime		Nighttime	
	Mean	SD	Mean	SD	Mean	SD	Mean	SD
Aw	1.3	2.7	1.4	1.2	311.0	7.0	296.2	4.0
BSh	−0.6	2.3	2.0	1.3	312.3	9.6	293.4	6.7
BSk	−0.1	2.1	1.9	1.3	296.5	17.0	279.7	13.0
Csa	0.0	2.0	1.7	1.2	303.6	11.0	287.8	7.3
Cfa	1.8	2.2	1.5	1.2	301.0	11.0	287.3	9.3
Cfb	1.7	1.9	1.2	1.1	292.1	11.4	280.5	6.9
Cwa	1.2	2.0	1.8	1.2	304.5	10.2	291.5	8.3
Cwb	1.2	3.1	2.2	1.6	305.5	7.2	286.0	5.1
Dfa	1.9	2.3	1.4	1.2	293.8	13.6	280.6	11.1
Dfb	1.5	1.9	1.5	1.4	288.2	17.0	276.3	11.5
Dwa	1.1	2.3	2.1	1.4	294.4	16.2	280.2	14.3

3.2. SUHII Seasonal Hysteresis

The daytime SUHII hysteresis loops of the examined climate classes are presented in Figure 3. Overall, their shape matches well that of the SUHII climatologies and provides a clearer view of how the SUHII and LST_{rural} vary within the year. They are the most different for temperate climates (Table 2), where each sub-class exhibits a distinct looping pattern. The daytime Cfa and Cfb hysteresis loops exhibit a concave-up pattern, as the model of Manoli et al. [42] suggests, with the SUHII and LST_{rural} peaking almost simultaneously. The shape of the Csa loop exhibits a weak concave-down pattern, while that of the Cwa exhibits a twisted concave-up pattern. The shape of the Cwb daytime loop is convex (triangle-like), as is the case for the Aw. For cities in the Dfa, Dfb, and Dwa continental climates, the daytime SUHII hysteresis shows a concave-up pattern that is flat when the LST_{rural} is below ~300 K and peaks rapidly as the LST_{rural} increases. Similarly, to the Cfa and Cfb climates, the SUHII and LST_{rural} of the Dfa, Dfb, and Dwa become maximum almost simultaneously. For the BSh and BSk semi-arid climates, the daytime SUHII loops are rather flat. This result does not agree well with the shapes of the individual city loops shown in Figure 3 (grey lines). Further investigations focusing on the BSh cities show that the shape of the individual hysteresis loops differs with geographic location and can take the form of concave-down, flat, twisted, and triangle-like loops (Figure 4). These differences are attributed primarily to differences in the characteristics of the surrounding rural areas, which exert a strong influence on the SUHII [24,25], and suggest that semi-arid cities should be classified into even finer groups. Our analysis also shows that the shape of the humid temperate and continental loops is more stable than that of the dry-climate loops. This observation corroborates the remark of Manoli et al. [42] that the shape of dry-climate loops is more susceptible to perturbations in the seasonality and the magnitude of rainfall. The direction of the daytime loops is clockwise in all cases, except for the Aw, BSh, Cwa, Cwb, and Dwa (Figure 3).

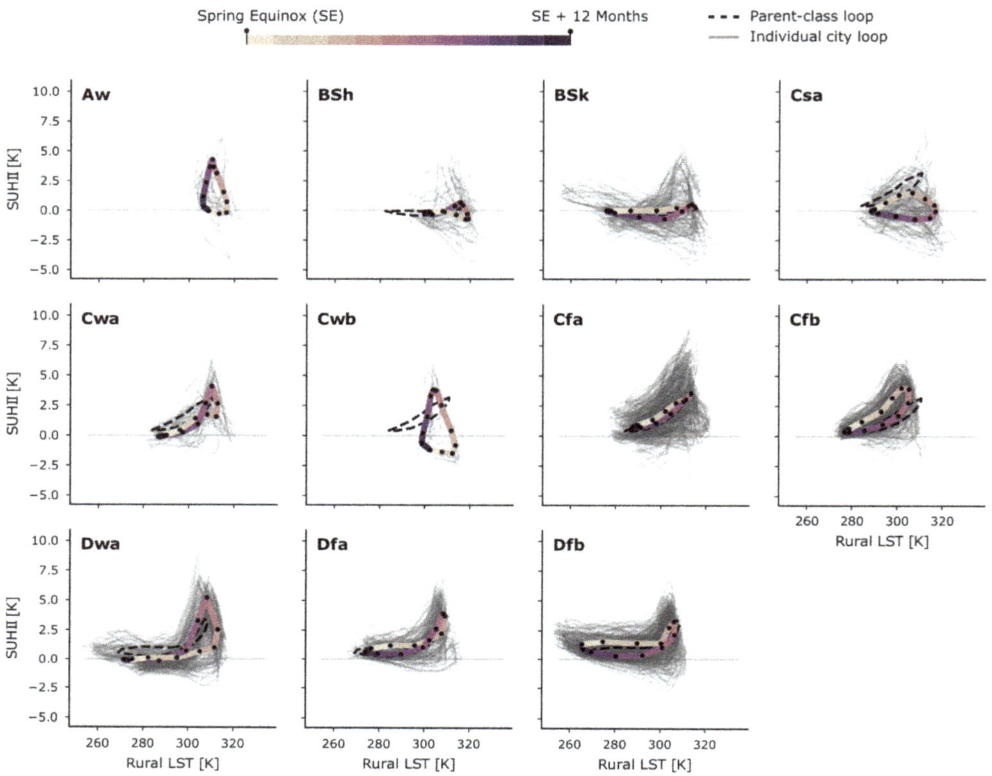

Figure 3. The daytime SUHII hysteresis loops for each Köppen–Geiger climate class (colored lines). Each black dot represents a month (the start and end points correspond to spring equinox), and the change in color indicates the direction of each loop. The solid grey loops represent the individual city loops per climate class, and the dashed one represents the corresponding parent-class loops.

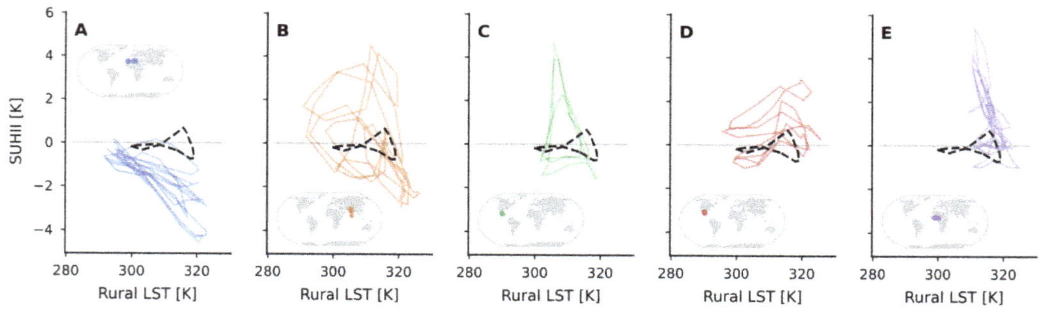

Figure 4. The individual daytime SUHII hysteresis loops for hot semi-arid (BSh) cities in North Africa (**A**), South Asia (**B**), Central America—Mexico (**C**), Central America—USA (**D**), and West Africa (**E**). The dashed line is the mean BSh loop from Figure 3.

In Figure 5, we present the seasonal variation of the daytime SUHII, SW, and precipitation for each Köppen–Geiger class examined in this work. We focus on these two variables because they have been shown to play a key role in the seasonality of the SUHII [42]. For the Aw cities we observe that the SUHII peaks when the SW and precipitation are almost maximum. This is also the case for the Cwb cities in our analysis, which are located mainly in elevated regions within the tropics and the subtropics (Figure 2H). In temperate and continental climates (Cfa, Cfb, Dfa, and Dfb), where the precipitation is rather constant throughout the year, the SUHII appears to vary mainly with the SW. This, however, is not the case for the Cwa and Dwa variates that exhibit a monsoonal tendency with a much higher precipitation in summer than in winter. In these climates, the SUHII intensifies as the summertime precipitation peaks, which suggests that monsoons influence the concave-up hysteresis of the SUHII in the Cwa and Dwa. The climate class with the most distinct behavior is the hot-summer Mediterranean (Csa), where the precipitation decreases as the SW increases. The anti-correlation between the SW and precipitation during spring and summer makes the Csa SUHII-SW-precipitation loop the only one with a clockwise direction (Figure 5). The daytime SUHII of the Csa cities peaks in late spring/early summer and then starts to drop as the precipitation approaches its minimum and SW its maximum. During this phase, a significant portion of the natural vegetation begins to dry due to water stress [54]. Under such water-limited conditions, the evapotranspiration of rural areas decreases, which impacts their ability to cool [11,42]. In dry-climate cities, the precipitation is low throughout the year and does not vary much with the SW (monsoon-influenced cities in India make the bulk of the examined BSh cities and are responsible for the precipitation peak in Figure 5). The SUHII of the BSh and BSk cities does not vary with the SW either; however, as discussed above, this result is a fluke caused by averaging dissimilar SUHII loops (see Figures 3 and 4). Overall, Figure 5 shows that the SUHII of tropical, temperature, and continental cities is generally strongest when the SW and precipitation peak. Under these conditions, the vegetation surrounding each city reaches peak greenness, which in turn suggests that the observed SUHII increase should not be attributed solely to an increase in the urban LST.

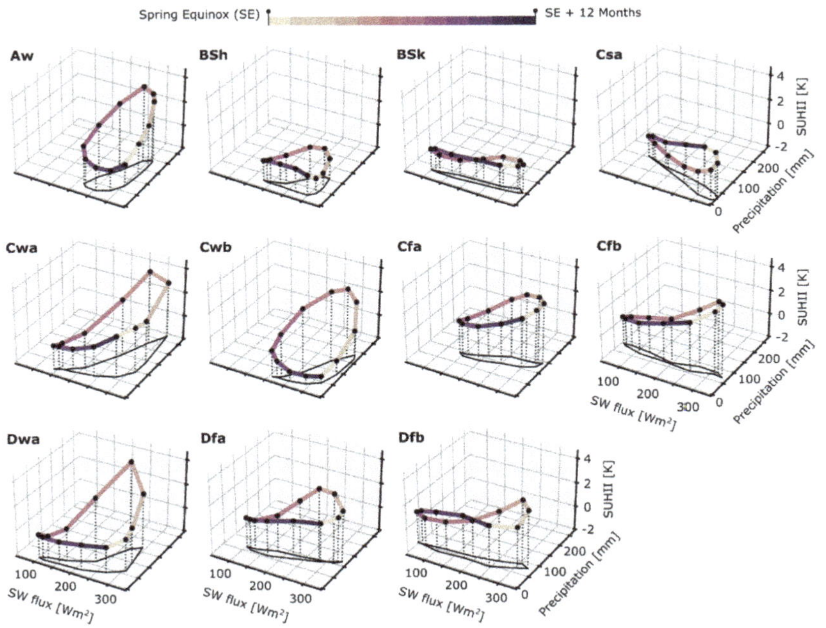

Figure 5. The seasonal variation of precipitation, clear-sky at-surface downwelling shortwave radiation flux (SW), and daytime SUHII for each Köppen–Geiger climate class examined in this work. Each dot represents a month (the start and end points correspond to spring equinox). The monthly means refer to the years 2000–2018 and the color of the line indicates the direction of each loop.

The corresponding nighttime hysteresis loops are rather similar and exhibit mostly flat and concave-up patterns (Figure 6). In humid temperate and continental climates, the SUHII increases and decreases almost in sync with the LST_{rural}, while in dry climates, the shape of the nighttime hysteresis loops is mainly flat. The looping direction is always clockwise and the classes with the most distinct nighttime loops are the BSh for the dry climates, the Cfb for the temperate, and the Dwa for the continental. Contrary to the daytime, the shape of the individual BSh and BSk nighttime hysteresis loops are more alike and better represented by the mean loop (Figure 6). In respect to the seasonal variation of the precipitation and SW, the nighttime SUHII of the Csa, Cfa, Cfb, Dfa, and Dfb cities is strongest when the SW peaks (Figure 7). In contrast, the nighttime SUHII of the Aw cities is weakest when the precipitation and SW peak.

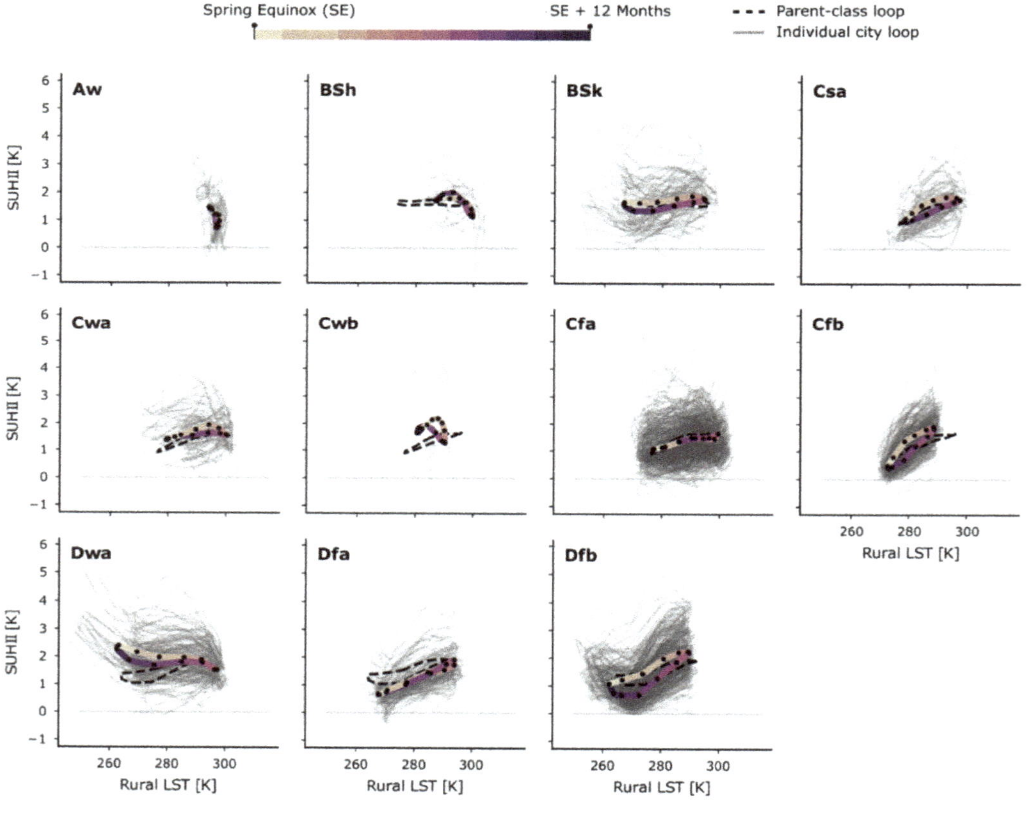

Figure 6. The nighttime SUHII hysteresis loops for each Köppen–Geiger climate class (colored lines). Each black dot represents a month (the start and end points correspond to spring equinox), and the change in color indicates the direction of each loop. The solid grey loops represent the individual city loops of each climate class, and the dashed one represents the corresponding parent-class loops.

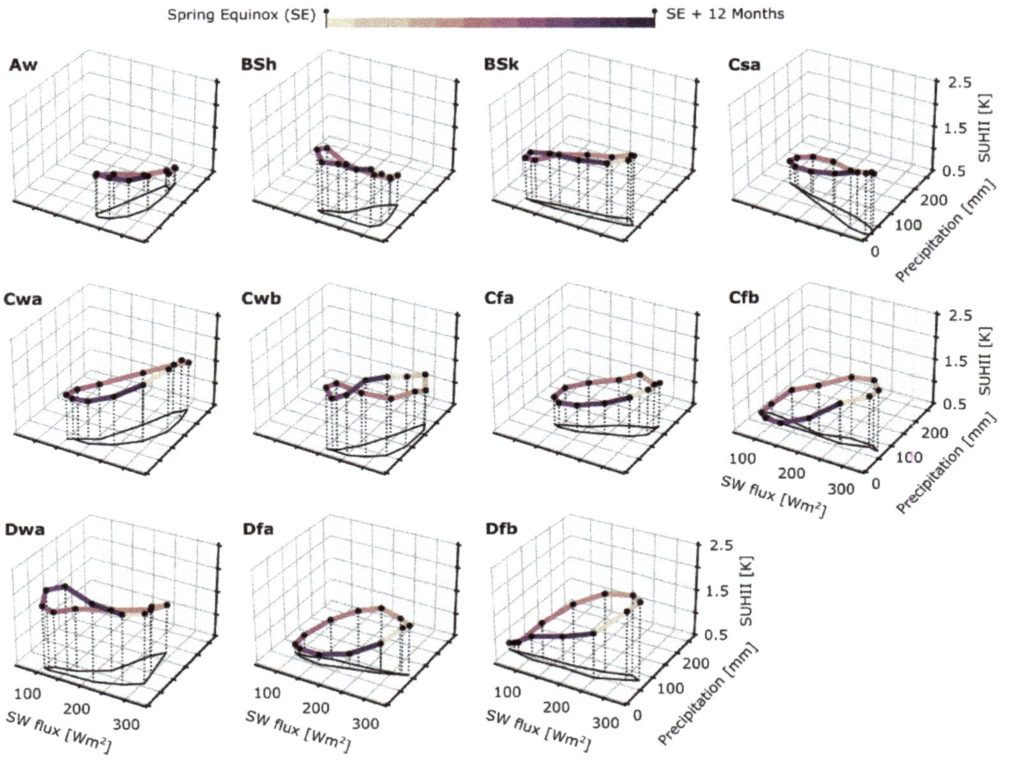

Figure 7. The seasonal variation of precipitation, clear-sky at-surface downwelling shortwave radiation flux (SW) and nighttime SUHII for each Köppen–Geiger climate class examined in this work. Each dot represents a month (the start and end points correspond to spring equinox). The monthly means refer to the years 2000–2018, and the color of the line indicates the direction of each loop.

In Figures 3 and 6, we also include the SUHII hysteresis of the dry, temperate, and continental parent classes (dashed lines) that we derive using the individual loops from all the relevant cities. The results support our thesis that aggregating multi-city data without considering the biome of each city can result in temporal means that fail to reflect the actual SUHII characteristics and show that the shape of each parent-class loop is determined by the climate sub-class with the most cities. This is particularly the case for the Cfa and Cfb, where the daytime temperate parent-class loop reflects their shape and is not representative of the other temperate sub-classes (e.g., Csa or Cwb, as seen in Figure 3).

3.3. Month of Minimum and Maximum SUHII

The month when the SUHII of each hysteresis loop is maximum and minimum in absolute values is shown with black dots in Figures 8 and 9 (the corresponding magnitudes are provided in Table 4). The colored dots refer to the individual city loops and indicate the variability in the examined cities. For the Aw cities, the daytime SUHII is strongest in September (4.3 ± 0.8 K) and weakest in February (−0.1 ± 0.7 K). The peak occurs four months later than that of the daytime LST_{rural} and one month later than that of the precipitation (Figure 8). The nighttime SUHII peaks in January (1.4 ± 0.2 K) and is least in September (0.8 ± 0.2 K), whereas the Aw LST_{rural} is maximum in May and minimum in December/January.

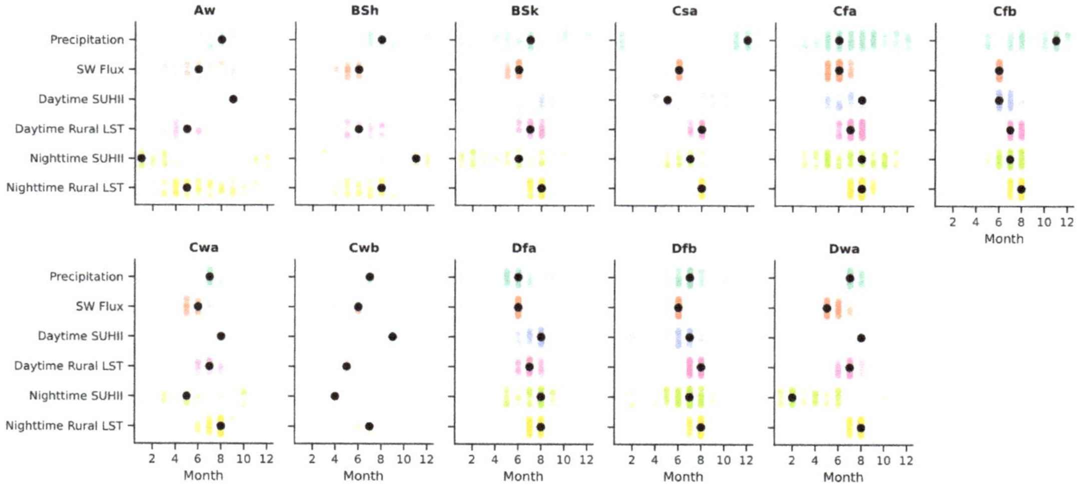

Figure 8. The month when SUHII, rural LST, SW, and precipitation are strongest per climate class. The colored dots represent the individual SUHII loops and the black dots the Köppen–Geiger climate-class mean loop (BSh and BSk are not included).

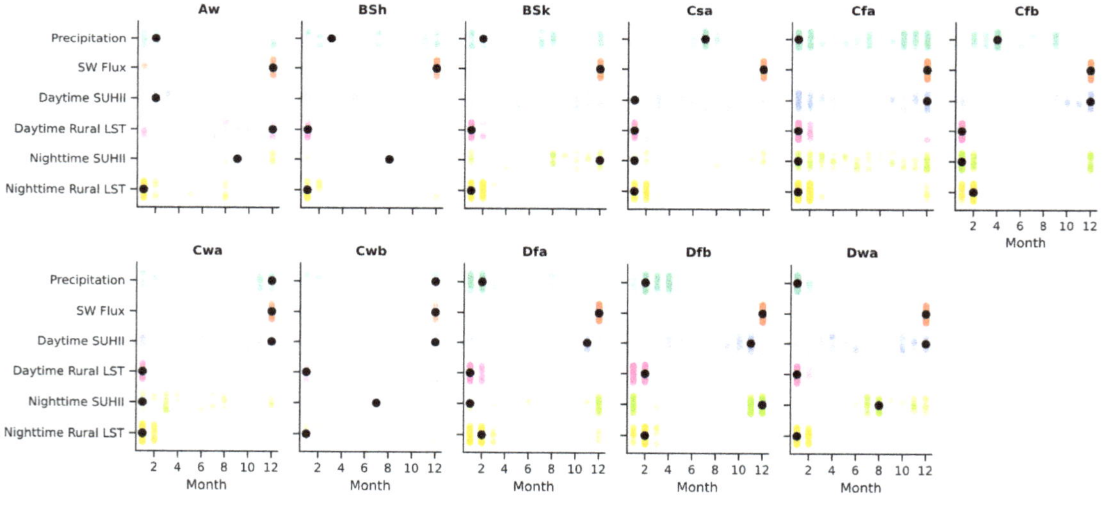

Figure 9. The month when SUHII, rural LST, SW flux, and precipitation are minimum. The colored dots represent the individual SUHII loops, and the black dots the Köppen–Geiger climate-class mean loop (BSh and BSk are not included).

Table 4. The minimum and maximum SUHII (and the 95% confidence intervals) for the Köppen–Geiger climate-class means shown in Figures 8 and 9 (black dots).

Climate	Maximum SUHII (K)		Minimum SUHII (K)	
	Daytime	Nighttime	Daytime	Nighttime
Aw	4.3 ± 0.8	1.4 ± 0.2	-0.1 ± 0.7	0.8 ± 0.2
BSh	-	1.9 ± 0.2	-	1.1 ± 0.3
BSk	-	2.0 ± 0.2	-	1.4 ± 0.2
Csa	1.5 ± 0.4	1.8 ± 0.2	0.0 ± 0.2	0.9 ± 0.2
Cfa	3.6 ± 0.3	1.7 ± 0.1	0.7 ± 0.1	1.1 ± 0.1
Cfb	4.0 ± 0.2	1.9 ± 0.1	0.4 ± 0.1	0.4 ± 0.1
Cwa	4.1 ± 0.5	1.9 ± 0.2	0.0 ± 0.2	1.4 ± 0.1
Cwb	3.8 ± 1.3	2.2 ± 1.1	0.2 ± 0.6	1.3 ± 1.1
Dfa	3.9 ± 0.4	1.9 ± 0.1	0.4 ± 0.1	0.7 ± 0.1
Dfb	3.3 ± 0.2	2.2 ± 0.1	0.3 ± 0.1	0.7 ± 0.1
Dwa	5.2 ± 0.3	2.4 ± 0.2	0.0 ± 0.1	1.5 ± 0.1

The daytime SUHII and LST_{rural} of semi-arid cities is generally strongest in summer. The nighttime SUHII peaks in November (1.9 ± 0.2 K) for the BSh and in June (2.0 ± 0.2 K) for the BSk, while the LST_{rural} peaks in August. In hot-Mediterranean (Csa) cities, the daytime SUHII is warmest in May (1.5 ± 0.4 K) and the LST_{rural} in August. In contrast, the nighttime SUHII peaks in July (1.8 ± 0.2 K) when the precipitation is minimum and the LST_{rural} is almost maximum. The month when the Csa SUHII and LST_{rural} are weakest is January (Figure 9). The Csa is the only temperate climate where the daytime SUHII peak occurs in spring and not in summer.

In wet temperate climates, the SUHII is strongest in summer. It peaks in August for the Cfa and in June/July for the Cfb. The corresponding SUHII magnitudes are 3.6 ± 0.3 K and 4.0 ± 0.2 K for the daytime and 1.7 ± 0.1 K and 1.9 ± 0.1 K for the nighttime (Table 4). The LST_{rural} and SW also peak in summer, while the precipitation is relative constant throughout the year (this explains the pronounced dispersion of the Cfa and Cfb colored dots in Figures 8 and 9). The SUHII is weakest in December/January for both climate classes, with the Cfa exhibiting a slightly greater magnitude (~0.9 K vs. 0.4 K). The daytime SUHII of the Cwa and Cwb cities is maximum in August/September and minimum in December. This is also the case for the precipitation, SW, and LST_{rural}. The maximum daytime SUHII is equal to 4.1 ± 0.5 K for the Cwa and 3.8 ± 1.3 K for the Cwb, while the corresponding minimums are 0.0 ± 0.2 K and 0.2 ± 0.6 K, respectively (Table 4). The nighttime SUHII peaks earlier than the LST_{rural} (in April/May vs. July/August) and is equal to 1.9 ± 0.2 K for the Cwa and 2.2 ± 1.1 K for the Cwb.

In continental climates (Dfa, Dfb, and Dwa), the SUHII and LST_{rural} are warmest in July/August. The only exception is the Dwa where the nighttime SUHII peaks in February (Figure 8). The precipitation also peaks in July, while the SW peaks in June. The maximum daytime SUHII is 3.9 ± 0.4 K for the Dfa, 3.3 ± 0.2 K for the Dfb, and 5.2 ± 0.3 K for the Dwa. The corresponding nighttime values are 1.9 ± 0.1 K, 2.2 ± 0.1 K, and 2.4 ± 0.2 K. The daytime SUHII minimums occur in November/December and the nighttime in January/December, except for the Dwa which occurs in August (Figure 9). The corresponding values are ~0 K for the daytime and 0.7 ± 0.1 K (Dfa, Dfb) and 1.5 ± 0.1 K (Dwa) for the nighttime. Figures 8 and 9 also show that the daytime and nighttime, minimum and maximum SUHII do not always occur in the same month. We attribute this to the different mechanisms that drive the SUHII during the day and night.

4. Discussion

In this work, we revisit the topic of SUHII seasonality and how it differs across climates. Instead of the typical tropical/dry/temperate/continental grouping, we describe Earth's climate using the Köppen–Geiger system that empirically maps Earth's biome distribution into 30 climate classes [44]. This climate classification system has proven to be a highly suitable means for aggregating complex climate gradients into simple but ecologically meaningful classes [44]. As such, it is regularly used across a range of disciplines for regionalizing variables [55]. We find this property particularly suitable for our work because it allows an indirect control of the biome of each city. Even though a global vegetation map would provide more accurate labels about the city biomes, this approach also leads to a sufficient discrimination between the examined tropical, temperate, and continental cities. This is because climate is the basis for most plant vegetation systems, and the regional extend of each biome is primarily determined by it.

Previous studies investigating the characteristics of the SUHII on a global level have largely neglected to control the biome of each city, despite the fact that it exerts a strong influence on the SUHII. The SUHII is a function of both urban and rural features [24,25], and the phenology of the rural surroundings can differ considerably between urban areas even within the same climate zone. This implies that failing to control this parameter when aggregating multi-city data can result in temporal means that do not reflect the actual SUHII dynamics of each group. Our findings support this thesis and show that the seasonality of tropical, dry, temperate, and continental SUHIIs differs considerably during the daytime. They also reveal that in dry and temperate climates, the SUHII seasonal dynamics exhibit considerable intra-class variations that cannot be represented by the parent class. The comparison between the SUHII characteristics (i.e., hysteresis, range, and month of minimum/maximum) of the temperate parent class and the corresponding sub-classes provides clear evidence of this and suggests that the parent-class characteristics reflect those of the dominant sub-class. This suggests that using parent-class summaries to make inferences about the SUHII characteristics of cities in non-dominant climate sub-classes should be avoided or exercised with caution. In contrast to daytime data, this issue does not appear to affect the analysis and aggregation of nighttime SUHIIs as much, particularly because they exhibit less inter- and intra-class annual variation.

The derived SUHII hysteresis loops reveal the strong influence that local climate conditions exert on daytime SUHIIs and suggest that almost every climate class exhibits a unique daytime looping pattern. For temperate climates, our results replicate the daytime concave-up and -down patterns observed in Europe [33,42] and present the convex and twisted hysteresis of temperate dry-winter cities in America and Asia. They also show that the daytime hysteresis of continental cities in Asia, Europe, and North America is always concave-up, with a rather constant SUHII when the LST_{rural} is below 300 K. For cities in dry semi-arid climates, we found a rather flat SUHII seasonality, like Chakraborty and Lee [23]. However, examining the individual daytime loops of semi-arid cities, we observed a variety of flat, twisted, triangle-like, and concave-up patterns that depend strongly on geographic location. This finding suggests that dry-climate SUHIIs should be grouped into even finer classes that describe the rural features in more detail (e.g., using the land cover fractions).

Overall, our results provide the most complete typology of daytime and nighttime hysteresis loops to date. This information improves our understanding about the global SUHII dynamics and can guide the analysis of multi-city SUHII data from different climates. The development of a consistent framework for analyzing global SUHI data has been the focus of many papers over the years [10,13,23,39]. Our work contributes to this goal by proposing a consistent method for delineating cities and demonstrating the impact of improper data aggregation. Future efforts should introduce further controls and investigate how the seasonal hysteresis of the SUHII varies as a function of urban form and function. These two attributes exhibit considerable heterogeneity [56,57] on a global level and have been shown to influence the SUHII [10,58]. Here, we overlook this issue primarily because we use coarse resolution data (~1 km) and a custom definition of urban areas that selects only pixels that form densely built-up clusters with almost no vegetation. The employed definition improves the consistency of the derived multi-city SUHII data by mitigating the influence of urban green. However, it also results in city polygons that are smaller, more fragmented, and with more gaps in comparison to city polygons retrieved from administrative or land cover data. This work is also one of the first to use data from the new ESA-CCI LST product that has been designed to meet the requirements of the Global Climate Observing System (GCOS) for climate applications.

In respect to other global SUHII studies [13,23], our study includes fewer cities. This is mainly due to the minimum city size threshold that we use to ensure that the employed cities are adequately resolved in the ~1 km LST data. A consequence of this decision is that our SUHII estimates are warmer than that from other studies (this is because the SUHII and city size are positive correlated [58,59]). We are also skeptical about the direct use of the findings presented here and in similar works for informing heat mitigation actions. This is due to the dependency of the SUHII on the characteristics of the reference rural areas [24,25] and also because the canopy-layer UHIs—which are the ones that should be mitigated—differ significantly from the SUHIs [12,17] and do not exhibit any pronounced seasonal hysteresis [26]. Nevertheless, this information can help us better understand how rural and urban land covers react to time-lags between radiation forcing and precipitation, which might provide further insights about urban heat.

5. Conclusions

The seasonal variation of the SUHII exhibits distinct hysteretic patterns that depend on the local climate conditions and in particular on the seasonal availability of energy and water. In this work, we use the new ESA-CCI LST data product for the Terra MODIS and characterize the seasonal hysteresis of the SUHII in almost every Köppen–Geiger climate class inhabited by humans. Our results advance the state-of-the-art and provide the most complete typology of daytime and nighttime SUHII hysteresis loops to date. They reveal that in addition to concave-up and -down shapes, the seasonal hysteresis of the daytime SUHII can also exhibit twisted, flat, and triangle-like patterns, and that nighttime loops are mostly flat and concave-up. They suggest that, in wet climates, the daytime SUHII hysteresis is almost universally concave-up but paint a more complex picture for dry climate cities, where the reference rural areas are more heterogenous and actual evapotranspiration is less than potential evapotranspiration. They also show that aggregating SUHII data from cities in different biomes results in temporal means that do not reflect the actual SUHII characteristics. Even though our results cannot and should not be used for urban heat mitigation, they improve our understanding about the global SUHI variability and the influence that rural surroundings exert on SUHII.

Author Contributions: Conceptualization, P.S. and B.B.; methodology, P.S. and B.B.; software, P.S.; validation, P.S.; formal analysis, P.S.; investigation, P.S. and B.B.; resources, M.P. and D.G.; data curation, P.S.; writing—original draft preparation, P.S.; writing—review and editing, P.S., B.B. and D.G.; visualization, P.S.; supervision, B.B.; project administration, P.S. and B.B.; funding acquisition, B.B. and D.G. All authors have read and agreed to the published version of the manuscript.

Funding: This work has received funding from the European Space Agency within the framework of the Land Surface Temperature project under the Climate Change Initiative, contract number 4000123553/18/I-NB, and the ENLIGHT project, funded by the German Research Foundation (DFG) under grant No. 437467569.

Data Availability Statement: The LST, LC, and climate data that support the findings of this study are openly available at http://gws-access.ceda.ac.uk/public/esacci_lst, https://cds.climate.copernicus.eu/portfolio/dataset/satellite-land-cover, and https://doi.org/10.6084/m9.figshare.6396959, respectively (all accessed on 5 May 2022). The SUHII and rural LST means that support the findings of this work are openly available in Zenodo at https://zenodo.org/record/6535187, reference number 6535187.

Conflicts of Interest: The authors declare no conflict of interest.

Abbreviations

The following abbreviations are used in this manuscript:

BGA	Boundary Generation Algorithm
BT	Brightness Temperatures
CIMSS	Cooperative Institute for Meteorological Satellite Studies
ESA-CCI	European Space Agency's Climate Change Initiative
GCOS	Global Climate Observing System
GPCC	Global Precipitation Climatology Centre
GSW	Generalized Split-Window
LC	Land Cover
LCZ	Local Climate Zones
LST	Land Surface Temperature
LST_cci	ESA-CCI project on Land Surface Temperature
MODIS	Moderate Resolution Imaging Spectroradiometer
RMSD	Root-Mean-Square-Deviation
SD	Standard Deviation
SUHI	Surface Urban Heat Island
SUHII	Surface Urban Heat Island Intensity
SW	At-surface clear-sky downwelling shortwave fluxes
UHI	Urban Heat Island
UNLCCS	United Nations Land Cover Classification System

Appendix A

Table A1 presents the agreement between the 2000–2018 SUHII data estimated from the LST_cci MODIS product (v.1.0) and the well-established MOD11A1 LST v.6.0 product. We retrieve the MOD11A1 SUHII data for all the focus areas of Figure 1 using the method described in Section 2.3. To quantify the agreement between the two SUHII datasets, we use the Root-Mean-Square-Deviation (RMSD) and Pearson's correlation coefficient. Our results show that the daytime and nighttime LST_cci and MOD11A1 SUHII data agree well, with the RMSD ranging from 0.5 to 1.1 K at daytime and 0.1 K to 0.7 K at nighttime. The correlation coefficient ranges from 0.90 to 0.97 and is generally higher during the daytime.

Table A1. The Root-Mean-Square-Deviation (RMSD) and the correlation coefficient between the 2000–2018 LST_cci and MOD11A1 (v.6.0) SUHII for each Köppen–Geiger climate class.

Climate	RMSD (K)		Correlation		Observations	
	Daytime	Nighttime	Daytime	Nighttime	Daytime	Nighttime
Aw	0.6	0.1	0.97	0.92	9297	31,831
BSh	0.8	0.5	0.96	0.91	3393	13,180
BSk	0.6	0.5	0.97	0.94	9392	25,613
Csa	0.7	0.5	0.94	0.91	2828	11,988
Cfa	0.8	0.5	0.94	0.90	19,401	64,639
Cfb	0.6	0.6	0.92	0.90	10,150	26,337
Cwa	0.6	0.4	0.97	0.92	5519	26,964
Cwb	0.6	0.4	0.95	0.95	469	974
Dfa	1.1	0.4	0.94	0.91	13,951	29,807
Dfb	0.6	0.6	0.92	0.94	20,873	44,667
Dwa	0.5	0.4	0.97	0.95	14,775	41,282

Appendix B

Figure A1 presents the characteristics (area, elevation, and percentage of inland/coastal cities) of the examined cities per Köppen–Geiger climate class. The mean city elevation ranges from 98 m in the Dwa to 1928 m in the Cwb. The pooled mean and the standard deviation (SD) of the elevation means is 400 $\pm$ 503 m, while the median of the means is 220 m. The climate classes with the most pronounced elevation differences are the BSk and Cwb, where the interquartile range (ΔQ) is 966 m and 611 m, respectively. The classes where the urban elevations differ the least are the Cfa ($\Delta Q = 132$ m), Cfb (101 m), and Dwa (117 m). The mean city size (Figure A1B) ranges from 32 km^2 in the Cfb and BSk to 159 km^2 in the Cwb. The pooled mean ($\pm$SD) is 57 ± 36 km^2 and the median of the means is 48 km^2. The size distributions are heavily skewed to the right and appear to follow Zipf's Law. The class with the largest cities ($\geq$500 km^2) is the Cfa (11), followed by the Cfb (3) and Dfa (3). The classes with the greatest variation in city size are the Csa (the SD is 396 km^2) and the Cwb (258 km^2), while the ones with the least variation are the BSk (43 km^2), Cwa (46 km^2), and Dfb (53 km^2). The percentage of inland and coastal cities for each climate is presented in Figure A1C (we assign a city as coastal if at least one point of its boundary is 10 km or less from the coast). Inland cities make up most of the urban areas in each class. The median and the mean ($\pm$SD) percentage for these classes is 84% and 82% ($\pm$12%), respectively. The only class with no coastal cities is the Cwb, which corresponds to subtropical highlands.

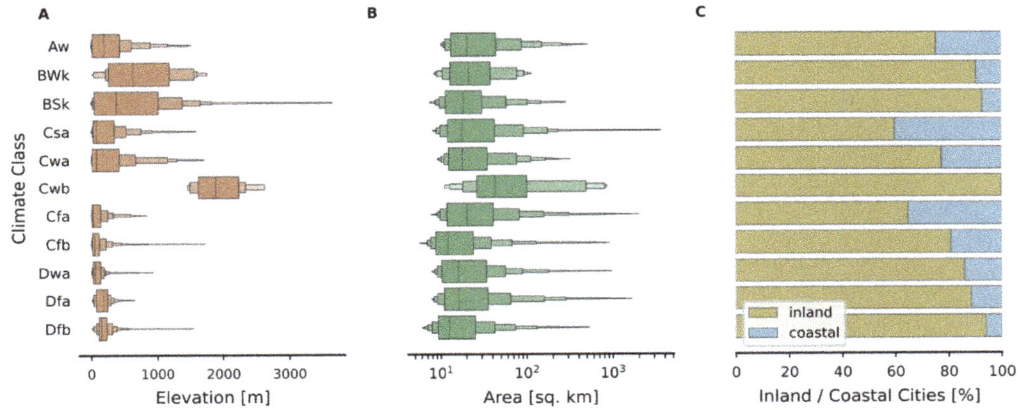

Figure A1. The distribution of city elevations (**A**) and sizes (**B**) and the percentage of inland and coastal cities (**C**) per Köppen–Geiger climate class.

References

1. Oke, T.R.; Mills, G.; Christen, A.; Voogt, J.A. *Urban Climates*, 1st ed.; Cambridge University Press: Cambridge, UK, 2017; ISBN 978-1-139-01647-6.
2. Santamouris, M. On the Energy Impact of Urban Heat Island and Global Warming on Buildings. *Energy Build.* **2014**, *82*, 100–113. [CrossRef]
3. Sismanidis, P.; Keramitsoglou, I.; Barberis, S.; Dorotić, H.; Bechtel, B.; Kiranoudis, C.T. PLANHEAT's Satellite-Derived Heating and Cooling Degrees Dataset for Energy Demand Mapping and Planning. *Remote Sens.* **2019**, *11*, 2048. [CrossRef]
4. Keramitsoglou, I.; Sismanidis, P.; Analitis, A.; Butler, T.; Founda, D.; Giannakopoulos, C.; Giannatou, E.; Karali, A.; Katsouyanni, K.; Kendrovski, V.; et al. Urban Thermal Risk Reduction: Developing and Implementing Spatially Explicit Services for Resilient Cities. *Sustain. Cities Soc.* **2017**, *34*, 56–68. [CrossRef]
5. Sailor, D.J.; Baniassadi, A.; O'Lenick, C.R.; Wilhelmi, O.V. The Growing Threat of Heat Disasters. *Environ. Res. Lett.* **2019**, *14*, 054006. [CrossRef]
6. Sarrat, C.; Lemonsu, A.; Masson, V.; Guedalia, D. Impact of Urban Heat Island on Regional Atmospheric Pollution. *Atmos. Environ.* **2006**, *40*, 1743–1758. [CrossRef]
7. Baccini, M.; Biggeri, A.; Accetta, G.; Kosatsky, T.; Katsouyanni, K.; Analitis, A.; Anderson, H.R.; Bisanti, L.; D'Ippoliti, D.; Danova, J.; et al. Heat Effects on Mortality in 15 European Cities. *Epidemiology* **2008**, *19*, 711–719. [CrossRef]
8. Zafeiratou, S.; Analitis, A.; Founda, D.; Giannakopoulos, C.; Varotsos, K.V.; Sismanidis, P.; Keramitsoglou, I.; Katsouyanni, K. Spatial Variability in the Effect of High Ambient Temperature on Mortality: An Analysis at Municipality Level within the Greater Athens Area. *Int. J. Environ. Res. Public Health* **2019**, *16*, 3689. [CrossRef]
9. Santamouris, M.; Paraponiaris, K.; Mihalakakou, G. Estimating the Ecological Footprint of the Heat Island Effect over Athens, Greece. *Clim. Chang.* **2007**, *80*, 265–276. [CrossRef]
10. Bechtel, B.; Demuzere, M.; Mills, G.; Zhan, W.; Sismanidis, P.; Small, C.; Voogt, J. SUHI Analysis Using Local Climate Zones—A Comparison of 50 Cities. *Urban Clim.* **2019**, *28*, 100451. [CrossRef]
11. Zhao, L.; Lee, X.; Smith, R.B.; Oleson, K. Strong Contributions of Local Background Climate to Urban Heat Islands. *Nature* **2014**, *511*, 216–219. [CrossRef]

12. Parlow, E. Regarding Some Pitfalls in Urban Heat Island Studies Using Remote Sensing Technology. *Remote Sens.* **2021**, *13*, 3598. [CrossRef]
13. Clinton, N.; Gong, P. MODIS Detected Surface Urban Heat Islands and Sinks: Global Locations and Controls. *Remote Sens. Environ.* **2013**, *134*, 294–304. [CrossRef]
14. Chrysoulakis, N.; Grimmond, S.; Feigenwinter, C.; Lindberg, F.; Gastellu-Etchegorry, J.-P.; Marconcini, M.; Mitraka, Z.; Stagakis, S.; Crawford, B.; Olofson, F.; et al. Urban Energy Exchanges Monitoring from Space. *Sci. Rep.* **2018**, *8*, 11498. [CrossRef] [PubMed]
15. Keramitsoglou, I.; Kiranoudis, C.T.; Ceriola, G.; Weng, Q.; Rajasekar, U. Identification and Analysis of Urban Surface Temperature Patterns in Greater Athens, Greece, Using MODIS Imagery. *Remote Sens. Environ.* **2011**, *115*, 3080–3090. [CrossRef]
16. Sismanidis, P.; Keramitsoglou, I.; Kiranoudis, C.T. A Satellite-Based System for Continuous Monitoring of Surface Urban Heat Islands. *Urban Clim.* **2015**, *14*, 141–153. [CrossRef]
17. Voogt, J.A.; Oke, T.R. Thermal Remote Sensing of Urban Climates. *Remote Sens. Environ.* **2003**, *86*, 370–384. [CrossRef]
18. Venter, Z.S.; Chakraborty, T.; Lee, X. Crowdsourced Air Temperatures Contrast Satellite Measures of the Urban Heat Island and Its Mechanisms. *Sci. Adv.* **2021**, *7*, eabb9569. [CrossRef]
19. Li, D.; Liao, W.; Rigden, A.J.; Liu, X.; Wang, D.; Malyshev, S.; Shevliakova, E. Urban Heat Island: Aerodynamics or Imperviousness? *Sci. Adv.* **2019**, *5*, eaau4299. [CrossRef]
20. Stewart, I.D.; Krayenhoff, E.S.; Voogt, J.A.; Lachapelle, J.A.; Allen, M.A.; Broadbent, A.M. Time Evolution of the Surface Urban Heat Island. *Earth's Future* **2021**, *9*, e2021EF002178. [CrossRef]
21. Sismanidis, P.; Keramitsoglou, I.; Kiranoudis, C.T. Diurnal Analysis of Surface Urban Heat Island Using Spatially Enhanced Satellite Derived LST Data. In Proceedings of the 2015 Joint Urban Remote Sensing Event (JURSE), Lausanne, Switzerland, 30 March–1 April 2015; pp. 1–4.
22. Peng, S.; Piao, S.; Ciais, P.; Friedlingstein, P.; Ottle, C.; Bréon, F.-M.; Nan, H.; Zhou, L.; Myneni, R.B. Surface Urban Heat Island Across 419 Global Big Cities. *Environ. Sci. Technol.* **2012**, *46*, 696–703. [CrossRef]
23. Chakraborty, T.; Lee, X. A Simplified Urban-Extent Algorithm to Characterize Surface Urban Heat Islands on a Global Scale and Examine Vegetation Control on Their Spatiotemporal Variability. *Int. J. Appl. Earth Obs. Geoinf.* **2019**, *74*, 269–280. [CrossRef]
24. Martilli, A.; Krayenhoff, E.S.; Nazarian, N. Is the Urban Heat Island Intensity Relevant for Heat Mitigation Studies? *Urban Clim.* **2020**, *31*, 100541. [CrossRef]
25. Yang, X.; Yao, L. Reexamining the Relationship between Surface Urban Heat Island Intensity and Annual Precipitation: Effects of Reference Rural Land Cover. *Urban Clim.* **2022**, *41*, 101074. [CrossRef]
26. Zhou, B.; Lauwaet, D.; Hooyberghs, H.; De Ridder, K.; Kropp, J.P.; Rybski, D. Assessing Seasonality in the Surface Urban Heat Island of London. *J. Appl. Meteorol. Climatol.* **2016**, *55*, 493–505. [CrossRef]
27. Haashemi, S.; Weng, Q.; Darvishi, A.; Alavipanah, S.K. Seasonal Variations of the Surface Urban Heat Island in a Semi-Arid City. *Remote Sens.* **2016**, *8*, 352. [CrossRef]
28. Bechtel, B. Robustness of Annual Cycle Parameters to Characterize the Urban Thermal Landscapes. *IEEE Geosci. Remote Sens. Lett.* **2012**, *9*, 876–880. [CrossRef]
29. Bechtel, B. Multitemporal Landsat Data for Urban Heat Island Assessment and Classification of Local Climate Zones. In Proceedings of the 2011 Joint Urban Remote Sensing Event, Munich, Germany, 11–13 April 2011; pp. 129–132.
30. Mathew, A.; Khandelwal, S.; Kaul, N. Investigating Spatial and Seasonal Variations of Urban Heat Island Effect over Jaipur City and Its Relationship with Vegetation, Urbanization and Elevation Parameters. *Sustain. Cities Soc.* **2017**, *35*, 157–177. [CrossRef]
31. Aslam, B.; Maqsoom, A.; Khalid, N.; Ullah, F.; Sepasgozar, S. Urban Overheating Assessment through Prediction of Surface Temperatures: A Case Study of Karachi, Pakistan. *IJGI* **2021**, *10*, 539. [CrossRef]
32. Zhao, Z.; Sharifi, A.; Dong, X.; Shen, L.; He, B.-J. Spatial Variability and Temporal Heterogeneity of Surface Urban Heat Island Patterns and the Suitability of Local Climate Zones for Land Surface Temperature Characterization. *Remote Sens.* **2021**, *13*, 4338. [CrossRef]
33. Zhou, B.; Rybski, D.; Kropp, J.P. On the Statistics of Urban Heat Island Intensity. *Geophys. Res. Lett.* **2013**, *40*, 5486–5491. [CrossRef]
34. Bechtel, B.; Sismanidis, P.; Voogt, J.; Zhan, W. Seasonal Surface Urban Heat Island Analysis. In Proceedings of the 2019 Joint Urban Remote Sensing Event (JURSE), Vannes, France, 21–24 May 2019; pp. 1–4.
35. Shastri, H.; Barik, B.; Ghosh, S.; Venkataraman, C.; Sadavarte, P. Flip Flop of Day-Night and Summer-Winter Surface Urban Heat Island Intensity in India. *Sci. Rep.* **2017**, *7*, 40178. [CrossRef] [PubMed]
36. Zhou, D.; Zhao, S.; Zhang, L.; Sun, G.; Liu, Y. The Footprint of Urban Heat Island Effect in China. *Sci. Rep.* **2015**, *5*, 11160. [CrossRef] [PubMed]
37. Chakraborty, T.; Hsu, A.; Manya, D.; Sheriff, G. A Spatially Explicit Surface Urban Heat Island Database for the United States: Characterization, Uncertainties, and Possible Applications. *ISPRS J. Photogramm. Remote Sens.* **2020**, *168*, 74–88. [CrossRef]
38. Imhoff, M.L.; Zhang, P.; Wolfe, R.E.; Bounoua, L. Remote Sensing of the Urban Heat Island Effect across Biomes in the Continental USA. *Remote Sens. Environ.* **2010**, *114*, 504–513. [CrossRef]
39. Schwarz, N.; Lautenbach, S.; Seppelt, R. Exploring Indicators for Quantifying Surface Urban Heat Islands of European Cities with MODIS Land Surface Temperatures. *Remote Sens. Environ.* **2011**, *115*, 3175–3186. [CrossRef]
40. Li, X.; Stringer, L.C.; Dallimer, M. The Role of Blue Green Infrastructure in the Urban Thermal Environment across Seasons and Local Climate Zones in East Africa. *Sustain. Cities Soc.* **2022**, *80*, 103798. [CrossRef]

41. Liu, Z.; Zhan, W.; Lai, J.; Bechtel, B.; Lee, X.; Hong, F.; Li, L.; Huang, F.; Li, J. Taxonomy of Seasonal and Diurnal Clear-Sky Climatology of Surface Urban Heat Island Dynamics across Global Cities. *ISPRS J. Photogramm. Remote Sens.* **2022**, *187*, 14–33. [CrossRef]
42. Manoli, G.; Fatichi, S.; Bou-Zeid, E.; Katul, G.G. Seasonal Hysteresis of Surface Urban Heat Islands. *Proc. Natl. Acad. Sci. USA* **2020**, *117*, 7082–7089. [CrossRef]
43. Manoli, G.; Fatichi, S.; Schläpfer, M.; Yu, K.; Crowther, T.W.; Meili, N.; Burlando, P.; Katul, G.G.; Bou-Zeid, E. Magnitude of Urban Heat Islands Largely Explained by Climate and Population. *Nature* **2019**, *573*, 55–60. [CrossRef]
44. Beck, H.E.; Zimmermann, N.E.; McVicar, T.R.; Vergopolan, N.; Berg, A.; Wood, E.F. Present and Future Köppen-Geiger Climate Classification Maps at 1-Km Resolution. *Sci. Data* **2018**, *5*, 180214. [CrossRef]
45. Wan, Z.; Dozier, J. A Generalized Split-Window Algorithm for Retrieving Land-Surface Temperature from Space. *IEEE Trans. Geosci. Remote Sens.* **1996**, *34*, 892–905. [CrossRef]
46. Seemann, S.W.; Borbas, E.E.; Knuteson, R.O.; Stephenson, G.R.; Huang, H.-L. Development of a Global Infrared Land Surface Emissivity Database for Application to Clear Sky Sounding Retrievals from Multispectral Satellite Radiance Measurements. *J. Appl. Meteorol. Climatol.* **2008**, *47*, 108–123. [CrossRef]
47. Ghent, D.; Veal, K.; Trent, T.; Dodd, E.; Sembhi, H.; Remedios, J. A New Approach to Defining Uncertainties for MODIS Land Surface Temperature. *Remote Sens.* **2019**, *11*, 1021. [CrossRef]
48. ESA. ESA CCI Land Cover Product User Guide. 2017; p. 105. Available online: http://maps.elie.ucl.ac.be/CCI/viewer/download/ESACCI-LC-Ph2-PUGv2_2.0.pdf (accessed on 5 May 2022).
49. van der Walt, S.; Schönberger, J.L.; Nunez-Iglesias, J.; Boulogne, F.; Warner, J.D.; Yager, N.; Gouillart, E.; Yu, T. Scikit-Image: Image Processing in Python. *PeerJ* **2014**, *2*, e453. [CrossRef]
50. Rozenfeld, H.D.; Rybski, D.; Gabaix, X.; Makse, H.A. The Area and Population of Cities: New Insights from a Different Perspective on Cities. *Am. Econ. Rev.* **2011**, *101*, 2205–2225. [CrossRef]
51. Stewart, I.D.; Oke, T.R. Local Climate Zones for Urban Temperature Studies. *Bull. Am. Meteorol. Soc.* **2012**, *93*, 1879–1900. [CrossRef]
52. Schneider, U.; Becker, A.; Finger, P.; Meyer-Christoffer, A.; Ziese, M.; Rudolf, B. GPCC's New Land Surface Precipitation Climatology Based on Quality-Controlled in Situ Data and Its Role in Quantifying the Global Water Cycle. *Theor. Appl. Climatol.* **2014**, *115*, 15–40. [CrossRef]
53. Doelling, D.R.; Sun, M.; Nguyen, L.T.; Nordeen, M.L.; Haney, C.O.; Keyes, D.F.; Mlynczak, P.E. Advances in Geostationary-Derived Longwave Fluxes for the CERES Synoptic (SYN1deg) Product. *J. Atmos. Ocean. Technol.* **2016**, *33*, 503–521. [CrossRef]
54. Gouveia, C.M.; Trigo, R.M.; Beguería, S.; Vicente-Serrano, S.M. Drought Impacts on Vegetation Activity in the Mediterranean Region: An Assessment Using Remote Sensing Data and Multi-Scale Drought Indicators. *Glob. Planet. Chang.* **2017**, *151*, 15–27. [CrossRef]
55. Peel, M.C.; Finlayson, B.L.; Mcmahon, T.A. Updated World Map of the Köppen-Geiger Climate Classification. *Hydrol. Earth Syst. Sci. Discuss.* **2007**, *11*, 1633–1644. [CrossRef]
56. Taubenböck, H.; Debray, H.; Qiu, C.; Schmitt, M.; Wang, Y.; Zhu, X.X. Seven City Types Representing Morphologic Configurations of Cities across the Globe. *Cities* **2020**, *105*, 102814. [CrossRef]
57. Demuzere, M.; Bechtel, B.; Middel, A.; Mills, G. Mapping Europe into Local Climate Zones. *PLoS ONE* **2019**, *14*, e0214474. [CrossRef] [PubMed]
58. Liu, H.; Huang, B.; Zhan, Q.; Gao, S.; Li, R.; Fan, Z. The Influence of Urban Form on Surface Urban Heat Island and Its Planning Implications: Evidence from 1288 Urban Clusters in China. *Sustain. Cities Soc.* **2021**, *71*, 102987. [CrossRef]
59. Zhou, B.; Rybski, D.; Kropp, J.P. The Role of City Size and Urban Form in the Surface Urban Heat Island. *Sci. Rep.* **2017**, *7*, 4791. [CrossRef] [PubMed]

remote sensing

Article

Reverse Thinking: The Logical System Research Method of Urban Thermal Safety Pattern Construction, Evaluation, and Optimization

Chunguang Hu [1,2,3] and He Li [1,*]

1 School of Geographic Mapping and Urban Planning, Jiangsu Normal University, Xuzhou 221116, China
2 School of Architecture and Urban Planning, Huazhong University of Science and Technology, Wuhan 430074, China
3 Hubei Engineering and Technology Research Center of Urbanization, Wuhan 430074, China
* Correspondence: lihe09@mails.ucas.ac.cn

Abstract: The acceleration of urbanization has significantly impacted the changing regional thermal environment, leading to a series of ecological and environment-related problems. A scientific evaluation of the urban thermal security pattern (TSPurban) strongly benefits the planning and layout of sustainable development and the construction of comfortable human settlements. To analyze the characteristics of the TSPurban under cross-regional differences and provide targeted solutions to mitigate the urban heat island effect in later stages, the logical system research framework of the TSPurban based on the "construction–evaluation–optimization" model was explored using reverse thinking. This study selected the Wuhan metropolitan area in China as the research object. First, a morphological spatial pattern analysis (MSPA) model was used to extract the top 30 core heat island patches, and Conefor 2.6 software was used for connection analysis to evaluate their importance. Second, based on the characteristics of various land cover types, the friction (cost) map of surface urban heat island (SUHI) diffusion was simulated. The spatial attributes of the heat island resistance surface were examined using a standard deviation ellipse and hot spot analysis. Finally, this paper used circuit theory to find 56 low-cost heat island links (corridors) and circuit scape software to find widely distributed vital nodes. The optimization of the TSPurban network was then investigated using a reverse thinking process. Heat island patches, corridors, and vital nodes are among the crucial components of the TSPurban. By obstructing corridor links and disturbing important nodes, it is possible to appropriately and proficiently reduce the TSPurban network's connection efficiency and stability, which will have a positive influence on regional climate mitigation and the heat island effect.

Keywords: urban thermal security pattern; surface temperature; circuit theory; research framework; Wuhan urban agglomeration; China

Citation: Hu, C.; Li, H. Reverse Thinking: The Logical System Research Method of Urban Thermal Safety Pattern Construction, Evaluation, and Optimization. *Remote Sens.* **2022**, *14*, 6036. https://doi.org/10.3390/rs14236036

Academic Editors: Yuji Murayama and Ruci Wang

Received: 21 October 2022
Accepted: 27 November 2022
Published: 29 November 2022

1. Introduction

As the world urbanizes, impervious surfaces such as cement, asphalt, and concrete are replacing the natural surface [1]. Heat builds up in metropolitan areas due to built structures and human activity. This build-up leads to a significant temperature difference between urban and suburban areas, resulting in the urban heat island (UHI) effect [2], which Howard originally discovered in 1833 [3]. Subsequently, numerous scholars have carried out a considerable number of studies on the form and structure [4,5], evolution and change [6,7], and mechanism and simulation [8,9] of UHIs. Relevant studies have considered the UHI effect to be one of the most critical issues influencing urban public health and sustainable development [10,11], regional climate [12], biodiversity [13], vegetation phenology [14], and quality of air and water [15]. Even human morbidity and mortality are impacted by UHIs [6].

The thermal environment security of urban agglomerations is facing unprecedented problems, particularly in light of the growing worldwide UHI intensity, the rapid expansion of urban development land, the ongoing growth in city populations, and the effects of global warming [7]. Academic research on heat mitigation is becoming increasingly important [5]. Therefore, it is extremely relevant for regional climate research to study the characteristics and optimization of the urban thermal security pattern model (TSPurban). Generally, atmospheric UHIs (AUHIs) and surface UHIs (SUHIs) are the two major types of UHIs [16]. We can calculate SUHIs by observing the land surface temperature (LST), which can be retrieved from satellite observations [17]. Additionally, we understand that SUHIs have convenient data acquisition characteristics and ranges of observable coverage [18]. Therefore, this study focuses on SUHIs.

Regarding how to effectively mitigate the security of the urban thermal environment, studies have shown that blue and green spaces play a significant role in cooling the Earth's surface compared to the surrounding thermal environment. Urban green infrastructure (UGI) can influence urban climate by increasing transpiration [19], providing shading or water features, and supporting heat exchange through the selective absorption and reflection of solar radiation [20,21]. However, these studies have mainly used forward thinking to reduce UHIs by directly selecting negative correlates that counteract the thermal environment. Furthermore, they have mainly focused on isolated spatial patches; less consideration has been given to the holistic and connected nature of the UHI mitigation effect [22,23].

Due to its abstract character, the term "reverse thinking" necessitates a thorough conceptual understanding before it can be properly applied to the text's internal logic. From the perspective of the term's connotation, "reverse thinking" refers to thinking backward to solve problems that are difficult to solve by conventional thinking. From the perspective of the TSPurban, we don't use large-scale cooling measures. Instead, we focus on correctly locating the important nodes and corridors of SUHIs and stopping nodes from connecting. This keeps a network of heat island patches from forming, which can do a lot of damage to the thermal environment in the region. Therefore, we choose to address the heat island patch problem in the real world and use "reverse thinking" to consider how to make an effective breakthrough based on the current TSPurban. In this paper, we will examine "reverse thinking" as a way to deal with TSPurban from two points of view: theoretical research and practical exploration. We'll talk about how reverse thinking works and what it means in real life.

In terms of theoretical research, on the one hand, studies have shown that the connectivity of heat island patches potentially enhances the regional UHI effect [24,25]. In this case, it is important to effectively block the connectivity between heat island patches, and this connectivity needs to be reflected at the specific spatial level of the targeted situation [26–28]. However, these studies have not received sufficient attention [29]. On the other hand, other researchers have focused on spatial connectivity, such as Peng et al. (2022), who focuses on enhancing the connectivity of future "cool islands" (CIs) by constructing a network, which is typical of forward positive thinking [30]. Based on the current situation, the heat island patch network problem in the TSPurban urgently needs an effective breakthrough. At the same time, since the connectivity of heat island patches has the function of enhancing the UHI effect [31,32], a new method to mitigate the UHI effect by interrupting the connectivity of heat island patch corridors is proposed for later stages.

On the one hand, in terms of practical exploration and due to real environmental costs, especially in the past, many investments have been made to mitigate the regional thermal environment [33]. Among them, urban blue/green landscape space is characterized by limited development and high investment costs [34]; with the overall economic slowdown, it is challenging to break through the limited funds available for mitigating urban thermal safety patterns. On the other hand, due to China's territorial spatial planning control, too many projects to mitigate the regional thermal environment will involve encroachment on many permanent basic agricultural land and construction development control areas [35,36]. Therefore, it will be more difficult to coordinate these projects under realistic conditions, which in turn will reduce the operability of regional thermal environment mitigation projects.

Therefore, constructing a specific distribution of heat island networks at the spatial level is an important technical point. Based on the analysis based on previous research, it was found that circuit theory, although it first originated as a related concept in physics, was gradually introduced to identify important heat island corridors and critical nodes related to heat exchange based on LST data. Studies have already been done that apply circuit theory to these networks, mainly by defining heat source areas and connecting corridors to build networks [30,32]. On the one hand, these corridors and critical nodes are similar to the effects of electric currents in that they have a random, "wandering" nature. On the other hand, this approach can assist us in pinpointing the key nodes and corridors of the SUHI. Thus, it is possible to identify important patch corridors and nodes in the TSPurban based on the movement patterns in complex cost resistance.

This paper constructed a regional heat island network based on the circuit theory simulation, i.e., the final construction of the three major elements of the TSPurban, namely, the heat source, the resistance surface, and the corridor. In the evaluation of the TSPurban, the heat source analyzed patch fragmentation through connectivity, the resistance surface revealed spatial directional characteristics via the standard deviation ellipse, and the corridor analyzed hot and cold spot clustering regions via spatial autocorrelation. The corridor also thought about how important it was and figured out where the most important heat flow nodes were. In the optimization of the TSPurban, this paper mainly relies on "reverse thinking" to mitigate the heat island effect through the following three dimensions: the first dimension is to weaken the heat source role of the main heat island patches; the second is to disrupt the linking effect by reducing the role of links in the heat island corridors; the last is to use the key areas identified on the corridor for targeted engineering projects that can mitigate the heat island effect. In this way, the current UHI effect can be alleviated through gradual weakening. In summary, precisely locating the key nodes and corridors in the urban heat island (SUHI) at the spatial level is the basis for the effective mitigation of SUHIs. We use "reverse thinking" to efficiently and accurately "block" or "destroy" the critical nodes and corridors in the network to effectively decrease the regional heat island effect.

Additionally, the administrative boundaries of cities determined by anthropogenic rules should not bind the natural heat island network. However, most current research on heat island networks lacks a cross-regional perspective [37], so it is difficult to fully consider the spatial organization of heat islands. The inspiration for regional TSPurban governance and optimization is limited. Even from a cross-regional perspective, the research mainly focuses on China's Yangtze River Delta, Pearl River Delta, and other open coastal city clusters [38]. Relevant research has not paid enough attention to the central metropolitan area. Therefore, taking the Wuhan urban agglomeration as the research object, this paper reveals the locational characteristics of urban heat island patches and the spatial pattern of collaborative optimization. Then, we look at the main problems caused by the urban heat island effect, and finally, we suggest ways to improve transregional efforts to reduce urban heat islands. Specifically, this paper proposes and attempts to explore the following research objectives: (1) to investigate the characteristics of the TSPurban in the context of cross-regional differences; (2) to accurately identify corridors and key nodes in the TSPurban and then propose targeted solutions; (3) to provide strong theoretical support for the construction of regionally integrated urban heat island effect mitigation solutions.

2. Research Overview

2.1. Research Area

As shown in Figure 1, the Wuhan Urban Agglomeration (Central China) refers to the area around Wuhan with a radius of approximately 100 km. The total area of Wuhan, Huangshi, Ezhou, Xiaogan, Huanggang, Xianning, Xiantao, Tianmen, and Qianjiang is 579×10^4 hm^2, accounting for 31.09% of the total area of Hubei Province. In 2021, the total resident population of the region exceeded 33 million, and the GDP exceeded 3 trillion. The Wuhan Urban Agglomeration is located at latitudes 29°02′~31°51′N and longitudes 112°33′~116°07′E. It belongs to the midlatitude zone. Solar radiation varies seasonally, and the area is far from the ocean. It has a typical subtropical East Asian continental climate, with four clear seasons and substantial amounts of light, heat, and rain.

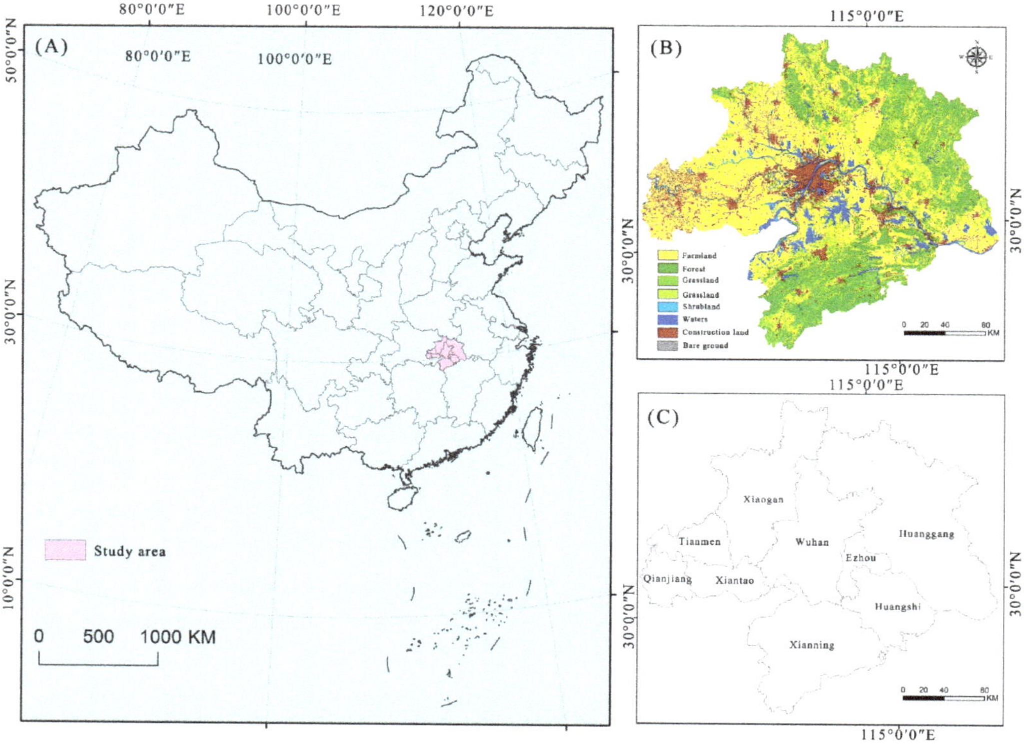

Figure 1. Location of the study area (**A**), land cover (**B**), and administrative division (**C**).

2.2. Research Framework and Data Sources

2.2.1. Research Framework

Some scholars have explored mitigation strategies for the UHI effect from the perspective of constructing spatial networks. For example, Yu et al. (2021) identified the critical nodes in a UHI network and then broke the network to effectively mitigate SUHI [32]. To reflect the contributory nature of the article in examining the overall logic of the TSPurban, in this research, as shown in Figure 2, we propose a TSPurban based on the new paradigm of "heat source–resistance surface–corridor" and alleviate the UHI effect by blocking connectivity. Finally, we creatively develop the research framework of the TSPurban's "construction–evaluation–optimization" logic system.

First, in terms of construction, the heat source was mainly defined by the climate characteristics of the Wuhan metropolitan area. Then, the average maximum value of that period was synthesized, and the morphological spatial pattern analysis (MSPA) model was used to further extract the core heat island patches. Surface resistance mainly combines different land use cover characteristics in the region and generates a friction (cost) map by giving a differential resistance coefficient. The corridor mainly uses circuit theory to identify the path with the least cumulative resistance by sharing random "wandering" attributes and finally identifies the important SUHI links and key nodes related to heat flow, forming a regional heat island network.

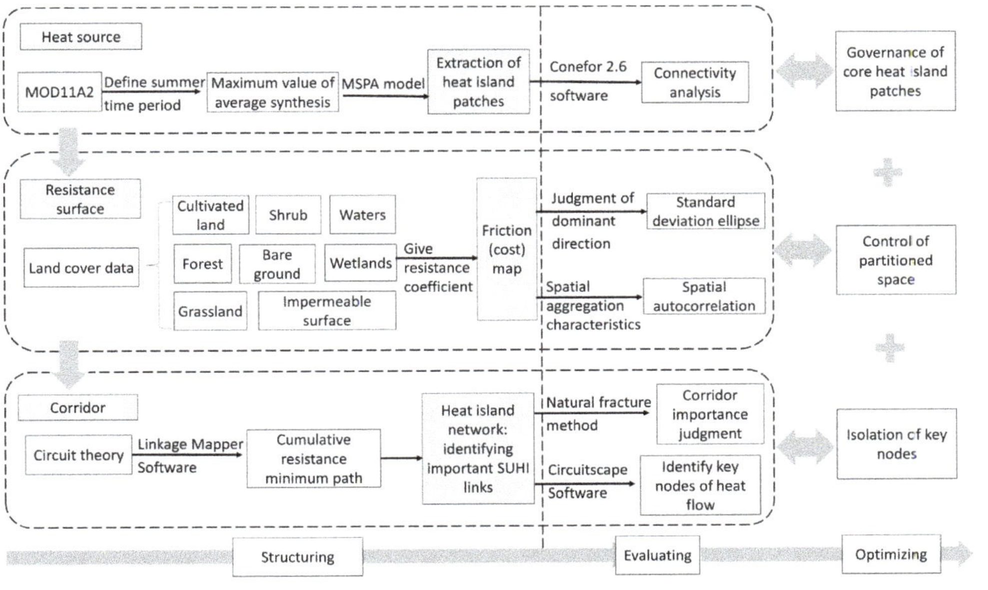

Figure 2. TSPurban's logical framework of "construction–evaluation–optimization".

Second, in the evaluation, Conefor software can quantitatively evaluate the connectivity of important patches as the basis for later maintenance or connectivity improvement [39,40]. Therefore, the heat source mainly uses Conefor 2.6 software to analyze the connectivity of the extracted core heat island patch. To measure the directional and element aggregation characteristics of the resistance surface, the standard deviation ellipse is used to judge the dominant direction, and the spatial autocorrelation is used to analyze the spatial aggregation characteristics. At the same time, in terms of corridor evaluation, the natural fracture method is used to judge the importance of corridors. In the circuit model, the random walk of electrons in the circuit is used to simulate the process of heat flow in regional diffusion [41,42]. Therefore, Circuitscape software can be employed to find the most important nodes, setting the stage for more precise heat island mitigation.

Finally, we investigate the precise and efficient mitigation strategies from the three aspects of heat islands (heat source, resistance surface, and heat corridor) based on the aforementioned construction and evaluation via a reverse thinking method. Then, we control the space between the partitions, manage the core heat island patches, and isolate the important nodes from the corridor.

2.2.2. Data Sources

MODIS remote sensing data are used as the data source for SUHIs in this project. The MOD11A2 remote sensing image is a 1 km^2 resolution image of the Earth's surface temperature in 2020 synthesized over eight days. The data can be obtained from NASA's official website (https://ladsweb.modaps.eos.dis.nasa.gov/search, accessed on 1 June 2022). The data sensor is on board the Terra satellite.

GlobeLand30 data (http://www.globallandcover.com/, accessed on 15 June 2022), developed by the National Basic Geographic Information Center, provided the land use data for this study. The data are multispectral images at 30 m resolution that use the POK method based on pixel classification, object extraction, and knowledge verification [37]. According to China's standard for land use classification, the study area is divided into eight categories: farmland, forest, grassland, shrubland, wetland, waters, construction land, and bare land.

In this study, all the spatial data are projected into the WGS_1984_UTM_Zone_50N coordinate system. We also use tools for resampling to turn the data into uniform raster data with a 30 m spatial resolution.

3. Method

3.1. Seasonal Division and Calculation of Land Surface Temperature (LST)

With an annual mean temperature of 17 °C and an annual temperature difference of 16 °C, the area has a typical subtropical monsoon humid climate. In most areas, the annual daily temperature exceeds 35 °C for approximately 10–30 days. The maximum temperature is above 30 °C for 60 to 80 days out of the year. The minimum temperature is below 0 °C for 80 to 110 days out of the year in most areas, decreasing from north to south. June, July, August, and September (defined as the summer months) are the four hottest successive months of the year, with average temperatures exceeding 22 °C. Therefore, we will calculate the average maximum of the 6- to 9-month LST composition.

3.2. Division of Urban Heat Island Intensity

Due to the difference in time and environment, the surface temperatures cannot be directly compared. However, the distribution of the relative intensity of the thermal field is not affected by other factors, so to analyze the relative spatiotemporal changes in the surface thermal environment, the urban surface thermal field is classified based on the surface temperature [12].

This study uses the mean-standard deviation method to classify the land surface thermal field. This method calculates the land surface temperature by multiple combinations of the mean and standard deviation and defines the urban thermal environment. The standard deviation reflects the deviation of the surface temperature from the mean temperature. Compared with the common method of equidistant density segmentation, the standard deviation can better reflect the classification characteristics of the urban heat island [11]. Based on the above theory, the surface heat field of the Wuhan Urban Agglomeration is divided into five grades with μ (mean), 0.5 std (standard deviation), and 1 std as the cutoff points. To analyze the heat island state, the high-temperature and medium-high-temperature areas are taken as the heat island range. Table 1 shows the specific rules for dividing the LST intensity by using the mean value and standard deviation.

Table 1. Division method of the mean-standard deviation thermal field.

Temperature Class	Division Method of Thermal Field
High-temperature zone	$Ts > \mu + std$
Medium-high temperature zone	$\mu + 0.5\ std \leq Ts \leq \mu + std$
Middle-temperature zone	$\mu - 0.5\ std \leq Ts \leq \mu + 0.5\ std$
Medium-low temperature zone	$\mu - std \leq Ts \leq \mu - 0.5\ std$
Low-temperature zone	$Ts \leq \mu - std$

3.3. MSPA-Based SUHI Mode

The morphological spatial pattern analysis method (MSPA) proposed by Vogt and other scholars can accurately classify grid images' spatial pattern function types [40]. It can identify patches that play an essential role in regional landscape connectivity at the pixel level [43] and provide a more scientific basis for the selection of ecological sources and ecological corridors [44]. According to the needs of the MSPA method, ArcGIS is used to convert the land use classification data after remote sensing interpretation into 30 m × 30 m binary TIFF grid data [45]. The foreground and background are two categories of research objects in the MSPA method. Therefore, to identify the types of heat island patches important for maintaining connectivity, the extracted heat island area is set as the foreground data (set to 2), and other areas are the background data (set to 1). We used GuidosToolbox2.8 software to analyze the landscape pattern. After a series of mathematical calculations, such as skeleton extraction and corrosion calculation, seven landscape types with different functions are obtained. Next, we count the analysis results for MSPA. The core area is basically a large heat island patch within the research area, which may be the primary locus where solar radiation is absorbed. From here, we can select the source of the core heat island.

3.4. Connectivity Analysis

Connectivity analysis was initially applied to the landscape field, which refers to the comprehensive effect of the ecological processes of energy, material exchange, and migration among landscape elements and the movement rate of patches [46]. Therefore, connectivity analysis can also measure the connectivity and importance of heat island patches. In this paper, the "core" SUHI patches in MSPA are extracted, and the patch importance index dPC is analyzed using Conefor 2.6 software to measure the degree of connectivity between the core heat island patches in the study area. The higher connectivity index of the heat island patches indicates that they are essential to the heat sources of SUHIs. The calculation formula is as follows:

$$PC = \frac{\sum_{i=1}^{n} \sum_{j=1}^{n} a_i \times a_j \times p_{ij}^*}{A_L^2} \quad (1)$$

$$\text{dPC} = \frac{PC - PC_{\text{remove}}}{PC} \times 100\% \quad (2)$$

where $i \neq j$; n represents the number of patches in the core area; a_i and a_j represent the areas of patches i and j, respectively; and p_{ij}^* represents the maximum product of all path probabilities between patch i and patch j. The value range of PC is 0 to 1. The larger the PC value is, the higher the connectivity between patches; PC_{remove} indicates the possible connectivity index of the remaining patches after removing a patch.

3.5. Building the Friction (Cost) Map

Previous studies have fully demonstrated the ability of land cover elements (such as forest, water, and grassland) to reduce temperature [37,47]. The resistance scores of different land use types in the heat island diffusion pathway are determined, and the larger the resistance scores are, the greater the heat island blocking force, patch migration, and diffusion.

Because of their large heat capacities, low heat conductivities, and slow rate of temperature rise, wetland and water areas typically have the lowest temperatures [48]; hence, the resistance of the wetland and water areas is 100, ranking first.

Forestland also plays essential roles in lowering the surface temperature, collecting dust, minimizing wind, fixing soil fixation, beautifying the area, and greening cities [49]. Forestland is thought to play a large part in maintaining the stability of the urban environment; hence, its resistance value is 75, ranking second.

Irrigation of arable land has the potential to reduce regional temperatures. Water evaporation can absorb a large amount of heat from the air via a phenomenon called the "irrigation cooling effect" [50]. Additionally, crops grown on the cultivated land create high levels of vegetation coverage, reducing the bare surface area. As a result, the surface temperature is significantly reduced [51]. Therefore, the resistance value of arable land is set at 50, ranking third. Additionally, the effect of shrubs on plant coverage was less than that of woodlands; hence, the value of resistance for shrubs is also 50.

The results show that the mean temperature of grassland is the highest among the vegetation types, which indicates that the effect of grassland on urban heat islands is not pronounced [52]. Hence, the resistance value of grassland is 25, ranking fourth.

In general, the average temperature of construction land is very high, mainly because of the high density of buildings within the land area, which is not conducive to air circulation. Building materials such as bricks, tiles, and cement have low thermal capacity and inertia, but their thermal conductivity and diffusivity are large. After receiving solar radiation, these materials diffuse heat quickly into the surrounding atmosphere, resulting in the surrounding temperature being higher than that of the vegetated areas [53,54]. Therefore, the resistance value of construction land is 5, ranking fifth.

The high surface temperature of unused land is due to the characteristics of bare land. Because there is no plant growth on bare land and solar radiation impacts bare land directly, the temperature changes considerably [55,56]. Therefore, we set the bare ground resistance value at 1, ranking it sixth.

3.6. Standard Deviation Ellipse and Spatial Autocorrelation Analysis

The standard deviation ellipse can measure the concentration, dispersion, and direct distribution of geographical elements, explore the distribution and development trend of geographical elements, and visualize the spatial pattern of geographical elements. Lefever first proposed this method in 1926, and it is widely used in many fields and studies, such as population, economy, and sociology [57,58]. Based on the analysis of the standard deviation ellipse of the heat island friction (cost) map for the Wuhan urban agglomeration, the direction and trend of the heat island resistance patch in the study area are reflected. Among them, the shape of the ellipse can, to some extent, reflect the characteristics of the geographical element distribution. It mainly includes the directional angle, center, ellipse X and Y axes, and standard deviation. The formulas are as follows:

$$H_x = \sqrt{\frac{\sum_{i=1}^{n}(x_i - \overline{x})^2}{n}} \quad (3)$$

$$H_y = \sqrt{\frac{\sum_{i=1}^{n}(y_i - \overline{y})^2}{n}} \quad (4)$$

where H_x and H_y represent the calculated ellipse variance, x_i and y_i are the coordinates of geographical element i, $\overline{x}$ and $\overline{y}$ are the arithmetic average centers of geographical elements, and n is the number of elements. The directional angle of an ellipse is calculated by:

$$\tan\theta = \frac{\sum_{i=1}^{n}\widetilde{x}_i^2 - \sum_{i=1}^{n}\widetilde{y}_i^2 + \sqrt{\left(\sum_{i=1}^{n}\widetilde{x}_i^2 - \sum_{i=1}^{n}\widetilde{y}_i^2\right)^2 + 4\left(\sum_{i=1}^{n}\widetilde{x}_i\widetilde{y}_i\right)^2}}{2\sum_{i=1}^{n}\widetilde{x}_i\widetilde{y}_i} \quad (5)$$

where θ is the angle that starts at 0° due north and rotates clockwise to the x-axis; $\widetilde{x}_i$, and $\widetilde{y}_i$ is the mean center deviation. The standard deviation of the x- and y-axes can be determined:

$$\left.\begin{array}{l} \sigma_x = \sqrt{2}\sqrt{\frac{\sum_{i=1}^{n}(\widetilde{x}_i\cos\theta - \widetilde{y}_i\cos\theta)^2}{n}} \\ \sigma_y = \sqrt{2}\sqrt{\frac{\sum_{i=1}^{n}(\widetilde{x}_i\sin\theta + \widetilde{y}_i\cos\theta)^2}{n}} \end{array}\right\} \quad (6)$$

where σ_x and σ_y are the standard deviations of the x- and y-axes, respectively.

Global spatial autocorrelation and local spatial autocorrelation are two aspects of spatial autocorrelation analysis. Moran's I index, derived from global spatial autocorrelation analysis, can comprehensively describe the spatial average correlation degree, spatial distribution pattern, and significance of specific variables or attributes of each unit in the region [59]. The formula is expressed as:

$$I = \frac{n\sum_{i=1}^{n}\sum_{j=1}^{n}W_{ij}\left(x_i - \overline{X}\right)\left(x_j - \overline{X}\right)}{\left(\sum_{i=1}^{n}\sum_{j=1}^{n}w_{ij}\right)\sum_{i=1}^{n}\left(x_i - \overline{X}\right)^2} \quad (7)$$

where x_i and x_j denote the spatial position of unit i and unit j, respectively, and $i \neq j$; $\overline{X}$ denotes the average value of the corresponding attribute values of n position units; W_{ij} denotes the spatial weight matrix, and n represents the number of cells. The significance of spatial autocorrelation was tested by the Z statistic of a normal distribution [60]. The value of Moran's I index is distributed between ±1; the absolute value indicates the strength of the spatial correlation of variables, the positive sign and the negative sign indicate the positive and negative correlation of variables in the region, respectively, and the trend to 0 indicates the noncorrelation and the random distribution.

Although the global Moran's I index can judge the global correlation degree of an attribute value in the region, it cannot determine the specific location of the aggregation. Due to spatial heterogeneity, the spatial autocorrelation degree of each location in the study area can be different, so local Getis-OrdGi* hot spot analysis is carried out to locate the specific location of the spatial aggregation of heat island resistance. Hot spot analysis is used to identify statistically significant high-value and low-value aggregation areas, that is, hot and cold spots [61]. It is expressed as:

$$G_i^* = \frac{\sum_{j=1}^{n} W_{ij}x_j - \overline{X}\sum_{j=1}^{n} W_{ij}}{S\sqrt{\frac{n\sum_{j=1}^{n} W_{ij} - \left(\sum_{j=1}^{n} W_{ij}\right)^2}{n-1}}} \tag{8}$$

where S is the standard deviation of the corresponding attributes of n units; the others are the same as in Formula (7). To facilitate comparison and research, G_i^* is normalized and expressed as:

$$Z(G_i^*) = \frac{1 - E(G_i^*)}{\sqrt{VAR(G_i^*)}} \tag{9}$$

where $E(G_i^*)$ is the expected value of the local Getis-OrdGi* index and $VAR(G_i^*)$ is the variance of the local Getis-OrdGi* index. $Z(G_i^*) > 1.96$ indicates a significant hot spot area; $1.65 < Z(G_i^*) < 1.96$ indicates a significant hot spot area; $-1.65 < Z(G_i^*) < 1.65$ indicates an area with insignificant aggregation; $-1.96 < Z(G_i^*) < -1.65$ indicates a significant cold spot area; and $Z(G_i^*) < -1.96$ indicates a significant cold spot area.

3.7. Construction of a Heat Island Network Based on Circuit Theory

In circuit theory, the movement of heat island patches is analogous to current, and the friction (cost) map is defined as a resistance surface. The higher the cumulative current value is, the better the connectivity of the circuit network, and vice versa [62,63]. In the process of simulating the regional diffusion of heat island patches based on circuit theory, some patches are first grounded, 1A current is input to other patches, the friction (cost) resistance surface is constructed, the current value between each pair of patches is calculated, and the current value can represent the probability of heat island patches moving along a certain path. Therefore, different calculation methods are used to determine how well different decentralized paths in the heat island network work based on how the electrons in the circuit move around randomly.

Based on circuit theory, this study used Circuitscape software and the Linkage Mapper tool for simulations. Circuitscape software provides four methods of calculation, of which "all-to-one" (many-to-one mode) is mainly used for the identification of essential patches, and "pairwise" (pair mode) is mainly used for the identification of corridors. In this study, "all-to-one" and "pairwise" were used to simulate the connectivity between heat island sources. Regarding how to scientifically set the resistance value, on the one hand, the theoretical circuit model creates the accumulated current value of migration or diffusion between patches through different resistance values [64]. Although different resistance values will produce different current densities, connectivity between patches can be observed [41] because a higher accumulated current value indicates that the connectivity between two patches in the region is better [40]. On the other hand, this paper combines the research results of relevant circuit theory and considers the propagation distance of heat island patches to comprehensively determine the resistance value [65,66]. The Linkage Mapper tool is used to construct the UGI network, and the search radius of the barrier Mapper tool is set to 500 m to identify obstacles in the network.

4. Results

4.1. Construction of the Urban Thermal Security Pattern

Figure 3A shows the average maximum value of LST synthesis in the Wuhan Urban Agglomeration from June to September 2020, where the minimum temperature was 23.33 °C, the maximum temperature was 48.17 °C, the mean value was 33.76 °C, and the standard deviation was 2.31. The LST showed an extensive range of changes or fluctuations. Five grades of heat island intensity in the Wuhan Urban Agglomeration are divided using the mean standard deviation. Figure 3B depicts the spatial distribution of heat island intensity change. The high-temperature area was 7774.29 km^2, accounting for 13.41% of the total study area. The area of the medium- and high-temperature areas was 8462.14 km^2, accounting for 14.59% of the total study area, indicating that the heat island covered 28.00% of the total study area. The proportion of the medium-temperature zone was as high as 40.09%, that of the medium- and low-temperature zones was 18.28%, and that of the low-temperature zone was 13.63%. Specifically, from the perspective of spatial distribution, the heat island patches in the Wuhan metropolitan area were mainly concentrated in the area of Wuhan's main urban regions. However, they tended to spread to the surrounding areas. In addition, the "1 + 8" urban circle of Tianmen (Zone b), Huanggang (Zone c), Huangshi (Zone d), Xianning (Zone e), and other urban centers also had apparent heat islands. The spatial pattern of the heat island was consistent with the urban spatial distribution. The background areas were rich in water and forests; thus, they were mainly low-temperature and medium-low-temperature areas.

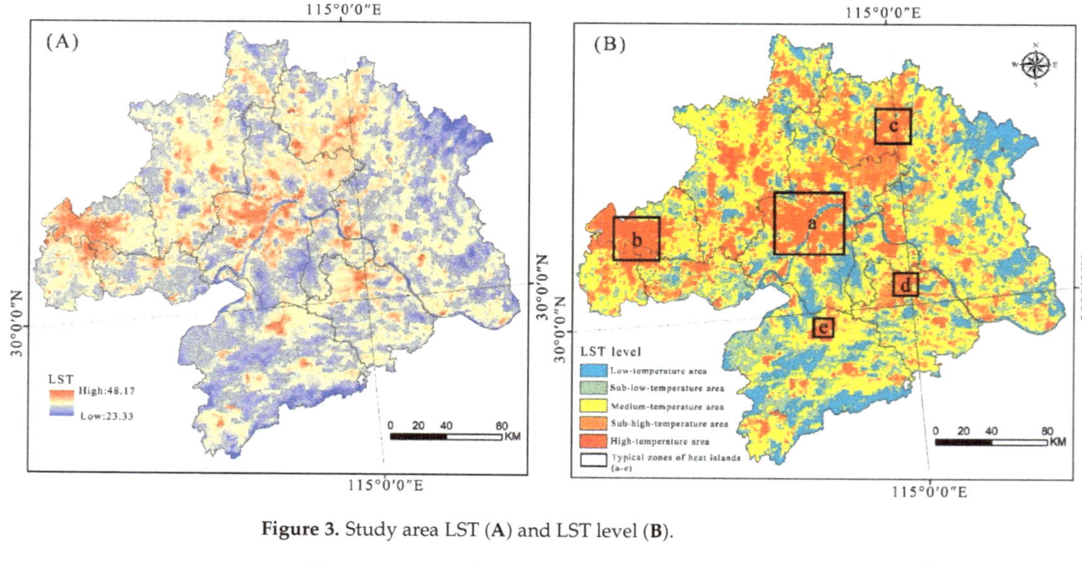

Figure 3. Study area LST (**A**) and LST level (**B**).

The prospective data area for MSPA analysis was 16,236.4356 km^2, as shown in Table 2 and Figure 4A, accounting for approximately 27.999912% of the total area of the study area. In the foreground data, the heat island area in the core area was the largest, accounting for 97.59243% of the total foreground area. In addition, as the transition patch between the core heat island area and the outside background area, the marginal area accounts for 2.25766% of the foreground area. The porosity was along the inner edge of the core region, which had the same edge effect as the edge region. The porosity accounted for 0.14583% of the foreground area, indicating that the core heat island region had a specific edge effect. Moreover, the rest of the measurement indicators accounted for less.

Table 2. MSPA classification statistics.

Types	Area/km^2	Percent in Foreground Area/%	Percent in Total Area/%
Core	15,845.5323	97.59243%	27.323795%
Loop	0.1377	0.00085%	0.000237%
Bridge	0.4896	0.00302%	0.000844%
Edge	366.5637	2.25766%	0.632143%
Islet	0.0027	0.00002%	0.000005%
Branch	0.0324	0.00020%	0.000056%
Perforation	23.6772	0.14583%	0.040832%

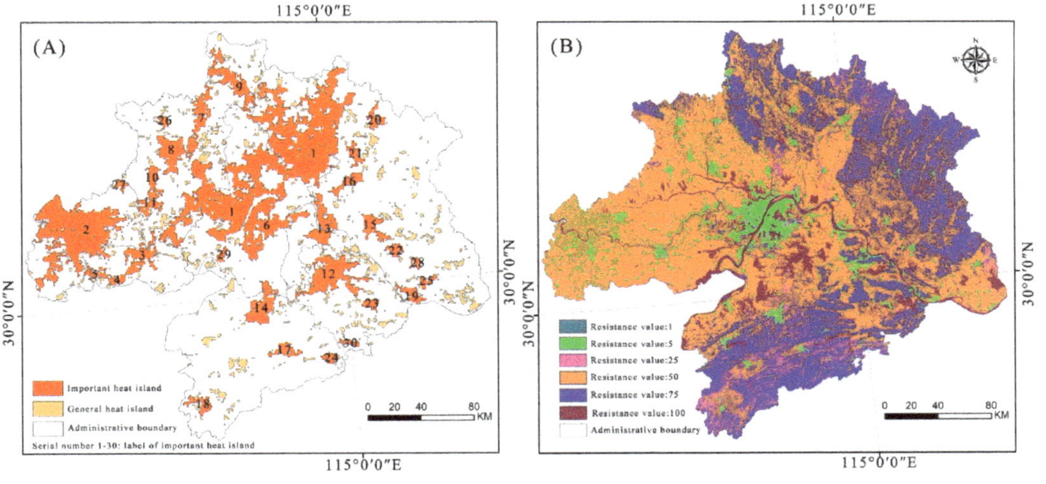

Figure 4. Heat island map (**A**) and friction (cost) map (**B**) in the study area.

Through the analysis of land cover resistance factors, Figure 4B shows the resistance surface of heat island propagation and diffusion in the Wuhan Urban Agglomeration. The study area's spatial distribution of heat island resistance generally presented a "high in the east, low in the west, low in the middle, and high in the north and south" pattern. The terrain significantly influenced the heat island resistance distribution. The resistance value was low on the riverside plain in eastern Hubei and the Hanjiang plain west of the circle. The resistance values in the vast northern, northeastern, and southern low mountain and hilly areas were generally on the high side. In addition, the dense distribution of construction land led to the distribution of low-value resistance, which promoted the flow of heat island energy and led to the better connectivity of regional heat islands.

Figure 5A depicts the study area's minimum cumulative resistance distribution map. The areas with larger cumulative resistance values were mainly distributed in the low mountain and hilly regions of the northeast and south. The distribution pattern was mainly related to the trend of the valleys. Therefore, a "depression" was formed due to the complex topography, convenient transportation, flat terrain, water distribution in more areas, more human activities, and heat island expansion resistance. As shown in Figure 5B, the minimum cost path for the starting and ending points was calculated by the minimum path tool from the minimum cumulative resistance model, simulating potential diffusion corridors that interconnect between patches of heat island sources. The results showed 56 channels in the Heat Island Patch, among which the northern corridor was sparse, and the eastern and southern corridors were more densely distributed, indicating that the exchange of matter and energy was more unimpeded in the Heat Island Patch.

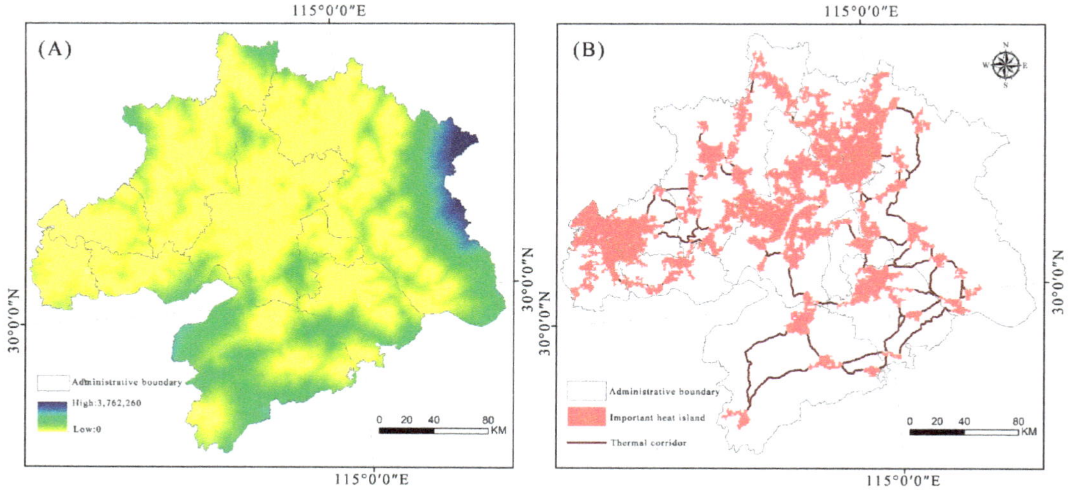

Figure 5. Distribution of minimum cumulative resistance (**A**) and the thermal corridor (**B**).

4.2. Evaluation of the Urban Thermal Safety Pattern

Table 3 measures the significance of the 30 heat sources selected in the previous stage using dPC. The total area of the heat island source area was 13,023.2295 km^2, which was distributed widely in space. The source area with a sequence number of 1 in the north was 5227.9668 km^2, and the patch importance index (dPC) was 87.1181, indicating that the well-connected source area was conducive to the exchange of heat islands. It was located at the source of the heat island with serial numbers 2–7 in the north, and its patch importance index (dPC) was greater than 10, so its connectivity was better than that of other areas. The heat island source with serial numbers 8–11 had a patch importance index (dPC) greater than one and medium connectivity; however, the connectivity of heat island patches with other serial numbers was small. In the Wuhan urban agglomeration, heat island sources were not connected in the same way, and there was considerable patch fragmentation.

Table 3. Importance of heat source dPC value.

Serial Number	Area/km^2	dPC Value
1	5227.9668	87.1181
2	2124.2952	23.8945
3	314.8272	22.0948
4	68.1552	18.6799
5	90.6255	18.2483
6	793.9566	17.1256
7	352.3500	10.6279
8	366.7203	6.7918
9	434.9169	6.6787
10	120.9843	2.2007
11	163.9728	1.3063
12	713.1960	0.7596
13	384.5322	0.2208
14	373.0311	0.2078
15	207.6714	0.0644
16	149.6457	0.0334
17	136.6938	0.0279
18	124.4772	0.0231
19	120.0708	0.0215

Table 3. *Cont.*

Serial Number	Area/km^2	dPC Value
20	116.4915	0.0203
21	108.6786	0.0176
22	85.7034	0.0110
23	69.7050	0.0073
24	68.1696	0.0069
25	62.8362	0.0059
26	51.4611	0.0040
27	51.1101	0.0039
28	48.0726	0.0035
29	46.6524	0.0033
30	46.2600	0.0032

Note: The above serial numbers are consistent with the serial numbers in Figure 4A.

To reveal the spatial characteristics of the heat island friction (cost) map of the Wuhan urban agglomeration, this paper measured the center of gravity shift and ellipse distribution of the standard deviation ellipse. The center of gravity of the heat island friction (cost) map corresponded to the center of mass of this geographic plane. Table 4 shows the specific parameters of the ellipse. Combined with the ellipse shape of the heat island friction (cost) map in Figure 6, the east-west semiaxis was more significant than the south-north semiaxis, showing an obvious east-west distribution pattern. The standard deviation ellipse azimuth θ of the friction (cost) map was approximately 92.13°, which basically followed a horizontal straight line.

Table 4. Standard deviation ellipse parameters for friction (cost) maps.

The Circumference of an Ellipse/km	The Area of an Ellipse/km^2	Center Point X Coordinates	Center Point Y Coordinates	The Length of the X-Axis of an Ellipse/km	The Length of the Y-Axis of the Ellipse/km	Azimuth/(°)
635.32	27,922.31	114°6′33″	30°24′4″	114.54	77.60	92.13

Based on the first-order spatial weight matrix of the Queen Standard, the global spatial autocorrelation analysis of the Wuhan urban heat island friction (cost) map was carried out. Furthermore, the global Moran's I index results were obtained. The global Moran's I index was 0.41, and all attributes passed the significance level test with a P value less than 0.01 and a z value greater than 2.58. The heat island friction (cost) map showed significant spatial autocorrelation. That is, the map of the Wuhan Urban Agglomeration's heat island friction (cost) showed a very significant spatial clustering effect, and counties with high (or low) heat island friction (cost) were close to each other.

To better understand the local spatial aggregation characteristics of the heat island friction (cost) map, the hot spots and cold spots of the heat island friction map were explored using the Getis-OrdGi* hot spot analysis method. Compared with the local spatial autocorrelation, the hot spot analysis was more evident for the aggregation state of heat island friction (cost). According to the hot spot analysis, as shown in Figure 6, there were hot and cold spots in the friction (cost) of the heat island. The friction (cost) spatial agglomeration of the Wuhan urban heat island was characterized by "cold in the west and hot in the east". Heat island friction (cost) hotspots were mainly located in lakes and nature reserves such as Long Lake and Hong Lake. At the same time, the eastern and southern regions contained low mountains and hills, the land use cover types were mostly woodland and grassland, the forest coverage rate was high, and the population was small, which was the secondary hot spot. The friction (cost) cold spots of heat islands were mainly found in the region's middle and west, where there was much human disturbance.

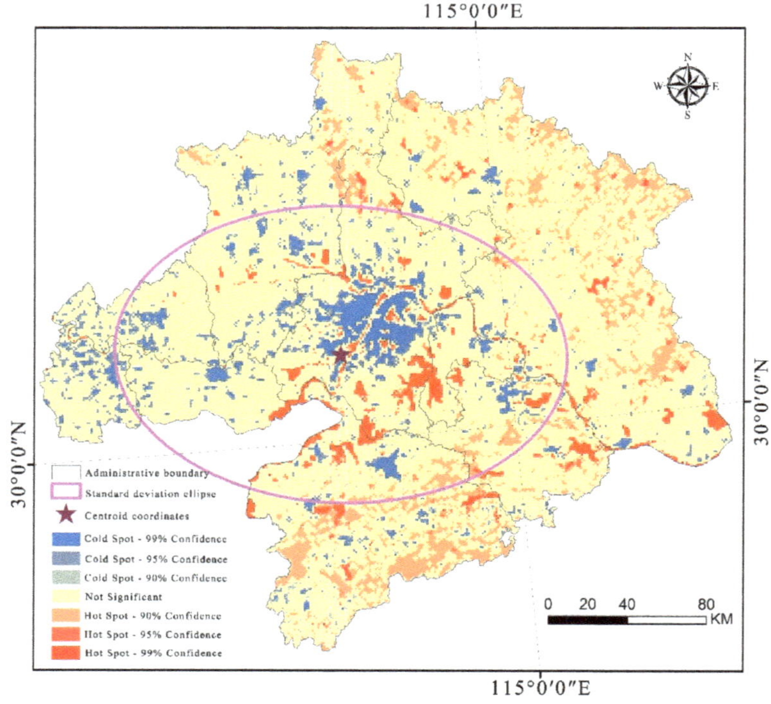

Figure 6. Standard deviation ellipse and local spatial autocorrelation analysis in the study area.

5. Discussion

5.1. Characteristics of the Urban Thermal Safety Pattern

In summary, the distribution law of the heat island patches in the Wuhan Urban Agglomeration is consistent with the urban spatial distribution. It gradually spreads from the main urban area of Wuhan to the surrounding urban areas, and the centers of other cities in the urban circle also have a noticeable heat island effect. Generally, the effect is mainly distributed around the developed areas of cities. In particular, as the economic activities of urban areas have accelerated and the population density has grown in recent years, the number of buildings has increased, and traffic congestion and other factors have likewise increased in tandem. These results are consistent with those of previous studies [67,68].

At the same time, heat island patch spreading resistance is mainly distributed in mountains, woodlands, and water areas where human activity is minimized, such as river basins. Because of their intrinsic resistance to heat island patch spreading, the resistance value of these localities is higher. Finally, the resistance values of these regions form the characteristic spatial pattern of "middle low, four sides high". Figure 7A shows that the corridors spatially connect the otherwise independent heat sources. Due to the limitations of geographical space and friction cost along corridors, the connection lengths between them are not the same. The total length of the corridor centerline is 1004.71 km. In the Wuhan Urban Agglomeration, a network of heat sources has grown, making it easier for each heat source to connect to other heat sources and strengthening the urban heat island as a whole.

As shown in Figure 7B, the critical nodes in the study area are located on the edge or corridor of the heat source, and more significant obstacles are found where the natural and artificial ecosystems intersect. The area of relatively small obstacles, on the other hand, appears in land use areas classified as woodland or water areas. These essential nodes, which are often tiny in size and distant from major heat sources, are thought to be crucial for the proper operation of heat island corridors. As a result, the focus of the TSPurban network optimization is the effective blocking of critical points, which plays an important role in reducing the structure and function of the regional heat island system.

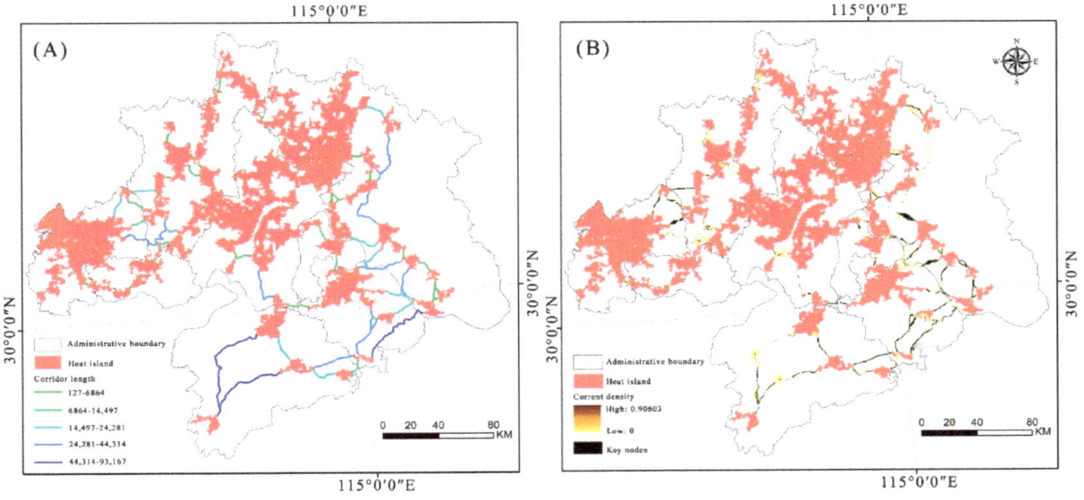

Figure 7. Locations of the heat island network (**A**) and key nodes (**B**) in the study area.

5.2. Optimization of the Urban Thermal Safety Pattern

5.2.1. Comprehensive "Destruction" of Patches and Corridors

As a reverse thinking process, one should think about how to reduce the stability and connectivity of the heat island patch network under the conditions of the status quo TSPurban. The first goal, from the perspective of the heat source itself, is to weaken the heat island source site. For this purpose, we consider natural and artificial drivers: on the one hand, based on natural drivers, the reduction in heat islands via green areas is extremely obvious [69]. Therefore, we should vigorously develop green urban construction. While improving the ecological and environmental function of green spaces, we should strictly implement ecological conservation measures, such as implementing ecological protection projects, returning farmland to forests, and closing mountains for afforestation. On the other hand, to address artificial drivers, local resource advantages should be used to their fullest during regional economic development. This means gradually reducing the production of high-energy-consuming businesses, adjusting the industrial structure and layout, and reducing the entry of artificial heat sources.

The second goal, from the perspective of heat source connectivity, is to destroy heat island corridors. To this end, two options focus on nodes and corridors. First, because the number and quality of nodes dictate the formation of heat island corridors, the destruction of corridors is contingent on the continuous reduction in heat island sources. Second, integrated heat island mitigation projects should be carried out for both the corridor and its perimeter. The mitigation projects for the corridor itself should mainly focus on minimizing the corridor width to reduce the contribution to network connectivity, while the mitigation projects for the corridor perimeter need to be integrated to increase the diffusion resistance of heat flow, a process that is largely influenced by the dominant influence of land cover [47,52].

Specifically, to address the main characteristics of the Wuhan Urban Agglomeration in this study, various methods must be implemented to destroy the heat island system within the TSPurban network and block heat island corridors at the key points. On the one hand, the regional heat island resistance barrier formed by the "green veins" (forests and grasslands) and "blue veins" (waters and wetlands) of the Wuhan Urban Agglomeration should be protected. On the other hand, the construction and protection of multiple groups of ecological "isolated green wedges" are required to gradually form green isolation zones within the Wuhan Urban Agglomeration, thus reducing the heat island effect.

5.2.2. Effective "Barrier" Based on Key Point Analysis

In the reverse thinking process, when the TSPurban network is disrupted by blocking the links and key pinch points at the spatial level, the SUHI connectivity is reduced. Thus the SUHI effect can be effectively mitigated. Since we have already discussed how connected links can be broken from the point of view of heat source connectivity, we know that the next step in reducing the effects of heat islands will be to pinpoint the priority mitigation zones. This will be done by implementing effective mitigation measures at key points in the TSPurban network.

Specifically, the above analysis shows that the corridor connections between heat island patches are where most critical points are located, and these areas require the most urgent improvement. Once mitigation measures become significant, the improvement effect becomes obvious. The heat island "barrier" in this area can directly reduce the overall connectivity efficiency of the TSPurban network. In this study, the critical nodes are mainly located in the eastern and southern regions of the Wuhan urban area, and planners and policymakers should focus on these areas when implementing future climate adaptation planning and SUHI mitigation measures. If the key nodes involve regional ecological and arable land protection areas, they can be adjusted by micro-means, such as optimizing spatial layout structures and updating building materials. At the same time, because of the wide distribution of these key points, "blocking" will be a long-term investment. Effectively "blocking" a key node is the cornerstone for significantly mitigating the formation of a regional TSPurban network.

5.3. Limitations and Deficiencies

This study also has some shortcomings. First, this paper mainly focuses on the urban heat island effect formed by LST during the daytime. However, the heat island effect is more potent at night [70] because daytime solar radiation induces a warming state in the nearby ground, while the whole region enters a cooling environment when the ground emits longwave radiation at night. The greater the heat escaping the ground, the lower the LST [71]. Therefore, in future work, we can compare nighttime data features and the daytime TSPurban network. Second, the heat island friction (cost) resistance surface is subject to the superposition of both artificial and natural factors [72,73]. The focus of this paper is to measure the integrated results of land cover; the consideration of artificial resistance drivers should be enhanced at a later stage.

6. Conclusions

This study selected the Wuhan Urban Agglomeration, an important strategic place in Central China, and constructed three major elements of the TSPurban based on the MSPA model and circuit theory "heat source, resistance surface, and corridor," and used "reverse thinking" to explore a variety of ways to optimize the TSPurban. The main conclusions were as follows: (1) regarding the heat source, heat island patches were extracted according to the MSPA model, and the patches in the top 30 important heat island source areas were evaluated using the connectivity analysis method. It was found that the spatial form of the heat island was consistent with the urban spatial distribution; (2) regarding the resistance surface, the resistance surface of heat island propagation and diffusion was constructed by combining the land cover resistance factors, and the resistance surface was evaluated by using the standard deviation ellipse and spatial autocorrelation analysis methods. It was found that the resistance surface was distributed in an "east–west" spatial pattern and that the gathering situation was characterized by "cold in the west and hot in the east"; (3) regarding the corridor, we identified the link (corridor) based on the circuit theory and used the Linkage Mapper software to connect the lowest-cost path of the SUHI "source," thus generating 56 heat island corridors with a total length of 1004.71 km. At the same time, Circuitscape software was used to identify key nodes to effectively control key points, and the ecological regulation function for alleviating the urban heat island effect was more prominent. In terms of optimizing the TSPurban, it is necessary to weaken the interconnection between heat island sources at the spatial level by "breaking" the links and key nodes in the TSPurban network. Ultimately, the focus is on increasing the fragmentation of heat island patches and their resistance to diffusion and reducing the connections between heat islands while weakening the heat source patches themselves. This study utilizes reverse thinking to effectively mitigate the urban thermal environment.

Author Contributions: C.H.: Writing—original draft, Data curation. H.L.: Data curation, Software, Supervision. All authors have read and agreed to the published version of the manuscript.

Funding: This work was supported by the Natural Science Foundation of China under grant 41807319 and the Fundamental Research Funds for the Central Universities (YCJJ202204012).

Data Availability Statement: The original contributions presented in the study are included in the article, and further inquiries can be directed to the corresponding author.

Conflicts of Interest: The authors declare that the research was conducted without commercial or financial relationships that could be construed as a potential conflict of interest.

References

1. Lo, C.P.; Quattrochi, D.A.; Luvall, J.C. Application of High-Resolution Thermal Infrared Remote Sensing and GIS to Assess the Urban Heat Island Effect. *Int. J. Remote Sens.* **1997**, *18*, 287–304. [CrossRef]
2. O'Malley, C.; Piroozfar, P.; Farr, E.R.; Pomponi, F. Urban Heat Island (UHI) Mitigating Strategies: A Case-Based Comparative Analysis. *Sustain. Cities Soc.* **2015**, *19*, 222–235. [CrossRef]
3. Howard, L. *The Climate of London: Deduced from Meteorological Observations Made in the Metropolis and at Various Places around It*; Harvey and Darton: London, UK, 1833; Volume 3.
4. Stone Jr, B.; Rodgers, M.O. Urban Form and Thermal Efficiency: How the Design of Cities Influences the Urban Heat Island Effect. *Am. Plan. Assoc. J. Am. Plan. Assoc.* **2001**, *67*, 186. [CrossRef]
5. Rizwan, A.M.; Dennis, L.Y.; Chunho, L.I.U. A Review on the Generation, Determination and Mitigation of Urban Heat Island. *J. Environ. Sci.* **2008**, *20*, 120–128. [CrossRef]
6. Zhou, D.; Xiao, J.; Bonafoni, S.; Berger, C.; Deilami, K.; Zhou, Y.; Frolking, S.; Yao, R.; Qiao, Z.; Sobrino, J. Satellite Remote Sensing of Surface Urban Heat Islands: Progress, Challenges, and Perspectives. *Remote Sens.* **2018**, *11*, 48. [CrossRef]
7. Chen, X.-L.; Zhao, H.-M.; Li, P.-X.; Yin, Z.-Y. Remote Sensing Image-Based Analysis of the Relationship between Urban Heat Island and Land Use/Cover Changes. *Remote Sens. Environ.* **2006**, *104*, 133–146. [CrossRef]
8. Li, H.; Zhou, Y.; Wang, X.; Zhou, X.; Zhang, H.; Sodoudi, S. Quantifying Urban Heat Island Intensity and Its Physical Mechanism Using WRF/UCM. *Sci. Total Environ.* **2019**, *650*, 3110–3119. [CrossRef] [PubMed]
9. Corburn, J. Cities, Climate Change and Urban Heat Island Mitigation: Localising Global Environmental Science. *Urban Stud.* **2009**, *46*, 413–427. [CrossRef]
10. Gao, J.; Yu, Z.; Wang, L.; Vejre, H. Suitability of Regional Development Based on Ecosystem Service Benefits and Losses: A Case Study of the Yangtze River Delta Urban Agglomeration, China. *Ecol. Indic.* **2019**, *107*, 105579. [CrossRef]
11. Sun, D.; Hu, C.; Wang, Y.; Wang, Z.; Zhang, J. Examining Spatio-Temporal Characteristics of Urban Heat Islands and Factors Driving Them in Hangzhou, China. *IEEE J. Sel. Top. Appl. Earth Obs. Remote Sens.* **2021**, *14*, 8316–8325. [CrossRef]
12. Wang, Z.; Sun, D.; Hu, C.; Wang, Y.; Zhang, J. Seasonal Contrast and Interactive Effects of Potential Drivers on Land Surface Temperature in the Sichuan Basin, China. *Remote Sens.* **2022**, *14*, 1292. [CrossRef]
13. Akbari, H.; Kolokotsa, D. Three Decades of Urban Heat Islands and Mitigation Technologies Research. *Energy Build.* **2016**, *133*, 834–842. [CrossRef]
14. Wu, J. Urban Ecology and Sustainability: The State-of-the-Science and Future Directions. *Landsc. Urban Plan.* **2014**, *125*, 209–221. [CrossRef]
15. Grimm, N.B.; Faeth, S.H.; Golubiewski, N.E.; Redman, C.L.; Wu, J.; Bai, X.; Briggs, J.M. Global Change and the Ecology of Cities. *Science* **2008**, *319*, 756–760. [CrossRef] [PubMed]
16. Oke, T.R.; Mills, G.; Christen, A.; Voogt, J.A. *Urban Climates*; Cambridge University Press: Cambridge, UK, 2017; ISBN 0-521-84950-0.
17. Li, X.; Zhou, Y.; Asrar, G.R.; Imhoff, M.; Li, X. The Surface Urban Heat Island Response to Urban Expansion: A Panel Analysis for the Conterminous United States. *Sci. Total Environ.* **2017**, *605*, 426–435. [CrossRef]
18. Zhou, D.; Zhao, S.; Liu, S.; Zhang, L.; Zhu, C. Surface Urban Heat Island in China's 32 Major Cities: Spatial Patterns and Drivers. *Remote Sens. Environ.* **2014**, *152*, 51–61. [CrossRef]
19. Saaroni, H.; Amorim, J.H.; Hiemstra, J.A.; Pearlmutter, D. Urban Green Infrastructure as a Tool for Urban Heat Mitigation: Survey of Research Methodologies and Findings across Different Climatic Regions. *Urban Clim.* **2018**, *24*, 94–110. [CrossRef]

20. Bowler, D.E.; Buyung-Ali, L.; Knight, T.M.; Pullin, A.S. Urban Greening to Cool Towns and Cities: A Systematic Review of the Empirical Evidence. *Landsc. Urban Plan.* **2010**, *97*, 147–155. [CrossRef]
21. Zhou, W.; Wang, J.; Cadenasso, M.L. Effects of the Spatial Configuration of Trees on Urban Heat Mitigation: A Comparative Study. *Remote Sens. Environ.* **2017**, *195*, 1–12. [CrossRef]
22. Chen, L.; Sun, R.; Lu, Y. A Conceptual Model for a Process-Oriented Landscape Pattern Analysis. *Sci. China Earth Sci.* **2019**, *62*, 2050–2057. [CrossRef]
23. Yu, Z.; Zhang, J.; Yang, G.; Schlaberg, J. Reverse Thinking: A New Method from the Graph Perspective for Evaluating and Mitigating Regional Surface Heat Islands. *Remote Sens.* **2021**, *13*, 1127. [CrossRef]
24. Zhou, D.; Bonafoni, S.; Zhang, L.; Wang, R. Remote Sensing of the Urban Heat Island Effect in a Highly Populated Urban Agglomeration Area in East China. *Sci. Total Environ.* **2018**, *628–629*, 415–429. [CrossRef] [PubMed]
25. Yu, Z.; Yao, Y.; Yang, G.; Wang, X.; Vejre, H. Spatiotemporal Patterns and Characteristics of Remotely Sensed Region Heat Islands during the Rapid Urbanization (1995–2015) of Southern China. *Sci. Total Environ.* **2019**, *674*, 242–254. [CrossRef] [PubMed]
26. Pan, J. Area Delineation and Spatial-Temporal Dynamics of Urban Heat Island in Lanzhou City, China Using Remote Sensing Imagery. *J. Indian Soc. Remote Sens.* **2016**, *44*, 111–127. [CrossRef]
27. Santamouris, M. Cooling the Cities–a Review of Reflective and Green Roof Mitigation Technologies to Fight Heat Island and Improve Comfort in Urban Environments. *Sol. Energy* **2014**, *103*, 682–703. [CrossRef]
28. Meerow, S.; Newell, J.P. Spatial Planning for Multifunctional Green Infrastructure: Growing Resilience in Detroit. *Landsc. Urban Plan.* **2017**, *159*, 62–75. [CrossRef]
29. Sun, R.; Xie, W.; Chen, L. A Landscape Connectivity Model to Quantify Contributions of Heat Sources and Sinks in Urban Regions. *Landsc. Urban Plan.* **2018**, *178*, 43–50. [CrossRef]
30. Peng, J.; Cheng, X.; Hu, Y.; Corcoran, J. A Landscape Connectivity Approach to Mitigating the Urban Heat Island Effect. *Landsc. Ecol.* **2022**, *37*, 1707–1719. [CrossRef]
31. Peng, J.; Hu, Y.; Dong, J.; Liu, Q.; Liu, Y. Quantifying Spatial Morphology and Connectivity of Urban Heat Islands in a Megacity: A Radius Approach. *Sci. Total Environ.* **2020**, *714*, 136792. [CrossRef]
32. Yu, Z.; Zhang, J.; Yang, G. How to Build a Heat Network to Alleviate Surface Heat Island Effect? *Sustain. Cities Soc.* **2021**, *74*, 103135. [CrossRef]
33. Qin, Y. A Review on the Development of Cool Pavements to Mitigate Urban Heat Island Effect. *Renew. Sustain. Energy Rev.* **2015**, *52*, 445–459. [CrossRef]
34. Guo, G.; Wu, Z.; Cao, Z.; Chen, Y.; Zheng, Z. Location of Greenspace Matters: A New Approach to Investigating the Effect of the Greenspace Spatial Pattern on Urban Heat Environment. *Landsc. Ecol.* **2021**, *36*, 1533–1548. [CrossRef]
35. Zhang, Y.; Odeh, I.O.; Ramadan, E. Assessment of Land Surface Temperature in Relation to Landscape Metrics and Fractional Vegetation Cover in an Urban/Peri-Urban Region Using Landsat Data. *Int. J. Remote Sens.* **2013**, *34*, 168–189. [CrossRef]
36. Gunawardena, K.R.; Wells, M.J.; Kershaw, T. Utilising Green and Bluespace to Mitigate Urban Heat Island Intensity. *Sci. Total Environ.* **2017**, *584*, 1040–1055. [CrossRef]
37. Hu, C.; Wang, Z.; Wang, Y.; Sun, D.; Zhang, J. Combining MSPA-MCR Model to Evaluate the Ecological Network in Wuhan, China. *Land* **2022**, *11*, 213. [CrossRef]
38. Liu, Y.; Fang, X.; Xu, Y.; Zhang, S.; Luan, Q. Assessment of Surface Urban Heat Island across China's Three Main Urban Agglomerations. *Theor. Appl. Climatol.* **2018**, *133*, 473–488. [CrossRef]
39. Saura, S.; Pascual-Hortal, L. A New Habitat Availability Index to Integrate Connectivity in Landscape Conservation Planning: Comparison with Existing Indices and Application to a Case Study. *Landsc. Urban Plan.* **2007**, *83*, 91–103. [CrossRef]
40. An, Y.; Liu, S.; Sun, Y.; Shi, F.; Beazley, R. Construction and Optimization of an Ecological Network Based on Morphological Spatial Pattern Analysis and Circuit Theory. *Landsc. Ecol.* **2021**, *36*, 2059–2076. [CrossRef]
41. McRae, B.H.; Dickson, B.G.; Keitt, T.H.; Shah, V.B. Using Circuit Theory to Model Connectivity in Ecology, Evolution, and Conservation. *Ecology* **2008**, *89*, 2712–2724. [CrossRef]
42. Dickson, B.G.; Albano, C.M.; Anantharaman, R.; Beier, P.; Fargione, J.; Graves, T.A.; Gray, M.E.; Hall, K.R.; Lawler, J.J.; Leonard, P.B. Circuit-theory Applications to Connectivity Science and Conservation. *Conserv. Biol.* **2019**, *33*, 239–249. [CrossRef]
43. Wang, S.; Wu, M.; Hu, M.; Fan, C.; Wang, T.; Xia, B. Promoting Landscape Connectivity of Highly Urbanized Area: An Ecological Network Approach. *Ecol. Indic.* **2021**, *125*, 107487. [CrossRef]
44. Huang, X.; Wang, H.; Shan, L.; Xiao, F. Constructing and Optimizing Urban Ecological Network in the Context of Rapid Urbanization for Improving Landscape Connectivity. *Ecol. Indic.* **2021**, *132*, 108319. [CrossRef]
45. Wang, T.; Li, H.; Huang, Y. The Complex Ecological Network's Resilience of the Wuhan Metropolitan Area. *Ecol. Indic.* **2021**, *130*, 108101. [CrossRef]
46. Zhao, S.; Ma, Y.; Wang, J.; You, X. Landscape Pattern Analysis and Ecological Network Planning of Tianjin City. *Urban For. Urban Green.* **2019**, *46*, 126479. [CrossRef]
47. Buyantuyev, A.; Wu, J. Urban Heat Islands and Landscape Heterogeneity: Linking Spatiotemporal Variations in Surface Temperatures to Land-Cover and Socioeconomic Patterns. *Landsc. Ecol.* **2010**, *25*, 17–33. [CrossRef]
48. Coutts, A.M.; Tapper, N.J.; Beringer, J.; Loughnan, M.; Demuzere, M. Watering Our Cities: The Capacity for Water Sensitive Urban Design to Support Urban Cooling and Improve Human Thermal Comfort in the Australian Context. *Prog. Phys. Geogr.* **2013**, *37*, 2–28. [CrossRef]

49. Chen, W.Y.; Jim, C.Y. Assessment and Valuation of the Ecosystem Services Provided by Urban Forests. In *Ecology, Planning, and Management of Urban Forests*; Springer: Berlin/Heidelberg, Germany, 2008; pp. 53–83.
50. Sacks, W.J.; Cook, B.I.; Buenning, N.; Levis, S.; Helkowski, J.H. Effects of Global Irrigation on the Near-Surface Climate. *Clim. Dyn.* **2009**, *33*, 159–175. [CrossRef]
51. Shi, W.; Tao, F.; Liu, J. Regional Temperature Change over the Huang-Huai-Hai Plain of China: The Roles of Irrigation versus Urbanization. *Int. J. Climatol.* **2014**, *34*, 1181–1195. [CrossRef]
52. Zou, Z.; Yan, C.; Yu, L.; Jiang, X.; Ding, J.; Qin, L.; Wang, B.; Qiu, G. Impacts of Land Use/Land Cover Types on Interactions between Urban Heat Island Effects and Heat Waves. *Build. Environ.* **2021**, *204*, 108138. [CrossRef]
53. Peng, J.; Jia, J.; Liu, Y.; Li, H.; Wu, J. Seasonal Contrast of the Dominant Factors for Spatial Distribution of Land Surface Temperature in Urban Areas. *Remote Sens. Environ.* **2018**, *215*, 255–267. [CrossRef]
54. Verbeke, S.; Audenaert, A. Thermal Inertia in Buildings: A Review of Impacts across Climate and Building Use. *Renew. Sustain. Energy Rev.* **2018**, *82*, 2300–2318. [CrossRef]
55. Jiang, J.; Tian, G. Analysis of the Impact of Land Use/Land Cover Change on Land Surface Temperature with Remote Sensing. *Procedia Environ. Sci.* **2010**, *2*, 571–575. [CrossRef]
56. Tran, D.X.; Pla, F.; Latorre-Carmona, P.; Myint, S.W.; Caetano, M.; Kieu, H.V. Characterizing the Relationship between Land Use Land Cover Change and Land Surface Temperature. *ISPRS J. Photogramm. Remote Sens.* **2017**, *124*, 119–132. [CrossRef]
57. Yuill, R.S. The Standard Deviational Ellipse; an Updated Tool for Spatial Description. *Geogr. Ann. Ser. B Hum. Geogr.* **1971**, *53*, 28–39. [CrossRef]
58. Lefever, D.W. Measuring Geographic Concentration by Means of the Standard Deviational Ellipse. *Am. J. Sociol.* **1926**, *32*, 88–94. [CrossRef]
59. Fan, C.; Myint, S. A Comparison of Spatial Autocorrelation Indices and Landscape Metrics in Measuring Urban Landscape Fragmentation. *Landsc. Urban Plan.* **2014**, *121*, 117–128. [CrossRef]
60. Liu, Y.; Tian, J.; Zheng, W.; Yin, L. Spatial and Temporal Distribution Characteristics of Haze and Pollution Particles in China Based on Spatial Statistics. *Urban Clim.* **2022**, *41*, 101031. [CrossRef]
61. Li, G.; Fang, C.; Wang, S. Exploring Spatiotemporal Changes in Ecosystem-Service Values and Hotspots in China. *Sci. Total Environ.* **2016**, *545*, 609–620. [CrossRef]
62. Xie, P.; Yang, J.; Wang, H.; Liu, Y.; Liu, Y. A New Method of Simulating Urban Ventilation Corridors Using Circuit Theory. *Sustain. Cities Soc.* **2020**, *59*, 102162. [CrossRef]
63. Huang, J.; Hu, Y.; Zheng, F. Research on Recognition and Protection of Ecological Security Patterns Based on Circuit Theory: A Case Study of Jinan City. *Environ. Sci. Pollut. Res.* **2020**, *27*, 12414–12427. [CrossRef]
64. Liu, Y.; Huang, T.-T.; Zheng, X. A Method of Linking Functional and Structural Connectivity Analysis in Urban Green Infrastructure Network Construction. *Urban Ecosyst.* **2022**, *25*, 909–925. [CrossRef]
65. Hathway, E.A.; Sharples, S. The Interaction of Rivers and Urban Form in Mitigating the Urban Heat Island Effect: A UK Case Study. *Build. Environ.* **2012**, *58*, 14–22. [CrossRef]
66. Wu, Z.; Zhang, Y. Water Bodies' Cooling Effects on Urban Land Daytime Surface Temperature: Ecosystem Service Reducing Heat Island Effect. *Sustainability* **2019**, *11*, 787. [CrossRef]
67. Xiong, Y.; Zhang, F. Effect of Human Settlements on Urban Thermal Environment and Factor Analysis Based on Multi-Source Data: A Case Study of Changsha City. *J. Geogr. Sci.* **2021**, *31*, 819–838. [CrossRef]
68. Meng, Q.; Zhang, L.; Sun, Z.; Meng, F.; Wang, L.; Sun, Y. Characterizing Spatial and Temporal Trends of Surface Urban Heat Island Effect in an Urban Main Built-up Area: A 12-Year Case Study in Beijing, China. *Remote Sens. Environ.* **2018**, *204*, 826–837. [CrossRef]
69. Xiao, X.D.; Dong, L.; Yan, H.; Yang, N.; Xiong, Y. The Influence of the Spatial Characteristics of Urban Green Space on the Urban Heat Island Effect in Suzhou Industrial Park. *Sustain. Cities Soc.* **2018**, *40*, 428–439. [CrossRef]
70. Sarrat, C.; Lemonsu, A.; Masson, V.; Guédalia, D. Impact of Urban Heat Island on Regional Atmospheric Pollution. *Atmos. Environ.* **2006**, *40*, 1743–1758. [CrossRef]
71. Oke, T.R.; Johnson, G.T.; Steyn, D.G.; Watson, I.D. Simulation of Surface Urban Heat Islands under 'Ideal'Conditions at Night Part 2: Diagnosis of Causation. *Bound.-Layer Meteorol.* **1991**, *56*, 339–358. [CrossRef]
72. Ulpiani, G. On the Linkage between Urban Heat Island and Urban Pollution Island: Three-Decade Literature Review towards a Conceptual Framework. *Sci. Total Environ.* **2021**, *751*, 141727. [CrossRef]
73. Wang, Z.; Meng, Q.; Allam, M.; Hu, D.; Zhang, L.; Menenti, M. Environmental and Anthropogenic Drivers of Surface Urban Heat Island Intensity: A Case-Study in the Yangtze River Delta, China. *Ecol. Indic.* **2021**, *128*, 107845. [CrossRef]

remote sensing

Article

Spatial Influence of Multifaceted Environmental States on Habitat Quality: A Case Study of the Three Largest Chinese Urban Agglomerations

Fei Liu [1,*], Yuji Murayama [2] and Yoshifumi Masago [1]

1 Center for Climate Change Adaptation, National Institute for Environmental Studies, 16-2 Onogawa, Tsukuba 305-8506, Ibaraki, Japan
2 Faculty of Life and Environmental Sciences, University of Tsukuba, Tsukuba 305-8572, Ibaraki, Japan
* Correspondence: liu.fei@nies.go.jp; Tel.: +81-29-850-2387

Abstract: Habitat structure and quality in the urban agglomeration (UA) are subject to multiple threats and pressures due to ongoing anthropogenic activities and call for comprehensively effective solutions. Many approaches, including cartographic comparison, correlation analysis, the local entropy model, and GeoDetector, were jointly used to clarify the interplay between habitat quality and multiple environmental issues. In response to the overlapped risks of diverse environmental systems, this study presented an integrated research framework to evaluate the spatial influences of multifaceted environmental situations on habitat quality. We conducted the case study in the three largest Chinese UAs: Beijing–Tianjin–Hebei (BTH), Greater Bay Area (GBA), and Yangtze River Delta (YRD). The evaluation results show that the three UAs shared similarities and differences in relationship/impact types and their strengths. In 2015, most of the three UAs' landscapes delivered low–medium magnitudes of habitat quality (score <0.7) and emerged with unevenly severe consequences over space across different environmental aspects, highlighting the importance of maintaining habitat safety. Overall, habitat quality scores were synergistic with NDVI, but antagonistic to surface heat island intensity (SHII), PM2.5 concentrations, and residential support. However, locally structured relationships exhibited geographical complexity and heterogeneity between habitat quality and environmental systems. Regarding GeoDetector evaluation, PM2.5 concentrations in BTH, SHII in GBA, and NDVI in YRD played a dominant role in single-factor and interaction analysis. More importantly, the synergistic effect of various environmental issues on habitats was manifested as mutually enhanced rather than independent or weakened interactive effects, implying the aggravation of compound effects and the necessity of prioritization schemes. This study could provide beneficial insights into the interconnections between habitats' sustainability and multifaceted environmental situations in UAs.

Keywords: air quality; GeoDetector; habitat quality; thermal environment; urban agglomeration

Citation: Liu, F.; Murayama, Y.; Masago, Y. Spatial Influence of Multifaceted Environmental States on Habitat Quality: A Case Study of the Three Largest Chinese Urban Agglomerations. *Remote Sens.* **2023**, *15*, 921. https://doi.org/10.3390/rs15040921

Academic Editor: Riccardo Buccolieri

Received: 4 December 2022
Revised: 3 February 2023
Accepted: 4 February 2023
Published: 7 February 2023

1. Introduction

Ongoing socioeconomic development has given rise to multifaceted problems and challenges for habitats. The Intergovernmental Panel on Climate Change (IPCC) highlighted the necessity of understanding the interrelationships and linkages between various environmental issues in the Sixth Assessment Report [1]. The natural ecosystem and human system are intertwined, and alterations in one component may stimulate others [1]. Thus, sustainable development strategies require a new understanding of compound environmental issues and overlapping consequences [2].

The United Nations declared that over half of the current world's dwellers reside in urban settlements in the World Urbanization Prospects [3]. This proportion will rise to two-thirds by the 2050s [3]. Accordingly, improving human well-being increasingly relies on upgrading the situations and services in human-dominated urban areas [4,5].

Urban agglomerations (UAs) are a newly evolved manifestation of spatially clustered regions with a common interest and fate [6]. Regional integration development and drastic urbanization have propelled the formation of UAs [7]. The UA system is a complex ensemble comprising multiple cities and their extended outskirts with large-scale urban fabrics and artificial landscapes. These artificial landscapes provide adequate space to guarantee social-economic boom and human well-being [6]. In parallel, natural habitats provide a majority of essential ecosystem services to sustain the ecological security of the entire UA region. However, with drastic economic and population growth, UA regions have manifested vigorous trade-offs between ecosystem services provision and human social development, giving rise to enormous ecological deficits, especially in the core megacities and highly populated areas [4,8]. As such, the habitats of UAs are diverse but fragile, and might be subjected to a multitude of risk factors from natural and human interference [4].

Existing studies on UAs are typically confined to the individual environmental problem, formation mechanism, and mitigation strategy, including the thermal [9], air [10], housing [11], landscape [12], and resource [13] spheres. A vast quantity of studies have identified, for example, the heat island phenomenon [14,15], air pollution [16,17], and urban containment [18]. Abundant solutions have continuously contributed to a future sustainable world, such as nature-based solutions [19] and ecological intensification interventions [20]. Against the backdrops of UAs, implementing measures that safeguard environmental health and human well-being across multiple aspects is warranted [21,22]. However, few studies have been concerned with the fate of compound environmental risks and their interactive relationships. The environmental crisis in the UAs may be magnified due to the overlap of multiple issues, their influences on habitats may be aggravated, and comprehensive carrying capacity might have further deteriorated [1,23]. Considering the growing demands for comprehensive issues and compound risks in the UAs, it is imperative to understand the associations between diffident environmental problems and seek co-benefits.

Unraveling the relationships between habitats and environmental issues is the foundation of effective, comprehensive policies. Yet, one of the key challenges is obtaining insights into the complex interplay of multiple environmental situations [1,4]. The impact of human interference on different environment systems varies broadly from slight, modest, to severe, but may share similar driving factors [8,24–26]. Additionally, the measurement of the geographical association of multiple variables relies on their locations, distances, neighboring settings, and scales [24,27,28]. However, conventional methods, such as cartographic comparison, correlation, and regression analysis, generally only mirror simple relationships [29]. Exponential, quadratic, and interactive relationships are difficult to represent using these conventional methods. The lack of knowledge about complex relationships has obstructed the achievement of multifaceted environmental evaluation [29]. In light of this, this study presents a framework that integrates multi-method and multi-source data to evaluate the associations between multifaceted environmental states and habitat quality in different UA settings. Our proposed framework incorporates cartographic comparison, correlation analysis, the local entropy model, and GeoDetector to conduct a spatially multi-perspective relationship profiling. The local entropy model is a novel method for examining the locally varying relationships over space, which can capture linear and polynomial complex associations [29]. GeoDetector is a powerful tool for identifying spatial stratified heterogeneities and interactive effects [30,31].

We designated the UAs of Beijing–Tianjin–Hebei (BTH), Guangdong–Hong Kong–Macao Greater Bay Area (GBA), and Yangtze River Delta (YRD) as the study regions. In recent decades, the Chinese government has been committed to ecological civilization construction and sustainability management in these three UAs based on a series of relevant initiatives and policies [32–34], which have been in response to habitat degradation from dramatic socioeconomic growth [13,35–38]. Moreover, many scientific investigations have contributed to sustainable development schemes for the three UAs [37,39–44]. The thermal environment, air quality, living space, and green landscapes have been the focal topics among environmental monitoring and assessments, which are also deeply discussed in this study. Thus, focusing on the three UAs of BTH, GBA, and YRD in 2015, this study consists of four goals, as follows: (1) Quantification of habitat quality; (2) Evaluation of diverse environmental situations from the thermal, air, living, and landscape spheres; (3) Identification of the interconnection and interplay between habitat quality and multi-dimensional environmental situations; and (4) Elucidation of a potential management pathway for multi-dimensional sustainability and security for the UAs.

2. Materials and Methods

2.1. Study Areas

This study focused on the three largest Chinese UAs: BTH, GBA, and YRD (Figure 1). The study regions were determined due to (1) their rapid socioeconomic development, (2) plentiful nature capitals, (3) critical geographical location, (4) urgent sustainability needs, and (5) superior policy support. BTH, GBA, and YRD are nationally and globally representative UA regions. These UA regions are not only primary engines and important pivots for China's socioeconomic development, but also the significant guarantee and forefront of China's ecological security. BTH UA (113°27′E to 119°50′E, 35°03′N to 42°40′N) is situated in Northeast China, encompassing 13 cities with 2.2% (2.2×10^7 ha) of the national territorial area. GBA UA (111°27′E to 115°42′E, 21°57′N to 24°40′N) has a scope of 9 cities and 2 special administrative regions in South China, accounting for 0.6% (5.6×10^6 ha) of the national territory. YRD UA (111°27′E to 115°42′E, 21°57′N to 24°40′N) is composed of 16 cities on the eastern coast of China, covering 2.2% of China's territory (2.1×10^7 ha).

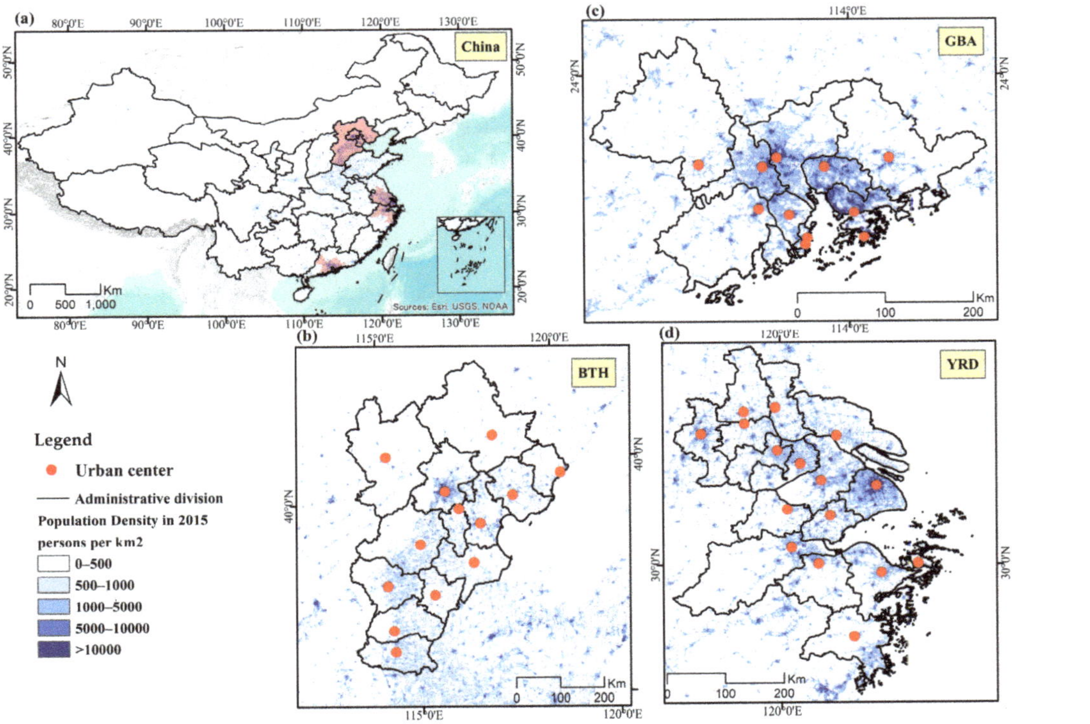

Figure 1. Study areas: (**a**) geographic location of the study areas in China; (**b**) Beijing–Tianjin–Hebei (BTH); (**c**) Guangdong–Hong Kong–Macao Greater Bay Area (GBA); and (**d**) Yangtze River Delta (YRD).

At the end of 2015, approximately 8%, 5%, and 11% of the national population resided in BTH, GBA, and YRD, respectively [45]. BTH, GBA, and YRD had 10.2%, 14%, and 18.5% of the respective national totals in the gross domestic product (GDP) [45]. These three UAs emerge as compound habitat systems of multiple urban cores and their peripheral countrysides under different climate schemes characterized by abundant land details. BTH, GBA, and YRD are configured by a variety of ecosystems, landscapes, and landforms. BTH is located in a temperate monsoon climate zone, while GBA and YRD are dominated by a monsoon subtropical climate context. Significant alterations and transformations in the environmental landscapes and socioeconomic scales of these UAs in recent decades [38,42,46,47] have given rise to habitat degradation, ecological fragility, and climate crisis [48–51], such as intensified heat island effects [52] and deteriorated air pollution [16]. Thus, national development initiatives also highlight the urgency and necessity of sustainability-related, ecological-oriented strategies and projects in BTH, GBA, and YRD [32–34].

2.2. Data Preparation and Treatment

Given the data availability, precision, and consistency of the three UAs, this study set 2015 as the year for the research. The year 2015 was an extraordinary time for economic growth and social boom in BTH, GBA, YRD, and even the entire territory of China due to a number of important development policies, including the 13th Five-Year Plan, Belt and Road initiatives, and regional collaboration plans [32–34,53].

The land cover/use maps of the BTH, GBA, and YRD in 2015 were cited and pruned from Climate Change Initiative (CCI) global land cover (LC) products, European Space Agency (ESA) (https://www.esa-landcover-cci.org/ (accessed on 30 November 2022)). ESA CCI-LC product is a 300 m spatial resolution of a time-series global surface coverage dataset generated based on multi-source satellite archives [54]. The overall weighted-area accuracy of the ESA CCI-LC map in 2015 was validated to be 71.1% [54]. Thus, the ESA CCI-LC map can deliver abundant land details and flexibly serve diverse land-oriented applications. Referring to the official classification scheme [54] and IUCN habitat-related suggestions [55,56], we rearranged the land cover/use thematic legend into eight categories: agricultural habitat (AH), forest habitat (FH), grassland habitat (GH), wetland habitat (WH), shrubland habitat (SH), urban fabric (UF), vacant land (VL), and water (WA).

The annual average daytime and nighttime land surface temperatures (LSTs) of the BTH, GBA, and YRD in 2015 were composited and preprocessed based on the 1 km resolution of Moderate Resolution Imaging Spectroradiometer (MODIS) daily/night land surface temperature/emissivity products (MOD11A1 and MYD11A1, https://modis.gsfc.nasa.gov/data/dataprod/mod11.php (accessed on 30 November 2022)) using the Google Earth Engine (GEE). The 2015 yearly average normalized difference vegetation index (NDVI) distributions in the BTH, GBA, and YRD were also GEE-derived and composed using MODIS 1 km 16-Day vegetation indices (MOD13A2, https://modis.gsfc.nasa.gov/data/dataprod/mod13.php (accessed on 30 November 2022)). The 1 km resolution spatial coverages of the 2015 average PM2.5 (particulate matter in the air with aerodynamic diameters smaller than 2.5 μm) concentrations in the BTH, GBA, and YRD were sourced from China High Air Pollutants (CHAP) datasets [57,58] (https://weijing-rs.github.io/product.html (accessed on 30 November 2022)). The China High PM2.5 datasets were developed by coupling in site PM2.5 data, MODIS aerosol optical depth (AOD) products, and various auxiliary data based on the space-time extra-trees model [57]. Various air-relevant studies have used the ChinaHighPM2.5 datasets due to their wide spatiotemporal range and high cross-validation accuracy (R^2 = 0.86–0.90) [57,58]. Spatially gridded population density data (1 km of spatial resolution) of the BTH, GBA, and YRD in 2015 were extracted from the high-profile WorldPop population products. WorldPop (https://www.worldpop.org/ (accessed on 30 November 2022)), developed by the University of Southampton, has been dedicated to supporting the open utilization of various spatial demographic studies [59]. China-relevant spatial data (e.g., administrative boundary) were obtained from the Resource and Environmental Science and Data Center, Chinese Academy of Sciences (https://www.resdc.cn/Default.aspx (accessed on 30 November 2022)).

2.3. Research Framework

This study aims to evaluate the interconnection and spatial heterogeneity between habitat quality and multiple environmental situations from multiple perspectives. Thus, the cascade diagram of our proposed framework is shown in Figure 2. The overall research framework is mainly composed of four modules: (I) data preparation and preprocessing; (II) correlation analysis; (III) local bivariate analysis; and (IV) GeoDetector analysis.

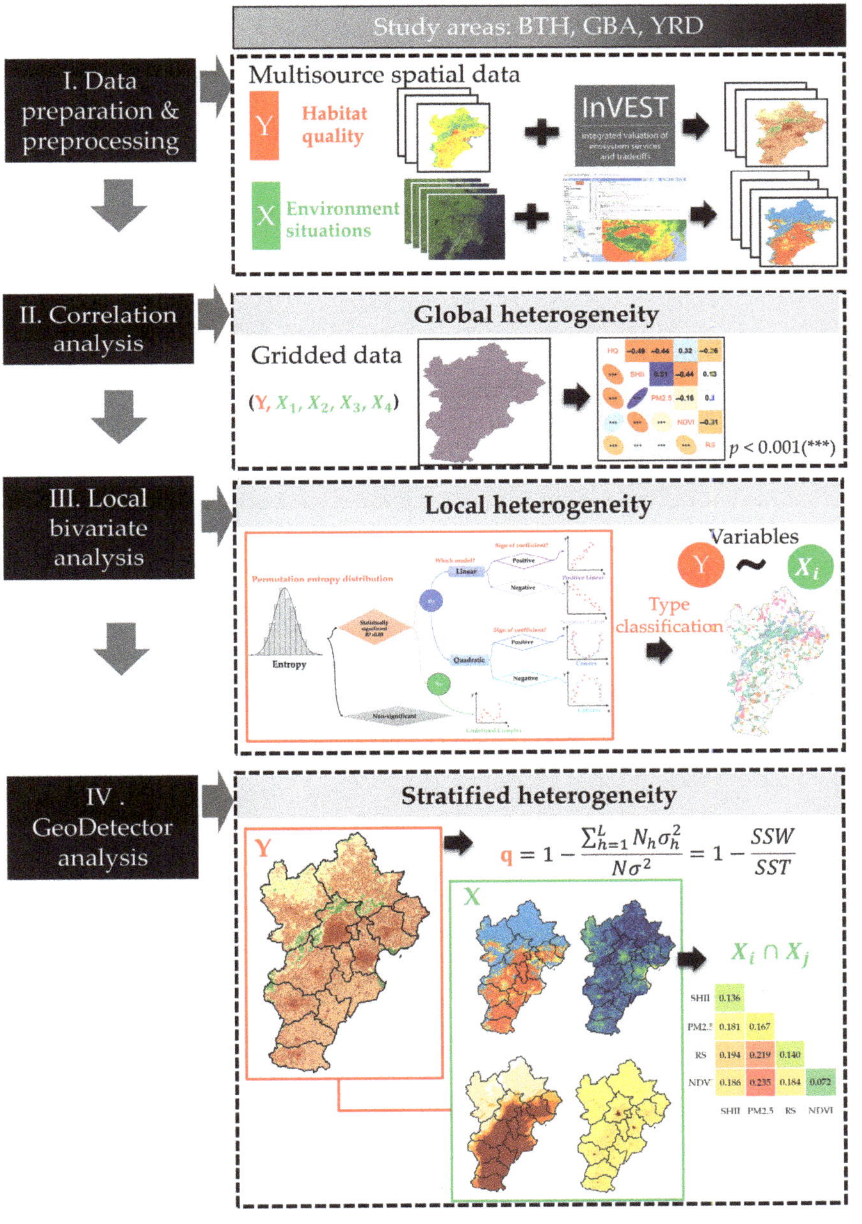

Figure 2. Overall research framework. Note: InVEST logo image sources from the website (https://naturalcapitalproject.stanford.edu/software/invest (accessed on 30 November 2022)).

2.4. Quantification of Habitat Quality

Habitat quality is geographically heterogeneous, which depends on habitat characteristics, suitability, and the presence and relative magnitudes of threat factors [60]. In light of this, the spatial patterns and magnitudes could be evaluated by the joint use of land cover/use and threats data [61,62]. The Integrated Valuation of Ecosystem Services and Tradeoffs–Habitat Quality (InVEST–HQ) model was developed by Natural Capital Projects in conjunction with Stanford University, the Nature Conservancy, and the World Wide Fund for Nature (https://naturalcapitalproject.stanford.edu/software/invest (accessed on 30 November 2022)). Given the low data demand, superior spatial visualization, and easy operation, the InVEST–HQ model is the dominant tool for habitat quality quantification. It has been widely applied to ecological studies across different regions and spatial scales [60]. Thus, we employed the InVEST–HQ model to estimate the habitat quality in the three UAs of BTH, GBA, and YRD. Habitat quality can be calculated using the following equations:

$$Q_{xj} = H_j(1 - (\frac{D_{xj}^z}{D_{xj}^z + k^z})) \tag{1}$$

where Q_{xj} represents the habitat quality of grid x in land cover/use type j; H_j refers to the habitat suitability of land cover/use type j; D_{xj} reflects the total threat level in grid cell x with land cover/use type j, which is judged by the types, intensities, and proximities of neighboring threats [62]; k is the half-saturation constant, which is equivalent to the half of the crest value of D_{xj}; and z is default at 2.5, which is a scaling parameter for mirroring spatial differentiation [60].

$$D_{xj} = \sum_{r=1}^{R}\sum_{y=1}^{Y_r}(\frac{w_r}{\sum_{r=1}^{R} w_r})r_y i_{rxy} \beta_x S_{jr} \tag{2}$$

where R denotes the number of threats; y indexes all grid cells on the raster layer of threat r; Y_r indicates the set of grid cells occupied by the threat on the raster layer; w_r and r_y reflect the weight of the threat and the interference intensity of the threat for grid cell y, respectively; i_{rxy} is the habitat interference level of grid cell x from threat r on the grid cell y; β_x refers to the reachability level (the anti-disturbance ability) of grid cell x; and S_{jr} represents the relative sensitivity of land cover/use type j to threat factor r. The habitat quality scores range from 0 to 1, which embodies serious to ideal situations for habitat quality [60]. In this paper, i_{rxy} was considered as follows:

$$i_{rxy} = 1 - \left(\frac{d_{xy}}{d_{r\ max}}\right) if\ linear\ attenuation \tag{3}$$

$$i_{rxy} = \exp\left(-\left(\frac{2.99}{d_{r\ max}}\right)d_{xy}\right) if\ exponential\ decay \tag{4}$$

where d_{xy} is the linear distance between grid cells x and y, and $d_{r\ max}$ is the maximum effective influence distance of threat r. More specific information is provided in the InVEST user guide [60] and the relevant literature [35,49,63–66].

We primarily prepared three categories of documentation as input parameters of the InVEST–HQ model: (1) raster layers of land cover/use, (2) threat data, sources, and their accessibility, and (3) suitable habitat types and corresponding threat sensitivities [60]. Threat sources in this study included cultivated lands, bare lands, and built-up lands due to data availability and consistency. The raster layers of threat sources were derived in ArcGIS Pro Version 3.0.1 (https://www.esrij.com/products/arcgis-pro/ (accessed on 30 November 2022)). The setting of threat factors and sensitivity parameters were based on previous literature in the study regions, expert knowledge, and the official user guide manual [35,49,60,63–66], as shown in the Supplementary Materials.

2.5. Evaluation of Multifaceted Environmental States

Land surface temperature (LST) is an essential key variable in surface energy balance, heat fluxes, and energy exchanges over different land cover types, playing a crucial role in geo-environmental and human health [28]. Thermal infrared (TIR) data from satellite platforms and retrieved LST are reliable barometers for diagnosing thermal circumstances in various fields [28,67]. Numerous studies have found different responses to daytime and nighttime LSTs and reported their different influences on environmental situations and human behaviors [28,67]. Thus, we utilized both daytime and nighttime LSTs. Instead of absolute LST observations, the intensity of the surface heat island (SHII) was estimated for evaluating the states of thermal environments in BTH, GBA, and YRD to minimize the effects and uncertainty of the climate background. The SHII was generated using the average daytime and nighttime surface heat island intensities, computed as Equations (5) and (6). The day and night SHII were, respectively, derived from the daytime and nighttime LSTs, which subtracts the corresponding mean LST value of non-UF category pixels from the LST value of each pixel in the study region [68]:

$$SHII = average\left(SHII_{day}, SHII_{night}\right) \tag{5}$$

$$SHII_{day\ or\ night} = LST_{urban} - LST_{rural} = LST_i - \overline{LST_{non-UF}} \tag{6}$$

where $SHII_{day}$ and $SHII_{night}$ are the surface heat island intensities for day and night, respectively. LST_{urban} is the LST value of urban pixels, while LST_{rural} indicates the LST value of rural pixels. LST_i denotes the LST value of pixel i within the individual study regions of BTH, GBA, and YRD. The non-UF category pixels are substitutions for rural pixels, encompassing the pixels from various non-UF land cover/use categories. Thus, $\overline{LST_{non-UF}}$ indicates the average LST values of non-UF pixels in the three UAs.

In this study, the yearly average concentration of PM2.5 was captured to reflect the air environmental situation and pollution in the three UAs. The values of the yearly average NDVI were used to mirror the environmental comfort and biological health degrees based on green spaces and vegetation situations.

Residential support (RS) refers to the spaces or services arranged for the necessities of human living and life, taking into consideration the land cover/use weights and population density [69]. BTH, GBA, and YRD are the most populated regions in China. Thus, residential support is an essential component of UA's environmental issues. The magnitude of RS in this study was evaluated based on the population distribution and built-up proportion, which is computed using the following equation [69]:

$$RS = Pop_{std} \times P_{BU} \tag{7}$$

where Pop_{std} indicates standardized population density. P_{BU} denotes the built-up proportion derived from the calculation of land cover/use.

2.6. Relationship Profiling

2.6.1. Correlation and Local Bivariate Analysis

Given the data-processing capacity and spatial autocorrelation, we generated a set of 3000 m × 3000 m grid cells as the appropriate analytical scale for the spatial analysis. Thus, this study rescaled the habitat quality and all environmental indicators into 3000 m spatial fishnets for evaluating the relationships and interactions between habitat quality and multifaceted environmental situations. All the spatial operations and modeling in this study were completed in the ArcGIS Pro Version 3.0.1 platform. Initially, we utilized the corrplot R package (under R v4.1.0 statistical software, https://www.r-project.org/ (accessed on 30 November 2022)) to visualize the correlation matrix for disclosing the global association between various indicators across the entire UA region. Next, we adopted the local entropy model to explore the local connection between habitat quality and various environmental situations.

The local spatial associations between habitat quality and environmental situations were judged based on a local entropy model. The local entropy model is a nonparametric approach proposed by Guo [29] for multivariate data. It integrates local spatial analysis, Rényi entropy calculation, permutation-based distribution estimation, and a set of statistical tests, allowing the interactive detection of significant local multivariate associations over space [29]. Entropy is a gauge for quantifying uncertainties and randomness, such as Rényi entropy and Shannon entropy [29,70,71]. Rényi and Shannon entropies are computed as the following equations:

$$H_\lambda = \frac{1}{1-\lambda} \log\left(\int f(x)^\lambda dx \right) \quad \lambda \geq 0, \lambda \neq 1 \tag{8}$$

$$H = -\int f(x) \log f(x) dx \tag{9}$$

where H_λ indicates Rényi entropy, x denotes a multi-dimensional vector in the data space, $f(x)$ refers to the probability density function, and λ (≥ 0) represents the order of the H_λ. When λ is closing to 1, H_λ converges to the Shannon entropy (H) in the data space [29]. However, the unknown probability density function is the principal challenge when estimating the entropy for exploratory data analysis [29,70,71]. The power-weighted minimum spanning tree (MST) method is an effective alternative for entropy estimation in multivariate data analysis [29]. Accordingly, the local entropy model can be conducted by the following analytical procedures: (1) multivariate Rényi entropy estimation and observation for the local neighborhood, (2) construction for empirical distributions of local joint entropy using permutations based on the null hypothesis and MST method, (3) test for statistically significant local relationships using permutations, and (4) examination and classification of the local relationships [29,70,71]. Notably, using the appropriate parameter values for the size of the local neighborhood and the edge power of the MST is vital in implementing this local entropy model. Specifically, the scaling factor can modulate the sensitivity to subtle relationships. The proper parameter of the neighboring size can improve the likelihood and efficiency of detecting significant relationships and patterns. Additionally, optimizing the number of permutations can permit a neutralization between precision and processing time [29].

The local entropy model proposed by Guo [29] did not assume a prior relationship form, which can estimate multivariate entropy distribution without a probability density function and implement local spatial analysis without a regression model [29]. Thus, there is a growing application of local entropy maps in geography-related studies [29,70–72]. We designed the research steps of local bivariate analysis based on Guo's approach, as shown in Figure 3. The average habitat quality score of each analytical grid was extracted as the dependent variable. The respective average values of SHII, PM2.5 concentration, NDVI, and RS for each analytical grid were estimated as explanatory variable parameters. Further, we conducted a series of local bivariate relationship analyses to investigate the local association between habitat quality and each explanatory variable. We specified a 95% confidence level of the hypothesis test for significant relationships. Six categories of relationships can be identified based on the local entropy model: not statistically significant, positive linear, negative linear, concave, convex, and undefined complex.

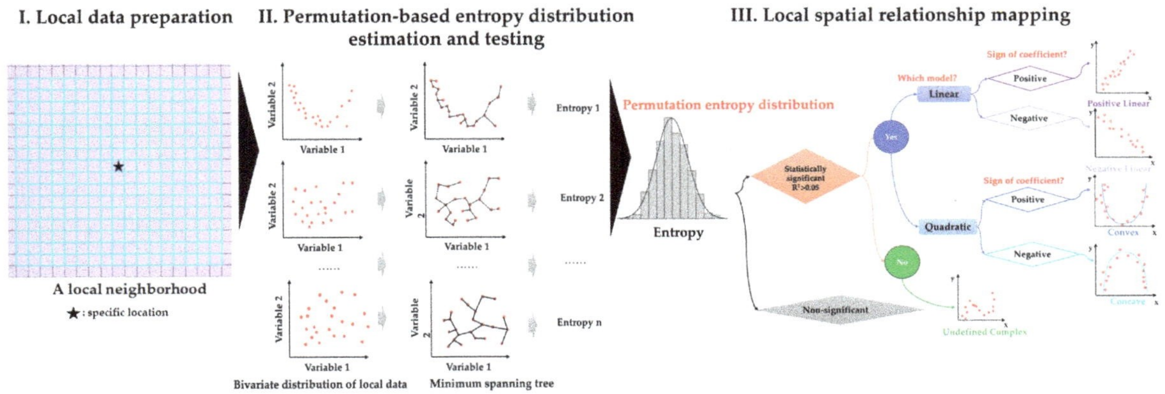

Figure 3. Research flowchart of local bivariate analysis. Note: The flowchart of local bivariate analysis was designed and revised based on Guo's study [29], consisting of three steps: (I) local data preparation; (II) permutation-based entropy distribution estimation and testing; and (III) local spatial relationship mapping.

2.6.2. Geographical Detector Model

The Geographical Detector Model (GeoDetector) is a novel method for capturing stratified spatial heterogeneity and related driving forces, consisting of risk, factor, ecology, and interaction detectors modules [30,31]. GeoDetector can quantitatively analyze the association and similarity of the spatial distributions between a geostatistical variable and a series of stratified independent variables [31]. The merits of GeoDetector are identifying interaction influences between multiple factors and comparing their differences [31].

The GeoDetector model assumes that the study area is composed of multiple region blocks. Spatial heterogeneity is detected if the total variance of the regions is greater than the sum of the variances of the subregions [31]. In contrast, there is a statistical association between the two variables when the spatial patterns of the two variables tend to be identical [31]. Due to no presuppositions and constraints, the GeoDetector analysis is universal and transplantable [31]. Thus, the GeoDetector model has been used in various research fields, such as the geoenvironmental and sustainability sciences [14,73,74].

Given the consistency of the three UAs, the four numerical indicators—RS, SHII, PM2.5 concentrations, and NDVI—were stratified and categorized using equal intervals classification in conjunction with Jenks natural breaks optimization methods. These four stratified indicators served as the explanatory variables (X) in the GeoDetector analysis. The estimated score of habitat quality was still taken as the dependent variable (Y). Among the four modules of GeoDetector, the risk detector showed the responses of habitat quality on each stratified explanatory variable. The ecological detector was used to examine the significant impact difference of stratified explanatory variables on habitat quality. These two modules can determine the suitable range or type of explanatory variables. However, we were more concerned about the outputs of factor and interaction detectors in this study.

Based on the factor detector module, the magnitude of stratified spatial heterogeneity for habitat quality can be quantified using q statistical measures [31].

$$q = 1 - \frac{\sum_{h=1}^{L} N_h \sigma_h^2}{N\sigma^2} = 1 - \frac{SSW}{SST} \tag{10}$$

where the dependent variable Y (habitat quality in this study) is divided into strata $\{h\}$, with $h = 1, \ldots, L$; N and N_h indicate the number of grids in the whole UA region and strata h, respectively; and σ^2 and σ_h^2 denote the variances of the whole region and strata h, respectively. SSW is the sum of squares within, while SST is the sum of squares total. q stands for the explanatory power of a specific explanatory variable X (RS, SHII, PM2.5, and NDVI) for habitat quality, ranging from 0 to 1. The greater the q value, the stronger the influences of the explanatory variable X and the more apparent spatial differentiation of habitat quality [31]. Remarkably, habitat quality exhibits a random spatial distribution when the q value is 0, whereas habitat quality shows complete spatial differentiation when the q value equals 1 [31]. The q value satisfies the non-central F distribution, where λ is the non-central parameter and Y_h is the mean of layer h [31].

$$F = \frac{N-L}{L-1}\frac{q}{1-q}F(\widetilde{L1,\ NL;}) \tag{11}$$

$$\lambda = \frac{1}{\sigma^2}[\sum_{h=1}^{L}\overline{Y_h}^2 - \frac{1}{N}(\sum_{h=1}^{L}\sqrt{N_h\overline{Y_h}})^2] \tag{12}$$

According to the equations above, the p-value can be calculated as a statistical significance test for the q statistics. The interaction detector module can discern the interactive effect on the habitat quality between two environmental situations. Firstly, the impact (q values) of two stratified environmental situations on habitat quality was estimated as $q(X_i)$ and $q(X_j)$, respectively. Next, we computed the q values of the interactive effect ($q(X_i \cap X_j)$). Ultimately, according to judging criteria (Table 1), the interaction type between two stratified environmental situations can be determined by comparing $q(X_i \cap X_j)$ with $q(X_i)$ and $q(X_j)$. More detailed information on the GeoDetector model can be obtained from the official website (www.geodetector.cn (accessed on 30 November 2022)).

Table 1. Definition of the interaction types in the GeoDetector model [31].

Interaction Types	Judging Criteria	Interaction Relationship Descriptions
Nonlinear-weakened	$q(X_i \cap X_j) < Min(q(X_i), q(X_j))$	The synergistic effect is nonlinearly weakened by the interplay of two variables.
Univariate-weakened	$Min(q(X_i), q(X_j)) < q(X_i \cap X_j) < Max(q(X_i), q(X_j))$	The synergistic effect is univariately weakened by the interplay of two variables.
Independent	$q(X_i \cap X_j) > Max(q(X_i), q(X_j))$	The effects of individual variables are independent.
Bivariate-enhanced	$q(X_i \cap X_j) = q(X_i) + q(X_j)$	The synergistic effect is mutually enhanced by the interplay of two variables.
Nonlinear-enhanced	$q(X_i \cap X_j) > q(X_i) + q(X_j)$	The synergistic effect is nonlinearly enhanced by the interplay of two variables.

3. Results

3.1. Spatial Characteristics of Habitat Quality Based on Land Cover/Use Evaluation

The thematic maps of land cover/use for three UAs were prerequisites for quantifying the habitat quality, as shown in Figure 4. The UF areas were highly urbanized areas dominated by large-scale built-up areas, highlighted in red. Natural habitat types are presented using different gradients of green and blue. AH referred to a semi-habitat system used for cultivation and farming activities, depicted in yellow.

Upon the completion of the InVEST–HQ analysis, the magnitudes of habitat quality for three UAs in 2015 were assessed, as shown in Figure 5. The habitat quality scores ranged from 0 to 1, representing the worst to perfect ecological situations. We assorted and mapped the habitat quality index into ten levels using an equal interval of 0.1, indicating the lowest (HL1: 0–0.1) to highest habitat quality (HL10: 0.9–1). The habitat qualities of the three UAs and their geographical variations differed by habitat types and suitability. Overall, the habitat situations of the three UAs in 2015 were not optimistic, and approximately one-third of the regions showed low magnitudes of habitat quality (HL1–HL3). The average habitat quality scores in BTH, GBA, and YRD were 0.3563, 0.3399, and 0.3336, respectively. The moderate-quality levels of habitats (HL4–HL6) made up the largest proportion of the three UA regions. However, the high habitat quality areas (HL7–HL10) added up to under 15%

(Figure 5d). The spatial layouts of habitat quality in the three UAs were highly consistent with the configuration characteristics of land cover/use. The scores of non-habitat areas (UF and VL) were estimated to be zero. In contrast, FH and WA showed higher quality scores than other habitat types in the three UAs. However, the habitat quality score of the WA type in GBA was much lower than that of BTH and YRD. There was a large margin of habitat quality score for the GH type between BTH and the other two UAs. The average scores of AH, FH, GH, and WA in 2015 are shown in Figure 6, as these four habitat types dominated the three UAs (over 85% of area shares).

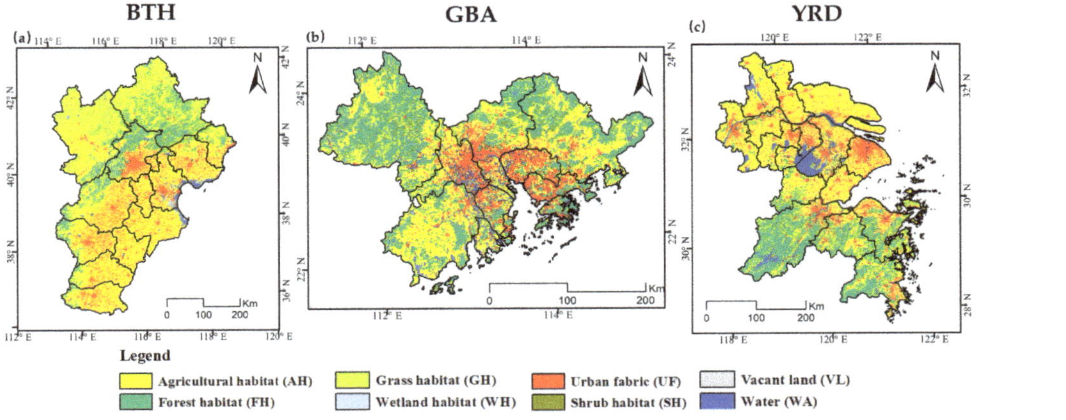

Figure 4. Thematic maps of land cover/use for three UAs: (**a**) BTH, (**b**) GBA, and (**c**) YRD.

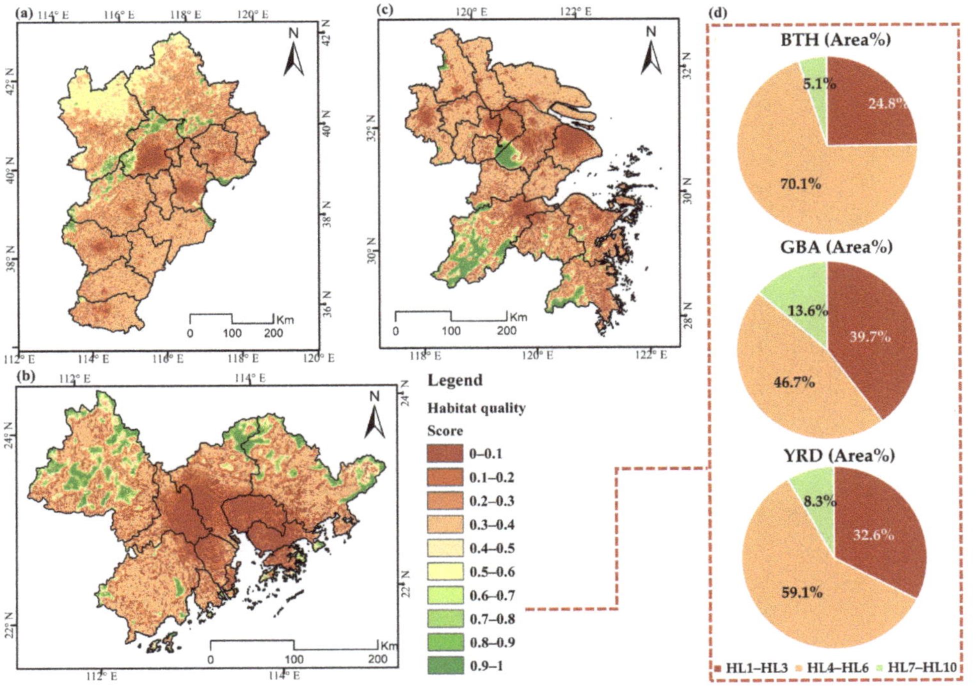

Figure 5. Spatial patterns and distributions of habitat quality in the three UAs: (**a**) BTH, (**b**) GBA, (**c**) YRD, and (**d**) area summary.

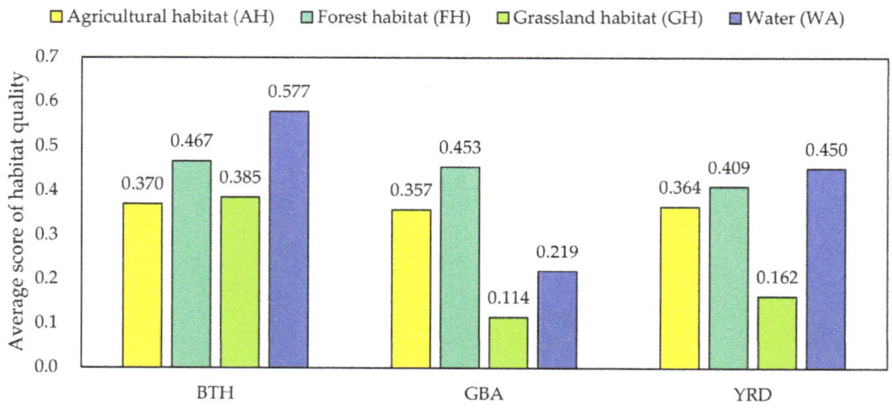

Figure 6. The scores of habitat quality for the four dominant land cover/use.

Geographically, the areas with low habitat suitability in the three UAs in 2015 were chiefly located in human settlements, namely, urban and peri-urban fringes, indicating the adverse influences of anthropogenic activities on the regional habitats. The northwest part of BTH showed high-quality scores due to mountain ridges and massive forests, while the most serious habitat fragile areas were detected in Beijing and Tianjin. The severe habitat delicate areas in GBA were roughly configured as a triangle-shaped region, extending from Guangzhou–Foshan outwards to Macao, Zhuhai, Shenzhen, and Hong Kong. Only Zhaoqing and Huizhou exhibited favorable circumstances for habitats due to their abundant natural capital. In YRD, the southern part and the Taihu region were prominent due to their superior habitat suitability. Shanghai, the south of the Jiangsu Province, and the north of the Zhejiang Province suffered from a challenging habitat situation in 2015 due to the sparse vegetation availability.

3.2. Spatial Stratification of Multi-dimensional Environmental Situations

Although the annual LST values showed a large difference in tendency and influence depending on the day and night, the spatial patterns of day and night heat island intensities in the three UAs were highly consistent. In light of this, this study utilized the mean value of yearly daytime and nighttime SHII to reflect the regional comprehensive thermal environment status in the three UAs. The spatial arrangements of daytime and nighttime thermal environments (both LST and SHII) in the three UAs are shown in the Supplementary Materials. This study divided the SHII index into five categories: ≤0, 0~1, 1~2, 2~3, >3 (unit: °C, Grades I–V) (Figure 7). The 2015 average values of daytime LST in BTH, GBA, and YRD were estimated to be 19.99 °C, 24.7 °C, and 20.83 °C, respectively, and those of respective nighttime LST were 21.84 °C, 26.79 °C, and 22.67 °C. The non-habitat areas and agricultural habitats dominated the areas with high SHII values. The area ratios of positive SHII regions were 56.36% in BTH, 57.12% in GBA, and 60.51% in YRD.

Significant spatial differentiations of yearly average PM2.5 concentrations were found in the three UAs. The spatial maps of the average PM2.5 concentrations in 2015 for the three UAs were stratified into five categories: ≤30, 30~40, 40~50, 50~60, >60 (unit: ug/m^3, Grades I–V), as shown in Figure 8. The BTH regions showed the worst air condition among the three UAs, and about 43.78% of the regional territory was exposed to severe air pollution (Grade V). Most of the GBA regions (94.26%) in 2015 were clustered in Grade II and III PM2.5 concentrations. Only 13.8% of YRD regions did not suffer from severe air pollution.

This study weighed the UAs' living supporting capacity based on the product of built-up proportion and population density. Regions without living supporting capacity were assigned the value of zero. The RS value was equal to one, indicating perfect residential suitability. We defined five grades for the RS index: 0 (without RS capacity), 0~0.1, 0.1~0.2, 0.2~0.3, >0.3 (unitless, Grades I–V). There were multiple cores with extremely high living supporting capacity in the three UAs, as shown in Figure 9. Regions with residential suitability in BTH, GBA, and YRD were estimated to be 26.75%, 26.27%, and 30.1%, respectively.

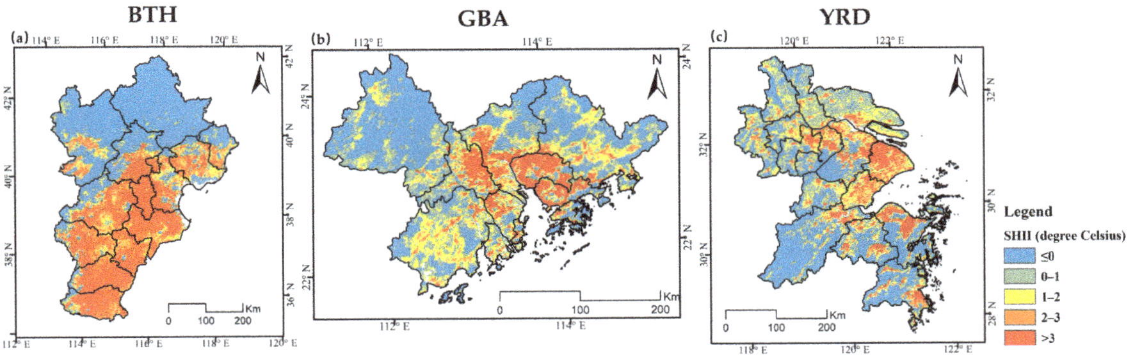

Figure 7. Spatial stratification of the thermal environmental situations in the three UAs: (**a**) BTH, (**b**) GBA, and (**c**) YRD.

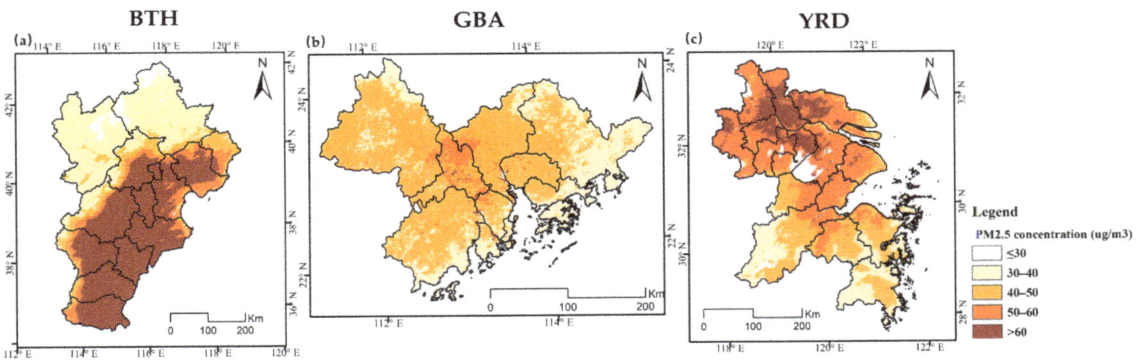

Figure 8. Spatial stratification of the air environmental situations in the three UAs: (**a**) BTH, (**b**) GBA, and (**c**) YRD.

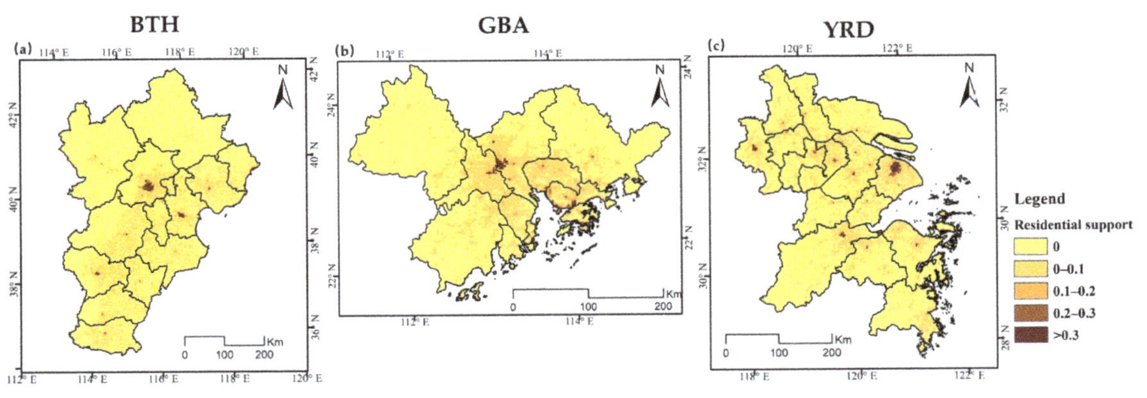

Figure 9. Spatial stratification of the residential situations in the three UAs: (**a**) BTH, (**b**) GBA, and (**c**) YRD.

NDVI represents the aesthetic, recreational, and regulating functions of natural landscapes. We mapped the spatial patterns of the NDVI index with an interval of 0.2 for the three UAs (unitless, Grades I–V, Figure 10). In 2015, the average values of NDVI were projected to be 0.7263 in BTH, 0.7196 in GBA, and 0.6798 in YRD. Approximately 84.77% of BTH regions, 75.33% of GBA regions, and 72.30% of YRD regions had high values of NDVI, exceeding 0.6 in 2015.

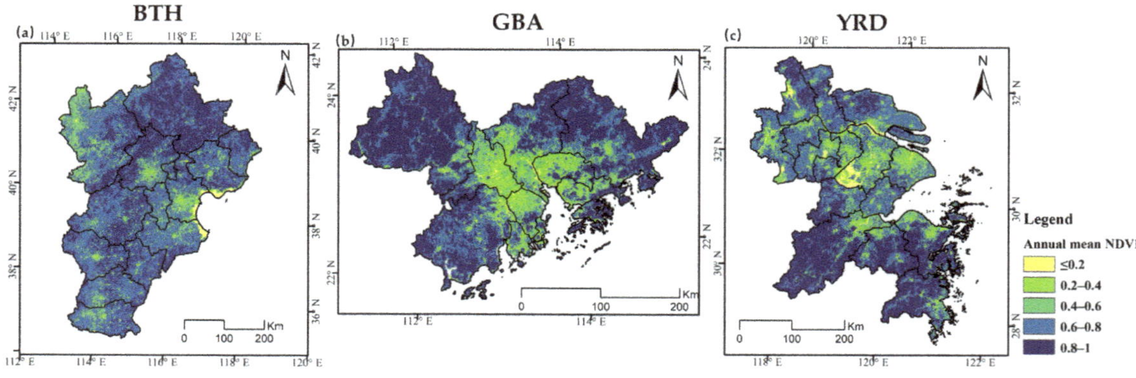

Figure 10. Spatial stratification of the biological health situations in the three UAs: (**a**) BTH, (**b**) GBA, and (**c**) YRD.

3.3. Correlations of Environmental States

We started with the correlation analysis between habitat quality and four environmental status indicators. In general, three UAs exhibited identical relation patterns in the entire regional scale, albeit with varying degrees. Habitat quality in the three UAs was negatively associated with the indicators of SHII ($R^2 \geq 0.49$, $p < 0.001$), PM2.5 concentrations ($R^2 \geq 0.38$, $p < 0.001$), and RS ($R^2 \geq 0.26$, $p < 0.001$), while positively related to NDVI ($R^2 \geq 0.3$, $p < 0.001$). The correlation coefficients between habitat quality and SHII were the highest in GBA ($R^2 = 0.73$) and the lowest in BTH ($R^2 = 0.49$). The association between habitat quality and NDVI degrees in GBA ($R^2 = 0.62$) was the strongest among the three UAs. The most salient impact of PM2.5 concentrations ($R^2 = 0.44$) on habitat was detected in BTH. YRD showed the highest coefficient between habitat quality and RS ($R^2 = 0.32$). As for the four environmental status indicators, there were positive correlations among SHII, PM2.5 concentration, and RS, but NDVI negatively varied with other indicators ($p < 0.001$). In terms of UA, the strongest associations were SHII–PM2.5 concentration in BTH ($R^2 = 0.81$) and SHII–NDVI in GBA ($R^2 = 0.74$) and YRD ($R^2 = 0.64$). In contrast, there were minor linkages between PM2.5 concentrations and RS in all the studied UAs ($R^2 \leq 0.12$). The relevant correlation coefficients are shown in Figure 11.

We identified six types of spatial structural relationships between habitat quality and four environmental status indicators based on the local entropy model. We assumed that there were local relationships between quality and individual environmental status indicators in the three UAs under the 95% confidence level. We set the 0.5 scaling factor, 30 neighborhoods, and 199 permutations to perform the hypothesis test for accommodating local spatial relationships.

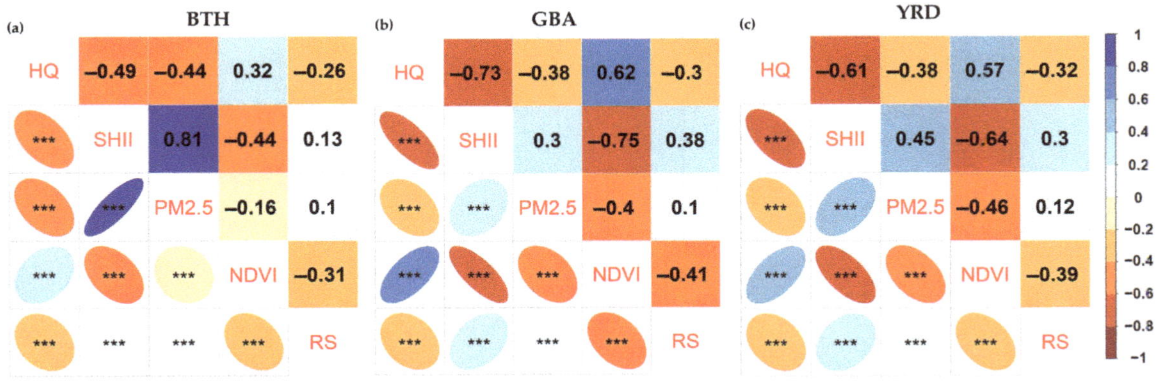

Figure 11. Correlograms of multifaceted environmental situations in the three UAs: (**a**) BTH, (**b**) GBA, and (**c**) YRD. Note: The asterisk symbols ('***') are representative of the statistical significance at the 0.1% level ($p < 0.001$).

The relationships were predicted and determined based on the adjusted R-square and the model's Modified Akaike Information Criterion (AICc). The classified local relationships in the three UAs are mapped in Figure 12, and the relevant categorical summary is shown in Table 2. We provided the entropy statistic and detailed information on the local bivariate analysis in the Supplementary Materials. The results of the local bivariate analysis displayed local structural relationships of various forms between habitat quality and environmental situations over space. Furthermore, the generated maps of local bivariate analysis differentiated the significance levels within a certain neighboring extent. Many regions showed statistically non-significant relationships in the local contexts of the three UAs, indicating a spatially random distribution pattern between variables presented in these regions. The spatially explicit local relationship maps provided an overview of the spatial diversities for significant local associations between habitat quality and environmental situations and navigated the analysis to focus on specific areas. The local bivariate analysis, thus, offered considerable interpretation clues. For example, about 4.86% of BTH regions showed a concave relationship between habitat quality and SHII, indicating that habitat quality in the corresponding BTH regions followed a concave curve as the SHII varied.

We found that the total area ratios of linear and quadratic models for the HQ–SHII relationship in BTH were 23.95% and 10.59%, respectively. This finding indicates that the polynomial and linear models can explain the 23.95% and 10.59% relationship between habitat quality and thermal environment in BTH, respectively. These explanations were also suitable for other estimated coefficients. Thus, the local bivariate analysis allowed us to capture the extent of local heterogeneity between habitat quality and different environmental situations: 27.37–43.14% in BTH, 23.7–65.37% in GBA, and 27.95–55.84% in YRD.

Comparing the performance of four environmental situation indicators, the spatial influences of SHII, NDVI, and RS on habitats in the neighboring local setting were more apt to be detected. However, understanding how the PM2.5 concentration locally influenced the UAs' habitat over space was not completed due to there being over 72% of non-significant areas. Looking at the statistically significant features, the negative linear relationship accounted for the largest share in the HQ–SHII, HQ–PM2.5, and HQ–RS results. In contrast, the positive linear relationship was the most remarkable in the HQ–NDVI results.

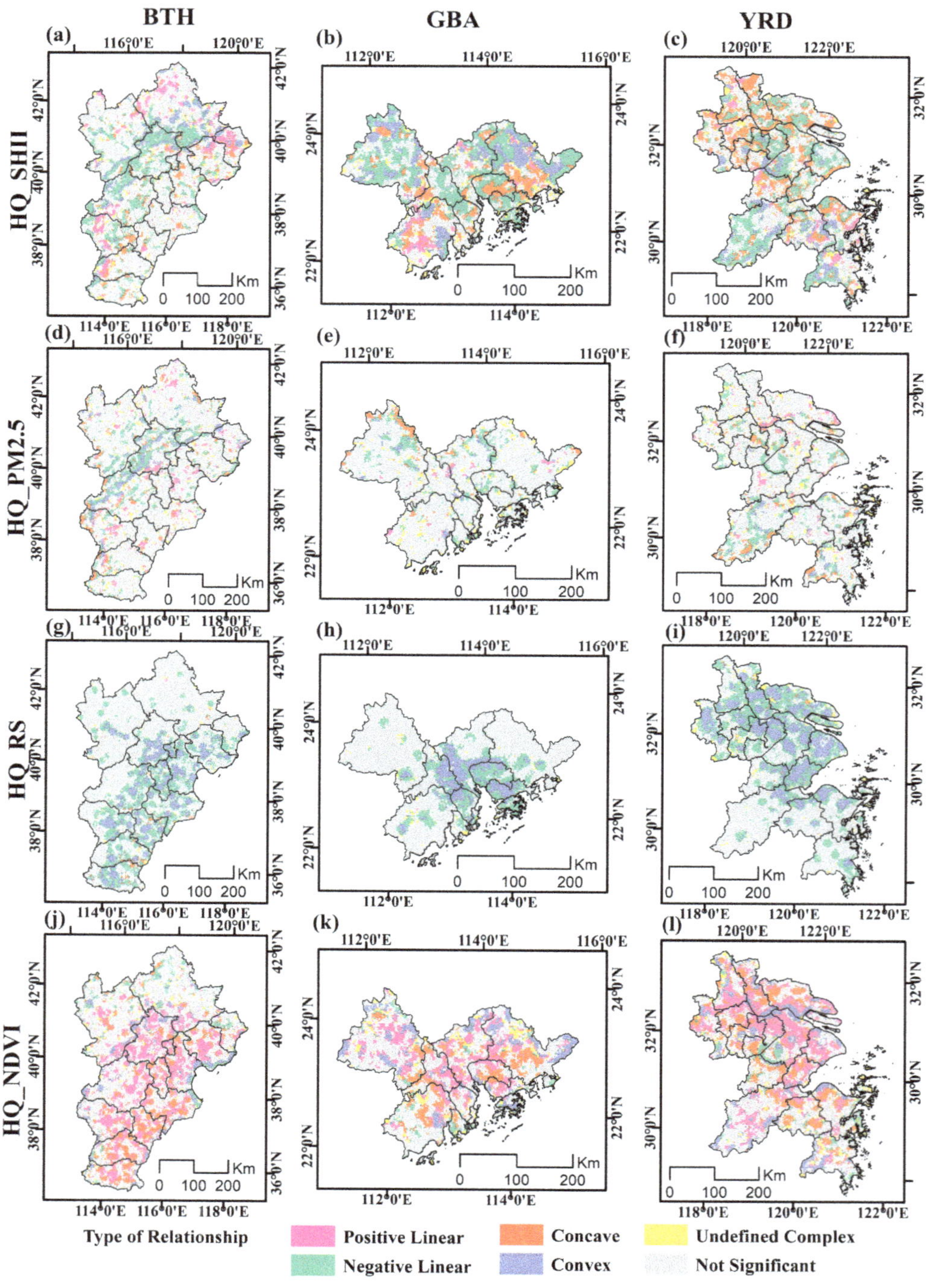

Figure 12. Local spatial relationship between habitat quality and multifaceted environmental states: (**a–c**) HQ–SHII for BTH, GBA, and YRD, respectively; (**d–f**) HQ–PM2.5 for BTH, GBA, and YRD, respectively; (**g–i**) HQ–RS for BTH, GBA, and YRD, respectively; and (**j–l**) HQ–NDVI for BTH, GBA, and YRD, respectively.

Table 2. Categorical summary of the local bivariate analysis.

BTH				
Type of Relationship	**HQ–SHII**	**HQ–PM2.5**	**HQ–RS**	**HQ–NDVI**
Positive Linear	6.99%	4.62%	0.11%	20.87%
Negative Linear	16.96%	10.47%	21.16%	4.27%
Concave	4.86%	3.77%	1.00%	9.87%
Convex	5.73%	4.56%	10.44%	5.48%
Undefined Complex	3.89%	3.95%	1.15%	2.66%
Not Significant	61.57%	72.63%	66.15%	56.86%
GBA				
Type of relationship	**HQ–SHII**	**HQ–PM2.5**	**HQ–RS**	**HQ–NDVI**
Positive Linear	4.17%	1.35%	0.01%	16.91%
Negative Linear	30.17%	10.48%	16.50%	4.78%
Concave	11.53%	3.23%	0.03%	9.28%
Convex	12.26%	1.91%	10.01%	10.81%
Undefined Complex	7.25%	6.72%	2.25%	7.43%
Not Significant	34.63%	76.30%	71.20%	50.80%
YRD				
Type of relationship	**HQ–SHII**	**HQ–PM2.5**	**HQ–RS**	**HQ–NDVI**
Positive Linear	4.16%	3.20%	0.01%	22.00%
Negative Linear	21.41%	11.36%	27.99%	5.46%
Concave	16.18%	4.08%	0.30%	11.43%
Convex	6.23%	2.11%	16.86%	10.53%
Undefined Complex	5.65%	7.20%	3.37%	6.42%
Not Significant	46.36%	72.05%	51.47%	44.16%

3.4. GeoDetector-Based Interactive Effects Assessment

Spatial stratified heterogeneity and interactive influence between habitat quality and multi-dimensional environmental status in the three UAs were captured using GeoDetector. The outputs show that the geographically stratified heterogeneities of habitat quality in the three UAs were, to a certain extent, affected by multifaceted environmental status. The habitat quality varied significantly over space for each graded environmental status indicator based on a series of t-tests at a significance level of 0.05. The related outputs of factor and ecological detector modules are shown in the Supplementary Materials. The stratified average values of habitat quality in each stratum of environmental situations for the three UAs are shown in Figure 13. The q statistic coefficients for four graded environmental status indicators in the three UAs under the factor detector module are shown in Figure 14. The higher q statistic coefficient indicated a greater explanatory power for the stratified heterogeneity of habitat quality. According to the coefficients of the q statistic, the most crucial influencing indicators on habitat quality in the three UAs were PM2.5 concentration (BTH, q = 0.17), SHII (GBA, q = 0.30), and NDVI (YRD, q = 0.21). Other indicators also showed a considerable impact on habitat quality. The outputs of the interaction detector module revealed the interactive influences from a combination of multiple environmental status indicators on habitat quality, as shown in Figure 15. With reference to the judging criteria of interaction (Table 1), we found that the q statistics for the bivariate interactions were greater than the q statistics for every single factor, indicating that the explanatory power of a single indicator could be mutually enhanced when interacting with another one. The cumulative effects of different environmental statuses could bring about more pressures and risks on habitat quality. Specifically, in BTH, the PM2.5 concentration played a dominant role in the single-factor influence analysis and strongly interacted with other indicators. The predominant interaction between PM2.5 concentration and NDVI showed the highest q statistic. A similar phenomenon was also found in the SHII indicator of GBA and the NDVI indicator of YRD.

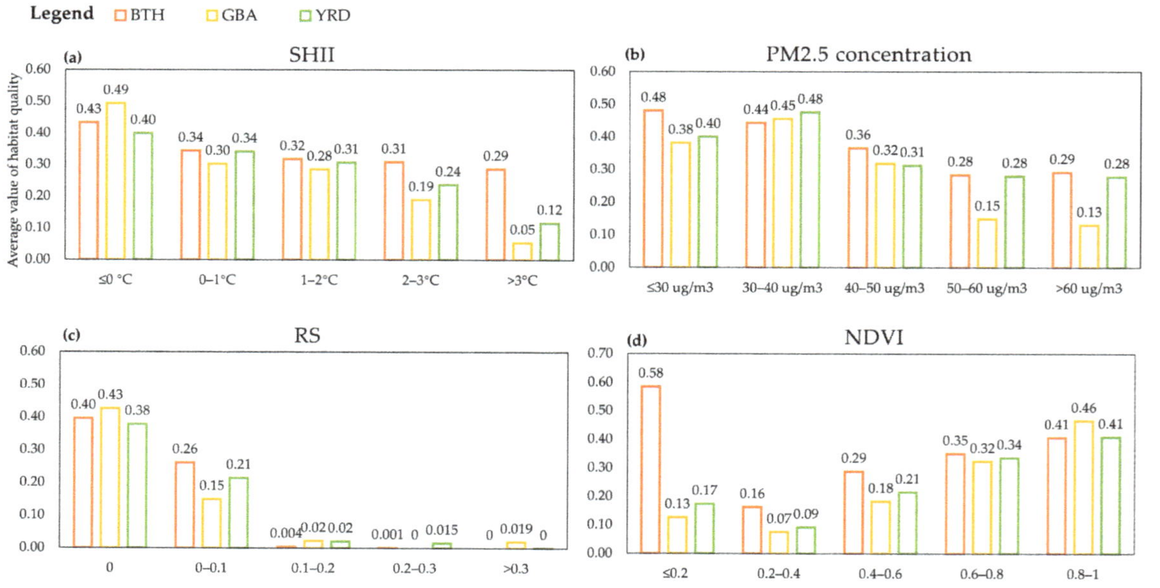

Figure 13. The performances of the graded environmental situations under the risk detector: (**a**) SHII, (**b**) PM2.5, (**c**) RS, and (**d**) NDVI.

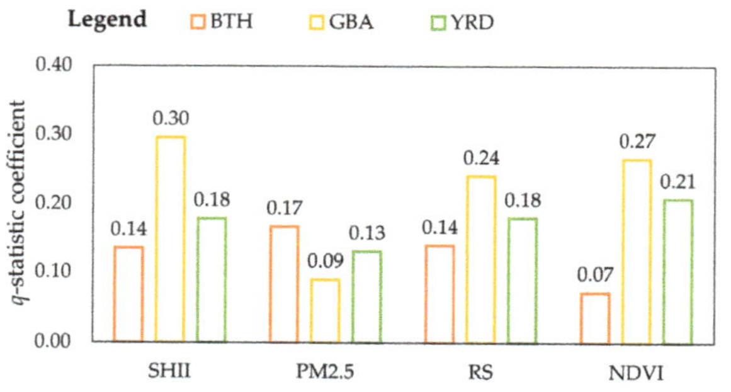

Figure 14. The q-statistic coefficients for the grading indicators in the three UAs.

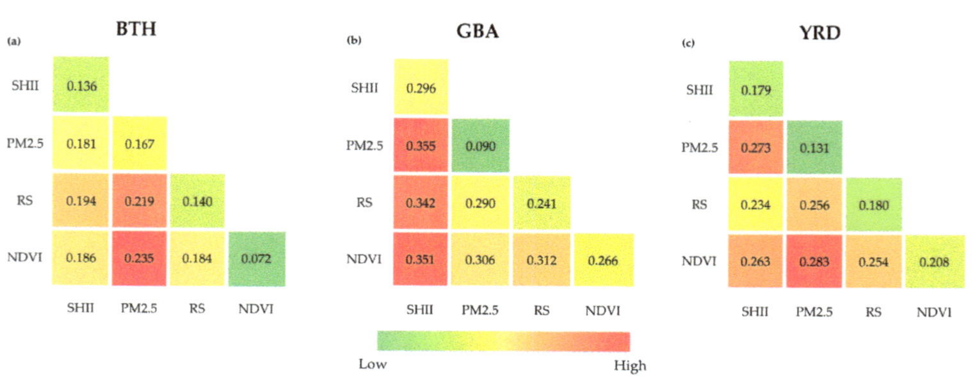

Figure 15. Interactive influences of different environmental situations on habitat quality: (**a**) BTH, (**b**) GBA, and (**c**) YRD.

4. Discussion

4.1. The Impacts of Multifaceted Environmental States on Habitat Quality in the Three Urban Agglomerations

Focusing on BTH, GBA, and YRD, this study spatially evaluated the relationships between habitat quality and multifaceted environmental states using various methods. The cartographic comparison allowed us to capture spatial differentiation and hotspots, the correlation analysis provided a global overview of the statistical relationships, the local entropy model inspected the local structural associations, and GeoDetector identified the spatial stratification heterogeneity and interactive influence. Through such incorporation of multi-perspective profiling, this study presents a comprehensive panorama of the interconnection of various environmental circumstances and their impacts on habitats for different UAs. As a series of biophysical and anthropogenic drivers can threaten the habitat quality and simultaneously trigger changes in different environmental mechanisms, there were similarities and discrepancies in the performances of the three UAs, reflected by the dominant impact and relationship strengths.

Based on the overlay of habitat quality and land/cover use maps in the three UAs, we found that more than four-fifths showed low–medium habitat suitability, spatially clumped in the human settlement and cultivation areas. Meanwhile, large-scale UF areas displayed prominent heat island phenomena, terrible air quality, over-intensive living space, and sparse green availability, albeit with varying degrees of three UAs. Habitat types with high greenness (e.g., forests) in the three UAs showed small gaps in habitat quality scores, ranging from 0.409 to 0.467. The scores of habitats with low greenness (e.g., grass, water) greatly varied across different UAs, depending on habitat types and extents (Figures 6 and 13d). Grounded on correlation analysis, SHII, PM2.5 concentrations, and RS showed an overall deteriorative impact, whereas NDVI had a beneficial influence on the habitat quality in the three UAs. This result was in line with previous environmental-related studies [10,14,15,44,49,74,75]. We also recognized that the most noticeable impacts on habitat were air pollution in BTH, thermal stress, and green capacity in GBA, and residential pressure in YRD. Thus, habitat quality tied in closely with natural assets' situations, indicating that habitat-related strategies depended on improving and maintaining the green and blue spaces.

Next, bivariate relationship profiling based on the local entropy model identified and delineated the local spatial influence of individual environmental status on habitat quality. The spatial mapping of local relationships showed the advantage of distinguishing diverse types of significant associations and highlighting their spatial patterns. Local spatial relationships were roughly in agreement with the global trend of the entire UA. Except for non-significant areas, the habitat quality scores of most regions in the three UAs were negatively related to SHII, PM2.5 concentrations, and RS, while positively associated with NDVI at the 95% confidence level. The local spatial influences of different environmental statuses on habitats were complicated with a rich mixture of relationship types, regardless of UAs. However, large shares of non-significant areas emerged in terms of HQ–PM2.5 (>72%) and HQ–RS (>51%). The PM2.5 concentrations used in this study were estimated based on in situ observations, AOD distribution, and auxiliary data (e.g., land cover/use and population) [57,58]. Our habitat quality scores were assessed using the land cover/use-based InVEST–HQ module [60]. It is difficult to simulate the habitat suitability and quality considering air-related threats due to model constraints, which interfere with the relationships captured between habitat and PM 2.5 concentrations. Regarding HQ–RS performance, the non-significant areas mainly belonged to natural habitats without residential function and service.

The upshots from GeoDetector analysis generally conformed with the foregoing paired relationship profiling and demonstrated the pile-up effect of any two different environmental issues. Relying on the GeoDetector investigation, this study emphasized spatial differences in the overlapping effects and interaction relationships of differently graded environmental strata on habitat quality. Using the interaction detector, we figured out that the impacts of the combination of different environmental issues on habitats in the three UAs were bivariate-enhanced rather than independent or weakened, indicating the aggravation and amplification of compound effect compared with the influence of a single environmental status. In terms of the BTH region, the cumulative effects of air pollution in conjunction with other environmental issues were more severe, coinciding with previous investigations [57,76,77]. BTH has struggled with the worst consequences of air pollution due to a chain of factors [16,78–80]. BTH is the national capital region of China, with over 110 million people [45]. The primary contributors were high energy consumption, vehicle emissions, biomass burning, and neighboring heavy industries [79,80]. As surrounded by mountains, BTH was a victim of its topography, where pollutants are trapped within the regional limits (mainly in the south and east of BTH).

In contrast, the overlapping influences of the heat island phenomenon and other issues in GBA have attracted more attention. We identified that the average annual daytime and nighttime LST values in GBA were the highest among the three UAs. On the other hand, a vast extent with low habitat quality scores in GBA was spatially congregated in the three contiguous megacity clusters (Dongguan–Guangzhou–Foshan, Zhuhai–Macao, and Shenzhen–Hong Kong). Meanwhile, these three contiguous megacity clusters exhibited forceful residential service capacities and low green gradients. However, GBA was superior to the other two UAs in the scale, extent, and gradient of air pollution. Most areas with over 50 ug/m^3 annual PM2.5 concentrations in GBA were merely spatial distributed in Foshan. Accordingly, we observed the lowest impact of PM2.5 concentrations on GBA's habitats. In the YRD case, the interactions between NDVI and other indicators stood out. Additionally, there were minor gaps in impact magnitudes between indicators for YRD in comparison to those for BTH and GBA. We noticed that the gradient of NDVI and habitat quality in YRD were inferior to that of the other two UAs. These performances of YRD can be partly ascribed to development gradients and geomorphic features. Owing to the fast-growing urban built-up and settlements in recent decades, YRD is the most populated, urbanized, and prosperous region in China, with numerous megacities and more than 80 million urban dwellers [45]. Most areas of YRD render a plain terrain, whereas uplands and forests are spatially configured in the south region. Shanghai, Hangzhou Bay, and Suxichang Metropolitan Areas are the most remarkable regions with high urbanness. Such geographical configuration of YRD could respond to its layout and composition of habitat quality, revealing the lag and importance of YRD's greenness and habitat security.

4.2. Road Ahead and Implication for the Spatial Management of Multi-Dimensional Environmental Issues

The uniqueness of this study was the multi-perspective profiling of numerous environmental issues and their interactions. Unlike other environmental studies, this study discerned the associations and interactive effects of different environmental situations instead of clarifying the mechanisms and driving forces of individual environmental problems. Initially, the circumstances of the designated onefold environment in the three UAs were separately evaluated using different indicators and methods. Subsequently, we constructed an integrated framework that combined multiple approaches to link the varying mixes and degrees of different environmental impacts on habitats. This framework propels the multi-objective sustainable management of natural-social capital and co-benefits-seeking policies.

The spatially explicit quantification, mapping, and profiling of various environmental impacts and their relationships serve as the chief references for development plans and policymaking. In recent years, numerous studies detected trade-offs and synergies among multiple environmental services, demonstrating how to move towards sustainability by optimizing the governance and service deliveries of different ecosystems [24,81]. Analogously, our understanding of multifaceted environmental situations and their influences on habitat quality can support work toward tailored sustainable pathways to specify (1) whether and where to adopt the countermeasures to safeguard nature and human habitats, (2) which sector and whose situation should be accounted for and take the corresponding precedence, and (3) how to reduce overlapped risks and advance the co-benefits, which we examined in the context of the three largest Chinese UAs.

Our evaluation results for these three UAs suggest the weightiness of embracing the spatial diversities and context heterogeneities of regional environmental situations for spatially targeted decision-making. Importantly, the landscapes delivering high habitat quality (score $\geq$0.7) in the three UAs occupied only under 15% of their respective areas. However, over half areas emerged with varying strengths of deteriorative environmental consequences across different aspects. This study focused on four environmental issues, namely, heat-related stress, air pollution, over-intensive living space, and limited urban afforestation, which have been the most common and urgent problems in UA regions and emphasized in national initiatives and local policies [23,25]. We identified that the compound impacts were amplified when one environmental issue overlapped with another. Since the interactive effects of different environmental issues were not independent, strategies to improve an onefold environment may contribute to ameliorating other environmental situations. It is imperative to take advantage of the interactive traits of environmental issues to safeguard and advance regional sustainability, preventing the formation and magnification of multifold risks from ecological, climate, and social aspects.

Furthermore, the motivation for UA formation is integrating regional resources, complementarity advantages, and beneficial cooperation [6]. The geographical heterogeneities and multi-dimensional gaps of UAs can offer opportunities for establishing rational resource allocation and cost-sharing mechanisms [6]. The core megacities and highly populated areas of the UA have been subjected to multiple environmental pressures. Due to the extremely constrained habitat availability and ecological carrying capacities, the supply of essential ecological services in these areas should be addressed by the enrichment and complementary of surrounding natural/semi-natural ecosystems rather than self-sufficiency [82]. Therefore, it is crucial to review the heterogeneities and relationships of the multifaceted environmental situations across the entire UA. By synthesizing different situations, the corresponding geographic areas should be identified to efficiently demarcate vulnerable environmental zones, implement the scopes of mitigation or restoration, and equitably assign environmental compensation [24,82].

Considering different hotspots and relationships between environmental issues, the spatial targeting of specific policies should be deliberated in potentially vulnerable areas to ensure effectiveness [38,83,84]. Spatially explicit mapping should visualize the performance of habitat quality and various environmental states, which can be more prone to identifying corresponding vulnerable locations. The most suitable strategies should be generated by integrating the different environmental profiles and minimizing overall negative feedback [23]. For example, since the most severe problem in BTH is air pollution, the interaction between air pollution and other issues can be aggravated. The mitigation of air pollution should be a considerable priority [19,79,80]. A series of strategies and interventions, including a clean energy plan, vast investments in urban greenery, the optimization of energy consumption structure and efficiency, and technological innovations [53], can take precedence to ameliorate the pollution condition of BTH. Simultaneously, thermal and biophysical environments in BTH can also be improved due to common management strategies. Similar arguments are suitable for GBA and YRD. A relief of the heat island phenomenon in GBA and the management intervention of urban afforestation in YRD should take precedence. Prioritization of decision-making in tandem with compound risk minimization sheds light on a promising amelioration pathway of both onefold and comprehensive UA environments, which can make the achievements of spatially comprehensive management and multi-dimensional sustainability possible and efficient from various aspects [19,23,82].

Although we are confident about the contributions of this study to sustainability-related policymaking processes, there is still enormous potential for our framework. Firstly, we are considering the improvement of the habitat quality quantification in further studies by incorporating more threat source data and models, such as traffic data and heavy industrial zone. Our habitat quality was mainly derived from land cover/use evaluation and, to some extent, affected the investigation of relationships between habitat quality and certain environmental situations (e.g., air quality). On the other hand, there can be a dynamic improvement when updating the indicators of environmental issues from a spatiotemporal perspective. In this study, restricted by data availability and consistency, we only evaluated the situations and impacts of four environmental issues on the spatial grid scale in 2015. The relationships between environmental issues and their influences may vary across different spatial scales and evolve over time. We desire to add more temporal dynamics and incorporate policy information to verify the proposed comprehensive research framework further. In particular, significant spatiotemporal disparities and trends may be reflected when comparing the situations before and after the COVID-19 pandemic. Moreover, we can endeavor to explore more environmental issues to enrich the multifaceted evaluation. Environmental profiles, such as energy emission, water purification, and food security, can be synthesized into our future research framework to produce more insights for sustainability support.

5. Conclusions

In tailoring sustainable development schemes for the three UAs of BTH, GBA, and YRD, we profiled the relationships and interactions between habitat quality and multi-environmental status indicators from various perspectives by integrating cartographic comparison, correlation analysis, local entropy map, and GeoDetector. The cartographic comparison highlighted geographical hotspots, the correlation analysis weighed global heterogeneities, the local entropy map differentiated the local spatial associations, and GeoDetector recognized interactive relationships.

The contributions of this study stem from four aspects: (1) spatial valuation of habitat quality for the three UAs; (2) spatial overview of the multifaceted environmental situations for the three UAs; (3) multi-perspective profiling of the relationships and interactive influences of different environmental situations on habitat quality for the three UAs; and (4) implications for supporting habitat sustainability and co-benefit-sought decisions. The critical discoveries are summarized as follows:

In 2015, most of the regional landscapes in the three UAs delivered low/medium habitat availability levels overlapped with threats from the thermal, air, living, and biological environmental spheres. The worst compound environmental risks were geographically aggregated in highly urbanized areas, showing an execrable habitat quality. As expected, the habitat security of the UAs was diminished by the heat island phenomenon, air pollution, and residential support. In contrast, enhancing afforestation in the UAs can be a promising pathway to maintain habitat health. The delineation of the local structural relationships between habitat quality and different environmental statuses highlighted the spatial complexity of the influences and interplay among various environmental systems, indicating potential environmental injustice, spatially heterogeneous changes, and context dependency. Strategies targeting specific environmental issues may induce spatially varying consequences on habitats and produce various trade-offs in other environmental systems. The interaction between various environmental issues manifested bivariate-enhanced effects, implying that the overlay of different environmental issues can magnify the influence of each environmental risk. The seriousness of the worst environmental issue would deteriorate when overlapping with others. Thus, the prioritization of decision-making processes may be required in different geographical contexts, such as air pollution removal in BTH, heat island mitigation in GBA, and green space improvement in YRD.

In conclusion, the findings of this study support the multi-dimensional sustainability and co-benefits sought by taking advantage of the relationships and interactions between habitats and diverse environmental issues in the UAs.

Supplementary Materials: The following supporting information can be downloaded at: https://www.mdpi.com/article/10.3390/rs15040921/s1, Table S1. Threat parameter settings for estimating habitat quality; Table S2. The sensitivity parameter settings of habitat types to threat factors; Table S3. Entropy results summary for local bivariate analysis; Figure S1. Spatial stratification of the land surface temperatures (LSTs) in the three UAs; Figure S2. Spatial stratification of the surface heat islands in the three UAs; Figure S3. Results for ecological detector module; Figure S4. Results for risk detector module.

Author Contributions: Conceptualization, F.L. and Y.M. (Yuji Murayama); Data curation, F.L.; Formal analysis, F.L.; Funding acquisition, Y.M. (Yuji Murayama); Investigation, F.L.; Methodology, F.L.; Project administration, Y.M. (Yuji Murayama); Resources, F.L., Y.M. (Yuji Murayama) and Y.M. (Yoshifumi Masago); Software, F.L.; Supervision, Y.M. (Yuji Murayama); Validation, F.L.; Visualization, F.L.; Writing—original draft, F.L.; Writing—review and editing, Y.M. (Yuji Murayama) and Y.M. (Yoshifumi Masago). All authors have read and agreed to the published version of the manuscript.

Funding: This research was partly supported by the Japan Society for the Promotion of Science (JSPS) grants of 21K01027 and 18H00763.

Data Availability Statement: The data that support the findings of this study are available upon reasonable request from the corresponding author.

Acknowledgments: The authors highly appreciate the editors and anonymous reviewers for their valuable and constructive comments and suggestions to improve the manuscript.

Conflicts of Interest: The authors declare no conflict of interest.

References

1. IPCC. *Climate Change 2022: Impacts, Adaptation, and Vulnerability. Contribution of Working Group II to the Sixth Assessment Report of the Intergovernmental Panel on Climate Change*; Pörtner, H.-O., Roberts, D., Tignor, M., Poloczanska, E., Mintenbeck, K., Alegría, A., Craig, M., Langsdorf, S., Löschke, S., Möller, V., et al., Eds.; Cambridge University Press: Cambridge, UK; New York, NY, USA, 2022.
2. Liu, J.; Hull, V.; Godfray, H.C.J.; Tilman, D.; Gleick, P.; Hoff, H.; Pahl-Wostl, C.; Xu, Z.; Chung, M.G.; Sun, J.; et al. Nexus Approaches to Global Sustainable Development. *Nat. Sustain.* **2018**, *1*, 466–476. [CrossRef]
3. United Nations, Department of Economic and Social Affairs, Population Division. *World Urbanization Prospects: The 2018 Revision (ST/ESA/SER.A/420)*; United Nations: New York, NY, USA, 2019.
4. Almenar, J.B.; Elliot, T.; Rugani, B.; Philippe, B.; Gutierrez, T.N.; Sonnemann, G.; Geneletti, D. Nexus between Nature-Based Solutions, Ecosystem Services and Urban Challenges. *Land Use Policy* **2021**, *100*, 104898. [CrossRef]
5. Cortinovis, C.; Geneletti, D. A Framework to Explore the Effects of Urban Planning Decisions on Regulating Ecosystem Services in Cities. *Ecosyst. Serv.* **2019**, *38*, 100946. [CrossRef]
6. Fang, C.; Yu, D. Urban Agglomeration: An Evolving Concept of an Emerging Phenomenon. *Landsc. Urban Plan.* **2017**, *162*, 126–136. [CrossRef]
7. Li, L.; Ma, S.; Zheng, Y.; Xiao, X. Integrated Regional Development: Comparison of Urban Agglomeration Policies in China. *Land Use Policy* **2022**, *114*, 105939. [CrossRef]
8. Yang, G.; Ge, Y.; Xue, H.; Yang, W.; Shi, Y.; Peng, C.; Du, Y.; Fan, X.; Ren, Y.; Chang, J. Using Ecosystem Service Bundles to Detect Trade-Offs and Synergies across Urban-Rural Complexes. *Landsc. Urban Plan.* **2015**, *136*, 110–121. [CrossRef]
9. Fu, X.; Yao, L.; Xu, W.; Wang, Y.; Sun, S. Exploring the Multitemporal Surface Urban Heat Island Effect and Its Driving Relation in the Beijing-Tianjin-Hebei Urban Agglomeration. *Appl. Geogr.* **2022**, *144*, 102714. [CrossRef]
10. Ouyang, X.; Wei, X.; Li, Y.; Wang, X.C.; Klemeš, J.J. Impacts of Urban Land Morphology on PM2.5 Concentration in the Urban Agglomerations of China. *J. Environ. Manag.* **2021**, *283*, 112000. [CrossRef]
11. Wen, Y.; Zhang, Z.; Liang, D.; Xu, Z. Rural Residential Land Transition in the Beijing-Tianjin-Hebei Region: Spatial-Temporal Patterns and Policy Implications. *Land Use Policy* **2020**, *96*, 104700. [CrossRef]
12. Zhang, D.; Huang, Q.; He, C.; Yin, D.; Liu, Z. Planning Urban Landscape to Maintain Key Ecosystem Services in a Rapidly Urbanizing Area: A Scenario Analysis in the Beijing-Tianjin-Hebei Urban Agglomeration, China. *Ecol. Indic.* **2019**, *96*, 559–571. [CrossRef]
13. Sun, M.; Wang, J.; He, K. Analysis on the Urban Land Resources Carrying Capacity during Urbanization—A Case Study of Chinese YRD. *Appl. Geogr.* **2020**, *116*, 102170. [CrossRef]
14. Feng, R.; Wang, F.; Wang, K.; Wang, H.; Li, L. Urban Ecological Land and Natural-Anthropogenic Environment Interactively Drive Surface Urban Heat Island: An Urban Agglomeration-Level Study in China. *Environ. Int.* **2021**, *157*, 106857. [CrossRef] [PubMed]
15. Zhang, H.; Yin, Y.; An, H.; Lei, J.; Li, M.; Song, J.; Han, W. Surface Urban Heat Island and Its Relationship with Land Cover Change in Five Urban Agglomerations in China Based on GEE. *Environ. Sci. Pollut. Res.* **2022**, *29*, 82271–82285. [CrossRef]

16. Wang, Y.; Yao, L.; Xu, Y.; Sun, S.; Li, T. Potential Heterogeneity in the Relationship between Urbanization and Air Pollution, from the Perspective of Urban Agglomeration. *J. Clean. Prod.* **2021**, *298*, 126822. [CrossRef]
17. Zhao, R.; Zhan, L.; Yao, M.; Yang, L. A Geographically Weighted Regression Model Augmented by Geodetector Analysis and Principal Component Analysis for the Spatial Distribution of PM2.5. *Sustain. Cities Soc.* **2020**, *56*, 102106. [CrossRef]
18. Mao, X.; Huang, X.; Song, Y.; Zhu, Y.; Tan, Q. Response to Urban Land Scarcity in Growing Megacities: Urban Containment or Inter-City Connection? *Cities* **2020**, *96*, 102399. [CrossRef]
19. Lafortezza, R.; Sanesi, G. Nature-Based Solutions: Settling the Issue of Sustainable Urbanization. *Environ. Res.* **2019**, *172*, 394–398. [CrossRef]
20. Tittonell, P.; Struik, P.C.; Kuyper, T.W. Ecological Intensification of Agriculture—Sustainable by Nature. *Curr. Opin. Environ. Sustain.* **2014**, *8*, 53–61. [CrossRef]
21. Bush, J.; Doyon, A. Building Urban Resilience with Nature-Based Solutions: How Can Urban Planning Contribute? *Cities* **2019**, *95*, 102483. [CrossRef]
22. Nesshöver, C.; Assmuth, T.; Irvine, K.N.; Rusch, G.M.; Waylen, K.A.; Delbaere, B.; Haase, D.; Jones-Walters, L.; Keune, H.; Kovacs, E.; et al. The Science, Policy and Practice of Nature-Based Solutions: An Interdisciplinary Perspective. *Sci. Total Environ.* **2017**, *579*, 1215–1227. [CrossRef]
23. Westman, L.; Patterson, J.; Macrorie, R.; Orr, C.J.; Ashcraft, C.M.; Broto, V.C.; Dolan, D.; Gupta, M.; van der Heijden, J.; Hickmann, T.; et al. Compound Urban Crises. *Ambio* **2022**, *51*, 1402–1415. [CrossRef] [PubMed]
24. Cord, A.F.; Bartkowski, B.; Beckmann, M.; Dittrich, A.; Hermans-Neumann, K.; Kaim, A.; Lienhoop, N.; Locher-Krause, K.; Priess, J.; Schröter-Schlaack, C.; et al. Towards Systematic Analyses of Ecosystem Service Trade-Offs and Synergies: Main Concepts, Methods and the Road Ahead. *Ecosyst. Serv.* **2017**, *28*, 264–272. [CrossRef]
25. Piracha, A.; Chaudhary, M.T. Urban Air Pollution, Urban Heat Island and Human Health: A Review of the Literature. *Sustainability* **2022**, *14*, 9234. [CrossRef]
26. Singh, N.; Singh, S.; Mall, R.K. Urban Ecology and Human Health: Implications of Urban Heat Island, Air Pollution and Climate Change Nexus. In *Urban Ecology*; Elsevier Inc.: Cambridge, MA, USA, 2020; pp. 317–334.
27. Niu, L.; Zhang, Z.; Peng, Z.; Liang, Y.; Liu, M.; Jiang, Y.; Wei, J.; Tang, R. Identifying Surface Urban Heat Island Drivers and Their Spatial Heterogeneity in China's 281 Cities: An Empirical Study Based on Multiscale Geographically Weighted Regression. *Remote Sens.* **2021**, *13*, 4428. [CrossRef]
28. Deilami, K.; Kamruzzaman, M.; Liu, Y. Urban Heat Island Effect: A Systematic Review of Spatio-Temporal Factors, Data, Methods, and Mitigation Measures. *Int. J. Appl. Earth Obs. Geoinf.* **2018**, *67*, 30–42. [CrossRef]
29. Guo, D. Local Entropy Map: A Nonparametric Approach to Detecting Spatially Varying Multivariate Relationships. *Int. J. Geogr. Inf. Sci.* **2010**, *24*, 1367–1389. [CrossRef]
30. Wang, J.F.; Li, X.H.; Christakos, G.; Liao, Y.L.; Zhang, T.; Gu, X.; Zheng, X.Y. Geographical Detectors-Based Health Risk Assessment and Its Application in the Neural Tube Defects Study of the Heshun Region, China. *Int. J. Geogr. Inf. Sci.* **2010**, *24*, 107–127. [CrossRef]
31. Wang, J.F.; Zhang, T.L.; Fu, B.J. A Measure of Spatial Stratified Heterogeneity. *Ecol. Indic.* **2016**, *67*, 250–256. [CrossRef]
32. Outline Development Plan for the Guangdong-Hong Kong-Macao Greater Bay Area. Available online: https://www.bayarea.gov.hk/en/outline/plan.html (accessed on 30 November 2022).
33. The Outline Plan of Beijing-Tianjin-Hebei Integrated Development. Available online: http://jjj.chinadevelopment.com.cn/ (accessed on 30 November 2022). (In Chinese).
34. The Yangtze River Delta Integration Plan. Available online: http://www.gov.cn/zhengce/2019-12/01/content_5457442.htm (accessed on 30 November 2022). (In Chinese)
35. Wu, L.; Sun, C.; Fan, F. Estimating the Characteristic Spatiotemporal Variation in Habitat Quality Using the Invest Model—A Case Study from Guangdong–Hong Kong–Macao Greater Bay Area. *Remote Sens.* **2021**, *13*, 1008. [CrossRef]
36. Lin, M.; Lin, T.; Sun, C.; Jones, L.; Sui, J.; Zhao, Y.; Liu, J.; Xing, L.; Ye, H.; Zhang, G.; et al. Using the Eco-Erosion Index to Assess Regional Ecological Stress Due to Urbanization—A Case Study in the Yangtze River Delta Urban Agglomeration. *Ecol. Indic.* **2020**, *111*, 106028. [CrossRef]
37. Dong, L.; Longwu, L.; Zhenbo, W.; Liangkan, C.; Faming, Z. Exploration of Coupling Effects in the Economy–Society–Environment System in Urban Areas: Case Study of the Yangtze River Delta Urban Agglomeration. *Ecol. Indic.* **2021**, *128*, 107858. [CrossRef]
38. Wang, J.; Zhou, W.; Pickett, S.T.A.; Yu, W.; Li, W. A Multiscale Analysis of Urbanization Effects on Ecosystem Services Supply in an Urban Megaregion. *Sci. Total Environ.* **2019**, *662*, 824–833. [CrossRef] [PubMed]
39. Gao, K.; Yang, X.; Wang, Z.; Zhang, H.; Huang, C.; Zeng, X. Spatial Sustainable Development Assessment Using Fusing Multisource Data from the Perspective of Production-Living-Ecological Space Division: A Case of Greater Bay Area, China. *Remote Sens.* **2022**, *14*, 2772. [CrossRef]
40. Jiang, H.; Peng, J.; Dong, J.; Zhang, Z.; Xu, Z.; Meersmans, J. Linking Ecological Background and Demand to Identify Ecological Security Patterns across the Guangdong-Hong Kong-Macao Greater Bay Area in China. *Landsc. Ecol.* **2021**, *36*, 2135–2150. [CrossRef]
41. Lin, G.; Jiang, D.; Fu, J.; Cao, C.; Zhang, D. Spatial Conflict of Production-Living-Ecological Space and Sustainable-Development Scenario Simulation in Yangtze River Delta Agglomerations. *Sustainability* **2020**, *12*, 2175. [CrossRef]

42. Fang, G.; Wang, Q.; Tian, L. Green Development of Yangtze River Delta in China under Population-Resources-Environment-Development-Satisfaction Perspective. *Sci. Total Environ.* **2020**, *727*, 138710. [CrossRef] [PubMed]
43. Yang, Y.; Bao, W.; Liu, Y. Coupling Coordination Analysis of Rural Production-Living-Ecological Space in the Beijing-Tianjin-Hebei Region. *Ecol. Indic.* **2020**, *117*, 106512. [CrossRef]
44. Peng, C.; Li, B.; Nan, B. An Analysis Framework for the Ecological Security of Urban Agglomeration: A Case Study of the Beijing-Tianjin-Hebei Urban Agglomeration. *J. Clean. Prod.* **2021**, *315*, 128111. [CrossRef]
45. National Bureau of Statistics of China (NBSC). Available online: http://www.stats.gov.cn/english/Statisticaldata/AnnualData/ (accessed on 30 November 2022).
46. Wang, X.; Yan, F.; Zeng, Y.; Chen, M.; Su, F.; Cui, Y. Changes in Ecosystems and Ecosystem Services in the Guangdong-Hong Kong-Macao Greater Bay Area since the Reform and Opening up in China. *Remote Sens.* **2021**, *13*, 1611. [CrossRef]
47. Zhou, M.; Lu, L.; Guo, H.; Weng, Q.; Cao, S.; Zhang, S.; Li, Q. Urban Sprawl and Changes in Land-Use Efficiency in the Beijing–Tianjin–Hebei Region, China from 2000 to 2020: A Spatiotemporal Analysis Using Earth Observation Data. *Remote Sens.* **2021**, *13*, 2850. [CrossRef]
48. Wu, J.; Li, X.; Luo, Y.; Zhang, D. Spatiotemporal Effects of Urban Sprawl on Habitat Quality in the Pearl River Delta from 1990 to 2018. *Sci. Rep.* **2021**, *11*, 1–15. [CrossRef] [PubMed]
49. Wang, H.; Tang, L.; Qiu, Q.; Chen, H. Assessing the Impacts of Urban Expansion on Habitat Quality by Combining the Concepts of Land Use, Landscape, and Habitat in Two Urban Agglomerations in China. *Sustainability* **2020**, *12*, 4346. [CrossRef]
50. He, C.; Gao, B.; Huang, Q.; Ma, Q.; Dou, Y. Environmental Degradation in the Urban Areas of China: Evidence from Multi-Source Remote Sensing Data. *Remote Sens. Environ.* **2017**, *193*, 65–75. [CrossRef]
51. Zhang, D.; Huang, Q.; He, C.; Wu, J. Impacts of Urban Expansion on Ecosystem Services in the Beijing-Tianjin-Hebei Urban Agglomeration, China: A Scenario Analysis Based on the Shared Socioeconomic Pathways. *Resour. Conserv. Recycl.* **2017**, *125*, 115–130. [CrossRef]
52. Lu, D.; Song, K.; Zang, S.; Jia, M.; Du, J.; Ren, C. The Effect of Urban Expansion on Urban Surface Temperature in Shenyang, China: An Analysis with Landsat Imagery. *Environ. Model. Assess.* **2015**, *20*, 197–210. [CrossRef]
53. Zhang, G.; Gao, Y.; Li, J.; Su, B.; Chen, Z.; Lin, W. China's Environmental Policy Intensity for 1978–2019. *Sci. Data* **2022**, *9*, 75. [CrossRef]
54. European Space Agency (ESA). Land Cover CCI Product User Guide Version Tech Rep. 2017. Available online: http://maps.elie.ucl.ac.be/CCI/viewer/download/ESACCI-LC-Ph2-PUGv2_2.0.pdf (accessed on 30 November 2022).
55. IUCN Habitats Classification Scheme. Available online: https://www.iucnredlist.org/resources/habitat-classification-scheme (accessed on 30 November 2022).
56. Jung, M.; Dahal, P.R.; Butchart, S.H.M.; Donald, P.F.; De Lamo, X.; Lesiv, M.; Kapos, V.; Rondinini, C.; Visconti, P. A Global Map of Terrestrial Habitat Types. *Sci. Data* **2020**, *7*, 1–8. [CrossRef]
57. Wei, J.; Li, Z.; Cribb, M.; Huang, W.; Xue, W.; Sun, L.; Guo, J.; Peng, Y.; Li, J.; Lyapustin, A.; et al. Improved 1km Resolution PM2.5 Estimates across China Using Enhanced Space-Time Extremely Randomized Trees. *Atmos. Chem. Phys.* **2020**, *20*, 3273–3289. [CrossRef]
58. Wei, J.; Li, Z.; Lyapustin, A.; Sun, L.; Peng, Y.; Xue, W.; Su, T.; Cribb, M. Reconstructing 1-Km-Resolution High-Quality PM2.5 Data Records from 2000 to 2018 in China: Spatiotemporal Variations and Policy Implications. *Remote Sens. Environ.* **2021**, *252*, 112136. [CrossRef]
59. Lloyd, C.T.; Sorichetta, A.; Tatem, A.J. High Resolution Global Gridded Data for Use in Population Studies. *Sci. Data* **2017**, *4*, 170001. [CrossRef]
60. Natural Capital Project. InVEST-Habitat Quality. Available online: https://storage.googleapis.com/releases.naturalcapitalproject.org/invest-userguide/latest/habitat_quality.html (accessed on 30 November 2022).
61. Hall, L.S.; Krausman, P.R.; Morrison, M.L. The Habitat Concept and a Plea for Standard Terminology Key Words Peer Refereed. *Wildl. Soc. Bull.* **1997**, *25*, 173–182.
62. Mckinney, M.L. Urbanization, Biodiversity, and Conservation. *Bioscience* **2002**, *52*, 883–890. [CrossRef]
63. Tang, F.; Fu, M.; Wang, L.; Song, W.; Yu, J.; Wu, Y. Dynamic Evolution and Scenario Simulation of Habitat Quality under the Impact of Land-Use Change in the Huaihe River Economic Belt, China. *PLoS ONE* **2021**, *16*, e0249566. [CrossRef] [PubMed]
64. Chen, T.; Feng, Z.; Zhao, H.; Wu, K. Dataset of Ecosystem Services in Beijing and Its Surrounding Areas. *Data Br.* **2020**, *29*, 105151. [CrossRef] [PubMed]
65. Xu, L.; Chen, S.S.; Xu, Y.; Li, G.; Su, W. Impacts of Land-Use Change on Habitat Quality during 1985-2015 in the Taihu Lake Basin. *Sustainability* **2019**, *11*, 3513. [CrossRef]
66. Li, F.; Wang, L.; Chen, Z.; Clarke, K.C.; Li, M.; Jiang, P. Extending the SLEUTH Model to Integrate Habitat Quality into Urban Growth Simulation. *J. Environ. Manag.* **2018**, *217*, 486–498. [CrossRef]
67. Zhou, D.; Xiao, J.; Bonafoni, S.; Berger, C.; Deilami, K.; Zhou, Y.; Frolking, S.; Yao, R.; Qiao, Z.; Sobrino, J.A. Satellite Remote Sensing of Surface Urban Heat Islands: Progress, Challenges, and Perspectives. *Remote Sens.* **2019**, *11*, 48. [CrossRef]
68. Liu, F.; Zhang, X.; Murayama, Y.; Morimoto, T. Impacts of Land Cover/Use on the Urban Thermal Environment: A Comparative Study of 10 Megacities in China. *Remote Sens.* **2020**, *12*, 307. [CrossRef]

69. Wang, Q.; Wang, H.; Zeng, H.; Chang, R.; Bai, X. Understanding Relationships between Landscape Multifunctionality and Land-Use Change across Spatiotemporal Characteristics: Implications for Supporting Landscape Management Decisions. *J. Clean. Prod.* **2022**, *377*, 134474. [CrossRef]
70. Naimi, B.; Hamm, N.A.S.; Groen, T.A.; Skidmore, A.K.; Toxopeus, A.G.; Alibakhshi, S. ELSA: Entropy-Based Local Indicator of Spatial Association. *Spat. Stat.* **2019**, *29*, 66–88. [CrossRef]
71. Kalantari, M.; Ghavagh, A.R.; Toomanian, A.; Dero, Q.Y. A New Methodological Framework for Crime Spatial Analysis Using Local Entropy Map. *Mod. Appl. Sci.* **2016**, *10*, 179. [CrossRef]
72. Yaakub, N.F.; Masron, T.; Marzuki, A.; Soda, R. GIS-Based Spatial Correlation Analysis: Sustainable Development and Two Generations of Demographic Changes. *Sustainability* **2022**, *14*, 1490. [CrossRef]
73. Wang, H.; Qin, F.; Xu, C.; Li, B.; Guo, L.; Wang, Z. Evaluating the Suitability of Urban Development Land with a Geodetector. *Ecol. Indic.* **2021**, *123*, 107339. [CrossRef]
74. Zhang, X.; Liao, L.; Xu, Z.; Zhang, J.; Chi, M.; Lan, S.; Gan, Q. Interactive Effects on Habitat Quality Using InVEST and GeoDetector Models in Wenzhou, China. *Land* **2022**, *11*, 630. [CrossRef]
75. Wang, Y.; Guo, Z.; Han, J. The Relationship between Urban Heat Island and Air Pollutants and Them with Influencing Factors in the Yangtze River Delta, China. *Ecol. Indic.* **2021**, *129*, 107976. [CrossRef]
76. Yang, Y.; Zheng, Z.; Yim, S.Y.L.; Roth, M.; Ren, G.; Gao, Z.; Wang, T.; Li, Q.; Shi, C.; Ning, G.; et al. PM2.5 Pollution Modulates Wintertime Urban Heat Island Intensity in the Beijing-Tianjin-Hebei Megalopolis, China. *Geophys. Res. Lett.* **2020**, *47*, e2019GL084288. [CrossRef]
77. Wu, H.; Wang, T.; Wang, Q.; Riemer, N.; Cao, Y.; Liu, C.; Ma, C.; Xie, X. Relieved Air Pollution Enhanced Urban Heat Island Intensity in the Yangtze River Delta, China. *Aerosol Air Qual. Res.* **2019**, *19*, 2683–2696. [CrossRef]
78. Wu, D.; Xu, Y.; Zhang, S. Will Joint Regional Air Pollution Control Be More Cost-Effective? An Empirical Study of China's Beijing-Tianjin-Hebei Region. *J. Environ. Manag.* **2015**, *149*, 27–36.
79. Wang, L.; Zhang, F.; Pilot, E.; Yu, J.; Nie, C.; Holdaway, J.; Yang, L.; Li, Y.; Wang, W.; Vardoulakis, S.; et al. Taking Action on Air Pollution Control in the Beijing-Tianjin-Hebei (BTH) Region: Progress, Challenges and Opportunities. *Int. J. Environ. Res. Public Health* **2018**, *15*, 306. [CrossRef]
80. Li, Y.; Huang, S.; Yin, C.; Sun, G.; Ge, C. Construction and Countermeasure Discussion on Government Performance Evaluation Model of Air Pollution Control: A Case Study from Beijing-Tianjin-Hebei Region. *J. Clean. Prod.* **2020**, *254*, 120072. [CrossRef]
81. Li, Y.; Zhang, L.; Qiu, J.; Yan, J.; Wan, L.; Wang, P.; Hu, N.; Cheng, W.; Fu, B. Spatially Explicit Quantification of the Interactions among Ecosystem Services. *Landsc. Ecol.* **2017**, *32*, 1181–1199. [CrossRef]
82. Shen, J.; Li, S.; Liang, Z.; Liu, L.; Li, D.; Wu, S. Exploring the Heterogeneity and Nonlinearity of Trade-Offs and Synergies among Ecosystem Services Bundles in the Beijing-Tianjin-Hebei Urban Agglomeration. *Ecosyst. Serv.* **2020**, *43*, 101103. [CrossRef]
83. Li, G.; Fang, C.; Wang, S. Exploring Spatiotemporal Changes in Ecosystem-Service Values and Hotspots in China. *Sci. Total Environ.* **2016**, *545–546*, 609–620. [CrossRef] [PubMed]
84. Hou, Y.; Li, B.; Müller, F.; Fu, Q.; Chen, W. A Conservation Decision-Making Framework Based on Ecosystem Service Hotspot and Interaction Analyses on Multiple Scales. *Sci. Total Environ.* **2018**, *643*, 277–291. [CrossRef] [PubMed]

remote sensing

MDPI

Article

Driving Factors of Land Surface Temperature in Urban Agglomerations: A Case Study in the Pearl River Delta, China

Wenxiu Liu [1,2], Qingyan Meng [1,2,3,*], Mona Allam [4], Linlin Zhang [1], Die Hu [1,2] and Massimo Menenti [5]

1 Aerospace Information Research Institute, Chinese Academy of Sciences, Beijing 100101, China; Liuwx@radi.ac.cn (W.L.); zhangll@aircas.ac.cn (L.Z.); hudie@radi.ac.cn (D.H.)
2 University of Chinese Academy of Sciences, Beijing 100101, China
3 Sanya Institute of Remote Sensing, Sanya 572029, China
4 Environment & Climate Changes Research Institute, National Water Research Center, El Qanater EI khairiya 13621/5, Egypt; monaallam@sci.asu.edu.eg
5 Geosciences and Remote Sensing Department, Delft University of Technology, Stevinweg, 12628 CN Delft, The Netherlands; m.menenti@tudelft.nl
* Correspondence: mengqy@radi.ac.cn

Abstract: Land surface temperature (LST) in urban agglomerations plays an important role for policymakers in urban planning. The Pearl River Delta (PRD) is one of the regions with the highest urban densities in the world. This study aims to explore the spatial patterns and the dominant drivers of LST in the PRD. MODIS LST (MYD11A2) data from 2005 and 2015 were used in this study. First, spatial analysis methods were applied in order to determine the spatial patterns of LST and to identity the hotspot areas (HSAs). Second, the hotspot ratio index (HRI), as a metric of thermal heterogeneity, was developed in order to identify the features of thermal environment across the nine cities in the PRD. Finally, the geo-detector (GD) metric was employed to explore the dominant drivers of LST, which included elevation, land use/land cover (LUCC), the normalized difference vegetation index (NDVI), impervious surface distribution density (ISDD), gross domestic product (GDP), population density (POP), and nighttime light index (NLI). The GD metric has the advantages of detecting the dominant drivers without assuming linear relationships and measuring the combined effects of the drivers. The results of Moran's Index showed that the daytime and nighttime LST were close to the cluster pattern. Therefore, this process led to the identification of HSAs. The HSAs were concentrated in the central PRD and were distributed around the Pearl River estuary. The results of the HRI indicated that the spatial distribution of the HSAs was highly heterogeneous among the cities for both daytime and nighttime. The highest HRI values were recorded in the cities of Dongguan and Shenzhen during the daytime. The HRI values in the cities of Zhaoqing, Jiangmen, and Huizhou were relatively lower in both daytime and nighttime. The dominant drivers of LST varied from city to city. The influence of land cover and socio-economic factors on daytime LST was higher in the highly urbanized cities than in the cities with low urbanization rates. For the cities of Zhaoqing, Huizhou, and Jiangmen, elevation was the dominant driver of daytime LST during the study period, and for the other cities in the PRD, the main driver changed from land cover in 2005 to NLI in 2015. This study is expected to provide useful guidance for planning of the thermal environment in urban agglomerations.

Keywords: land surface temperature; spatial analysis; urban agglomeration; driving factors; geo-detector metric

Citation: Liu, W.; Meng, Q.; Allam, M.; Zhang, L.; Hu, D.; Menenti, M. Driving Factors of Land Surface Temperature in Urban Agglomerations: A Case Study in the Pearl River Delta, China. *Remote Sens.* **2021**, *13*, 2858. https://doi.org/10.3390/rs13152858

Academic Editors: Yuji Murayama and Ruci Wang

Received: 16 June 2021
Accepted: 18 July 2021
Published: 21 July 2021

1. Introduction

The urban heat island (UHI) is a global phenomenon caused by urbanization [1]. UHI affects air quality [2], threatens the health of urban residents [3,4], influences building energy consumption, and leads to the risk of overheating in outdoor thermal environments [5]. Generally, UHI can be assessed by using the air temperature or land surface temperature

(LST). Urban agglomerations, which represent groups of cities that have a compact spatial organization and close economic connections, have become the most prominent feature of global urbanization in recent decades [6,7]. Especially in China, urban agglomerations have become the major form of national urbanization [8]. Therefore, it is important to understand the spatial distribution patterns of LST and the main driving factors affecting LST in order to formulate informed urban policies for urban agglomerations in the future [9].

Most previous studies focused on the spatial distribution characteristics and driving factors of LST [10–18]. The spatial distribution of high LSTs is significant for understanding the landscape patterns of LST [19,20]. According to previous studies, the potential factors of LST can be divided into three types: (1) land cover factors, which include land use/land cover (LUCC) [21], the normalized difference vegetation index (NDVI) [22–24], and the normalized difference built-up index (NDBI) [25,26]. This type also includes the impervious surface area (ISA) and the impervious surface distribution density (ISDD), which indicate the urbanization intensity of an area from the perspective of impervious surfaces [27,28]. (2) Socio-economic factors, which include the gross domestic product (GDP) [29], population density (POP) [30,31], and industrial production activities [32]. In addition, the nighttime light index (NLI) can indicate the human density [33], energy consumption [34], built-up area [35], and socioeconomic dynamics [36]. However, few studies have explored the relationship between NLI and LST. (3) Physiographical factors, which include elevation and slope, have also been studied [37,38].

The correlations between LST and various factors should be clarified in order to determine the main drivers of LST. Several methods, including the use of Pearson correlation coefficient [39], linear regression analysis [8,30], and stepwise multiple-linear regression [40], have been widely used in previous studies. LST is usually affected by multiple factors, and their interactions are quite complex. However, the above-mentioned methods cannot adequately measure the non-linear correlations between LST and the various drivers [9], or they cannot deal with categorical data [41]. In addition, few studies have taken the combined effect of different factors into consideration when determining the dominant factors of LST. Therefore, this study applied the geo-detector (GD) metric [42] to explore the correlations between LST and three types of drivers in order to avoid the previous limitations. The GD metric is applicable to the non-linear relations between the drivers and LST, and can also measure the combined impact of different drivers on LST. Due to its good applicability, the GD metric has been widely used in studies of drivers, including air pollution [43–45], public health [41,46,47], land use [48,49], regional economies [50,51], and urban environment [9,52].

This study was conducted in the Pearl River Delta (PRD) urban agglomeration, China. The PRD is one of the most important economic centers in China, and is also one of the most highly urbanized and populous regions in the world [53]. The cities included in the PRD were merged into an urban agglomeration in 2005 [54]. The PRD contributed about 9.12% of the national gross domestic product in 2015 [55], and is known as an "economic miracle" and the "world factory". Its rapid urbanization and industrialization led to significant land cover changes and brought a series of environmental problems, including urban flood risk, air pollution, and the UHI effect in the PRD [56–58]. The PRD has a subtropical climate and is located along the coast of the South China Sea. The dual effects of rapid urbanization and climate make the PRD highly susceptible to extreme heatwave events [20]. However, the current studies on the thermal environment in the PRD are still not adequate. Most previous studies focused solely on individual factors of LST, such as LUCC, ISA, or greenspace in the PRD [19,20,37,59–62]. In addition, more attention has been paid to the core cities, such as Guangzhou, Dongguan, Shenzhen, and Zhongshan cities [19,63–66], while the cities in the east and west of the PRD have received less attention.

To address the above problems, this study aimed to explore the spatial distribution characteristics of LST and to detect its main driving factors by using the GD metric in the PRD. First, we studied the spatial distribution of LST and used Moran's Index to indicate the presence of hotspot areas (HSAs). Then, the HSAs were determined by using hotspot analysis in a GIS environment. Second, the Hotspot Ratio Index (HRI) was developed to evaluate and rank the thermal gradient of each city in the urban agglomeration. Finally, the GD metric was used to measure the influence of each factor and the combined influence of the different factors on LST. Seven factors were selected based on the literature review and available data. The effects of the seven factors on the LST in the nine cities were compared to reveal the complex mechanism of LST in the PRD. Our study contributes to a better understanding of the spatial distribution characteristics of LST and its main driving factors, which is helpful in providing insights into the optimization of the urban thermal environment.

2. Study Area and Data Sources

2.1. Study Area

The PRD urban agglomeration is one of the regions with the highest urban densities in the world and is one of the most developed regions in China. It is located in the south-central coastal region of Guangdong Province, and it includes nine cities: Guangzhou, Foshan, Dongguan, Shenzhen, Zhongshan, Zhuhai, Huizhou, Zhaoqing, and Jiangmen (Figure 1). The PRD is surrounded by hills and mountains to the north, west, and east. The total area of the PRD is 55,000 km^2 [61]. The PRD experienced rapid population growth and economic growth from 2005 to 2015, with its population increasing by 13.27 million, and its GDP increased by about 3.8 trillion yuan (http://stats.gd.gov.cn, accessed on 15 September 2020). The PRD became the largest mega-region in the world according to the World Bank Group (2015). The average annual temperature ranges from 21 to 23 °C, and the average annual precipitation is over 1500 mm [60,67]. The PRD was selected for this study because its thermal environment became a serious problem due to the acceleration of its urbanization. Table 1 shows some key attributes of the nine cities in the PRD. The air temperature data were obtained from the National Meteorological Data Center (http://data.cma.cn, accessed on 10 April 2021). The permanent population and electricity consumption data were derived from the Statistical Yearbook of Guangdong Province.

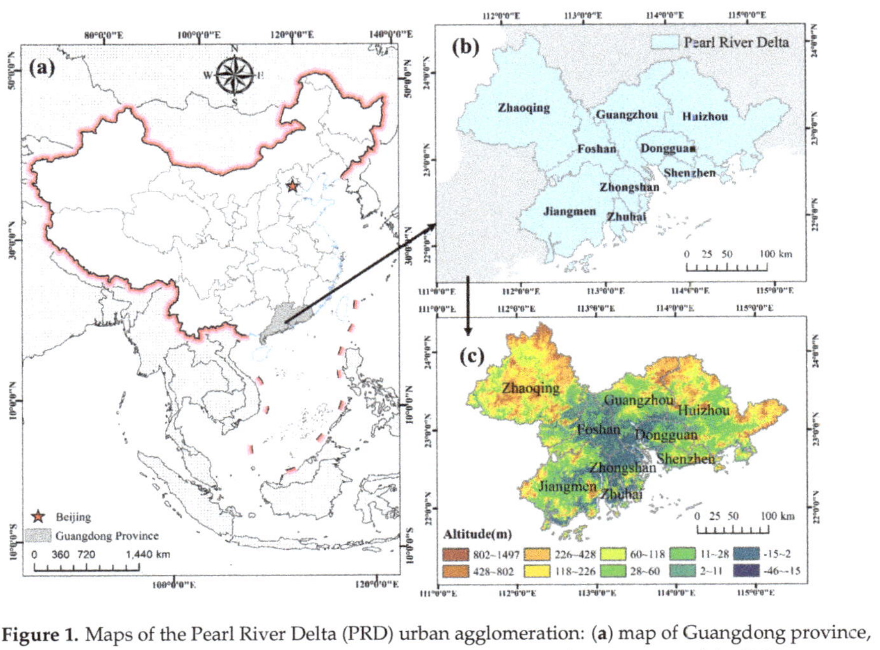

Figure 1. Maps of the Pearl River Delta (PRD) urban agglomeration: (**a**) map of Guangdong province, China; (**b**) PRD urban agglomerations, including nine cities; (**c**) elevation map of the PRD.

Table 1. Absolute values of selected constant attributes and rates of change for the selected changeable attributes from 2005 to 2015.

Variables	Unit	Guangzhou	Dongguan	Shenzhen	Foshan	Zhongshan	Zhuhai	Huizhou	Zhaoqing	Jiangmen
Constant variables										
Administrative area	km^2	7215.27	2449.57	1942.73	3795.64	1744.12	1579.48	11,319.16	14,898.67	9365.96
Mean Elevation	m	110.96	42.86	90.25	24.50	23.51	38.21	167.26	207.32	77.80
Changeable variables										
Average air temperature	°C	+0.18	+2.69	+3.04	+1.55	+2.32	+3.94	+2.12	+2.97	+3.07
Permanent population	10,000 persons	+42.16	+25.81	+37.47	+28.11	+31.83	+15.43	+28.29	+10.44	+10.15
Electricity consumption	100 million kw/h	+83.08	+58.84	+83.25	+85.85	+98.58	+136.07	+176.20	+213.50	+108.69
Cultivated land percentage	%	−6.18	−18.54	−11.50	−7.78	−7.49	−4.45	−3.68	−1.33	−2.31
Woodland percentage	%	−2.05	−9.40	−4.42	−3.45	−2.75	−1.43	−1.00	−0.73	−1.28
Grassland percentage	%	−3.05	−5.37	−3.31	−5.88	0.00	−20.00	+9.72	+25.63	+10.09
Water area percentage	%	−4.00	-8.53	−14.48	−7.87	−5.56	−12.70	+1.07	−1.65	−1.74
Construction land percentage	%	+19.43	+14.81	+9.95	+21.65	+16.55	+22.99	+25.68	+16.73	+18.89
Unused land percentage	%	0.00	0.00	0.00	0.00	0.00	0.00	0.00	0.00	−33.33
Gross domestic product (GDP)	1000 yuan/km^2	+284.58	+266.61	-27.47	+225.57	+329.39	+226.23	+123.42	+253.53	+196.55
Population density (POP)	Person/km^2	+85.52	+385.87	+389.66	+104.63	+121.56	+65.99	+66.93	+12.07	+19.09
Nighttime light index (NLI)	—	+66.05	+27.98	+32.53	+49.68	+39.19	+71.48	+96.06	+123.03	+117.74
Normalized difference vegetation index (NDVI)	—	+8.20	+9.09	+17.65	+2.00	−2.00	+1.82	+6.76	+3.85	+2.78
Impervious surface distribution density (ISDD)	—	+35.71	+28.57	+22.58	+28.00	+40.91	+66.67	+66.67	+100.00	+25.00

Note: "+" indicates the positive changes from 2005 to 2015, and "−" indicates the negative changes from 2005 to 2015; the rate of change is calculated with the formula: (T2015-T2005)/T2005*100.

2.2. Data Sources

In this study, the LST data for 2005 and 2015 were obtained from the 8-days (version 6) MODIS Aqua LST composite products (MYD11A2) (https://search.earthdata.nasa.gov, accessed on 6 June 2019). The MYD11A2 LST product was validated with in situ measurements to yield a bias of <0.5 K [68]. We used the annual mean LST observed in the daytime/nighttime at 13:30/01:30 local solar time [37]. The potential driving factors were selected according to previous research and available data, including the elevation, LUCC, NDVI, ISDD, GDP, POP, and NLI (Table 2). The elevation, LUCC, NDVI, POP, and GDP data were obtained from the Resource and Environment Science and Data Center (www.resdc.cn, accessed on 15 June 2019). The LUCC data included six types: cultivated land, woodland, grassland, water area, construction, and unused land. The NDVI product was computed from the continuous time series of the SPOT/VEGETATION NDVI remote sensing dataset. The GDP and POP were grid data with a 1 km resolution. In addition, the impervious surface data were obtained from published global urban datasets (https://doi.org/10.6084/m9.figshare.11513178.v1, accessed on 18 June 2020). The calculation of the ISDD is detailed in [27]. The NLI was obtained from nighttime light imageries, which were downloaded from the Resource and Environment Science and Data Center (www.resdc.cn, accessed on 15 June 2019) and the National Geophysical Data Center (http://ngdc.noaa.gov, accessed on 15 June 2019). All data were resampled to a 1 km resolution to keep the consistence of the spatial resolution of the data analyzed. The sources of the driver data are summarized in Table 2.

Table 2. Influencing drivers and data sources.

Drivers	Variables	Original Resolution/Resample Resolution	Time	Source
Physiographical factor	Elevation	1 km/1 km	/	www.resdc.cn
Land cover factors	Land use/land cover (LUCC)	1 km/1 km	2005, 2015	www.resdc.cn
	NDVI	1 km/1 km	2005, 2015	www.resdc.cn
	ISDD	30 m/1 km	2005, 2015	https://doi.org/10.6084/m9.figshare.11513178.v1
Socio-economic factors	GDP	1 km/1 km	2005, 2015	www.resdc.cn
	POP	1 km/1 km	2005, 2015	www.resdc.cn
	NLI	1 km/1 km	2005	www.resdc.cn
		750 m/1 km	2015	http://ngdc.noaa.gov

3. Methods

3.1. Spatial Variability of LST

The MYD11A2 LST products were used to get the annual average daytime LST and annual average nighttime LST for 2005 and 2015. The MODIS Reprojection tool was used to preprocess the MYD11A2 LST data, which included format conversion, re-projection, and clipping. The mask tool was used to remove cloud pixels; only pixels with a high quality (average LST error < = 1 K) were used to calculate the annual average LST for 2005 and 2015.

3.1.1. Global Moran's Index

Spatial distribution analysis is often used to explore geographical phenomena. The Global Moran's Index was used to analyze the spatial distribution of the LST. The value of the Global Moran's Index is between −1.0 and 1.0, where 1 indicates perfect positive spatial autocorrelation, −1 indicates perfect negative spatial autocorrelation, and 0 indicates a perfect spatial randomness [69]. The Global Moran's Index is calculated as follows:

$$I = \frac{n \sum_{i=i}^{n} \sum_{j=1}^{n} W_{ij}(x_i - \overline{x})(x_j - \overline{x})}{\sum_{i=1}^{n} \sum_{j=1}^{n} W_{ij} \sum_{i=1}^{n} (x_i - \overline{x})^2} \tag{1}$$

where n is the number of pixels; x_i and x_j are the LSTs at pixels i and j; $\overline{x}$ is the mean LST; w_{ij} is the spatial weight determined with the spatial correlation of the LSTs at pixels i and j.

3.1.2. Hotspot Analysis

Many previous methods for delineating high-temperature regions have applied numerical classification [70], Gaussian surface fitting [71–73], radial sampling [74,75], and hotspot analysis [76,77]. Hotspot analysis can be used to identify the spatial clusters of high LSTs (hotspots) and low LSTs (cold spots) by using only LST data, thus avoiding the subjective influence that may be caused by having too many parameter settings. Therefore, hotspot analysis was applied to identify the areas with high (hotspot) or low (cold spot) LSTs in this study. Based on the clustering patterns of the spatial distribution of the LST, the Getis–Ord Gi (*Gi**) index was applied to measure the degree of clustering for each pixel [77]. For each pixel, the hotspot analysis returns a z-score. A higher positive z- score indicates a higher degree of clustering of high LSTs (hot spots) [78]. In this study, the areas with z-scores $\geq$ 2.58 (corresponding to the 99% confidence level) were defined as HSAs [76]. The *Gi** is defined as:

$$G_i^* = \frac{\sum_{j=1}^{n} \omega_{ij} x_j - \overline{X} \sum_{j=1}^{n} \omega_{ij}}{S\sqrt{\frac{\left[n \sum_{j=1}^{n} \omega_{ij}{}^2 - \left(\sum_{j=1}^{n} \omega_{ij}\right)^2\right]}{n-1}}} \tag{2}$$

$$\overline{X} = \frac{\sum_{j=1}^{n} x_j}{n} \tag{3}$$

$$S = \sqrt{\frac{\sum_{j=1}^{n} x_j{}^2}{n} - \left(\overline{X}\right)^2} \tag{4}$$

where, x_j is the LST value of pixel j, ω_{ij} represents the spatial weight between pixels i and j, and n is the total number of pixels; the spatial weight is determined according to the Queen's adjacency connectivity matrix [76].

3.2. Hotspot Ratio Index

The urban-heat-island ratio index was applied in previous studies to evaluate and rank thermal environments [79,80]. It is effective in measuring the characteristics of a thermal environmental, but its calculation requires urban or rural boundary data [80,81]. In this study, the hotspot ratio index (HRI) was used to evaluate and rank the thermal environments of the nine cities. The HRI gives scores and ranks the areas according to the value of the LST. The HRI is calculated as the weighted sum of the percentage of HSAs with different classes in the selected region. The HRI reflects the thermal gradient in the selected region. A higher HRI indicates that a city has a higher thermal gradient, which implies that the city is more likely to experience greater thermal stress. The HRI is calculated as follows:

$$HRI = \sum_{i=1}^{n} i * P_i \tag{5}$$

where n is the total number of classes. To avoid skewed data, the quantile classification method was applied to classify the HSAs into 5 classes according to the LSTs. P_i is the ratio of the area of class i to the total study area. i is the class index, i.e., i = 1, 2, 3, 4, and 5. A higher class index refers to a higher LST. To calculate the HRI, the areas of 1st, 2nd, 3rd,

4th, and 5th classes of HSAs were divided by the total study area, then multiplied by the class index.

3.3. Geo-Detector Metric

The geo-detector (GD) metric is an effective metric for revealing the drivers of geographical phenomena [42]. The main idea assumes that if an independent variable affects the dependent variable, then their spatial distributions should be similar. The GD metric includes four detectors: factor, interaction, risk, and ecological. We used the factor detector to measure the influence of each driver on the LST. We also used the interaction detector to measure the combined influence of the interactions between drivers on the LST. The GD metric is freely available from www.geodetector.cn, accessed on 18 July 2021.

3.3.1. Factor Detector

The factor detector signifies the effects of factors on the LST. The greater the probability distribution (q) value is, the greater the influence of the factors on the LST will be. The value of q is between 0 and 1; in extreme cases, the q value equals 1, indicating that factor X completely affects the spatial distribution of Y, and a q value of 0 indicates that factor X has nothing to do with Y. The value of q is calculated as follows [42]:

$$q = 1 - \frac{1}{N\sigma^2} \sum_{h=1}^{L} N_h \sigma_h^2 \tag{6}$$

where the LST (Y) and the drivers (X) are composed of L classes ($h = 1, 2, \ldots L$); N and N_h represent the numbers of cells in the entire area and the h class, and σ^2 and σ_h^2 represent the variance in the LST.

3.3.2. Interaction Detector

The interaction detector was used to measure the influence of the interactions between different factors on the LST. It identifies whether the interactions between drivers (X1 and X2) have an effective influence on the LST or not. First, we separately calculate the influence of each driver (q(X1) and q(X2)) on the LST; then, the influence of the interaction between X1 and X2 (q(X1∩X2)) on the LST is calculated based on an overlay of the two drivers (X1 and X2). According to the values of q(X1), q(X2), and q(X1∩X2), the effect of the interaction between the two drivers on the LST can be determined (Table 3) [9].

Table 3. The effects of interactions between two drivers.

Description	Interaction Effect
q(X1∩X2) > Min (q(X1), q(X2))	Enhance
q(X1∩X2) > Max (q(X1), q(X2))	Bi-enhance
q(X1∩X2) > q(X1) + q(X2)	Enhance, nonlinear
q(X1∩X2) = q(X1) + q(X2)	Independent
q(X1∩X2) < q(X1) + q(X2)	Weaken
q(X1∩X2) < Max (q(X1), q(X2))	Weaken, un-enhance
q(X1∩X2) < Min (q(X1), q(X2))	Weaken, nonlinear

3.3.3. Data Preparation

The GD metric captures the spatial heterogeneity of the attributes, and it requires the discretization of input data. Many methods have been applied for discretization [82]. The natural breaks method depends on the principle of the "maximum and minimum distance between classes" that can allow the original characteristics of the data to be kept [83]. This method has provided good results [43,52]. Accordingly, we applied natural breaks classification in order to classify the drivers. Sample points were generated in ArcGIS according to the 1 km grids over the entire study area. Therefore, the numbers of sample points in each city depended on the area of the city. The values of each driver and mean LST were extracted for the sample points. The natural breaks method was applied to classify all drivers (except LUCC, as it was already classified into cultivated land, woodland, grassland, water areas, construction, and unused land). For the GD metric, LST was the dependent variable, while the classified drivers were independent variables (Table 2).

4. Results

4.1. Spatial Distribution of LST

As seen in Figure 2, in 2005, higher daytime LSTs were mostly concentrated in the central area of the PRD, including the cities of Dongguan, Shenzhen, Guangzhou, Foshan, and Huizhou. By contrast, in 2015, higher daytime LSTs were concentrated in Dongguan, Shenzhen, Foshan, Zhongshan, Jiangmen, and Guangzhou. The daytime LSTs decreased in the cities located in the northern part of the PRD, including the cities of Zhaoqing, Guangzhou, and Huizhou. For the nighttime LST, the higher LSTs were also concentrated in the central area of the PRD in 2005. In 2015, the highest nighttime LSTs increased in the cities of Dongguan, Shenzhen, Foshan, Zhongshan, Jiangmen, and Guangzhou.

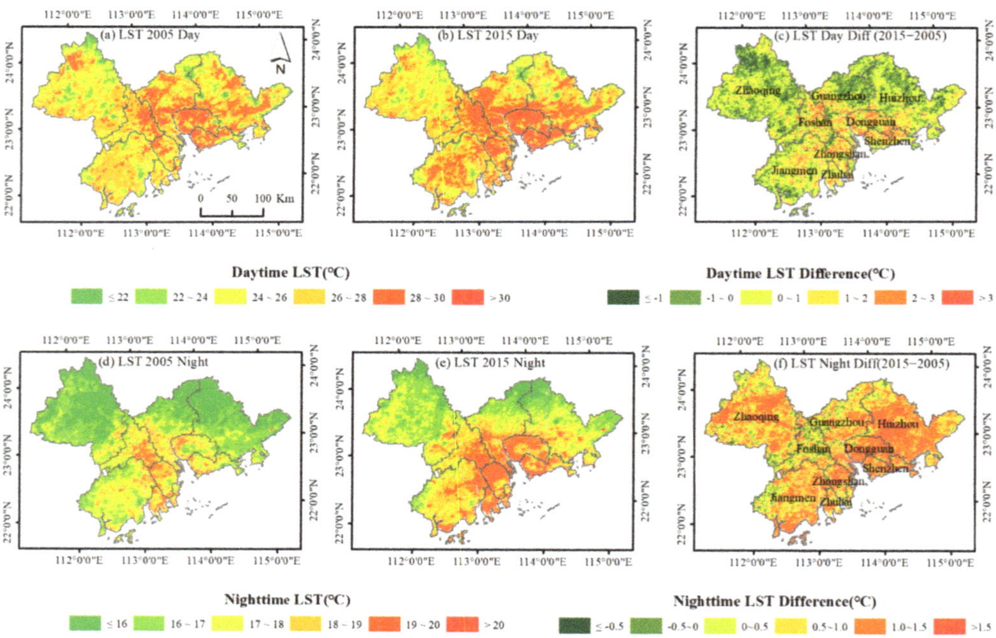

Figure 2. Daytime and nighttime land surface temperature (LST) maps of the PRD:(**a**) daytime LST in 2005; (**b**) daytime LST in 2015; (**c**) daytime LST difference between 2005 and 2015; (**d**) nighttime LST in 2005; (**e**) nighttime LST in 2015; (**f**) nighttime LST difference between 2005 and 2015.

4.1.1. General Spatial Pattern of LST

The spatial pattern of the LST was estimated by using the Global Moran's Index. The Global Moran's Indexes for the daytime LST were 0.903 and 0.915 at the 99% significance level in 2005 and 2015, respectively. For nighttime LST, the Global Moran's Indexes were 0.940 and 0.941 at the 99% significance level in 2005 and 2015, respectively. The results for the Moran's Index indicate that the LST distribution was a clustered pattern—areas had aggregations of high and low temperatures—and there was less than a 1% likelihood that the results were random, which indicates the presence of HSAs in the study area.

4.1.2. Spatial Distribution of Hotspot Areas (HSAs)

Hotspot analysis was applied to identify the HSAs in the PRD. For each LST pixel, the hotspot analysis returned a z-score. We defined areas with z-scores ≥ 2.58 as HSAs according to [78] (Table 4). For the daytime LST, the HSAs pattern was consistent with the pattern of the high-temperature regions (Figures 2 and 3).

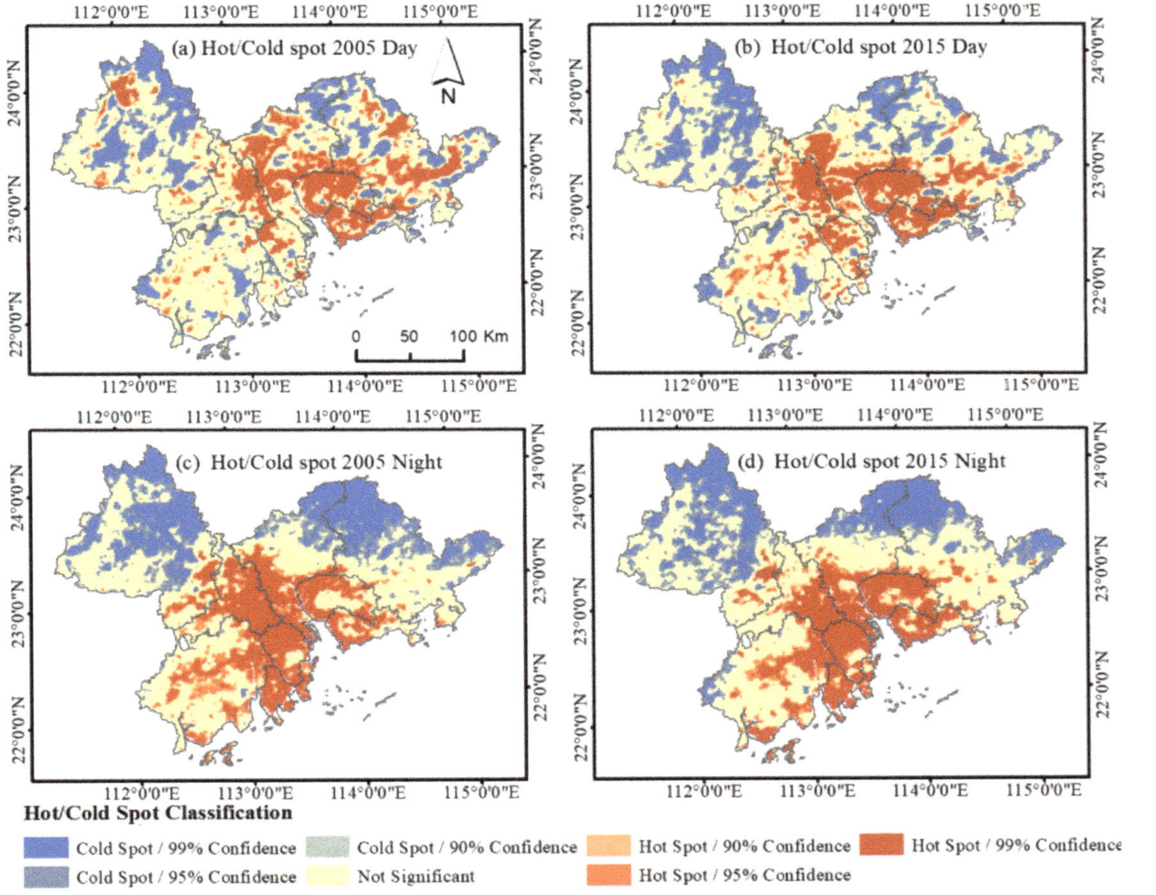

Figure 3. Hot-/cold-spot distributions in the PRD: (**a**) hot-/cold-spots for daytime LST in 2005; (**b**) hot-/cold-spots for daytime LST in 2015; (**c**) hot-/cold-spots for nighttime LST in 2005; (**d**) hot-/cold-spots for nighttime LST in 2015.

Three patterns of HSAs in the PRD were observed—stable, reduced, and expanded—by comparing the HSAs maps for 2015 with those for 2005 (Figure 4). For the daytime LST, the stable HSAs distribution indicates that the HSAs tended to be clustered in the center of the PRD. The HSAs expanded towards the central and southwestern parts of the PRD. A considerable growth of HSAs was observed in Jiangmen compared with Huizhou and Zhaoqing. In contrast, the HSAs were significantly reduced in the northwest and northeast of the PRD.

Table 4. Average LSTs and the percentage of hotspot areas (HSAs) in each city during the study periods.

Variables		Guangzhou	Dongguan	Shenzhen	Foshan	Zhongshan	Zhuhai	Huizhou	Zhaoqing	Jiangmen
Daytime LST (°C)	2005	27.04	28.66	28.00	27.27	27.29	26.44	26.56	25.76	26.31
	2015	27.47	30.28	29.38	28.30	28.64	27.56	26.92	25.93	27.04
	variation	0.43	1.62	1.38	1.03	1.35	1.12	0.36	0.17	0.73
Ratio of daytime HSAs to city area	2005	0.26	0.63	0.51	0.25	0.26	0.06	0.19	0.04	0.03
	2015	0.22	0.74	0.58	0.34	0.39	0.14	0.11	0.01	0.06
	variation	−0.04	0.10	0.07	0.09	0.13	0.07	−0.08	−0.03	0.03
Nighttime LST (°C)	2005	16.76	17.88	17.76	18.12	18.48	18.58	16.06	16.08	17.44
	2015	17.61	19.31	19.02	18.68	19.64	19.49	17.27	17.00	18.38
	variation	0.85	1.43	1.26	0.56	1.16	0.91	1.21	0.92	0.94
Ratio of nighttime HSAs to city area	2005	0.23	0.50	0.39	0.59	0.78	0.83	0.02	0.04	0.20
	2015	0.19	0.71	0.53	0.36	0.79	0.76	0.04	0.02	0.22
	variation	−0.04	0.21	0.15	−0.23	0.01	−0.07	0.02	−0.01	0.03

Note: HSAs were delineated at the 99% significance level.

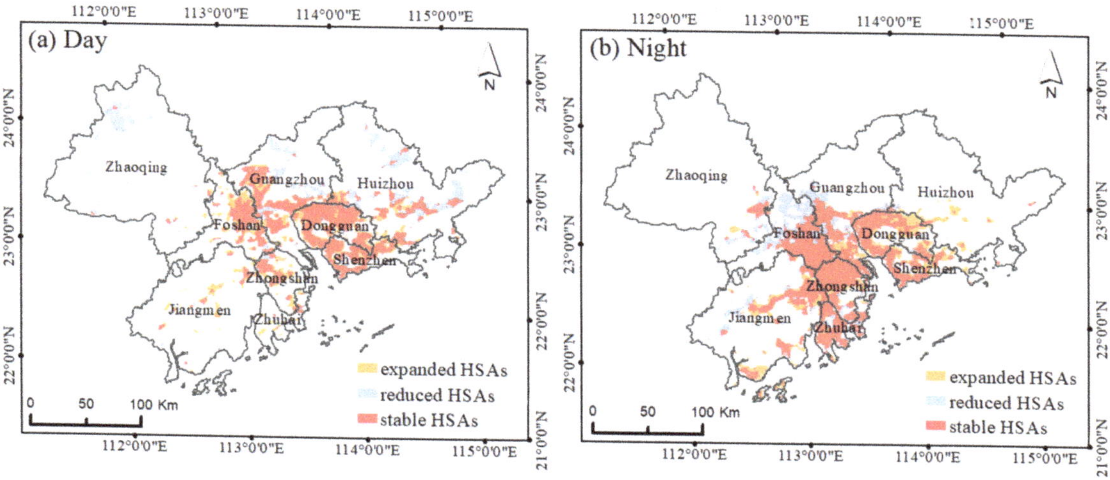

Figure 4. Maps of changes in the spatial distributions of hotspot areas (HSAs) at the 99% significance level in the PRD: (**a**) the changes in daytime HSAs in the PRD; (**b**) the changes in nighttime HSAs in the PRD (the "blank" areas did not include HSAs).

For the nighttime LST, the stable HSAs were bigger and more continuously distributed than those in daytime. Most of the stable HSAs were located near the coast and the Pearl River estuary, including in Zhuhai, Zhongshan, Foshan, Guangzhou, Shenzhen, and Dongguan.

4.2. Hotspot Ratio Index in the Nine Cities

The HSAs were classified according to [79] (Table 5 and Figure 5). The values of the HRI varied across cities in the daytime (Figure 6). Low HRI values were observed in Zhaoqing, Jiangmen, Zhuhai, and Huizhou (Figure 6). The HRI values of Dongguan, Shenzhen, Foshan, and Zhongshan were higher than those of the other cities.

Table 5. Criteria for HSAs classification.

HSAs Level	LST Range (°C)	
	Daytime	Nighttime
Level 1	<=28.78	<=18.36
Level 2	28.78~29.42	18.36~18.90
Level 3	29.42–30.19	18.90~19.30
Level 4	30.19~31.22	19.30~19.76
Level 5	>31.22	>19.76

The HRI also varied across cities in the nighttime. The highest HRI values were recorded in Zhongshan and Zhuhai. The HRI values of Zhaoqing and Huizhou were minimal. The higher HRI values were recorded in Dongguan, Shenzhen, Zhongshan, Zhuhai, and Foshan cities. These results indicate that the spatial distribution of HSAs was highly heterogeneous among the cities in both daytime and nighttime.

4.3. Dominant Drivers of LST in the PRD

The factor detector quantifies the influence of each driver on the LST by calculating the q values (Table 6). The q value shows that driver X explains (100*q) % of the LST. The q values at the 95% significance level are shown in Table 6. The results show the differences in the influence of each driver on the LST in 2005 and 2015.

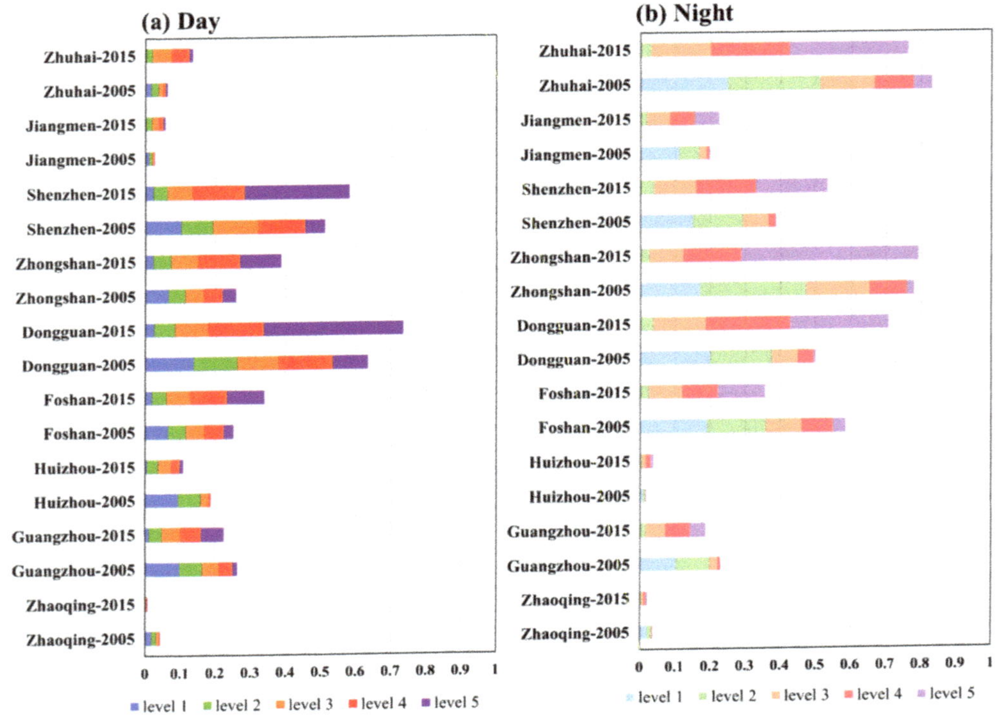

Figure 5. HSAs classification in each city in 2005 and 2015: (**a**) the percentage of different levels of HSAs in the total area of each city in the daytime; (**b**) the percentage of different levels of HSAs in the total area of each city in the nighttime.

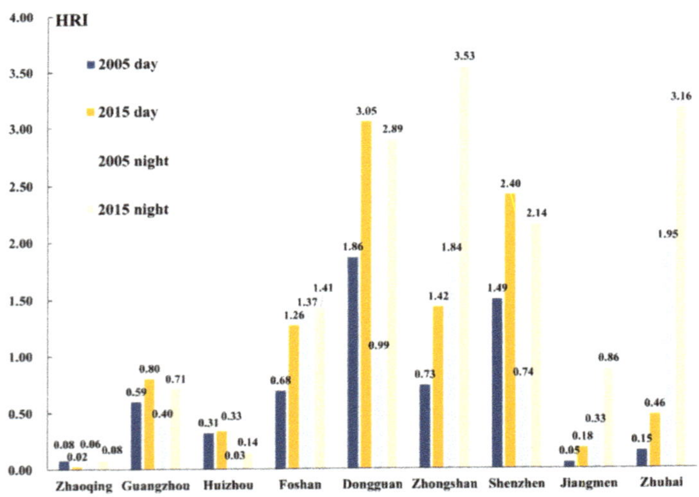

Figure 6. Hotspot Ratio Index (HRI) in the nine cities.

Table 6. The q values of the LST drivers in 2005 and 2015.

Year	Region	Elevation	LUCC	NDVI	ISDD	POP	GDP	NLI
2005 day	PRD	0.50	0.35	0.40	0.30	0.26	0.25	0.36
2015 day	PRD	0.50	0.43	0.50	0.44	0.37	0.31	0.58
2005 night	PRD	0.41	0.30	0.44	0.32	0.45	0.35	0.45
2015 night	PRD	0.43	0.31	0.45	0.34	0.33	0.29	0.50

For the daytime LST, in 2005, elevation had the greatest influence on LST (50%), followed by the NDVI (40%). In 2015, the NLI was the dominant driver of LST, which indicated the increase in the influence of socio-economic drivers on daytime LST due to the continuous development of the urban agglomeration [84]. For the nighttime LST, in 2005, the NLI (45%) and POP (45%) had the greatest influence, followed by the NDVI (44%) (Table 6). In 2015, the NLI (50%) also had the greatest influence, followed by the NDVI (45%) (Table 6). The results indicate that the NLI and NDVI were important factors for the nighttime LST.

The explanation rate of the NLI for the daytime LST increased from 36% in 2005 to 58% in 2015, and the explanation of the NLI for the nighttime LST increased from 45% in 2005 to 50% in 2015, which indicated the important influence of the NLI factor on LST in the PRD.

4.4. Dominant Drivers of LST in the Nine Cities

4.4.1. Factor Detector Analysis

The results for the factor detector for each city are presented in Figures 7 and 8, where a color-coding scheme was applied to facilitate the interpretation. Lower q values are shown in a lighter red color, while darker red colors indicate higher q values. There are two ways to explain these results. When read from left to right, the differences in the influences of all drivers in one city are clear. When reading from top to bottom, the differences across cities can be explained for each driver. The results show that the influence of the selected drivers on LST varied from city to city (Figures 7 and 8).

For Huizhou, Zhaoqing, and Jiangmen, the elevation had the greatest influence on daytime LST, as there are many mountains and forests in these three cities. Similarly, in Guangzhou, the elevation also exhibited the highest influence rate (61%) in 2005. Socio-economic factors also showed great influences on nighttime LST in the above four cities (Figure 8).

For Dongguan and Shenzhen, the NDVI had the greatest influence on daytime LST in 2005, while in 2015 the NLI was the dominant driver. The NLI was the most influential factor for nighttime LST in Dongguan and Shenzhen in 2015, which indicated the importance of socio-economic development in these areas and its effect on LST. The NDVI also had the high explanation rate for nighttime LST in these two cities during the study period.

For Foshan and Zhongshan, the NLI was the dominant driver of daytime LST during the study period (Figure 7). The NDVI showed the greatest influence on nighttime LST in these two cities (Figure 8). The influences of the drivers on the nighttime LST in Guangzhou and Zhuhai are shown in Figure 8. None of the seven drivers had a significant influence on nighttime LST in Zhuhai, but all seven drivers had a strong influence on nighttime LST in Guangzhou.

The socio-economic drivers (GDP, POP, and NLI) were highly correlated with daytime LST in the central cities in the PRD, including Dongguan, Shenzhen, Guangzhou, Foshan, and Zhongshan (Figure 7). The socio-economic drivers had a low level of correlation with the daytime LST in Huizhou, Zhaoqing, and Jiangmen, where elevation was the dominant driver of daytime LST (Figure 7). Similarly, the land cover drivers, i.e., the LUCC, ISDD, and NDVI, were highly correlated with daytime LST in Dongguan, Shenzhen, Guangzhou, Foshan, and Zhongshan, while the correlation was weaker in Huizhou, Zhaoqing, and Jiangmen. Due to topographical issues, urbanization is concentrated in the central region of the PRD [37]. For the nighttime LST, the NDVI showed a strong influence in most of the cities during the study period (Figure 8).

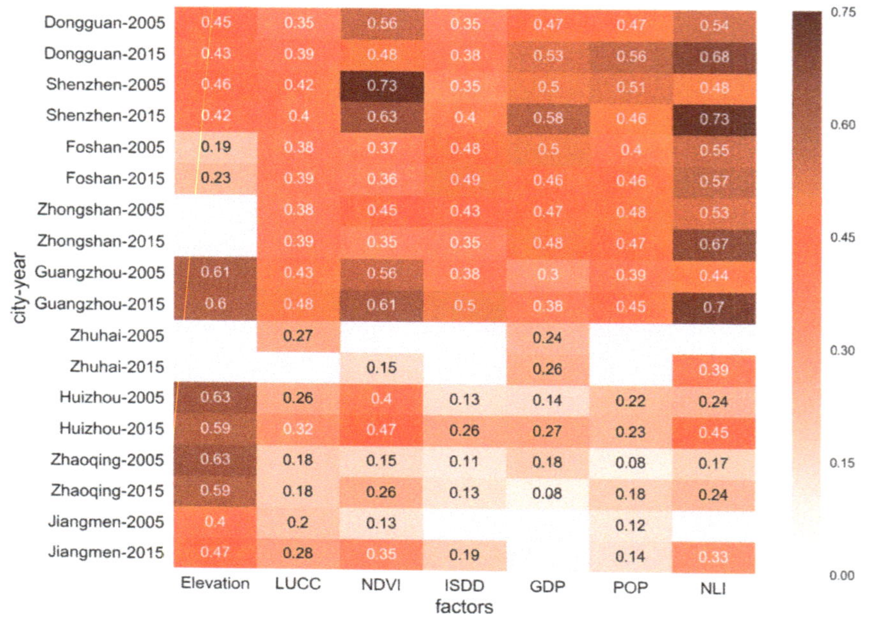

Figure 7. The q values of drivers for each city in the daytime (the blank values indicate that the results of corresponding factors are not significant at a p-value of 0.05).

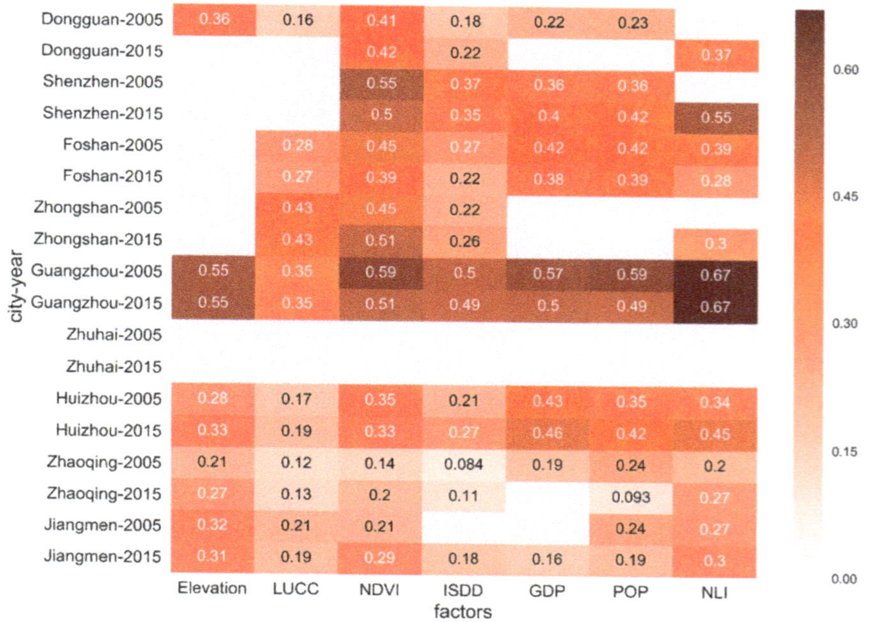

Figure 8. The q values of drivers for each city in the nighttime (the blank values indicate that the results of corresponding factors are not significant at a p-value of 0.05).

4.4.2. Interaction Detector Analysis

In this study, the interaction detector was calculated for certain drivers: elevation, LUCC, NDVI, ISDD, GDP, POP, and NLI. Strong interactions were observed among all seven drivers (Figures 9–12). The interactions between elevation and the other drivers were significantly strong in Huizhou, Jiangmen, Zhaoqing, and Guangzhou (Figures 9 and 10). The greatest effects of the drivers' interactions on daytime LST were Elevation ∩ NDVI (68%), Elevation ∩ LUCC (46%), Elevation ∩ GDP (68%), and Elevation ∩ NDVI (71%) in Huizhou, Jiangmen, Zhaoqing, and Guangzhou, respectively, in 2005. In 2015, the greatest effects of drivers' interactions on daytime LST were Elevation ∩ NLI in Huizhou (69%), Jiangmen (57%), and Guangzhou (78%). The explanation rate of Elevation ∩ LUCC, Elevation ∩ ISDD, and Elevation ∩ POP for daytime LST in Zhaoqing was 63%, which illustrates that elevation had the greatest influence on daytime LST in this city.

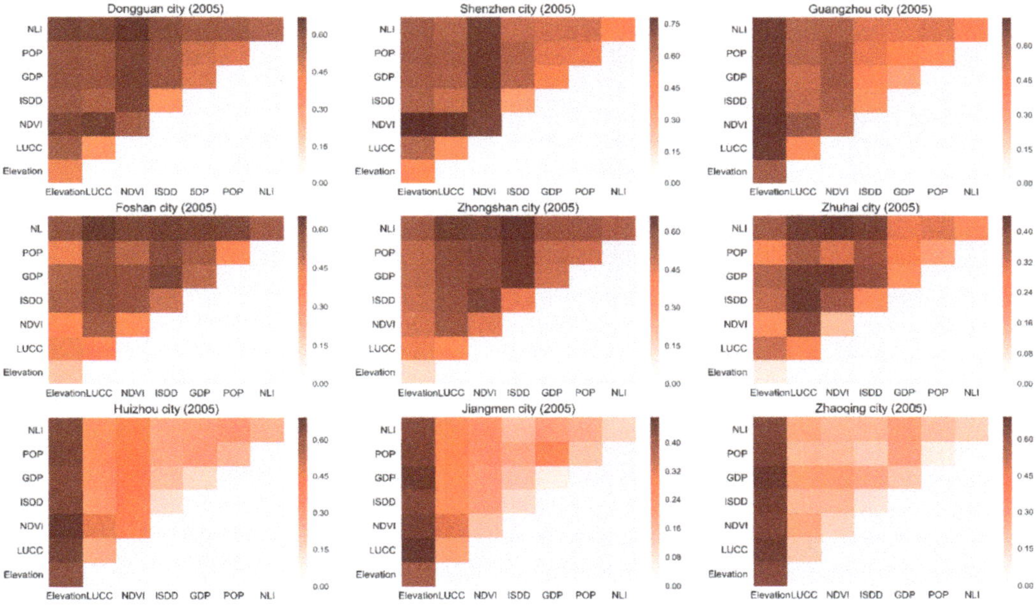

Figure 9. The interactions among drivers with respect to daytime LST in each city (2005).

The interactions between elevation and the socio-economic factors showed a high explanation rate for nighttime LST in Zhaoqing, Jiangmen, Huizhou, and Guangzhou (Figures 11 and 12). The explanation rates of the interactions between elevation and the three socio-economic factors exceeded 30% (in Zhaoqing), 40% (in Jiangmen), 42% (in Huizhou), and 72% (in Guangzhou) in 2005. In 2015, the highest combined explanations of elevation and the socio-economic factors reached 39% (in Zhaoqing), 43% (in Jiangmen), 57% (in Huizhou), and 72% (in Guangzhou).

In Dongguan and Shenzhen, in 2005, the interactions between the NDVI and the three socio-economic factors exceeded 63% and 74%, respectively (Figure 9). In 2015, the greatest influence of the interactions on daytime LST in Dongguan and Shenzhen was from NLI ∩ LUCC (72%) and NLI ∩ NDVI (79%), respectively (Figure 10). The results of the interaction analysis for daytime LST in Dongguan and Shenzhen were consistent with the results of the factor detector analysis for daytime LST in these two cities. For the nighttime LST in Dongguan and Shenzhen, the predominant interactions between drivers were NDVI ∩ Elevation, NDVI ∩ NLI, NDVI ∩ POP, and NDVI ∩ GDP (Figures 11 and 12). This result indicates the important influence of the NDVI and socio-economic factors on the nighttime LST in these two cities.

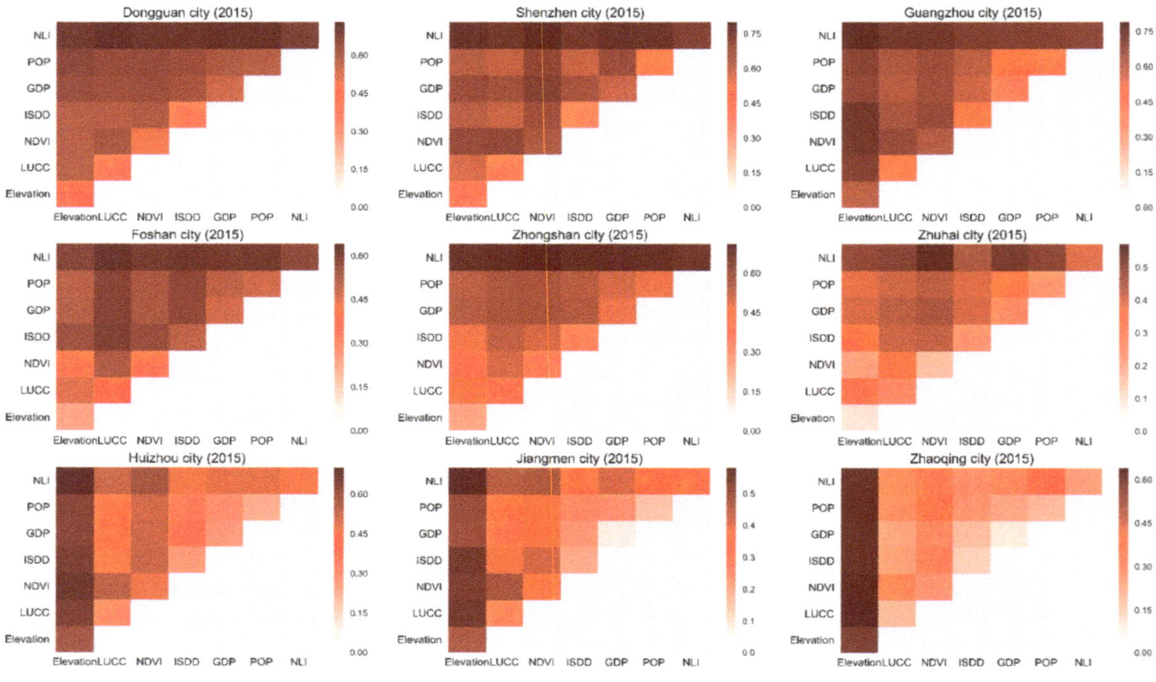

Figure 10. The interactions among drivers with respect to daytime LST in each city (2015).

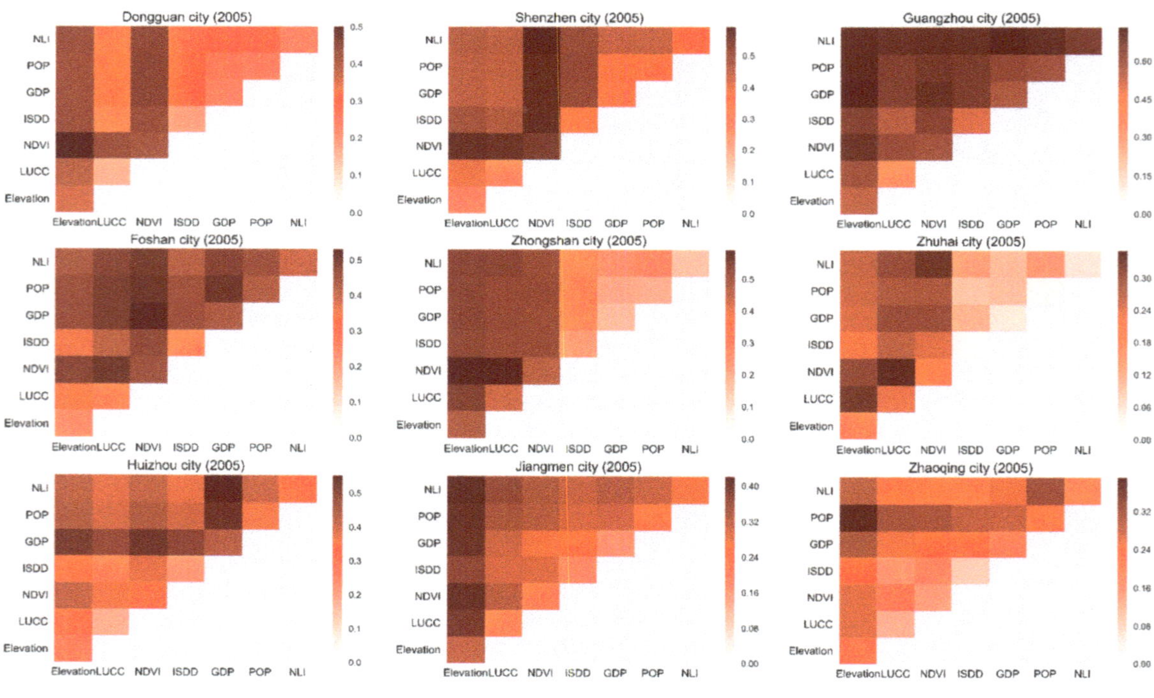

Figure 11. The interactions among drivers with respect to nighttime LST in each city (2005).

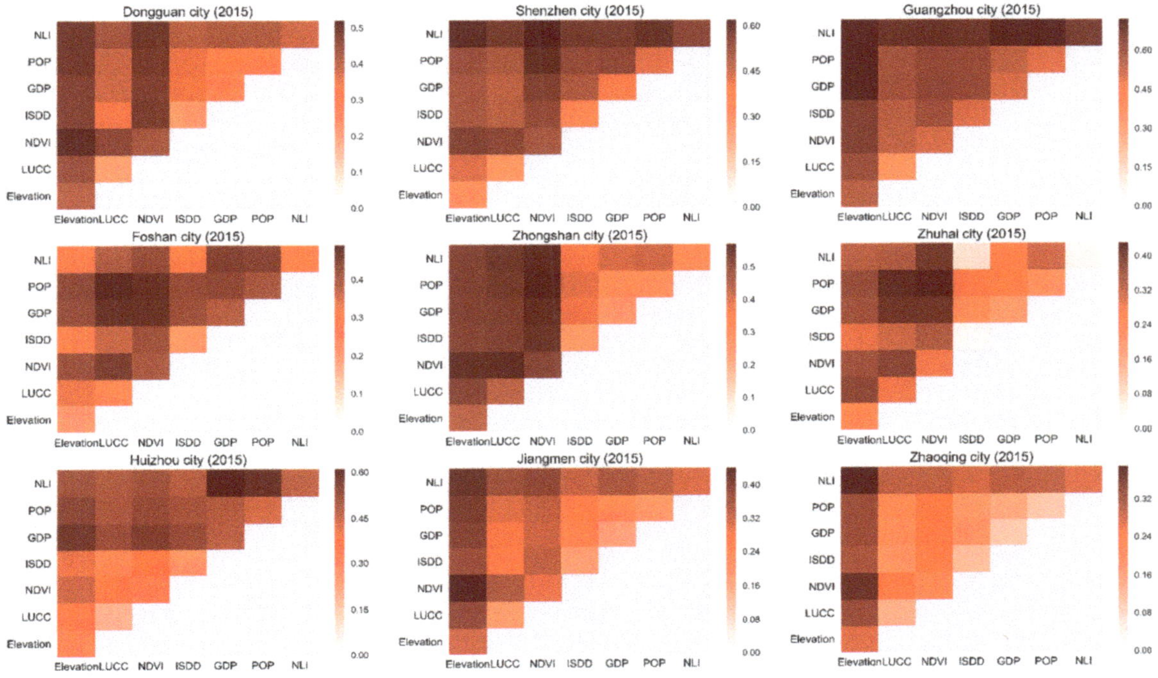

Figure 12. The interactions among drivers with respect to nighttime LST in each city (2015).

Similarly, the interactions between land cover and the socio-economic drivers were most predominant in Foshan, Zhongshan, and Zhuhai. NLI ∩ LUCC had the greatest influence on daytime LST in Foshan with 64% in 2005 and 63% in 2015. NLI ∩ NDVI had the greatest influence on daytime LST in Zhuhai with 43% in 2005 and 56% in 2015. The strongest interaction in Zhongshan was between the drivers NLI ∩ ISDD (65%) in 2005, and the interactions between NLI and the three land cover drivers explained 70% in 2015 (Figures 9 and 10). For the nighttime LST in Foshan, Zhongshan, and Zhuhai, the interactions between the NDVI and socio-economic factors were predominant during the study period (Figures 11 and 12).

5. Discussion

5.1. Spatial Patterns and Drivers of LST

This paper aimed to study the spatial distribution characteristics and main drivers of LST in the PRD urban agglomerations. The HSAs were mainly located on both sides of the Pearl River estuary. Our findings agree with those of [37]. The HSAs were mainly distributed in Shenzhen, Dongguan, Guangzhou, Foshan, Zhongshan, and Zhuhai, while there were fewer HSAs in Huizhou, Jiangmen, and Zhaoqing (Figure 4). The urbanization rates of the nine cities are as follows: Shenzhen > Guangzhou > Zhuhai > Dongguan > Foshan > Zhongshan > Huizhou > Jiangmen > Zhaoqing [85]. The HSAs results indicate that high temperatures are more likely to be observed in areas with high urbanization rates. There was a remarkable decrease in the daytime HSAs in the northern PRD (Figure 4). There are many mountains and forests in the northern PRD, where the environment has been improved significantly since 2006. A series of ecological projects have continuously promoted the increase in vegetation cover in these areas (http://gz.gov.cn, accessed on 15 September 2020), which decreased the LST in the northern PRD. This finding is consistent with the findings of [20]. There was a difference between the spatial distributions of daytime and nighttime HSAs. This could be explained by the impacts of impervious surfaces and human activities on the LST in areas with higher urbanization rates during the day, while the impact of water became significant at night due to its high thermal capacity and properties [10]. In other words, the differences between daytime and nighttime HSAs may be related to the different driving forces of daytime and nighttime LST [86].

The HRI was developed to quantitatively evaluate and rank the thermal environment of each city. The results for the HRI indicated that Dongguan and Shenzhen, which were characterized by high urbanization rates, were the most thermally cities in day and night (Figure 6). This can be explained by the fact that Dongguan is an industrial city, while Shenzhen has a big port with intensive traffic, as it is a coastal city, which increased the LSTs in both cities [66]. The daytime HRI index also increased more in Foshan and Zhongshan than in other cities (Figure 6), which can be due to the impervious surfaces, which obviously increased in these cities from 2005 to 2015 [37]. However, Zhaoqing, Jiangmen, and Huizhou, which were characterized by the lowest urbanization rates, showed the lowest HRI values. These findings can provide guidelines for the improvement of the thermal environment in urban agglomerations.

The drivers of LST were analyzed by applying the GD metric. All of the drivers showed significant influences on daytime and nighttime LST in the PRD. However, the main driving factors varied from city to city. Vegetation has a cooling effect through evapotranspiration and by shading other urban facets. Vegetation can absorb a large fraction of solar radiation and use a large fraction of it for evaporation and transpiration, thereby reducing sensible heat transfer to the boundary layer, i.e., reducing the temperatures of both the vegetation and urban air [19,38]. The shade from vegetation protects other urban surfaces from solar radiation and prevents increases in air and surface temperatures [87]. Our results showed that the NDVI had a strong correlation with daytime LST in Shenzhen and Guangzhou. Our results agree with those of [19,30]. The NDVI also showed a strong influence on nighttime LST in Dongguan, Shenzhen, Foshan, Zhongshan, and Guangzhou. This result is in agreement with the findings of [88].

Variations in LST due to elevation differences should be considered in LST studies in large areas where the terrain is not flat [89]. A negative correlation was observed between LST and elevation [90]. In our study, the elevation factor had a stronger influence on daytime LST in Huizhou, Zhaoqing, Jiangmen, Guangzhou, and Shenzhen. The elevation values of these five cities are higher than those of the other cities in the PRD (Figure 1 and Table 1). The high density of vegetation in the mountainous areas led to a great impact on temperature [91]. In addition, the low urbanization rates in Zhaoqing, Jiangmen, and Huizhou indicated the weak influences of the ISDD, GDP, POP, and NLI factors on daytime LST, which was in line with our results (Figure 7). LST is usually affected by multiple factors, and the higher correlation between the elevation factor and LST in these three cities may be the result of the weak influences of the other factors that were related to the urbanization rate on daytime LST. Similarly, the differences in correlation intensity between elevation and LST in 2005 and 2015 can also be partially explained by the variations in the influences of other factors on LST. The influence of elevation on nighttime LST in the PRD was lower than that on daytime LST in the PRD (Table 6, Figures 7 and 8); this could be due to the location of the PRD. The influence of the ocean at night makes the driving mechanism of nighttime LST more complex. This finding is in agreement with those of [37].

In this study, POP had a significant influence on daytime LST in Dongguan, Shenzhen, Foshan, Zhongshan, and Guangzhou, which were characterized by high urbanization rates (Figure 7). However, a case study in the Yangtze River Delta urban agglomeration found that there was no significant correlation between the surface UHI intensity and population density [92]. The statistical population census data aggregated by the district were used in the study of the Yangtze River Delta, while gridded population data with a 1 km spatial resolution were used in our study, and these were consistent with the spatial resolution of the LST data used in our study. Population indirectly reflects the development of an area and the complexity of the urban surface. An increasing population leads to rapid transformation of natural land cover into impervious surfaces, such as buildings, streets, and other human-made features, which can reduce thermal admittance and increase the surface temperature through a modification of the energy balance [31]. In addition, a high population density can directly (by metabolic heating) or indirectly (by anthropogenic heat emissions) affect urban surface temperature.

In 2005, the socio-economic drivers explained most of the observed variations in daytime LST in Dongguan, Shenzhen, Guangzhou, Foshan, and Zhongshan, which were characterized by high urbanization rate. The daytime LST is affected by landscape changes and anthropogenic activities [9]. Socio-economic activities are usually associated with energy consumption, which increases anthropogenic heat and local surface temperature [93]. The change rates of the population and electricity consumption were positive in these cities (Table 1), which indicated the possible increase in the intensity of the population and socio-economic activities. The NLI represents the intensity of nighttime lighting, and it is an indicator of human density [86]. Anthropogenic heat emissions caused by human activities contribute to daytime LST increases. A low intensity of nighttime lighting is the result of agricultural activities, especially in underdeveloped areas with low urbanization rates. In addition, the brightest areas in nighttime light images could represent factories or industries, which also consume energy and generate heat emissions in the daytime, causing daytime surface temperature changes [94]. Consequently, it can be inferred that the greater influence of NLI on the daytime LST in Dongguan and Shenzhen cities was related to the high urbanization rates and industrialization characteristics of these two cities. Our results agree with those of [95]. The results for the drivers in these nine cities further revealed the mechanisms that drive the LST.

5.2. Management of Urban Agglomerations

The results of the HSAs and HRI showed that the spatial distribution of LST in the PRD urban agglomeration was highly heterogeneous among the nine cities in both daytime and nighttime. Therefore, it is recommended that more attention and resources be committed to improving the thermal environments of cities that are characterized by higher thermal stress in daytime, such as Dongguan, Shenzhen, Zhongshan, and Foshan. Human-made surfaces absorb solar radiation and lead to heat storage in urban areas [39]. As the urbanization is greater in Dongguan and Shenzhen compared to Huizhou, Jiangmen, and Zhaoqing, the thermal stress in daytime is higher in Dongguan and Shenzhen than in Huizhou, Jiangmen, and Zhaoqing. To reduce the thermal stress, the spatial expansion of urban areas should be controlled in these cities. On the other hand, increasing green vegetation areas is an effective way to mitigate the increasing LSTs in regions with high thermal stress. The significant reduction of HSAs in the northern part of the PRD is due to the implementation of greening measures by local governments [20]. Therefore, increasing the green areas may also be an effective way to alleviate the thermal stress in Shenzhen, Dongguan, Guangzhou, and Huizhou, as the NDVI explained over 40% of the variations in the daytime LST in these cities (Figure 7). In addition, the NDVI also showed a strong influence on nighttime LST in Dongguan, Shenzhen, Foshan, Zhongshan, and Guangzhou. The dominant drivers of daytime LST in the central cities of the PRD were the socio-economic ones in 2015, including in Dongguan, Shenzhen, Foshan, Zhongshan, Guangzhou, and Zhuhai. The HRI greatly increased in Dongguan, Shenzhen, Foshan, and Zhongshan, which could be due to the increased intensity of socio-economic activities. Therefore, it is suggested that the heat emissions caused by socio-economic activities should be considered in the urban planning in order to reduce the thermal stress in these cities. The results of this study can provide useful guidance for planners towards better management of the thermal environment.

5.3. Limitations and Future Work

This study provided a comprehensive framework for identifying the spatial distribution characteristics and main driving factors of the LST at different spatial scales in the PRD. There are still some limitations. First, this study analyzed the spatial patterns and the influencing factors of the LST for just two years: 2005 and 2015. Therefore, further studies are needed in order to address this aspect by applying the most recent multi-temporal remote sensing data in the future. Second, the annual scale was selected in this study, as it provided good results in many previous studies [31,37,96]. Seasonal changes also need to be studied in the future. Third, we considered five levels of LST classification, but more classification criteria need to be considered in future studies. Fourth, the combined effects of two drivers on the LST were measured in this study. However, the combined effects of multiple variables on the LST are important and need to be studied in future works. Finally, the spatial resolution of the remote sensing data must be improved by using higher-resolution remote sensing data in the future.

6. Conclusions

This study took the PRD urban agglomeration, a typical urban agglomeration in Southeast China, as a case study and focused on the spatial distribution and main driving factors of the LST in 2005 and 2015. A spatial autocorrelation analysis, hotspot analysis, and the HRI were used to quantitatively analyze the spatial distribution characteristics of the LST. Seven drivers of LST were studied, including physiographical, land cover, and socio-economic drivers. The GD metric was applied to explore the explanation rate of the driving factors for the LST. The main conclusions from this study can be summarized as follows:

1. From 2005 to 2015, the daytime HSAs were concentrated towards the center of the PRD, while they decreased in the northern PRD. The stable daytime HSAs were concentrated and distributed on both sides of the Pearl River estuary.
2. The rankings of the HRI values of the nine cities showed that, during the study period, the highest daytime stress on the thermal environment among all cities was recorded in Dongguan and Shenzhen. The nighttime stress on the thermal environment recorded in Zhongshan, Zhuhai, Dongguan, and Shenzhen was higher than that in the other cities in 2015, while the lowest HRI values were observed in Zhaoqing, Jiangmen, and Huizhou, which were characterized by the lowest urbanization rates. This finding indicates that highly urbanized cities are more likely to experience severe thermal environments than cities with low urbanization rates.
3. The influence of land cover and socio-economic factors on daytime LST was higher in the relatively highly urbanized cities than in the cities with low urbanization rates. This finding indicates that human activities greatly contributed to the variations in LST in highly urbanized areas.
4. In 2015, the NLI factor exhibited the strongest influence on daytime LST in Shenzhen, Dongguan, Guangzhou, Foshan, Zhongshan, and Zhuhai. However, for the marginal cities of Zhaoqing, Jiangmen, and Huizhou, the influence of elevation was much higher than that of the other factors. This finding indicates that the influence of socio-economic activities on daytime LST was higher in highly urbanized areas, and even exceeded the influence of land cover. Controlling the anthropogenic heat released due to socio-economic activities is an important step in improving the thermal environment in highly urbanized areas with the development of urban agglomerations.
5. The NDVI showed an important influence on nighttime LST in most of the cities during the study period. Some factors had no significant effects on nighttime LST in some cities, suggesting that the driving mechanisms of nighttime LST are more complex than those of daytime LST.
6. LST is the result of the combined effects of multiple factors. The combined effects of different factors on the LST are greater than the independent effects of single factors. The combined effects of different drivers are important in studies of the driving mechanisms of LST.

This study is expected to provide useful guidance for the optimization of the thermal environment in PRD urban agglomerations. In addition, the research framework used in this study can also be applied in other urban agglomerations. The results of this study revealed the complex mechanisms of the variability of the LST among the nine cities in the PRD urban agglomeration, and corresponding improvement measures have been recommended to help urban planners in better articulating urban agglomeration management strategies.

Author Contributions: Conceptualization, W.L. and Q.M.; methodology, W.L. and M.A.; software, data curation, writing-original draft, W.L.; writing—review, Q.M., L.Z., D.H. and M.M.; writing-review and editing, W.L, M.A. and M.M.; visualization, M.A. and D.H.; investigation, L.Z.; supervision, Q.M.; funding acquisition, Q.M. All authors have read and agreed to the published version of the manuscript.

Funding: This research was funded by Tsien Hsue-shen Urbanology Award of Hangzhou International Urbanology Research Center & Zhejiang Urban Governance Studies Center, grant number no.20QXS007; the Major Projects of High-resolution Earth Observation Systems of National Science and Technology, grant number 05-Y30B01-9001-19/20-1.

Institutional Review Board Statement: Not applicable.

Informed Consent Statement: Not applicable.

Data Availability Statement: The data presented in this study are available on request from the corresponding website.

Conflicts of Interest: The authors declare no conflict of interest. The funders had no role in the design of the study; in the collection, analyses, or interpretation of data; in the writing of the manuscript, or in the decision to publish the results.

Abbreviations

GD	Geo-detector
GDP	Gross domestic product
HRI	Hotspot ratio index
HSAs	Hotspot areas
ISDD	Impervious surface distribution density
LST	Land surface temperature
LUCC	Land use/land cover
NDVI	Normalized difference vegetation index
NLI	Nighttime light index
POP	Population density
PRD	Pearl River Delta
UHI	Urban heat island

References

1. Oke, T.R. The energetic basis of the urban heat island. *Q. J. R. Meteorol. Soc.* **1982**, *108*, 1–24. [CrossRef]
2. Tao, M.; Chen, L.; Wang, Z.; Wang, J.; Tao, J.; Wang, X. Did the widespread haze pollution over China increase during the last decade? A satellite view from space. *Environ. Res. Lett.* **2016**, *11*, 054019. [CrossRef]
3. Curriero, F.C.; Heiner, K.S.; Samet, J.M.; Zeger, S.L.; Strug, L.; Patz, J.A. Temperature and mortality in 11 cities of the eastern United States. *Am. J. Epidemiol.* **2002**, *155*, 80–87. [CrossRef]
4. Patz, J.A.; Campbell-Lendrum, D.; Holloway, T.; Foley, J.A. Impact of regional climate change on human health. *Nature* **2005**, *438*, 310–317. [CrossRef] [PubMed]
5. Hwang, R.-L.; Lin, T.-P.; Lin, F.-Y. Evaluation and mapping of building overheating risk and air conditioning use due to the urban heat island effect. *J. Build. Eng.* **2020**, *32*, 101726. [CrossRef]
6. Fang, C. Important progress and future direction of studies on China's urban agglomerations. *J. Geogr. Sci.* **2015**, *25*, 1003–1024. [CrossRef]
7. Fang, C.; Yu, D. Urban agglomeration: An evolving concept of an emerging phenomenon. *Landsc. Urban Plan.* **2017**, *162*, 126–136. [CrossRef]
8. Li, W.; Han, C.; Li, W.; Zhou, W.; Han, L. Multi-scale effects of urban agglomeration on thermal environment: A case of the Yangtze River Delta Megaregion, China. *Sci. Total Environ.* **2020**, *713*, 136556. [CrossRef] [PubMed]
9. Hu, D.; Meng, Q.; Zhang, L.; Zhang, Y. Spatial quantitative analysis of the potential driving factors of land surface temperature in different "Centers" of polycentric cities: A case study in Tianjin, China. *Sci. Total Environ.* **2020**, *706*, 135244. [CrossRef] [PubMed]
10. Imhoff, M.L.; Zhang, P.; Wolfe, R.E.; Bounoua, L. Remote sensing of the urban heat island effect across biomes in the continental USA. *Remote Sens. Environ.* **2010**, *114*, 504–513. [CrossRef]

11. Estoque, R.C.; Murayama, Y.; Myint, S.W. Effects of landscape composition and pattern on land surface temperature: An urban heat island study in the megacities of Southeast Asia. *Sci. Total Environ.* **2017**, *577*, 349–359. [CrossRef] [PubMed]
12. Yao, R.; Wang, L.; Huang, X.; Niu, Z.; Liu, F.; Wang, Q. Temporal trends of surface urban heat islands and associated determinants in major Chinese cities. *Sci. Total Environ.* **2017**, *609*, 742–754. [CrossRef]
13. Kedia, S.; Bhakare, S.P.; Dwivedi, A.K.; Islam, S.; Kaginalkar, A. Estimates of change in surface meteorology and urban heat island over northwest India: Impact of urbanization. *Urban Clim.* **2021**, *36*, 100782. [CrossRef]
14. Sultana, S.; Satyanarayana, A. Assessment of urbanisation and urban heat island intensities using landsat imageries during 2000–2018 over a sub-tropical Indian City. *Sustain. Cities Soc.* **2020**, *52*, 101846. [CrossRef]
15. El Kenawy, A.M.; Hereher, M.; Robaa, S.M.; McCabe, M.F.; Lopez-Moreno, J.I.; Domínguez-Castro, F.; Gaber, I.M.; Al-Awadhi, T.; Al-Buloshi, A.; Al Nasiri, N. Nocturnal Surface Urban Heat Island over Greater Cairo: Spatial Morphology, Temporal Trends and Links to Land-Atmosphere Influences. *Remote Sens.* **2020**, *12*, 3889. [CrossRef]
16. Mohammad, P.; Goswami, A.; Bonafoni, S. The impact of the land cover dynamics on surface urban heat island variations in semi-arid cities: A case study in Ahmedabad City, India, using multi-sensor/source data. *Sensors* **2019**, *19*, 3701. [CrossRef]
17. Haashemi, S.; Weng, Q.; Darvishi, A.; Alavipanah, S.K. Seasonal variations of the surface urban heat island in a semi-arid city. *Remote Sens.* **2016**, *8*, 352. [CrossRef]
18. Liu, X.; Tian, G.; Feng, J.; Ma, B.; Wang, J.; Kong, L. Modeling the warming impact of urban land expansion on hot weather using the weather research and forecasting model: A case study of Beijing, China. *Adv. Atmos. Sci.* **2018**, *35*, 723–736. [CrossRef]
19. Guo, G.; Wu, Z.; Chen, Y. Complex mechanisms linking land surface temperature to greenspace spatial patterns: Evidence from four southeastern Chinese cities. *Sci. Total Environ.* **2019**, *674*, 77–87. [CrossRef]
20. Yu, Z.; Yao, Y.; Yang, G.; Wang, X.; Vejre, H. Spatiotemporal patterns and characteristics of remotely sensed region heat islands during the rapid urbanization (1995–2015) of Southern China. *Sci. Total Environ.* **2019**, *674*, 242–254. [CrossRef]
21. Chen, X.-L.; Zhao, H.-M.; Li, P.-X.; Yin, Z.-Y. Remote sensing image-based analysis of the relationship between urban heat island and land use/cover changes. *Remote Sens. Environ.* **2006**, *104*, 133–146. [CrossRef]
22. Yao, L.; Li, T.; Xu, M.; Xu, Y. How the landscape features of urban green space impact seasonal land surface temperatures at a city-block-scale: An urban heat island study in Beijing, China. *Urban For. Urban Green.* **2020**, *52*, 126704. [CrossRef]
23. Dwivedi, A.; Mohan, B.K. Impact of green roof on micro climate to reduce Urban Heat Island. *Remote Sens. Appl. Soc. Environ.* **2018**, *10*, 56–69. [CrossRef]
24. Singh, P.; Kikon, N.; Verma, P. Impact of land use change and urbanization on urban heat island in Lucknow city, Central India. A remote sensing based estimate. *Sustain. Cities Soc.* **2017**, *32*, 100–114. [CrossRef]
25. Dai, Z.; Guldmann, J.-M.; Hu, Y. Spatial regression models of park and land-use impacts on the urban heat island in central Beijing. *Sci. Total Environ.* **2018**, *626*, 1136–1147. [CrossRef] [PubMed]
26. Morabito, M.; Crisci, A.; Guerri, G.; Messeri, A.; Congedo, L.; Munafò, M. Surface urban heat islands in Italian metropolitan cities: Tree cover and impervious surface influences. *Sci. Total Environ.* **2021**, *751*, 142334. [CrossRef] [PubMed]
27. Meng, Q.; Zhang, L.; Sun, Z.; Meng, F.; Wang, L.; Sun, Y. Characterizing spatial and temporal trends of surface urban heat island effect in an urban main built-up area: A 12-year case study in Beijing, China. *Remote Sens. Environ.* **2018**, *204*, 826–837. [CrossRef]
28. Yuan, F.; Bauer, M.E. Comparison of impervious surface area and normalized difference vegetation index as indicators of surface urban heat island effects in Landsat imagery. *Remote Sens. Environ.* **2007**, *106*, 375–386. [CrossRef]
29. Li, Y.; Sun, Y.; Li, J.; Gao, C. Socioeconomic drivers of urban heat island effect: Empirical evidence from major Chinese cities. *Sustain. Cities Soc.* **2020**, *63*, 102425. [CrossRef]
30. Peng, J.; Jia, J.; Liu, Y.; Li, H.; Wu, J. Seasonal contrast of the dominant factors for spatial distribution of land surface temperature in urban areas. *Remote Sens. Environ.* **2018**, *215*, 255–267. [CrossRef]
31. Dewan, A.; Kiselev, G.; Botje, D.; Mahmud, G.I.; Bhuian, M.H.; Hassan, Q.K. Surface urban heat island intensity in five major cities of Bangladesh: Patterns, drivers and trends. *Sustain. Cities Soc.* **2021**, *71*, 102926. [CrossRef]
32. Portela, C.I.; Massi, K.G.; Rodrigues, T.; Alcântara, E. Impact of urban and industrial features on land surface temperature: Evidences from satellite thermal indices. *Sustain. Cities Soc.* **2020**, *56*, 102100. [CrossRef]
33. Lo, C. Modeling the population of China using DMSP operational linescan system nighttime data. *Photogramm. Eng. Remote Sens.* **2001**, *67*, 1037–1047.
34. Chand, T.K.; Badarinath, K.; Elvidge, C.; Tuttle, B. Spatial characterization of electrical power consumption patterns over India using temporal DMSP-OLS night-time satellite data. *Int. J. Remote Sens.* **2009**, *30*, 647–661. [CrossRef]
35. Yu, B.; Tang, M.; Wu, Q.; Yang, C.; Deng, S.; Shi, K.; Peng, C.; Wu, J.; Chen, Z. Urban built-up area extraction from log-transformed NPP-VIIRS nighttime light composite data. *IEEE Geosci. Remote Sens. Lett.* **2018**, *15*, 1279–1283. [CrossRef]
36. Bennett, M.M.; Smith, L.C. Advances in using multitemporal night-time lights satellite imagery to detect, estimate, and monitor socioeconomic dynamics. *Remote Sens. Environ.* **2017**, *192*, 176–197. [CrossRef]
37. Ma, Y.; Zhang, S.; Yang, K.; Li, M. Influence of spatiotemporal pattern changes of impervious surface of urban megaregion on thermal environment: A case study of the Guangdong–Hong Kong–Macao Greater Bay Area of China. *Ecol. Indic.* **2021**, *121*, 107106. [CrossRef]
38. Mathew, A.; Khandelwal, S.; Kaul, N. Investigating spatial and seasonal variations of urban heat island effect over Jaipur city and its relationship with vegetation, urbanization and elevation parameters. *Sustain. Cities Soc.* **2017**, *35*, 157–177. [CrossRef]

39. Huang, X.; Wang, Y. Investigating the effects of 3D urban morphology on the surface urban heat island effect in urban functional zones by using high-resolution remote sensing data: A case study of Wuhan, Central China. *ISPRS J. Photogramm. Remote Sens.* **2019**, *152*, 119–131. [CrossRef]
40. Zhou, D.; Bonafoni, S.; Zhang, L.; Wang, R. Remote sensing of the urban heat island effect in a highly populated urban agglomeration area in East China. *Sci. Total Environ.* **2018**, *628*, 415–429. [CrossRef]
41. Cui, C.; Wang, B.; Ren, H.; Wang, Z. Spatiotemporal variations in gastric cancer mortality and their relations to influencing factors in S County, China. *Int. J. Environ. Res. Public Health* **2019**, *16*, 784. [CrossRef]
42. Wang, J.F.; Li, X.H.; Christakos, G.; Liao, Y.L.; Zhang, T.; Gu, X.; Zheng, X.Y. Geographical detectors-based health risk assessment and its application in the neural tube defects study of the Heshun Region, China. *Int. J. Geogr. Inf. Sci.* **2010**, *24*, 107–127. [CrossRef]
43. Ding, Y.; Zhang, M.; Qian, X.; Li, C.; Chen, S.; Wang, W. Using the geographical detector technique to explore the impact of socioeconomic factors on PM2. 5 concentrations in China. *J. Clean. Prod.* **2019**, *211*, 1480–1490. [CrossRef]
44. Zhou, C.; Chen, J.; Wang, S. Examining the effects of socioeconomic development on fine particulate matter (PM2. 5) in China's cities using spatial regression and the geographical detector technique. *Sci. Total Environ.* **2018**, *619*, 436–445. [CrossRef]
45. Zhao, R.; Zhan, L.; Yao, M.; Yang, L. A geographically weighted regression model augmented by Geodetector analysis and principal component analysis for the spatial distribution of PM2. 5. *Sustain. Cities Soc.* **2020**, *56*, 102106. [CrossRef]
46. Liao, Y.; Zhang, Y.; He, L.; Wang, J.; Liu, X.; Zhang, N.; Xu, B. Temporal and spatial analysis of neural tube defects and detection of geographical factors in Shanxi Province, China. *PLoS ONE* **2016**, *11*, e0150332.
47. Huang, J.; Wang, J.; Bo, Y.; Xu, C.; Hu, M.; Huang, D. Identification of health risks of hand, foot and mouth disease in China using the geographical detector technique. *Int. J. Environ. Res. Public Health* **2014**, *11*, 3407–3423. [CrossRef] [PubMed]
48. Ren, Y.; Deng, L.; Zuo, S.; Luo, Y.; Shao, G.; Wei, X.; Hua, L.; Yang, Y. Geographical modeling of spatial interaction between human activity and forest connectivity in an urban landscape of southeast China. *Landsc. Ecol.* **2014**, *29*, 1741–1758. [CrossRef]
49. Ju, H.; Zhang, Z.; Zuo, L.; Wang, J.; Zhang, S.; Wang, X.; Zhao, X. Driving forces and their interactions of built-up land expansion based on the geographical detector–a case study of Beijing, China. *Int. J. Geogr. Inf. Sci.* **2016**, *30*, 2188–2207. [CrossRef]
50. Ding, Y.; Cai, J.; Ren, Z.; Yang, Z. Spatial disparities of economic growth rate of China's National-level ETDZs and their determinants based on geographical detector analysis. *Prog. Geogr.* **2014**, *33*, 657–666.
51. Wang, Y.; Wang, S.; Li, G.; Zhang, H.; Jin, L.; Su, Y.; Wu, K. Identifying the determinants of housing prices in China using spatial regression and the geographical detector technique. *Appl. Geogr.* **2017**, *79*, 26–36. [CrossRef]
52. Ren, Y.; Deng, L.-Y.; Zuo, S.-D.; Song, X.-D.; Liao, Y.-L.; Xu, C.-D.; Chen, Q.; Hua, L.-Z.; Li, Z.-W. Quantifying the influences of various ecological factors on land surface temperature of urban forests. *Environ. Pollut.* **2016**, *216*, 519–529. [CrossRef]
53. Bank, W. *East Asia's Changing Urban Landscape: Measuring a Decade of Spatial Growth*; The World Bank: Washington, DC, USA, 2015.
54. Zhang, L.; Weng, Q. Annual dynamics of impervious surface in the Pearl River Delta, China, from 1988 to 2013, using time series Landsat imagery. *ISPRS J. Photogramm. Remote Sens.* **2016**, *113*, 86–96. [CrossRef]
55. Liu, B.; Peng, S.; Liao, Y.; Long, W. The causes and impacts of water resources crises in the Pearl River Delta. *J. Clean. Prod.* **2018**, *177*, 413–425. [CrossRef]
56. Chan, F.K.S.; Yang, L.E.; Scheffran, J.; Mitchell, G.; Adekola, O.; Griffiths, J.; Chen, Y.; Li, G.; Lu, X.; Qi, Y. Urban flood risks and emerging challenges in a Chinese delta: The case of the Pearl River Delta. *Environ. Sci. Policy* **2021**, *122*, 101–115. [CrossRef]
57. Hou, H.; Liu, K.; Li, X.; Chen, S.; Wang, W.; Rong, K. Assessing the urban heat island variations and its influencing mechanism in metropolitan areas of Pearl River Delta, South China. *Phys. Chem. Earth Parts A/B/C* **2020**, *120*, 102953. [CrossRef]
58. Wu, Z.; Zhang, Y.; Zhang, L.; Huang, M.; Zhong, L.; Chen, D.; Wang, X. Trends of outdoor air pollution and the impact on premature mortality in the Pearl River Delta region of southern China during 2006–2015. *Sci. Total Environ.* **2019**, *690*, 248–260. [CrossRef]
59. Hu, M.; Wang, Y.; Xia, B.; Huang, G. Surface temperature variations and their relationships with land cover in the Pearl River Delta. *Environ. Sci. Pollut. Res.* **2020**, *27*, 37614–37625. [CrossRef]
60. Nichol, J.E.; Choi, S.Y.; Wong, M.S.; Abbas, S. Temperature change and urbanisation in a multi-nucleated megacity: China's Pearl River Delta. *Urban Clim.* **2020**, *31*, 100592. [CrossRef]
61. Wang, R.; Cai, M.; Ren, C.; Bechtel, B.; Xu, Y.; Ng, E. Detecting multi-temporal land cover change and land surface temperature in Pearl River Delta by adopting local climate zone. *Urban Clim.* **2019**, *28*, 100455. [CrossRef]
62. Zhang, Q.; Wu, Z.; Yu, H.; Zhu, X.; Shen, Z. Variable Urbanization Warming Effects across Metropolitans of China and Relevant Driving Factors. *Remote Sens.* **2020**, *12*, 1500. [CrossRef]
63. Chen, Y.; Yu, S. Impacts of urban landscape patterns on urban thermal variations in Guangzhou, China. *Int. J. Appl. Earth Obs. Geoinf.* **2017**, *54*, 65–71. [CrossRef]
64. Cao, Z.; Wu, Z.; Liu, L.; Chen, Y.; Zou, Y. Assessing the relationship between anthropogenic heat release warming and building characteristics in Guangzhou: A sustainable development perspective. *Sci. Total Environ.* **2019**, *695*, 133759. [CrossRef]
65. Wang, W.; Liu, K.; Tang, R.; Wang, S. Remote sensing image-based analysis of the urban heat island effect in Shenzhen, China. *Phys. Chem. Earth Parts A/B/C* **2019**, *110*, 168–175. [CrossRef]
66. Li, W.; Cao, Q.; Lang, K.; Wu, J. Linking potential heat source and sink to urban heat island: Heterogeneous effects of landscape pattern on land surface temperature. *Sci. Total Environ.* **2017**, *586*, 457–465. [CrossRef] [PubMed]

67. Wei, C.; Taubenböck, H.; Blaschke, T. Measuring urban agglomeration using a city-scale dasymetric population map: A study in the Pearl River Delta, China. *Habitat Int.* **2017**, *59*, 32–43. [CrossRef]
68. Wan, Z. New refinements and validation of the collection-6 MODIS land-surface temperature/emissivity product. *Remote Sens. Environ.* **2014**, *140*, 36–45. [CrossRef]
69. Wang, C.; Li, Y.; Myint, S.W.; Zhao, Q.; Wentz, E.A. Impacts of spatial clustering of urban land cover on land surface temperature across Köppen climate zones in the contiguous United States. *Landsc. Urban Plan.* **2019**, *192*, 103668. [CrossRef]
70. Shen, H.; Huang, L.; Zhang, L.; Wu, P.; Zeng, C. Long-term and fine-scale satellite monitoring of the urban heat island effect by the fusion of multi-temporal and multi-sensor remote sensed data: A 26-year case study of the city of Wuhan in China. *Remote Sens. Environ.* **2016**, *172*, 109–125. [CrossRef]
71. Quan, J.; Chen, Y.; Zhan, W.; Wang, J.; Voogt, J.; Wang, M. Multi-temporal trajectory of the urban heat island centroid in Beijing, China based on a Gaussian volume model. *Remote Sens. Environ.* **2014**, *149*, 33–46. [CrossRef]
72. Streutker, D.R. Satellite-measured growth of the urban heat island of Houston, Texas. *Remote Sens. Environ.* **2003**, *85*, 282–289. [CrossRef]
73. Yang, Q.; Huang, X.; Tang, Q. The footprint of urban heat island effect in 302 Chinese cities: Temporal trends and associated factors. *Sci. Total Environ.* **2019**, *655*, 652–662. [CrossRef] [PubMed]
74. Peng, J.; Hu, Y.; Dong, J.; Liu, Q.; Liu, Y. Quantifying spatial morphology and connectivity of urban heat islands in a megacity: A radius approach. *Sci. Total Environ.* **2020**, *714*, 136792. [CrossRef] [PubMed]
75. Qiao, Z.; Wu, C.; Zhao, D.; Xu, X.; Yang, J.; Feng, L.; Sun, Z.; Liu, L. Determining the boundary and probability of surface urban heat island footprint based on a logistic model. *Remote Sens.* **2019**, *11*, 1368. [CrossRef]
76. Feyisa, G.L.; Meilby, H.; Jenerette, G.D.; Pauliet, S. Locally optimized separability enhancement indices for urban land cover mapping: Exploring thermal environmental consequences of rapid urbanization in Addis Ababa, Ethiopia. *Remote Sens. Environ.* **2016**, *175*, 14–31. [CrossRef]
77. Ord, J.K.; Getis, A. Local spatial autocorrelation statistics: Distributional issues and an application. *Geogr. Anal.* **1995**, *27*, 286–306. [CrossRef]
78. Songchitruksa, P.; Zeng, X. Getis–Ord spatial statistics to identify hot spots by using incident management data. *Transp. Res. Rec.* **2010**, *2165*, 42–51. [CrossRef]
79. Shirani-Bidabadi, N.; Nasrabadi, T.; Faryadi, S.; Larijani, A.; Roodposhti, M.S. Evaluating the spatial distribution and the intensity of urban heat island using remote sensing, case study of Isfahan city in Iran. *Sustain. Cities Soc.* **2019**, *45*, 686–692. [CrossRef]
80. Xu, H.-Q.; Chen, B.-Q. Remote sensing of the urban heat island and its changes in Xiamen City of SE China. *J. Environ. Sci.* **2004**, *16*, 276–281.
81. Qiao, Z.; Tian, G.; Zhang, L.; Xu, X. Influences of urban expansion on urban heat island in Beijing during 1989–2010. *Adv. Meteorol.* **2014**, *2014*, 187169. [CrossRef]
82. Cao, F.; Ge, Y.; Wang, J.-F. Optimal discretization for geographical detectors-based risk assessment. *GISci. Remote Sens.* **2013**, *50*, 78–92. [CrossRef]
83. De Smith, M.J.; Goodchild, M.F.; Longley, P. *Geospatial Analysis: A Comprehensive Guide to Principles, Techniques and Software Tools*; Troubador Publishing Ltd: Kibworth Harcourt, UK, 2007.
84. Chen, W.; Zhang, Y.; Pengwang, C.; Gao, W. Evaluation of urbanization dynamics and its impacts on surface heat islands: A case study of Beijing, China. *Remote Sens.* **2017**, *9*, 453. [CrossRef]
85. Li, H.; Xu, E.; Zhang, H. Examining the coupling relationship between urbanization and natural disasters: A case study of the Pearl River Delta, China. *Int. J. Disaster Risk Reduct.* **2021**, *55*, 102057. [CrossRef]
86. Peng, S.; Piao, S.; Ciais, P.; Friedlingstein, P.; Ottle, C.; Bréon, F.o.-M.; Nan, H.; Zhou, L.; Myneni, R.B. Surface urban heat island across 419 global big cities. *Environ. Sci. Technol.* **2012**, *46*, 696–703. [CrossRef] [PubMed]
87. Aram, F.; García, E.H.; Solgi, E.; Mansournia, S. Urban green space cooling effect in cities. *Heliyon* **2019**, *5*, e01339. [CrossRef]
88. Logan, T.; Zaitchik, B.; Guikema, S.; Nisbet, A. Night and day: The influence and relative importance of urban characteristics on remotely sensed land surface temperature. *Remote Sens. Environ.* **2020**, *247*, 111861. [CrossRef]
89. Khandelwal, S.; Goyal, R.; Kaul, N.; Mathew, A. Assessment of land surface temperature variation due to change in elevation of area surrounding Jaipur, India. *Egypt. J. Remote Sens. Space Sci.* **2018**, *21*, 87–94. [CrossRef]
90. Goyal, R.; Khandelwal, S.; Kaul, N. Analysis of relative importance of parameters representing vegetation, urbanization and elevation with land surface temperature using ANN. In Proceedings of the Geospatial World Forum Hyderabad, Hyderabad, India, 18–21 January 2011.
91. Mathew, A.; Khandelwal, S.; Kaul, N. Spatial and temporal variations of urban heat island effect and the effect of percentage impervious surface area and elevation on land surface temperature: Study of Chandigarh city, India. *Sustain. Cities Soc.* **2016**, *26*, 264–277. [CrossRef]
92. Du, H.; Wang, D.; Wang, Y.; Zhao, X.; Qin, F.; Jiang, H.; Cai, Y. Influences of land cover types, meteorological conditions, anthropogenic heat and urban area on surface urban heat island in the Yangtze River Delta Urban Agglomeration. *Sci. Total Environ.* **2016**, *571*, 461–470. [CrossRef]
93. McCarthy, M.P.; Best, M.J.; Betts, R.A. Climate change in cities due to global warming and urban effects. *Geophys. Res. Lett.* **2010**, *37*, L09705. [CrossRef]

4. Tripathy, B.R.; Sajjad, H.; Elvidge, C.D.; Ting, Y.; Pandey, P.C.; Rani, M.; Kumar, P. Modeling of electric demand for sustainable energy and management in India using spatio-temporal DMSP-OLS night-time data. *Environ. Manag.* **2018**, *61*, 615–623. [CrossRef] [PubMed]

5. Li, J.; Wang, F.; Fu, Y.; Guo, B.; Zhao, Y.; Yu, H. A Novel SUHI Referenced Estimation Method for Multicenters Urban Agglomeration using DMSP/OLS Nighttime Light Data. *IEEE J. Sel. Top. Appl. Earth Obs. Remote Sens.* **2020**, *13*, 1416–1425. [CrossRef]

6. Liu, H.; Huang, B.; Zhan, Q.; Gao, S.; Li, R.; Fan, Z. The influence of urban form on surface urban heat island and its planning implications: Evidence from 1288 urban clusters in China. *Sustain. Cities Soc.* **2021**, *71*, 102987. [CrossRef]

remote sensing

MDPI

Article

A Remote Sensing Approach for Surface Urban Heat Island Modeling in a Tropical Colombian City Using Regression Analysis and Machine Learning Algorithms

Julián Garzón [1,2,*], Iñigo Molina [1], Jesús Velasco [1] and Andrés Calabia [3]

1 Department of Surveying and Cartography Engineering, Universidad Politécnica de Madrid, 28031 Madrid, Spain; inigo.molina@upm.es (I.M.); jesus.velasco@upm.es (J.V.)
2 Programa de Ingeniería Topográfica y Geomática, Universidad del Quindío, Armenia 630004, Colombia
3 School of Remote Sensing and Geomatics Engineering, Nanjing University of Information Science and Technology, Nanjing 210044, China; andres@calabia.com
* Correspondence: j.garzonb@alumnos.upm.es; Tel.: +57-318-621-5040

Abstract: The Surface Urban Heat Islands (SUHI) phenomenon has adverse environmental consequences on human activities, biophysical and ecological systems. In this study, Land Surface Temperature (LST) from Landsat and Sentinel-2 satellites is used to investigate the contribution of potential factors that generate the SUHI phenomenon. We employ Principal Component Analysis (PCA) and Multiple Linear Regression (MLR) techniques to model the main temporal and spatial SUHI patterns of Cartago, Colombia, for the period 2001–2020. We test and evaluate the performance of three different emissivity models to retrieve LST. The fractional vegetation cover model using Sentinel-2 data provides the best results with $R^2 = 0.78$, while the ASTER Global Emissivity Dataset v3 and the land surface emissivity model provide $R^2 = 0.27$ and $R^2 = 0.26$, respectively. Our SUHI model reveals that the factors with the highest impact are the Normalized Difference Water Index (NDWI) and the Normalized Difference Build-up Index (NDBI). Furthermore, we incorporate a weighted Naïve Bayes Machine Learning (NBML) algorithm to identify areas prone to extreme temperatures that can be used to define and apply normative actions to mitigate the negative consequences of SUHI. Our NBML approach demonstrates the suitability of the new SUHI model with uncertainty within 95%, against the 88% given by the Support Vector Machine (SVM) approach.

Keywords: Surface Urban Heat Island (SUHI); Land Surface Temperature (LST); Principal Component Analysis (PCA); Multiple Linear Regression (MLR); Machine Learning; Naïve Bayes

Citation: Garzón, J.; Molina, I.; Velasco, J.; Calabia, A. A Remote Sensing Approach for Surface Urban Heat Island Modeling in a Tropical Colombian City Using Regression Analysis and Machine Learning Algorithms. *Remote Sens.* **2021**, *13*, 4256. https://doi.org/10.3390/rs13214256

Academic Editors: Yuji Murayama and Ruci Wang

Received: 16 September 2021
Accepted: 19 October 2021
Published: 22 October 2021

1. Introduction

Urban expansion transforms natural areas into surfaces covered with concrete, asphalt, and buildings (highly impervious materials), reducing evapotranspiration and decreasing the cooling capacity of the air, which in turn helps to reduce the impacts of high urban surface temperature on the urban surface. Due to the existing urban growth, the climate in these areas becomes warmer than the regional areas of the suburban and rural regions, resulting in the phenomenon of Urban Heat Islands (UHI) [1]. The UHI refers to a phenomenon in which urban areas tend to have higher air or surface temperatures than their surroundings [2]. Traditionally, terrestrial observation methods, such as ground meteorological stations that record specific values of air temperature, have been used to model UHI [3]. The difference between air temperature measurements recovered from urban and rural meteorological stations is a direct method used to model UHI [4]. However, the high heterogeneity in urban areas makes temperature spatially diverse, making it difficult for a small number of stations to realistically represent the real variability [5]. When the UHI phenomenon is monitored by remote sensing, it is referred to as Surface Urban Heat Island (SUHI). The reason is that the parameter considered here is the Land

Surface Temperature (LST), which differs from studies of air temperature [6]. Therefore, LST is an essential variable to characterize SUHI, which has been listed as an essential climate variable of the World Meteorological Organization program. This variable is an important indicator of the energy balance between the atmosphere and the surface of the Earth [7]. Zhou et al. [8] presented a broad review of the SUHI phenomenon and suggested several methods of analysis through the integration of remote sensing data, thermal trends, field observations, and numerical modeling. Sekerteking and Zadbagher [9] suggest that to model and simulate LST, it would be important to investigate the performance of various Machine Learning methods associated with statistical and numerical models. Li et al. [10] and Mei et al. [11] pointed out that one of the problems to be solved in inference models of LST or geographic variables is to evaluate the influence of contributing factors.

Land-use land-cover changes (LULC) and spectral indices from satellite data, such as, e.g., Normalized Difference Vegetation Index (NDVI) or Normalized Difference Water Index (NDWI), have been used extensively to investigate the relationships between urban and biophysical systems, as well as their impact on surface temperature [12–15]. LULCs are due to complex interactions between the urban system and the biophysical environment that produce significant changes in local temperatures. In a recent work published by Shi et al. [16], eight parameters were referred to as urban design factors, where thermal properties of building materials, vegetation, vegetation cover ratio, and ground emissivity were taken into account. The authors suggested the need to detect more potential factors affecting this phenomenon.

Understanding and quantifying urban temperatures in space and time are significantly relevant for city planners in defining policies that generate adaptation strategies to mitigate the SUHI effects. A very useful tool is Principal Component Analysis (PCA), a multivariate statistical technique that aims to preserve the total variance and reduce the dimensionality of the data set, while eliminating redundancy in the data [17]. Several authors have used this technique to detect spatial patterns of biophysical factors by synthesizing information from a set of images [18–20]. Multiple Linear Regression (MLR) analysis is an approach used to evaluate the relationship between independent and dependent factors [21]. This method has also been widely used to determine the relationship of various environmental factors [22–24].

Advanced nonlinear analysis techniques, such as Machine Learning, have been applied in numerous studies that require analyzing variables related to urban thermal changes. Some examples are population density, land cover, and urbanization [25–29]. Voelkel and Shandas [30] implemented a UHI model to detect a daily distribution of temperatures. Their results revealed that a random forest (RF) model performed better in predicting temperature. Furthermore, Zumwald et al. [31] developed a model to create high-resolution air temperature maps. This model makes predictions by integrating an RF algorithm with low-cost weather stations. It is important to note that the behavior of SUHI varies over time and is associated with factors such as human development and changes in land use. In this sense, Kafy et al. [32] formulated a seasonal thermal prediction influenced by LULC through Cellular Automata and Artificial Neural Network algorithms. Their findings indicated that by 2039, the urban growth of Cumilla, and Bangladesh, plus the decrease in land cover, will cause 30% of cities to experience temperatures above 33 °C. Shi et al. [33] noted that the use of time series and Machine Learning techniques is a growing trend in SUHI research.

Several studies have widely documented the influence of spectral indices such as the NDBI, NDVI and Normalized Difference Water Index (NDWI) on the SUHI phenomenon [34,35]. However, none of these authors considered the weighted contribution of the factors to temperature changes, while it is an interesting analysis that can identify which factors generate the greatest influence on SUHI. On the basis of these factors, specific adaptation measures to thermal change can be defined and applied. The main novelty of the approach proposed in this work is the application of a weighted Naïve Bayes

Machine Learning (NBML) algorithm to segment the geographical space into regions of different thermal intensity, not explored in previous literature.

Understanding and quantifying urban temperatures in space and time is significantly relevant for city planners in defining policies that generate adaptation strategies in the face of adverse effects of SUHI. Here we study the application and assessment of modeling procedures that allow evaluating the contribution of various factors to SUHI. For this purpose, a combination of PCA and MLR techniques applied along with Machine Learning Algorithms is used to detect high thermal intensity patterns in the tropical Colombian Andean city of Cartago. Although SUHI is a derived quantity, expressed as the difference between urban and rural LST, the delimitation of thermal zones using LST ranges allows establishing comparisons with other zones, e.g., rural areas, and classifies the space into zones with greater or lesser thermal activity. In this study, the LST ranges are taken from Wang et al. [36], as they are based on statistical criteria, and they appear to conveniently reflect the LST differences of urban areas with their surroundings. The spatial patterns of the SUHI phenomenon can be represented through LST ranges, which, combined with the weights of the involved variables, are further classified using Machine Learning algorithms.

The methodology suggested in this article establishes an effective method for assessing SUHI patterns, locally, and attempts to draw several recommendations for planning sustainable urban development and for the regeneration of areas with thermal excesses.

2. Materials and Methods

2.1. Study Area

The city of Cartago is located in the south-west area of Colombia in the Andean region at an altitude of 917 m above sea level. It has an extension of 279 km^2 with moderate topographic relief. The geodetic coordinates of the city center are 4.75° N and 75.9° W. This area belongs to Valle del Cauca Department and is surrounded by the Cauca and La Vieja rivers.

The climate in this area is tropical dry and the average air temperature is 23.8 °C, with an annual rainfall of 1578 mm. March is the warmest month with an average temperature of 24.3 °C, while October is the coldest with 23.3 °C. According to official reports, the urban population growth rate during 2001–2020 was 12.3%, while the rural population decreased by 44% [37]. The population density is 464 inhabitants per km^2. The appearance of new urban units (red oval areas in Figure 1) denotes the urban growth from the city center to the north-east, near the La Vieja River, as well as to the south-east and south-west. Temperatures in tropical zones show small changes throughout the year. According to official reports from the study area, the difference between the average temperature of the warmest month and that of the coldest month is 1 °C [38].

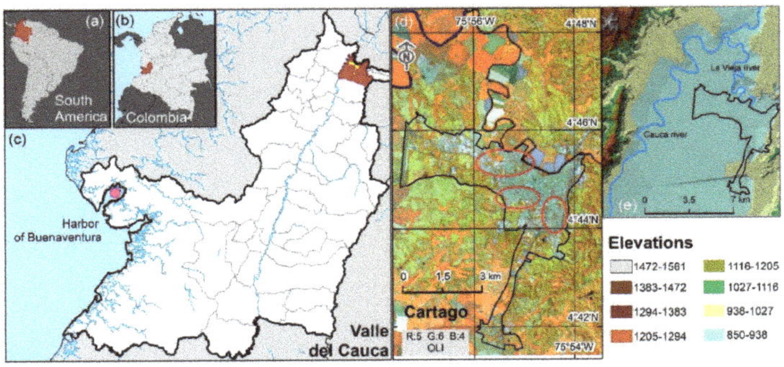

Figure 1. Location of the study area. (**a**) South America: Colombia highlighted. (**b**) Location of Valle del Cauca in Colombia. (**c**) Cartago in Department of Valle del Cauca. (**d**) Cartago, Landsat 8 OLI band combination (R:5, G:6, B:4). (**e**) Digital Elevation Model.

2.2. Data

Data used for the study area (Figure 1a,b) were freely acquired from ESRI World Countries (https://hub.arcgis.com/, accessed on 20 October 2021). The base cartography for the construction of thematic maps and the topographic model is available at https://geoportal.dane.gov.co/, accessed on 20 October 2021 and https://geoportal.igac.gov.co/, accessed on 20 October 2021. The primary information sources used are satellite Earth images from the Thematic Mapper (TM) instrument onboard the Landsat 5 and 7 satellites (L5TM, and L7ETM+), and from the Operational Land Imager (OLI) and Thermal Infrared Sensor (TIRS) onboard the Landsat 8 satellite (L8OLI/TIRS). The images used in this study are sparsely distributed within the period 2001–2020, as shown in Figure 2. Each Landsat product contains separated spectral bands in GeoTIFF format and is referenced to the WGS84 datum in the UTM (18N) cartographic projection. L5TM, L7ETM+ and L8OLI VIS and NIR bands have a spatial resolution of 30 m, while for the TIR satellite instruments, the resolutions are 120, 60, and 100 m, respectively. We employ a total of 37 Landsat scenes (satellite path 009 and row 057) including 2 images of L5TM, 20 of L7ETM+, and 15 of L8OLI/TIRS. In addition to the Landsat products, 11 multispectral Level-2A atmospheric corrected images from the Sentinel-2 Multispectral Instrument (S2-MSI) were also used to extract the so-called Fractional Vegetation Cover (F_{cover}) biophysical variable. S2-MSI offers a different spatial resolution; the three visible and the near infrared bands have 10 m spatial resolution. The three Red Edge bands, an NIR band, and two SWIR S2-MSI bands have 20 m spatial resolution. These data are very appropriate for the retrieval of geophysical surface parameters. Meanwhile, the three other S2-MSI bands (coastal aerosol, water vapor, and SWIR-Cirrus) have a resolution of 60 m resolution. The reflectance S2-MSI products are freely available on the European Space Agency (ESA) DataHUB server (ESA, https://scihub.copernicus.eu/, accessed on 20 October 2021). Details on the retrieval of F_{cover} are given in Section 2.3.2. The estimation of LST from the Thermal Infrared Sensor (TIRS) is highly dependent on the intrinsic properties of the coverage, such as the emissivity of the land surface. The emissivity retrieval method based on the F_{cover} is very suitable due to its ease of application. The performance shown in works such as Sobrino et al. [39] and Valor and Caselles [40].

Figure 2. Temporal distribution of Landsat and Sentinel 2.

2.3. Methods

The proposed methodology comprises five processing steps: (1) data calibration, (2) extraction of contributing factors, (3) estimation of temperature and emissivity, (4) validation of temperatures, and (5) modeling of the SUHI phenomenon. These are described in the following sections.

2.3.1. Data Calibration

The conversion of image digital values to top of atmosphere radiance (LTOA) was carried out using the gain and offset parameters included in the product metadata file. We use the radiance models provided by the USGS website [41]. Subsequently, the images were corrected from atmospheric effects to minimize the radiance scattering and absorption errors caused by water vapor, dust particles, and aerosols. We employ the Fast Line-of-sight Atmospheric Analysis of Spectral Hypercubes (FLAASH) module of ENVI®, which incorporates the MODerate Resolution Atmospheric TRANsmission (MODTRAN) model [42]. Given the geographic location of the area of interest, the tropical atmospheric model is applied to the Landsat products. FLAASH solves the radiative transfer equation by determining the water vapor for each pixel in the image. Water vapor content (WVC) retrieval is not a straightforward solution for Landsat bands, so this parameter was taken from a standard atmospheric model. Regarding the aerosol concentration or aerosol optical depth (AOD), the dark vegetation reflectance algorithm of Kaufman et al. [43] was applied. Finally, all images were subset to fit the boundaries of the study area.

2.3.2. Definition and Extraction of Contributing Factors

Spectral indices such as NDBI, NDVI, and NDWI are used to examine the underlying properties of SUHI formation. Analytical expressions of these indices can be found in Zha et al. [44], Tucker [45], and Gao [46]. Moreover, the components of the tasseled cap (TC) components (brightness, greenness, and wetness) are also computed [47]. The rationale for the selection of these biophysical indices is as follows.

- Energy exchange between latent and sensible heat is related to NDBI, since it detects impervious surfaces that reduce humidity and increase the average temperature of the environment [48].
- Temperature and vegetation maintain a spatially dependent relationship [49]. Vegetation reduces surface irradiation and increases humidity through physiological processes that allow energy exchange, while producing a cooling effect. In this sense, an index for measuring this photosynthetic activity is the NDVI.
- The presence of water bodies has a cooling effect on urban temperature [50]. In this scheme, the NDWI quantifies the water content in the vegetation, while suggesting a significant effect in reducing SUHI. Likewise, rivers play an important role as thermal regulators of urban climate, increasing the cooling potential through evaporation and facilitating airflow. Given that the urban center is the main point for the development of socioeconomic activities, two additional variables were considered to describe the expression of the proximity, i.e., the proximity map of the water body (PW) and the proximity map (PW) and the city center (PUC). A greater distance would imply a lower thermal intensity [51]. The proximity indices are computed by means of a Euclidean distance using the inverse weight distance operator in ArcGIS® (https://esri.com/, accessed on 20 October 2021).

The above indices conform to the contributing factors to our proposed SUHI model. To compute the emissivity values required to retrieve LST from Landsat thermal bands, a novel method is proposed through extracting F_{cover} biophysical variable, although this information can be derived indirectly from NDVI, Leaf area index (LAI), or other biophysical variables [52–54]. Bacour et al. [55] proposed a robust procedure based on the Neural Network training of the PROSAIL (PROSPECT leaf optical properties model and SAIL canopy bidirectional reflectance) model. This F_{cover} variable is implemented in the ESA's Sentinel Application Platform (SNAP (https://step.esa.int/main/toolboxes/snap/, accessed on 20 October 2021), and requires S2-MSI images. Detailed descriptions of this scheme are available in Weiss and Baret [56]. The F_{cover} variable provides the emissivity values necessary to compute LST with the L8OLI/TIRS thermal band 10. Compared to traditional methods based on NDVI, this new approach for extracting the emissivity is suitable for thermal radiation models. Due to temporal synchronization between S2-MSI and L8OLI/TIRS images, this method is only applicable since 2015.

2.3.3. Estimation of Land Surface Temperature and Emissivity

Land surface temperatures are retrieved from L5TM, L7ETM+, and L8OLI/TIRS. For L8OLI/TIRS, only band 10 is used, since band 11 has large uncertainties, as reported by the USGS [57]. The consistency of Landsat 5, 7 and 8 satellite thermal instruments in recovering LST was compared by Sekertekin and Bonafoni [58] and validated with in situ LST measurements. The RMSE values were 2.39 °C, 2.57 °C and 2.73 °C, respectively, resulting in an average difference of 0.2 °C between the sensors. The uncertainty values are adequate uncertainty for the purpose of this study. In Figure 3, our model to retrieve LST is presented in a flow chart. Temperatures are derived using the radiative components implemented by Barsi et al. [59] for single-channel algorithms. This method simulates the attenuation effects of the atmosphere that disturb the TIR signal.

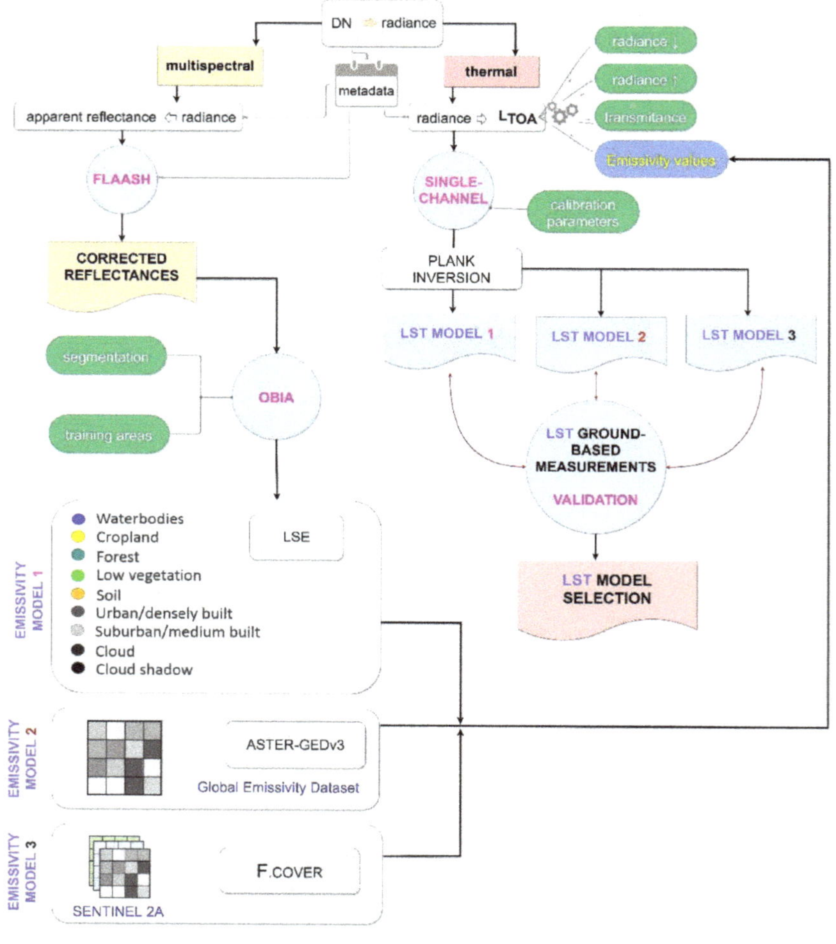

Figure 3. Flowchart for LST estimation and assessment of the emissivity models used in this study.

Radiance and transmissivity values are available at https://atmcorr.gsfc.nasa.gov/, accessed on 20 October 2021. The data is a compendium of atmospheric transmissivity values, along with upwelling and downwelling radiances for a given geographical location.

The radiative values can be used in atmospheric correction models, e.g., Equation (1), taking also into account the correction of spectral emissivity.

$$L_{TOA} = \tau \cdot \varepsilon \cdot L_T + L_{\lambda\uparrow} + \tau \cdot (1 - \varepsilon) \cdot L_{\lambda\downarrow} \quad (1)$$

In this equation, L_{TOA} is the spectral radiance at the top of the atmosphere (registered by the sensor), τ is the atmospheric transmittance, ε is the spectral emissivity, L_T is the spectral radiance of a black-body target of kinetic temperature T, $L_{\lambda\uparrow}$ and $L_{\lambda\downarrow}$ are the upwelling atmospheric path radiance and the downwelling or sky radiance, respectively.

Implementing Equation (1) requires the supply of adequate emissivity values for a suitable estimation of LST. Since different land covers emit thermal radiation differently, spectral emissivity corrections are necessary [60]. In this work, three emissivity models are tested to accurately estimate LST. First, the field-measured LSE (land surface emissivity) values are obtained from different authors, and are listed in Table 1. Then the emissivity data of the ASTER-GEDv3 product [61] were considered.

Then, the F_{cover} model of Valor and Caselles [40] is applied; this model allows the calculation of the emissivity in the Landsat 8 thermal band, considering the F_{cover} index and the minimum and maximum values of the emissivity in the corresponding spectral band. Finally, the three LST models are compared and validated. In this study, land use features are categorized into seven classes. These are water bodies, cropland, forest, low vegetation, bare soil, urban/densely built, and suburban/medium built. We applied this scheme following the land cover classes proposed by Park et al. [62]. Since impervious surfaces exhibit a large spectral variation [63], two classes are used to represent artificial surfaces: urban/dense and suburban/medium. These classes are particularly identified by the impact on emissivity.

For this purpose, an object-based classification is carried out using Trimble' s eCognition Developer software (https://geospatial.trimble.com/products-and-solutions/ecognition, accessed on 20 October 2021).

Table 1. Reference values for the LSE model with L8OLI/TIRS band 10.

Land Cover	Emissivity	Reference
Waterbodies	0.992	FROM-GLC cited by [64]
Cropland	0.971	FROM-GLC cited by [64]
Forest	0.995	FROM-GLC cited by [64]
Low vegetation	0.986	Tan et al. [65]
Soil	0.972	Tan et al. [65]
Urban/densely built	0.973	FROM-GLC cited by [64]
Suburban/medium built	0.971	Tan et al. [65]

The second emissivity dataset in this work is the ASTER Global Emissivity Dataset v3 (ASTER-GEDv3) [61], (https://emissivity.jpl.nasa.gov/aster-ged, accessed on 20 October 2021). This method was developed by the NASA Jet Propulsion Laboratory (JPL) as an algorithm based on temperature and emissivity separation along with an atmospheric correction model. More details can be found in Hulley and Hook [66].

The third emissivity model requires knowledge of the F_{cover} variable [40]. This method provides the emissivity of a heterogeneous surface as follows:

$$\varepsilon = \varepsilon_v \cdot F_{cover} + \varepsilon_g \cdot (1 - F_{cover}) + 4 \cdot \langle d\varepsilon \rangle \cdot F_{cover} \cdot (1 - F_{cover}) \quad (2)$$

In this equation, $\varepsilon_v = 0.985$ and $\varepsilon_g = 0.960$ are reference vegetation and bare soil emissivity, respectively. '$d\varepsilon$' is the cavity effect associated with the indirect radiance emitted due to internal reflections between the interfaces. Here, F_{cover} is obtained from S2-MSI Level 2A products (see Section 2.3.2). The procedure to retrieve the F_{cover} variable differs from the NDVI methods [67,68], and is presented as a novel alternative for thermal modeling with Landsat data. In tropical areas, throughout the year, vegetation dynamics does not exhibit abrupt changes, and this implies that F_{cover} lacks significant seasonal variations. For the L5TM and L7ETM+ thermal instruments, the emissivity model of Equation (3) [69] is used. This last method to obtain F_{cover} is based on the NDVI parameter.

$$\varepsilon = \varepsilon_{nonveg}\,(1 - F_{cover}) \cdot \varepsilon_{veg} \cdot F_{cover} \tag{3}$$

In this equation, ε_{nonveg} and ε_{veg} are the reference emissivity values for nonvegetated and vegetated areas, being 0.97 and 0.99, respectively [70]. In this work, the F_{cover} variable is recovered using the NDVI, as it effectively reflects the conditions of vegetation cover [42]. This is estimated by Equation (4).

$$F_{cover} = \left[\frac{NDVI - NDVI_s}{NDVI_v - NDVI_s}\right]^2 \tag{4}$$

In this equation, $NDVI_s$ is the NDVI value of pure soil, and $NDVI_v$ is the NDVI of pure vegetation obtained from the NDVI image.

This method is based on the Carlson and Ripley [69] model. Finally, the conversion from LTOA to LST is estimated by using the constants for sensor calibration and the inversion of the Planck equation [71].

2.3.4. Assessment of the Land Surface Temperature Retrieved from L8TIRS B10

Landsat-retrieved LST was verified with in situ measurements. Unfortunately, due to the Landsat overpasses schedule, starting in 2001, it was impossible to undertake a validation by means of field surveys for the entire Landsat time series. Due to these inconveniences, the comparison was limited only to two overpasses of L8OLI/TIRS band 10. The L5TM and L7ETM+ data, the Carlson and Ripley results [69] can be used as a reference. In situ temperatures were measured using 30 thermometers assembled into DS18B20 digital sensors. The direct calibration method was applied, which consists in recording the readings of the test and standard thermometers. The latter are preserved in an isothermal medium. This calibration procedure produced standard deviations of ±0.5 °C. The field survey consists of distributing 30 devices, as shown in Figure 4. LST measurement coincident with the two L8OLI/TIRS overpasses were recorded on 22 January and 9 September 2019. Each device recorded the temperature values by means of a probe in contact with the ground surface. The ground sensors were placed in areas with homogeneous land cover to minimize the spatial thermal variation caused by different emissivity values. These records will be used to contrast the LST values derived from satellite measurements. In Section 3.1, we perform a sensitivity analysis for the three emissivity models.

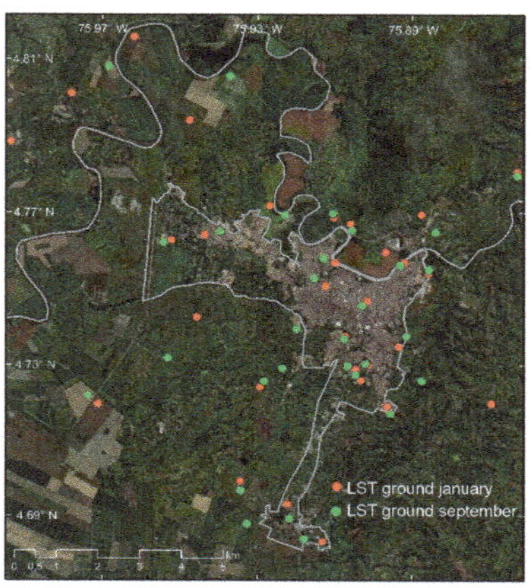

Figure 4. Location of in situ ground LST measurements.

2.3.5. Modelling the SUHI Phenomenon

According to Rasul et al. [72], SUHI modeling consists of identifying the spatial variation in time of thermal features in urban areas. Here, through the combination of thermal images from remote sensing and sparse measurements on field, our SUHI model employs the PCA to analyze space-time data. The PCA is a multivariate statistical technique that preserves the total variance of a dataset while reducing its dimensionality [73]. In this way, the PCA can retrieve the main spatial patterns of variability in a time-series. The application of PCA provides a generalization of the changes that characterize the variability patterns in a time series of images [18].

Then, the impacts of the eight factors considered in this study are assessed using a MLR approach. The MLR technique is a parametric model that adjusts the relationship between explanatory variables, that is, the contributing factors, and the response variable, e.g., LST. The inclusion or elimination of predictors depends on the significance of these variables within the model, which is defined by a test hypothesis based on the coefficients associated with the response variable. When using MLR techniques, it is important to examine the key assumptions of autocorrelation, normality of residuals, and multicollinearity. These factors determine the reliability of the model [74]:

- Autocorrelation of a variable represents its self-dependence and implies redundant information that makes the estimator lose efficiency. The Durbin-Watson statistic is used to measure autocorrelation [75].
- The normality of a residuals guarantees a satisfactory representation of the model.
- Multicollinearity occurs when the predictor variables are highly correlated. Multicollinearity increases the variance, causing instability of the regression and thus increasing the standard error [76]. Multicollinearity is measured with the Variance Inflation Factor (VIF).

Finally, outliers can also alter the modelling approach, causing problems with regression assumptions [77,78], and these must be controlled or removed from the dataset. Here, our MLR analysis is an equation capable of describing the thermal intensity depending on the contributing factors. To verify the relative importance of each individual predictor of the LST model, a normalization procedure was previously performed to standardize the coefficients. We use the deviation of the mean values, which is divided by the standard deviation of the response variable in LST. This allows us to derive the standardized coefficients [79]. Subsequently, the contribution of each variable to LST is obtained by weighting the absolute value of each variable. The resulting weights are further used for assessing the subsequent Machine Learning procedure that derives the multitemporal intensity of the SUHI model. This provides a technical basis for analyzing the factors that influence the thermal environment, which is of great significance for rational urban planning and sustainability.

The methodological workflow in Figure 5 shows the spatiotemporal model followed to characterize the impact of environmental factors on the thermal changes. First, the multitemporal factors, such as LST, spectral indices, and other variables, are derived from the Landsat 2001–2020 dataset. Then, the PCA technique is applied to extract the main patterns of variability. Subsequently, all the variables involved are included in the MLR scheme to model the possible dependences on LST. The MLR is implemented with the software R Studio (https://rstudio.com/, accessed on 20 October 2021).

Finally, the SUHI phenomenon is segmented into different zones depending on the thermal intensity. Thermal value ranges follow the categories of Wang et al. [36], which consider the average temperature of the land surface and its standard deviation (SD). Segmentation provides a definitive SUHI product that categorizes the urban environment according to specific conditions. Here we test two different Machine Learning methods for classification; Support Vector Machine (SVM) and Naïve Bayes Machine Learning (NBML). Both SVM and NBML methods have shown in previous research their robustness for the characterization of various types of geospatial data [80,81]. The SVM method defines a separate hyperplane in a higher-dimensional space that optimally classifies the data. This method is particularly useful for solving nonlinear relations [82], and is available as open-source software in Orfeo ToolBox (OTB) at https://www.orfeo-toolbox.org/, accessed on 20 October 2021. The NBML technique is based on the Bayes theorem for conditional probability and assumes independence between predictors, variables, or features [83].

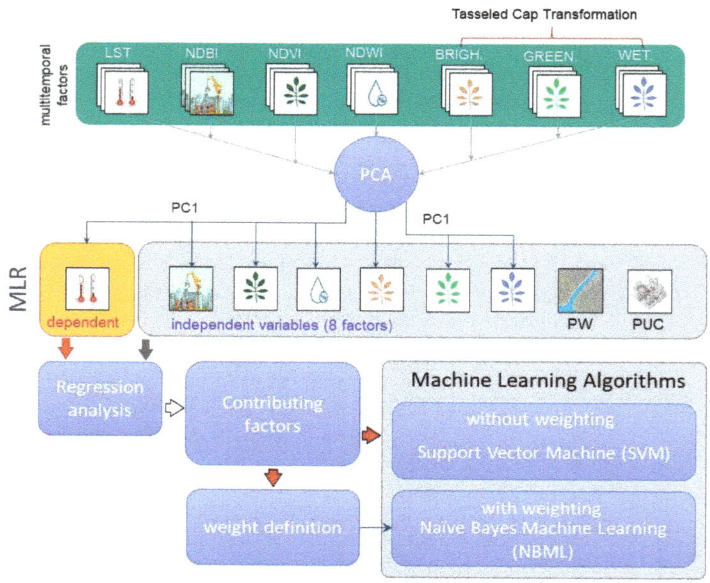

Figure 5. Flowchart of the proposed SUHI model.

NBML is often referred to as the maximum a posteriori decision rule [84], and its code can be easily written in any programming language. NBML assigns the most likely class to a certain observation by estimating the probability density of the training classes [85]. An observation is classified as a certain class when the posterior probability reaches the maximum value according to the following expression:

$$k(x) = argmax_k \mathrm{p}(\mathrm{C_k}) \prod_{\mathrm{i}=1}^{\mathrm{n}} \mathrm{p}(\mathrm{x_i} | \mathrm{C_k}) \quad (5)$$

In this equation, $k(x)$ is the maximum a posteriori of $\mathrm{x_i}$ for the class labeled as $\mathrm{C_k}$, $\mathrm{p}(\mathrm{C_k})$ is the prior probability for class $\mathrm{C_k}$, $\mathrm{p}(\mathrm{x_i}|\mathrm{C_k})$ represents the conditional probability distribution of $\mathrm{x_i}$ given $\mathrm{C_k}$, and $(\mathrm{w_i})$ is a particular weight applied to each factor. Usually, the independence assumption is not fulfilled, and the weighting of the features involved in the assignment process can satisfy the required assumptions [86]. Here, each feature or factor is affected by a particular weight $(\mathrm{w_i})$, which can be formally defined by:

$$k(x) = argmax_k \mathrm{p}(\mathrm{C_k}) \prod_{\mathrm{i}=1}^{\mathrm{n}} \mathrm{p}(\mathrm{x_i} | \mathrm{C_k})^{\mathrm{w_i}} \quad (6)$$

In this equation, $\mathrm{w_i}$ denotes the weight value of the ith attribute, with values restricted to the range [0, 1]. In this work, attributes are the contributing factors involved in the SUHI phenomenon, while the $\mathrm{C_k}$ classes are the seven temperature categories defined by Wang et al. [36]. These are described in Table 2.

Table 2. Range of LST intervals. T_s represents land surface temperature; T_a is the average land surface temperature. SD is the standard deviation.

Temperature Grade	Range
Extreme high temperature (EHT)	$T_S > T_a + 2SD$
High temperature (HT)	$T_a + SDT_S \leq T_a + 2SD$
Sub-high temperature (SHT)	$T_a + SD/2T_S \leq T_a + SD$
Medium temperature (MT)	$T_a - SD/2T_S \leq T_a + SD/2$
Sub-medium temperature (SMT)	$T_a - SDT_S \leq T_a - SD/2$
Low temperature (LT)	$T_a - 2SDT_S \leq T_a - SD$
Sub-low temperature (SLT)	$T_S < T_a + 2SD$

The prior $p(C_k)$ and conditional probabilities $p(x_i|C_k)$ are determined through a training process. Then, Equation (6) becomes:

$$k(x) = argmax_k \hat{p}(C_k) \prod_{i=1}^{n} \hat{p}(x_i|C_k)^{w_i} \quad (7)$$

In this equation, $\hat{p}(C_k)$ and $\hat{p}(x_i|C_k)$ are estimates of the probabilities density functions (PDFs). These are derived from the frequency of their respective arguments in the training sample. Here, $\hat{p}(C_k)$ can also be estimated from a preliminary outcome of a SVM process.

Equation (7) allows us to weight each environmental factor to generate the final SUHI product. The resulting map is generated according to the architecture shown in Figure 6, which is based on the NB decision rule. This approach categorizes the urban environment according to a specific condition and assigns a specific type of action based on each temperature category. This analytical procedure allows one to obtain a map that delimits the areas of different thermal intensities. The resulting areas are based on the spatiotemporal trends of the contributing factors, facilitating the management and application of measures to mitigate/adapt the SUHI phenomenon.

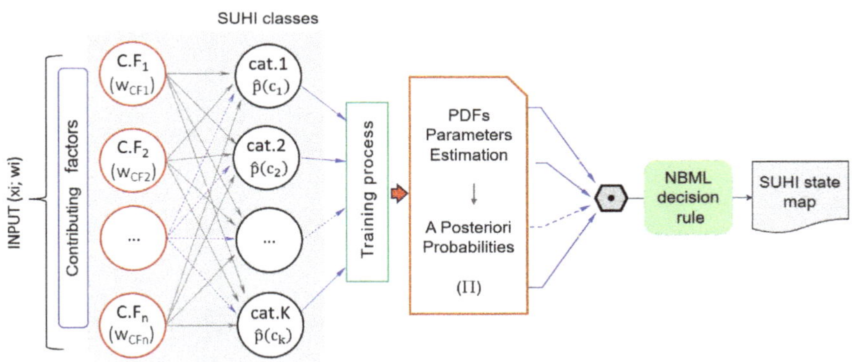

Figure 6. Architecture of the NBML modelling for generating the SUHI product.

3. Results

3.1. Land Surface Temperature

Sensitivity analysis of the three emissivity models is performed prior to retrieving land surface temperatures from the 37 Landsat images; the results of the assessment for the LST retrieved from L8/TIRS B10 (described in Section 2.3.4) are presented here. Figure 7 shows the differences between LST derived from the three models evaluated in this study (Fcover, AS-TER-GEDv3, and LSE), and these are compared with in situ LST. In this figure, the minimum, maxima, median and mean values are shown for (a) January 2019 and (b) September 2019. In both cases, the lowest differences agree with the LST values from our F_{cover} emissivity model.

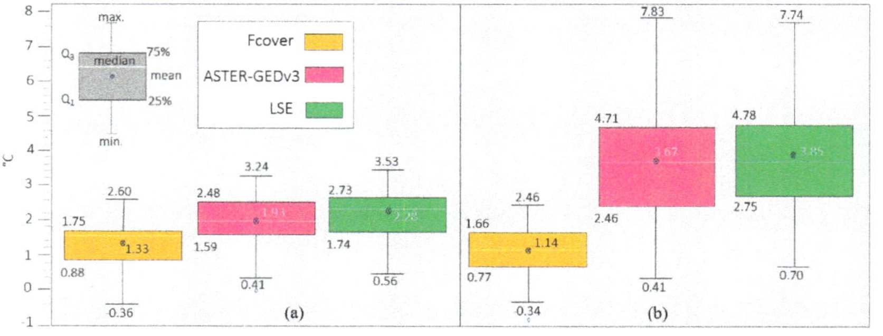

Figure 7. Boxplots of LST results from Fcover, ASTER-GEDv3, and LSE. The white horizontal line in each box is the median. (**a**) January 2019; (**b**) September 2019.

The interquartile ranges show a narrower dispersion for the F_{cover} model compared to the ASTER-GEDv3 and the LSE model. This feature is obvious for the campaign in September 2019. The regression analysis between the LST from ground-based sensors and that of the L8OLI/TIRS band 10 is shown in Figure 8. The dark gray areas represent the confidence boundaries of 95%, while the solid lines represent the line of best fit between the computed and in situ LST. The best determination coefficient is given by F_{cover} with R^2 = 0.78 and SD = 0.73 °C (Figure 8a). For the other two cases, the coefficients are R^2 = 0.27 and R^2 = 0.26 for the ASTER-GEDv3 and LSE models, respectively.

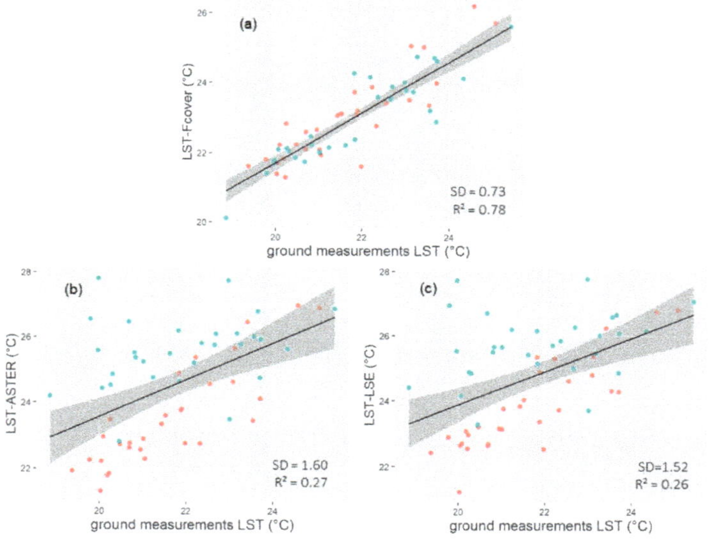

Figure 8. Regression analysis between LST from L8OLI/TIRS and that from ground measurements. Color code: Orange, January; Green, September (**a**) F_{cover}; (**b**) ASTER-GEDv3; (**c**) LSE model.

3.2. Principal Component Analysis

The PCA was carried out using all contributing factors during the period 2001–2020 (37 images for each variable). Figure 9 shows the contribution to the total variance of each PCA component. We can observe that the first PCA component (PCA1) of T-cap Brightness and T-cap Wetness provide a lower contribution to the total variance, with 54% and 64%, respectively. On the other hand, the rest of the variables show larger patterns of variability with only the first PCA component (above 75%).

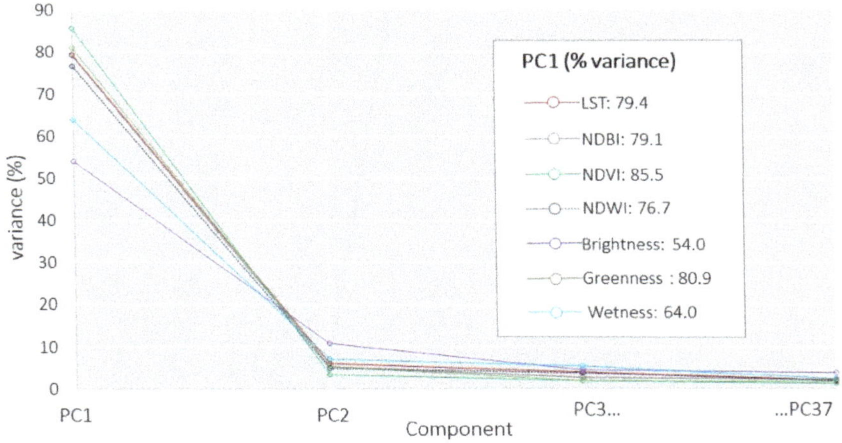

Figure 9. Contribution of the main PCA components to the total variance for the different variables used in this study.

Regarding the explained variance (%) of the second principal component or the T-cap Brightness, it is observed that it still retains a large amount of variance (12%), when compared to other factors. Since the goal of the PCA is to reduce the set of variables, in the case of the T-cap Brightness, the former dataset cannot be strictly explained by the first principal component, as it is the case of the remaining factors. This implies that further analysis is addressed towards investigating the second or even third components of the T-cap Brightness.

We employ the Jenks Natural Breaks grouping model [87] to identify the main groups and the inherent patterns that minimize the deviation of each class with respect to the mean value of the other groups. This method reduces the variance within the classes and maximizes the variance between classes. In this scheme, we obtain four groups for each factor, representing the spatiotemporal trends between 2001–2020. Figure 10 shows the resulting maps where the results for LST (Figure 10a) show the maximum concentration of temperature in densely populated areas, similar to the results of NDBI (Figure 10b). The LST results show gradual variations from low to high temperatures near the perimeter of urban areas. Regions with lower temperatures are mainly located in areas close to water bodies and dense vegetation. Vegetation areas can be identified in the NDVI results (Figure 10c). The NDVI and T-Cap Greenness maps Figure 10c,f have high similarity, while the T-Cap Brightness and Wetness maps Figure 10e,g lack spatial correlation with the thermal phenomenon. In the next section, we analyze the spatial correlations in more detail.

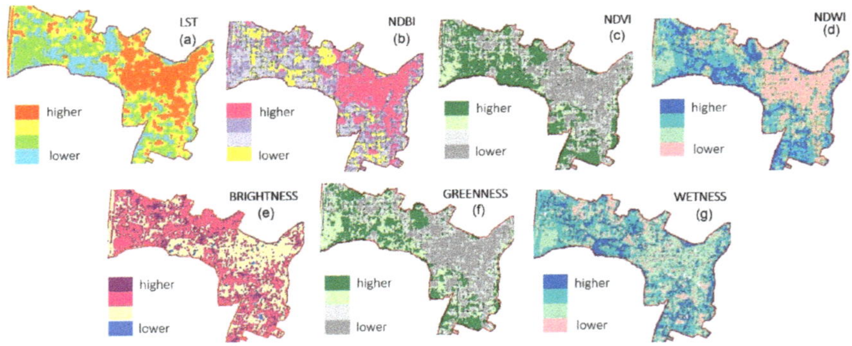

Figure 10. First PCA components for the main factors during the period 2001–2020: (**a**) land surface temperature; (**b**) normalized difference built-up index; (**c**) normalized difference vegetation index; (**d**) normalized difference water index; (**e**) T-cap brightness; (**f**) T-cap greenness; (**g**) T-cap wetness.

3.3. Multiple Linear Regression

A large number of outliers were identified and removed in the brightness and humidity factors to avoid introducing "noise" into the MLR analysis. All residuals greater than 3σ standard deviation from the mean value are considered outliers and thus removed. Since NDVI and Greenness factors are highly redundant ($R^2 = 0.99$), the latter was excluded. Concerning the Brightness factor, under the special circumstances observed in Section 3.2, the two first principal components only explain 66% of the total variance. Moreover, the low spatial correlation with LST (Figure 10) suggests excluding this variable.

Then, in the Fisher hypothesis test for the PW factor is larger than 0.1, and it was removed. Finally, the scrutiny explanatory variables are NDBI, NDVI, NDWI, and PUC, and the resulting MLR model outcomes as follows:

$$LST_{trend} = 0.29 + 0.48\ NDBI_{trend} + 0.21\ NDVI_{trend} - 0.61\ NDWI_{trend} - 0.51\ PUC \quad (8)$$

The regression analysis coefficients are shown in Table 3. In this table, p (>∣t = 0.05∣) represents the probability of observing any value larger than t. In our model, all p-values are below the significance level (0.05). This implies that NDBI, NDVI, NDWI and PUC are statistically significant predictors. The model has a high coefficient of determination ($R^2 = 0.82$), this means that these variables explain 82% of the variability observed in the LST.

Table 3. Multiple Linear Regression coefficients.

Factors	Estimate	SD	t Value	p (>t ∣0.05∣)
(Intercept)	0.29	0.01	34.79	<0.001
NDBI	0.48	0.05	9.91	<0.001
NDVI	0.21	0.02	13.21	<0.001
NDWI	−0.61	0.03	−23.65	<0.001
PUC	−0.51	0.01	−39.60	<0.001

The p-values of the regression analysis are shown in Table 3. All p-values are smaller than 0.05, indicating that the relationships between independent and dependent variables are statistically significant. Finally, to support the validity of the model, the following key assumptions were verified: autocorrelation, normality, and multicollinearity. The resulting values are given in Table 4.

Table 4. Fulfillment of the Assumptions.

Autocorrelation		Normality		Multicollinearity (VIF)			
D-W	*p*-Value	K-S	*p*-Value	NDBI	NDVI	NDWI	PUC
2.00	0.80	0.02	<0.001	45.03	9.12	45.75	1.26

In this table, we can appreciate that the NDBI and NDWI VIF values are greater than 10, thus exceeding the tolerance. This implies that these two variables should be disregarded. As stated by Szymanowski and Kryza [88], the variables that exceed this tolerance may be considered to improve a regression model. Moreover, these two predictors are very important variables in many UHI studies [89–91]. In the UHI study by Cruz et al. [92], after performing a multicollinearity test, explanatory variables with VIFs between 50 and 70 were selected for their multiregression analysis. These were considered an important component for modeling this phenomenon. These are the reasons for maintaining the NDBI and NDWI as explanatory variables in this study.

The independence between residuals was verified using the Durbin–Watson statistic (D-W) with a value of 2.0, which falls within the critical values of 1.5 < D-W < 2.5, indicating the absence of autocorrelation. The normality of the residuals was proved by applying a Kolmogorov–Smirnov (K-S) test, which confirms the normal distribution. Figure 11 shows the scatter diagrams, the histograms, and the correlation values for each pair of explanatory variables in the model. To verify our model assumptions, four scatterplots of residuals against fitted values are investigated. Figure 12 suggests that the data are randomly distributed around zero, with constant variability. There are no patterns that indicate that the assumptions of the model are fulfilled for the dataset.

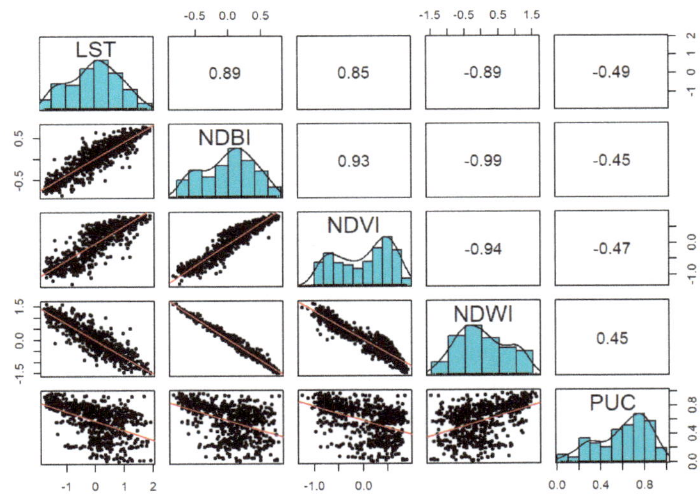

Figure 11. Histograms of the model variables (i.e., LST, NDBI, NDVI, NDWI, and PUC) in the main diagonal. Scattergrams between the model variables (below main diagonal), and the corresponding Pearson correlation (above main diagonal).

In this figure, good correlation of NDBI and NDWI with LST is observed.

The contribution of each variable (w_i) was obtained through the standardized regression coefficients ($\hat{w}_i$), which are weighted means absolute value. The resulting standardized regression coefficients and the contribution of the factors to the model are presented in Table 5. In this table, we can observe that the main contributing factors are the variables NDWI and NDBI, followed by NDVI and PUC. The derived weights are used in the next section to derive the SUHI model.

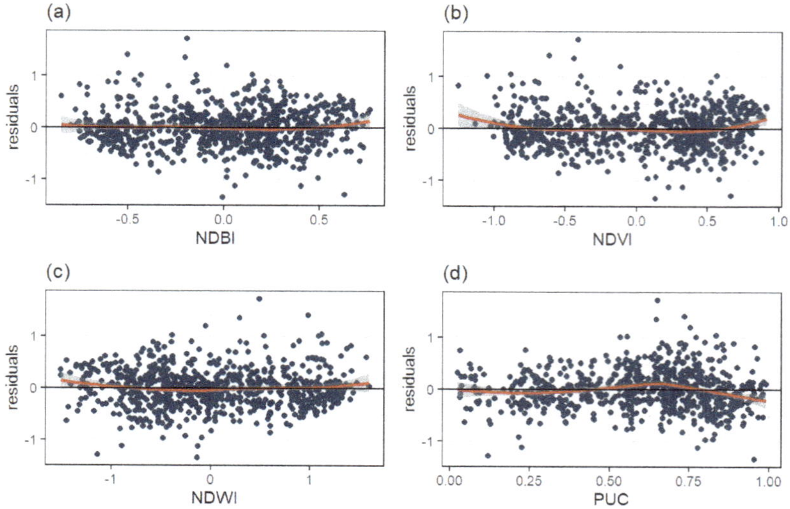

Figure 12. Performance of the linearity of the model. Graphs of the regression analysis residuals vs. fitted lines. (**a**) NDBI vs. residuals. (**b**) NDVI vs. residuals. (**c**) NDWI vs. residuals. (**d**) PUC vs. residuals.

Table 5. Standardized Regression Coefficients and Contribution of each Factor.

Factor	Standardized Coefficients $\hat{w}_i$	Weighted Contribution w_i (%)
NDBI	0.21	21.38
NDVI	0.13	12.84
NDWI	−0.51	51.46
PUC	−0.14	14.32

3.4. *SUHI Modeling*

The SUHI phenomenon depends on the properties of land cover properties which, combined with their energy absorption capacity, produce a thermal increase on the surface and represent a threat to the thermal regime of urban ecosystems. Our modeling approach is based on segmentation through the identification of potential thermal areas. Here we employ the seven temperature zones defined by Wang et al. [36]. The definition of the training areas is achieved with the LST variable (Figure 10a). The thermal ranges for the training process are those defined in Table 2. These ranges are based on LST averages and the standard deviations. We test both the SVM and NBML algorithms. The application of the NBML method requires estimating the conditional probability functions for each contributing factor. The Gaussian and Logistic probabilities density functions showed the best results for the respective training frequencies of observation/category. Each conditional probability was weighted according to Table 5. Moreover, the weighting capability of the NBML method allows taking into account the relevance of each factor for deriving the SUHI product. This feature is not possible with the SVM method.

The segmentation results for both methods are shown in Figure 13. In both cases, the results were validated with the criteria established in Table 2. Table 6 reports the Kappa index for SVM and NBML are approximately 88% and 94%, and the overall accuracy are 88% and 95%, respectively.

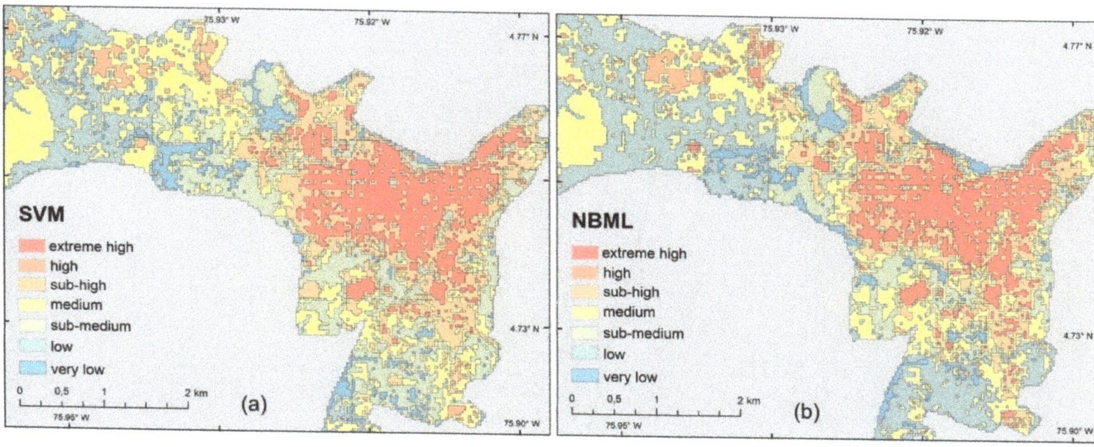

Figure 13. Temperature classification results. (**a**) SVM; (**b**) naïve Bayes.

Table 6. Kappa Index and Precision of SVM and NBML.

Algorithm	Kappa Index	Overall Accuracy
SVM	0.88	0.88
NBML	0.94	0.95

Figures 13 and 14 show the final SUHI map that categorizes the urban environment according to a specific SUHI state and assigns a specific type of action based on temperature. The proposed actions are: *intervene*, *monitor*, *strengthen*, and *preserve*. The *intervene* action is directly related to the SUHI areas exposed to the maximum thermal concentration. These areas need to be immediately intervened in and are considered an 'Extreme-high' class. The *monitor* action groups 'High' and 'Sub-high' categories, and points to the SUHI areas that should be kept under observation and intervened in a medium term. The *strengthen* action classifies the 'Medium' and 'Sub-medium' classes into SUHI areas that have gradually presented a temporary thermal trend increase. The *preserve* action contains the 'Low' and 'Very-low' classes and comprises the SUHI areas that must be preserved.

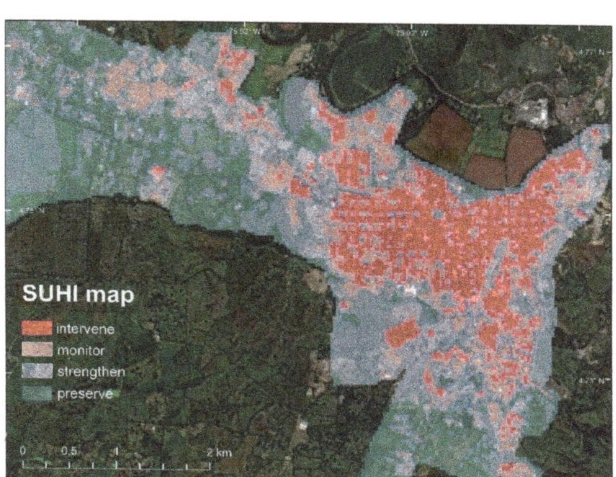

Figure 14. Final SHUI product from NBML. The legend recommendations are specific types of action based on temperature warnings.

4. Discussion

4.1. Sensitivity Analysis

The results of the January 2019 campaign (Figure 7a) suggest that the amplitude of the errors in recovering LST is similar between the models evaluated. The September 2019 measurement records (Figure 7b) indicated that the F_{cover} model provides the smallest deviation with a mean error of 1.14 °C. This is very obvious compared to ASTER-GEDv3 and the LSE model, which shows mean deviations of 3.67 °C and 3.85 °C, respectively. As shown in the Results section, the F_{cover} model exhibited better performance with a mean error of 1.33 °C. Data reported by Duan et al. [93], and Malakar et al. [94] showed differences for L5TM, L7ETM+, and L8OLI/TIRS among recovered LST and in situ LST between 0.7 and 1.2 °C. Furthermore, we observed mean differences between 1.1 °C and 1.3 °C. Authors such as Chen and Zhang [14] and Liu and Li [95] have analyzed the SUHI phenomena with similar differences. In this work, the F_{cover} model provided the smallest errors in LST recovery among all the tested schemes, and it is considered the most suitable for this kind of studies.

4.2. Statistical Analyses

The PCA was applied to derive the time trend of each variable and to analyze the LST variation. Then, the main PCA component was employed in the MLR. We achieve a coefficient of determination of approximately R^2 = 0.82. These results are in agreement with recent studies that have used regression models to quantify the impact of contributing factors on LST [16,96]. Moreover, the combination of these factors defines how the different types of land cover absorb temperatures. These absorptions manifest themselves with the corresponding increase in emissivity and surface temperature. Our findings confirm results of earlier studies, such as those of Rasul et al. [72], who modeled with the MLR method the spatiotemporal trend of temperature data, and provided robust results in determining SUHI areas.

Regarding the conditions for ensuring the validity of our proposed approach, several considerations must be addressed. First, the multicollinearity of the predictor variables and their effect on the model need to be validated with the VIF. Our results show that two of the VIF parameters exceeded the value of 10, which would exclude two of the explanatory variables of the prediction model. However, it is found that NDBI and NDWI are the factors with the highest contribution, while NDVI and PUC have lower VIF values, contributing to a lesser extent. Strong correlations, 0.89 and −0.89, were found between the LST and NDBI and NDWI, respectively. A similar correlation was found between NDVI and NDWI. It is important to note that removing highly correlated variables can benefit the overall result and simplify the approach. However, having high contributing predictor variables such as NDBI and NDWI may indeed improve prediction products, as noted in [88]. Although some collinearity was presented, the Pearson correlation indices and the fulfillment of independence and normality of the residuals have denoted a very reliable model.

Regarding the direct relationships of LST with the different physical variables and contributing factors, a strong correlation with NDBI was observed: building construction. This justifies why impervious areas have high caloric retention capacity and low water storage capacity, in turn reducing humidity. Previous studies have demonstrated strong correlations between LST and NDBI [97,98]. In contrast, a strong correlation was found between LST and NDVI/NDWI. Please note that temperature decrease follows increases in vegetation and humidity. The higher the vegetation cover, the lower the surface temperature becomes. The reason for this may be strongly related to the soil moisture content in vegetated areas, which alters the energy balance and causes variation effects from solar radiation. These results are in line with those obtained by Ibrahim and Rasul Faqe [99], who reported a strong negative correlation between these variables. Our results show urban planners that the identification of factors and their contribution to the SUHI phenomenon serves as a support to define adaptation measures to cities for thermal change, allowing them to adapt with other territorial planning priorities.

A moderate correlation was found between LST and PUC, indicating that LST increases moderately according to the proximity to the urban center. This correlation may be related to urban density distribution and road infrastructure, which, compared to the distance variable, are responsible for generating a complex structure not well represented by linear models. Bonafoni and Keeratikasikorn [100] also implemented a ring-based method and analyzed LST as a function of building density and proximity to urban centers. This issue is to be addressed in future research.

4.3. The SUHI Model

To reveal the multitemporal intensity of the SUHI phenomenon, two Machine Learning techniques were tested, the SVM and the NBML. Both algorithms performed satisfactorily, with Kappa indices of 89% and 93%, respectively. Better performance was observed for certain categories for the NBML algorithm (Figure 13b). Although both procedures are able to detect high-density urban areas affected by extremely high temperatures, NBML allows coupling criteria to assign individual weights to each class, increasing the quality of the results. Conventional NBML classifiers consider the model to be applicable when the Gaussian probability density function is present in the data set [84]. Molina et al. [101] showed that the combination of the best-fit distribution model (not necessarily Gaussian), and the weights of each variable led to satisfactory results.

The SUHI phenomenon is a complex system that occurs as a result of the interaction of various factors [102]. This interactivity produced by anthropic effects generates thermal imbalances requiring intervention, monitoring, strengthening, and preservation, as a fundamental expression between causes and effects of urban/rural ecosystems. Our results show that the highest temperatures are concentrated in the central area of the city and gradually decrease toward the periphery. The characterization of the space through the four proposed classes of actions (intervene, monitor, strengthen, and preserve) makes it possible to regulate the conditions that could mitigate the SUHI effect. The areas designated as *intervene* correspond to the center of the city and tend to have a higher population density and old buildings. It is recommended to change black roofs to less thermic roofs that have reduced solar energy absorption and increased energy savings, as suggested by Alshayeb and Chang [103]. Since these areas do not have appropriate physical space to create green areas, an alternative might be the use of road dividers to plant trees with large foliage and roots that do not weaken existing infrastructure. An interesting measure that allows the reduction of anthropogenic heat is to restrict the transit of private vehicles and limit access to specific areas. The access methods can be substituted for public transportation or cycling. These measures have already been implemented in many locations. The areas identified as *monitor* should implement small tree-lined sites and natural corridors to refresh the space. Rainwater irrigation channeled through sewerage systems can be used as a contribution to the restoration of urban wetlands. The areas indicated as *strengthen* show less thermal intensity than the above areas and are associated with urban growth. Within this policy, the morphology of these areas should integrate green spaces that allow increased water infiltration and cooling [104]. In general, the use of highly reflective building materials is recommended, reducing the amount of solar radiation absorbed by the surface, such as, for example, the use of cool pavements suggested by the U.S. Environmental Protection Agency [105]. Finally, areas marked as *preserve* have the highest vegetation cover and play an important role in the urban ecosystem. They reduce carbon dioxide emissions, becoming spaces that reduce the radiant load produced by various economic activities, and generate thermal regulation. In addition, these have a great potential for ecotourism.

Further development of this research can be undertaken by applying simulation techniques with Machine Learning Algorithms that allow the integration of weights to the variables involved in the predictive model, and that allow characterization of future thermal scenarios associated with the spatiotemporal trends of the explanatory variables.

The interactions produced by biophysical factors and the geometric changes that are transforming cities make the relationships between objects and phenomena increasingly complex. In this sense, it would be very pertinent to further explore the classification of local climatic zones in tropical cities in countries such as Colombia. These are highly vulnerable to climate change.

5. Conclusions

The results of this work demonstrate that emissivity data have a large impact on the retrieval of LST. Here, LST is obtained from L8OLI/TIRS band 10 and LSE from Sentinel-2. Both sources are more accurate and homogeneous than using traditional ground-based methods. Our innovative approach proposes quantifying the SUHI phenomenon from a set of contributing factors. We first employ the PCA to retrieve the main spatiotemporal variations in the initial data. Then, MLR is applied to integrate the dependencies and to analyze their impacts on SUHI. According to our regression model, the most influential factors in the SUHI are NDWI with a contribution of 52%, NDBI with 21%, NDVI with 13%, and PUC with a 14%. Finally, the integration of these predictors within an SVM and a NBML approaches confirms the existence of coupling mechanisms between each variable. The satisfactory results of the NBML confirm the suitability of the proposed approach, with an overall accuracy of 95%. We expect to improve the results of the model with future upgrades associated with structural complexity of the landscapes. The spatial variation of SUHI points out an enhanced phenomenon towards areas of high urban density. Our research demonstrates the suitability of Machine Learning Algorithms for mapping SUHI intensities, providing spatially explicit descriptions of urban heat distribution. The derived products are crucial for defining sustainable urban planning policies, as well as for adequate responses to thermal risks. These actions will in turn make it possible to define mitigation and adaptation strategies.

Author Contributions: J.G. and I.M. provided the main ideas, developed the methodology model, conceived and performed the comparison experiments, and analyzed the results. Both of them contributed equally. A.C. performed manuscript editing and revision tasks. J.V. supervised the whole process and reviewed the manuscript. All authors have read and agreed to the published version of the manuscript.

Funding: This research received no external funding.

Institutional Review Board Statement: Not applicable.

Informed Consent Statement: Not applicable.

Data Availability Statement: The data presented in this study are available on request from the corresponding author.

Conflicts of Interest: The authors declare no conflict of interest.

References

1. Oke, T.R. The energetic basis of the urban heat island. *Q. J. R. Meteorol. Soc.* **1982**, *108*, 1–24. [CrossRef]
2. Voogt, J.A.; Oke, T.R. Thermal remote sensing of urban climates. *Remote Sens. Environ.* **2003**, *86*, 370–384. [CrossRef]
3. Schwarz, N.; Schlink, U.; Franck, U.; Großmann, K. Relationship of land surface and air temperatures and its implications for quantifying urban heat island indicators—An application for the city of Leipzig (Germany). *Ecol. Indic.* **2012**, *18*, 693–704. [CrossRef]
4. Gubler, M.; Christen, A.; Remund, J.; Brönnimann, S. Evaluation and application of a low-cost measurement network to study intra-urban temperature differences during summer 2018 in Bern, Switzerland. *Urban Clim.* **2021**, *37*, 100817. [CrossRef]
5. Chakraborty, T.C.; Lee, X.; Ermida, S.; Zhan, W. On the land emissivity assumption and Landsat-derived surface urban heat islands: A global analysis. *Remote Sens. Environ.* **2021**, *265*, 112682. [CrossRef]
6. Sobrino, J.A.; Oltra-Carrió, R.; Sòria, G.; Jiménez-Muñoz, J.C.; Franch, B.; Hidalgo, V.; Mattar, C.; Julien, Y.; Cuenca, J.; Romaguera, M.; et al. Evaluation of the surface urban heat island effect in the city of Madrid by thermal remote sensing. *Int. J. Remote Sens.* **2013**, *34*, 3177–3192. [CrossRef]
7. WMO Essential Climate Variables. Available online: https://public.wmo.int/en/programmes/global-climate-observing-system/essential-climate-variables (accessed on 18 October 2019).

8. Zhou, D.; Xiao, J.; Bonafoni, S.; Berger, C.; Deilami, K.; Zhou, Y.; Frolking, S.; Yao, R.; Qiao, Z.; Sobrino, J.A. Satellite remote sensing of surface urban heat islands: Progress, challenges, and perspectives. *Remote Sens.* **2019**, *11*, 48. [CrossRef]
9. Sekertekin, A.; Zadbagher, E. Simulation of future land surface temperature distribution and evaluating surface urban heat island based on impervious surface area. *Ecol. Indic.* **2021**, *122*, 107230. [CrossRef]
10. Li, S.; Zhao, Z.; Miaomiao, X.; Wang, Y. Investigating spatial non-stationary and scale-dependent relationships between urban surface temperature and environmental factors using geographically weighted regression. *Environ. Model. Softw.* **2010**, *25*, 1789–1800. [CrossRef]
11. Mei, C.-L.; Wang, N.; Zhang, W.X. Testing the importance of the explanatory variables in a mixed geographically weighted regression model. *Environ. Plan. A* **2006**, *38*, 587–598. [CrossRef]
12. Senanayake, I.P.; Welivitiya, W.D.D.P.; Nadeeka, P.M. Remote sensing based analysis of urban heat islands with vegetation cover in Colombo city, Sri Lanka using Landsat-7 ETM + data. *Urban Clim.* **2013**, *5*, 19–35. [CrossRef]
13. Bokaie, M.; Zarkesh, M.K.; Arasteh, P.D.; Hosseini, A. Assessment of Urban Heat Island based on the relationship between land surface temperature and Land Use/Land Cover in Tehran. *Sustain. Cities Soc.* **2016**, *23*, 94–104. [CrossRef]
14. Chen, X.; Zhang, Y. Impacts of urban surface characteristics on spatiotemporal pattern of land surface temperature in Kunming of China. *Sustain. Cities Soc.* **2017**, *32*, 87–99. [CrossRef]
15. Hereher, M.E. Effect of land use/cover change on land surface temperatures—The Nile Delta, Egypt. *J. Afr. Earth Sci.* **2017**, *126*, 75–83. [CrossRef]
16. Shi, Y.; Xiang, Y.; Zhang, Y. Urban design factors influencing surface urban heat island in the high-density city of guangzhou based on the local climate zone. *Sensors* **2019**, *19*, 3459. [CrossRef] [PubMed]
17. Song, J.; Wang, J.; Xia, X.; Lin, R.; Wang, Y.; Zhou, M.; Fu, D. Characterization of urban heat islands using city lights: Insights from modis and viirs dnb observations. *Remote Sens.* **2021**, *13*, 3180. [CrossRef]
18. Parmentier, B. Characterization of land transitions patterns from multivariate time series using seasonal trend analysis and principal component analysis. *Remote Sens.* **2014**, *6*, 12639–12665. [CrossRef]
19. Firozjaei, M.K.; Alavipanah, S.K.; Liu, H.; Sedighi, A.; Mijani, N.; Kiavarz, M.; Weng, Q. A PCA-OLS model for assessing the impact of surface biophysical parameters on land surface temperature variations. *Remote Sens.* **2019**, *11*, 2094. [CrossRef]
20. Lemus-Canovas, M.; Martin-Vide, J.; Moreno-Garcia, M.C.; Lopez-Bustins, J.A. Estimating Barcelona's metropolitan daytime hot and cold poles using Landsat-8 Land Surface Temperature. *Sci. Total Environ.* **2020**, *699*, 134307. [CrossRef]
21. Shams, S.R.; Jahani, A.; Kalantary, S.; Moeinaddini, M.; Khorasani, N. Artificial intelligence accuracy assessment in NO2 concentration forecasting of metropolises air. *Sci. Rep.* **2021**, *11*, 1805. [CrossRef]
22. Wicki, A.; Parlow, E. Multiple regression analysis for unmixing of surface temperature data in an urban environment. *Remote Sens.* **2017**, *9*, 684. [CrossRef]
23. Zheng, Y.; Li, Y.; Hou, H.; Murayama, Y.; Wang, R.; Hu, T. Quantifying the cooling effect and scale of large inner-city lakes based on landscape patterns: A case study of hangzhou and nanjing. *Remote Sens.* **2021**, *13*, 1526. [CrossRef]
24. Deng, Y.; Chen, R.; Xie, Y.; Xu, J.; Yang, J.; Liao, W. Exploring the impacts and temporal variations of different building roof types on surface urban heat island. *Remote Sens.* **2021**, *13*, 2840. [CrossRef]
25. Oliveira, A.; Lopes, A.; Niza, S.; Soares, A. An urban energy balance-guided machine learning approach for synthetic nocturnal surface Urban Heat Island prediction: A heatwave event in Naples. *Sci. Total Environ.* **2022**, *805*, 150130. [CrossRef] [PubMed]
26. Hassan, T.; Zhang, J.; Prodhan, F.A.; Pangali Sharma, T.P.; Bashir, B. Surface urban heat islands dynamics in response to lulc and vegetation across south asia (2000–2019). *Remote Sens.* **2021**, *13*, 3177. [CrossRef]
27. Núñez-Peiró, M.; Mavrogianni, A.; Symonds, P.; Sánchez-Guevara Sánchez, C.; Neila González, F.J. Modelling long-term urban temperatures with less training data: A comparative study using neural networks in the city of Madrid. *Sustainability* **2021**, *13*, 8143. [CrossRef]
28. Kwak, Y.; Park, C.; Deal, B. Discerning the success of sustainable planning: A comparative analysis of urban heat island dynamics in Korean new towns. *Sustain. Cities Soc.* **2020**, *61*, 102341. [CrossRef]
29. Yoo, S. Investigating important urban characteristics in the formation of urban heat islands: A machine learning approach. *J. Big Data* **2018**, *5*, 2. [CrossRef]
30. Voelkel, J.; Shandas, V. Towards systematic prediction of urban heat islands: Grounding measurements, assessing modeling techniques. *Climate* **2017**, *5*, 41. [CrossRef]
31. Zumwald, M.; Knüsel, B.; Bresch, D.N.; Knutti, R. Mapping urban temperature using crowd-sensing data and machine learning. *Urban Clim.* **2021**, *35*, 100739. [CrossRef]
32. Kafy, A.-A.; Faisal, A.A.; Rahman, M.S.; Islam, M.; Al Rakib, A.; Islam, M.A.; Khan, M.H.H.; Sikdar, M.S.; Sarker, M.H.S.; Mawa, J.; et al. Prediction of seasonal urban thermal field variance index using machine learning algorithms in Cumilla, Bangladesh. *Sustain. Cities Soc.* **2021**, *64*, 102542. [CrossRef]
33. Shi, H.; Xian, G.; Auch, R.; Gallo, K.; Zhou, Q. Urban Heat Island and its regional impacts using remotely sensed thermal data—A review of recent developments and methodology. *Land* **2021**, *10*, 867. [CrossRef]
34. Alves, E.; Anjos, M.; Galvani, E. Surface urban heat island in middle city: Spatial and temporal characteristics. *Urban Sci.* **2020**, *4*, 54. [CrossRef]
35. Chakraborty, T.; Hsu, A.; Manya, D.; Sheriff, G. A spatially explicit surface urban heat island database for the United States: Characterization, uncertainties, and possible applications. *ISPRS J. Photogramm. Remote Sens.* **2020**, *168*, 74–88. [CrossRef]

36. Wang, H.; Zhang, Y.; Tsou, J.Y.; Li, Y. Surface urban heat island analysis of shanghai (China) based on the change of land use and land cover. *Sustainability* **2017**, *9*, 1538. [CrossRef]
37. DANE Departamento Administrativo Nacional de Estadística. Available online: https://www.dane.gov.co/ (accessed on 12 December 2020).
38. Municipio de Cartago Valle del Cauca—Alcaldía de Cartago. Available online: http://www.cartago.gov.co/pot-vigente (accessed on 15 January 2018).
39. Sobrino, J.A.; Raissouni, N.; Li, Z. A Comparative study of land surface emissivity retrieval from NOAA data. *Remote Sens. Environ.* **2001**, *75*, 256–266. [CrossRef]
40. Valor, E.; Caselles, V. Mapping land surface emissivity from NDVI: Application to European, African, and South American areas. *Remote Sens. Environ.* **1996**, *57*, 167–184. [CrossRef]
41. USGS Landsat 8 (L8) Data Users Handbook. Available online: https://www.usgs.gov/media/files/landsat-8-data-users-handbook (accessed on 7 January 2020).
42. Tarawally, M.; Xu, W.; Hou, W.; Mushore, T.D. Comparative analysis of responses of land surface temperature to long-term land use/cover changes between a coastal and Inland City: A case of Freetown and Bo Town in Sierra Leone. *Remote Sens.* **2018**, *10*, 112. [CrossRef]
43. Kaufman, Y.J.; Tanré, D.; Gordon, H.R.; Nakajima, T.; Lenoble, J.; Frouin, R.; Grassl, H.; Herman, B.M.; King, M.D.; Teillet, P.M. Passive remote sensing of tropospheric aerosol and atmospheric correction for the aerosol effect. *J. Geophys. Res. Atmos.* **1997**, *102*, 16815–16830. [CrossRef]
44. Zha, Y.; Gao, J.; Ni, S. Use of normalized difference built-up index in automatically mapping urban areas from TM imagery. *Int. J. Remote Sens.* **2003**, *24*, 583–594. [CrossRef]
45. Tucker, C.J. Red and photographic infrared linear combinations for monitoring vegetation. *Remote Sens. Environ.* **1979**, *8*, 127–150. [CrossRef]
46. Gao, B.-C. Naval Research Laboratory, 4555 Overlook Ave. *Remote Sens. Environ.* **1996**, *58*, 257–266. [CrossRef]
47. Baig, M.H.A.; Zhang, L.; Shuai, T.; Tong, Q. Derivation of a tasselled cap transformation based on Landsat 8 at-satellite reflectance. *Remote Sens. Lett.* **2014**, *5*, 423–431. [CrossRef]
48. Zhang, Y.; Odeh, I.O.A.; Han, C. Bi-temporal characterization of land surface temperature in relation to impervious surface area, NDVI and NDBI, using a sub-pixel image analysis. *Int. J. Appl. Earth Obs. Geoinf.* **2009**, *11*, 256–264. [CrossRef]
49. Karnieli, A.; Ohana-Levi, N.; Silver, M.; Paz-Kagan, T.; Panov, N.; Varghese, D.; Chrysoulakis, N.; Provenzale, A. Spatial and seasonal patterns in vegetation growth-limiting factors over Europe. *Remote Sens.* **2019**, *11*, 2406. [CrossRef]
50. Yang, C.; He, X.; Yu, L.; Yang, J.; Yan, F.; Bu, K.; Chang, L.; Zhang, S. The cooling effect of urban parks and its monthly variations in a snow climate city. *Remote Sens.* **2017**, *9*, 1066. [CrossRef]
51. Hathway, E.A.; Sharples, S. The interaction of rivers and urban form in mitigating the Urban Heat Island effect: A UK case study. *Build. Environ.* **2012**, *58*, 14–22. [CrossRef]
52. Olioso, A. Simulating the relationship between thermal emissivity and the Normalized Difference Vegetation Index. *Int. J. Remote Sens.* **1995**, *16*, 3211–3216. [CrossRef]
53. Wittich, K.-P. Some simple relationships between land-surface emissivity, greenness and the plant cover fraction for use in satellite remote sensing. *Int. J. Biometeorol.* **1997**, *41*, 58–64. [CrossRef]
54. Mallick, J.; Singh, C.K.; Shashtri, S.; Rahman, A.; Mukherjee, S. Land surface emissivity retrieval based on moisture index from LANDSAT TM satellite data over heterogeneous surfaces of Delhi city. *Int. J. Appl. Earth Obs. Geoinf.* **2012**, *19*, 348–358. [CrossRef]
55. Bacour, C.; Baret, F.; Béal, D.; Weiss, M.; Pavageau, K. Neural network estimation of LAI, fAPAR, fCover and LAI × Cab, from top of canopy MERIS reflectance data: Principles and validation. *Remote Sens. Environ.* **2006**, *105*, 313–325. [CrossRef]
56. Weiss, M.; Baret, F. S2ToolBox Level 2 Products: LAI, FAPAR, FCOVER Version 1.1. 2016. Available online: https://step.esa.int/docs/extra/ATBD_S2ToolBox_L2B_V1.1.pdf (accessed on 15 December 2019).
57. USGS Landsat 8 Thermal Infrared Sensor (TIRS) Calibration Notices. Available online: https://www.usgs.gov/land-resources/nli/landsat/landsat-8-oli-and-tirs-calibration-notices (accessed on 18 June 2020).
58. Sekertekin, A.; Bonafoni, S. Land surface temperature retrieval from Landsat 5, 7, and 8 over rural areas: Assessment of different retrieval algorithms and emissivity models and toolbox implementation. *Remote Sens.* **2020**, *12*, 294. [CrossRef]
59. Barsi, J.A.; Schott, J.R.; Palluconi, F.D.; Hook, S.J. Validation of a web-based atmospheric correction tool for single thermal band instruments. In Proceedings of the Earth Observing Systems X, SPIE, San Diego, CA, USA, 22 August 2005; Volume 5882, p. 58820E.
60. Yue, W.; Xu, J.; Tan, W.; Xu, L. The relationship between land surface temperature and NDVI with remote sensing: Application to Shanghai Landsat 7 ETM + data. *Int. J. Remote Sens.* **2007**, *28*, 3205–3226. [CrossRef]
61. Hulley, G.C.; Hook, S.J.; Abbott, E.; Malakar, N.; Islam, T.; Abrams, M. The aster global emissivity dataset (ASTER GED): Mapping Earth's emissivity at 100 meter spatial scale. *Geophys. Res. Lett.* **2015**, *42*, 7966–7976. [CrossRef]
62. Park, J.; Jang, S.; Hong, R.; Suh, K.; Song, I. Development of land cover classification model using AI based fusionnet network. *Remote Sens.* **2020**, *12*, 3171. [CrossRef]
63. Lu, D.; Hetrick, S.; Moran, E. Land cover classification in a complex urban-rural landscape with quickbird imagery. *Photogramm. Eng. Remote Sens.* **2010**, *76*, 1159–1168. [CrossRef]

64. Du, C.; Ren, H.; Qin, Q.; Meng, J.; Zhao, S. A practical split-window algorithm for estimating land surface temperature from landsat 8 data. *Remote Sens.* **2015**, *7*, 647–665. [CrossRef]
65. Tan, K.; Liao, Z.; Du, P.; Wu, L. Land surface temperature retrieval from Landsat 8 data and validation with geosensor network. *Front. Earth Sci.* **2017**, *11*, 20–34. [CrossRef]
66. Hulley, G.C.; Hook, S.J. Generating consistent land surface temperature and emissivity products between ASTER and MODIS data for earth science research. *IEEE Trans. Geosci. Remote Sens.* **2011**, *49*, 1304–1315. [CrossRef]
67. Sobrino, J.A.; Jiménez-Muñoz, J.C.; Sòria, G.; Romaguera, M.; Guanter, L.; Moreno, J.; Plaza, A.; Martínez, P. Land surface emissivity retrieval from different VNIR and TIR sensors. *IEEE Trans. Geosci. Remote Sens.* **2008**, *46*, 316–327. [CrossRef]
68. Jiménez-Muñoz, J.C.; Sobrino, J.A.; Skokovi, D.; Mattar, C.; Cristóbal, J.; Bands, A.L.-T. Land surface temperature retrieval methods from Landsat-8 thermal infrared sensor data. *IEEE Geosci. Remote Sens. Lett.* **2014**, *11*, 1840–1843. [CrossRef]
69. Carlson, T.N.; Ripley, D.A. On the relation between NDVI, fractional vegetation cover, and leaf area index. *Remote Sens. Environ.* **1997**, *62*, 241–252. [CrossRef]
70. Jiménez-Muñoz, J.C.; Cristobal, J.; Sobrino, J.A.; Sòria, G.; Ninyerola, M.; Pons, X. Revision of the single-channel algorithm for land surface temperature retrieval from landsat thermal-infrared data. *IEEE Trans. Geosci. Remote Sens.* **2009**, *47*, 339–349. [CrossRef]
71. Li, Z.-L.; Tang, B.-H.; Wu, H.; Ren, H.; Yan, G.; Wan, Z.; Trigo, I.F.; Sobrino, J.A. Satellite-derived land surface temperature: Current status and perspectives. *Remote Sens. Environ.* **2013**, *131*, 14–37. [CrossRef]
72. Rasul, A.; Balzter, H.; Smith, C.; Remedios, J.; Adamu, B.; Sobrino, J.; Srivanit, M.; Weng, Q. A review on remote sensing of urban heat and cool islands. *Land* **2017**, *6*, 38. [CrossRef]
73. Machidon, A.L.; Del Frate, F.; Picchiani, M.; Machidon, O.M.; Ogrutan, P.L. Geometrical approximated principal component analysis for hyperspectral image analysis. *Remote Sens.* **2020**, *12*, 1698. [CrossRef]
74. Adame-Campos, R.L.; Ghilardi, A.; Gao, Y.; Paneque-Gálvez, J.; Mas, J.F. Variables selection for aboveground biomass estimations using satellite data: A comparison between relative importance approach and stepwise Akaike's information criterion. *ISPRS Int. J. Geo-Inf.* **2019**, *8*, 245. [CrossRef]
75. Hanssens, D.M.; Parsons, L.J.; Schultz, R.L. Parameter estimation and model testing. In *International Series in Quantitative Marketing*; Springer: Boston, MA, USA, 2002; pp. 183–248.
76. Harrell, F. *Regression Modeling Strategies: With Applications to Linear Models, Logistic and Ordinal Regression and Survival Analysis*, 2nd ed.; Springer: Berlin/Heidelberg, Germany, 2015; ISBN 9783319194240.
77. Rahman, S.M.A.K.; Sathik, M.M.; Kannan, K.S. Multiple linear regression models in outlier detection. *Int. J. Res. Comput. Sci.* **2012**, *2*, 23–28. [CrossRef]
78. Zhao, X.; Zhang, Y.; Xie, S.; Qin, Q.; Wu, S.; Luo, B. Outlier detection based on residual histogram preference for geometric multi-model fitting. *Sensors* **2020**, *20*, 3037. [CrossRef]
79. Wilks, D.S. *Statistical Methods in the Atmospheric Sciences*, 4th ed.; Elsevier: Amsterdam, The Netherlands, 2019; ISBN 9780128158234.
80. Noi, P.T.; Kappas, M. Comparison of random forest, k-nearest neighbor, and support vector machine classifiers for land cover classification using Sentinel-2 Imagery. *Sensors* **2017**, *18*, 18. [CrossRef]
81. Barca, E.; Castrignanò, A.; Ruggieri, S.; Rinaldi, M. A new supervised classifier exploiting spectral-spatial information in the Bayesian framework. *Int J. Appl Earth Obs. Geoinf.* **2020**, *86*, 101990. [CrossRef]
82. Tian, S.; Zhang, X.; Tian, J.; Sun, Q. Random forest classification of wetland landcovers from multi-sensor data in the arid region of Xinjiang, China. *Remote Sens.* **2016**, *8*, 954. [CrossRef]
83. Lv, Z.Y.; He, H.; Benediktsson, J.A.; Huang, H. A generalized image scene decomposition-based system for supervised classification of very high resolution remote sensing imagery. *Remote Sens.* **2016**, *8*, 814. [CrossRef]
84. Park, D. Image classification using naïve bayes classifier. *Int. J. Comput. Sci. Electron. Eng.* **2016**, *4*, 135–139.
85. Judah, A.; Hu, B. The integration of multi-source remotely-sensed data in support of the classification of wetlands. *Remote Sens.* **2019**, *11*, 1537. [CrossRef]
86. Zhang, H.; Jiang, L.; Yu, L. Class-specific attribute value weighting for Naive Bayes. *Inf. Sci.* **2020**, *508*, 260–274. [CrossRef]
87. Jenks, G.F. The data model concept in statistical mapping. *Int. Yearb. Cartogr.* **1967**, *7*, 186–190.
88. Szymanowski, M.; Kryza, M. Local regression models for spatial interpolation of urban heat island-an example from Wrocław, SW Poland. *Theor. Appl. Climatol.* **2012**, *108*, 53–71. [CrossRef]
89. Ogashawara, I.; Bastos, V. A quantitative approach for analyzing the relationship between Urban Heat Islands and Land Cover. *Remote Sens.* **2012**, *4*, 3596–3618. [CrossRef]
90. Zhang, X.; Estoque, R.C.; Murayama, Y. An urban heat island study in Nanchang City, China based on land surface temperature and social-ecological variables. *Sustain. Cities Soc.* **2017**, *32*, 557–568. [CrossRef]
91. Guha, S.; Govil, H.; Diwan, P. Analytical study of seasonal variability in land surface temperature with normalized difference vegetation index, normalized difference water index, normalized difference built-up index, and normalized multiband drought index. *J. Appl. Remote Sens.* **2019**, *13*, 024518. [CrossRef]
92. Cruz, J.A.; Santos, J.A.; Blanco, A. Spatial disaggregation of Landsat-derived land surface temperature over a heterogeneous urban landscape using planetscope image derivatives. *Int. Arch. Photogramm. Remote Sens. Spat. Inf. Sci.* **2020**, *43*, 115–122. [CrossRef]

93. Duan, S.-B.; Li, Z.-L.; Wang, C.; Zhang, S.; Tang, B.-H.; Leng, P.; Gao, M.F. Land-surface temperature retrieval from Landsat 8 single-channel thermal infrared data in combination with NCEP reanalysis data and ASTER GED product. *Int. J. Remote Sens.* **2018**, *40*, 1763–1778. [CrossRef]
94. Malakar, N.K.; Hulley, G.C.; Hook, S.J.; Laraby, K.; Cook, M.; Schott, J.R. An operational land surface temperature product for landsat thermal data: Methodology and validation. *IEEE Trans. Geosci. Remote Sens.* **2018**, *56*, 5717–5735. [CrossRef]
95. Liu, C.; Li, Y. Spatio-temporal features of urban heat island and its relationship with land use/cover in mountainous city: A case study in Chongqing. *Sustainability* **2018**, *10*, 1943. [CrossRef]
96. Nill, L.; Ullmann, T.; Kneisel, C.; Sobiech-Wolf, J.; Baumhauer, R. Assessing spatiotemporal variations of landsat land surface temperature and multispectral indices in the Arctic Mackenzie Delta Region between 1985 and 2018. *Remote Sens.* **2019**, *11*, 2329. [CrossRef]
97. Dos-Santos, A.R.; de Oliveira, F.S.; da Silva, A.G.; Gleriani, J.M.; Gonçalves, W.; Moreira, G.L.; Silva, F.G.; Branco, E.R.F.; Moura, M.M.; da Silva, R.G.; et al. Spatial and temporal distribution of urban heat islands. *Sci. Total Environ.* **2017**, *605–606*, 946–956. [CrossRef]
98. Guha, S.; Govil, H.; Dey, A.; Gill, N. Analytical study of land surface temperature with NDVI and NDBI using Landsat 8 OLI and TIRS data in Florence and Naples city, Italy. *Eur. J. Remote Sens.* **2018**, *51*, 667–678. [CrossRef]
99. Ibrahim, G.R.F. Urban land use land cover changes and their effect on land surface temperature: Case study using Dohuk City in the Kurdistan Region of Iraq. *Climate* **2017**, *5*, 13. [CrossRef]
100. Bonafoni, S.; Keeratikasikorn, C. Land surface temperature and urban density: Multiyear modeling and relationship analysis using modis and landsat data. *Remote Sens.* **2018**, *10*, 1471. [CrossRef]
101. Molina, I.; Martinez, E.; Morillo, C.; Velasco, J.; Jara, A. Assessment of data fusion algorithms for earth observation change detection processes. *Sensors* **2016**, *16*, 1621. [CrossRef] [PubMed]
102. Renard, F.; Alonso, L.; Fitts, Y.; Hadjiosif, A.; Comby, J. Evaluation of the effect of urban redevelopment on surface urban heat islands. *Remote Sens.* **2019**, *11*, 299. [CrossRef]
103. Alshayeb, M.J.; Chang, J.D. Variations of PV panel performance installed over a vegetated roof and a conventional black roof. *Energies* **2018**, *11*, 1110. [CrossRef]
104. Grilo, F.; Pinho, P.; Aleixo, C.; Catita, C.; Silva, P.; Lopes, N.; Freitas, C.; Santos-reis, M.; Mcphearson, T.; Branquinho, C. Using green to cool the gre: Modelling the cooling effect of green spaces with a high spatial resolution. *Sci. Total Environ.* **2020**, *724*, 138182. [CrossRef] [PubMed]
105. U.S. Environmental Protection Agency. "Cool Pavements". Reducing Urban Heat Islands: Compendium of Strategies. 2012. Available online: https://www.epa.gov/heat-islands/heat-island-compendium (accessed on 1 June 2021).

MDPI
St. Alban-Anlage 66
4052 Basel
Switzerland
www.mdpi.com

Remote Sensing Editorial Office
E-mail: remotesensing@mdpi.com
www.mdpi.com/journal/remotesensing

www.ingramcontent.com/pod-product-compliance
Lightning Source LLC
Chambersburg PA
CBHW061559220926
51850CB00005B/371
9783036592749